Micro-Mesoporous Metallosilicates

Micro-Mesoporous Metallosilicates

Synthesis, Characterization, and Catalytic Applications

Edited by Peng Wu and Hao Xu

WILEY-VCH

Editors

Prof. Peng Wu
East China Normal University
North Zhongshan Rd. No. 3663
Shanghai, 200062
China

Prof. Hao Xu
East China Normal University
North Zhongshan Rd. No. 3663
Shanghai, 200062
China

Library of Congress Card No.: applied for

British Library Cataloguing-in-Publication Data
A catalogue record for this book is available from the British Library.

Bibliographic information published by the Deutsche Nationalbibliothek
The Deutsche Nationalbibliothek lists this publication in the Deutsche Nationalbibliografie; detailed bibliographic data are available on the Internet at <http://dnb.d-nb.de>.

Print ISBN: 978-3-527-35094-0
ePDF ISBN: 978-3-527-83936-0
ePub ISBN: 978-3-527-83937-7
oBook ISBN: 978-3-527-83938-4

Cover Image: Courtesy of Dr. Hao Xu

Typesetting Straive, Chennai, India
Printing and Binding CPI Group (UK) Ltd, Croydon, CR0 4YY

C9783527350940_230224

Contents

Preface

Zeolites have experienced tremendous progress in the last 60 years as key materials for a variety of applications, ranging from adsorbents to heterogeneous catalysts. The metallosilicates or heteroatom-containing microporous and mesoporous materials established a universally acknowledged milestone, with the most representative TS-1 titanosilicate as selective and clean oxidation catalyst using hydrogen peroxide as an oxidant. The discovery of TS-1 shattered our limited understanding that zeolites were solely confined to being heterogeneous solid-acid catalysts used for gas-phase hydrocarbon transformations. Innovative selective liquid-phase oxidation processes have been incessantly commercialized based on the TS-1/hydrogen peroxide catalytic system since its emergence in the 1980s, including phenol hydroxylation, ketone ammoximation, and propylene epoxidation. A great progress has occurred in the domain of these specific zeolites in past decades, resulting in the successful synthesis of other metallosilicates with either different structure topologies or changeable heteroatoms. The metallosilicates diversify in structure, composition, and multifunctional uses, substantially endowing these types of novel porous materials with new interesting properties and functionalities. They are not only useful for selective oxidations but also promising for dehydrogenation of propane and biomass conversion. This book is edited to comprehensively introduce the chemistry and catalytic technologies of metallosilicates, with a focus on design synthesis, active site construction, structural modification, characterization of active sites, and catalytic applications of titanosilicates, stannosilicates, germanosilicates, etc. This book will attract broad interests from both academia and industry, and it is also intended to bridge the technology gap between basic research and industrial requirements.

This book consists of 16 chapters. Chapter 1 concentrates on recent advances in the synthesis of titanosilicates, including hydrothermal synthesis, dry-gel conversion, fluoride-assisted synthesis, and post-synthesis methods. This is followed by a chapter by H. Xu et al. (Chapter 2), which provides the synthesis of metallosilicates with two-dimensional lamellar structures and their structural modifications as well as applications in selective oxidation reactions. The approaches to the synthesis of Sn and Zr zeolites and their applications in catalysis are summarized by Z. Zhu (Chapter 3). The synthesis of germanosilicates with specially designed organic structure-directing agents is presented by J. Jiang (Chapter 4). Germanosilicates are structurally unstable in humid environment, which inversely endows them with structurally modifiable properties. O. Veselý et al. demonstrate the application of germanosilicates as starting compounds for preparation of materials with highly controlled chemical and structural properties via methods of intercalation, pillaring, or the ADOR transformations (Chapter 5). In addition, the synthesis and catalytic applications of heteroatom-containing mesoporous silica are also included in this book. K. Zhang et al. (Chapter 6) specifically focus on the dendritic mesoporous silica nanoparticles with unique wrinkled center-radial structures. Post-modifications are effective in regulating the microenvironment of Ti active sites to enhance the catalytic activity and selectivity of titanosilicates. Y. Liu et al. (Chapter 7) present an overview of current developments in chemical post-modifications of titanosilicates in terms of the effects on transfer, adsorption/desorption, and surface reactions. The characterizations of heteroatom species in metallosilicates are critical to obtain structure–performance relationships and the accurate design synthesis of high-performance metallosilicate catalysts. J. Xu and F. Deng (Chapter 8) focus on the X-ray-based techniques, ultraviolet–visible–near infrared spectroscopy, Raman spectroscopy, and solid-state NMR spectroscopy, which provide information about the coordination, valence states, and locations of heteroatoms in zeolites, as well as their host–guest interactions with zeolite frameworks. W. Chen and A. Zheng (Chapter 9) stress the importance of theoretical calculation as an effective tool and supplement to understand the catalytic active center, structural character, and Brønsted/Lewis acidity of heteroatom-containing zeolites, and further they clarify the reaction mechanism catalyzed by metallosilicates. The remaining chapters are exclusively devoted to the application of metallosilicates in catalysis. P. Wu et al. (Chapter 10) concentrate on the application of titanosilicates in ammoximation reaction of ketones and aldehydes and the successful construction of large-scale plants for clean production of cyclohexanone ammoximation with NH_3 and H_2O_2. After this, C. Xia (Chapter 11) presents the application of titanosilicates in the liquid-phase epoxidation reaction of propylene and propylene chloride to corresponding epoxides as well as the ethylene epoxidation to ethylene oxide/ethyl glycol using H_2O_2 as the oxidant. The usage of cumene hydroperoxide as the oxidant and titanosilicates as the catalysts can also construct a highly efficient catalytic system for propylene oxide production, and the developments in both academic and industrial fields are outlined by L. Xu in Chapter 12. W. Fan et al. (Chapter 13) describe the applications of metallosilicates in the hydroxylation reactions of aromatics. The in situ formation of H_2O_2 is more appealing for the alkene epoxidation reactions due to the high cost and explosive risk of H_2O_2. The gas-phase propylene epoxidation with H_2 and O_2 over the bifunctional Au–Ti catalysts is presented by X. Duan and X. Zou in Chapter 14. The authors focus on the effects of particle sizes, oxidation state, and location sites of Au nanoparticles, and the Ti-containing materials with different structures and hydrophobicity on the epoxidation performance. The bifunctional metallosilicates have also found applications in the activation and conversion of low-carbon alkanes as demonstrated by L. Wang and F. Xiao in Chapter 15. The final chapter by the authors from the group of H. Yamashita presents possibilities of application of metallosilicates in photocatalysis.

These excellent chapters review and describe the cutting edge of basic research and R&D practices of metallosilicates, enlightening an optimistic view for their bright future. The number and type of new metallosilicates are ever increasing with the advancement of synthetic chemistry. Direct synthesis still dominates the preparation of metallosilicates. Post-synthesis proves to be useful to develop the materials with novel pore and crystalline structures as well as with desirable active sites that are difficult to simultaneously realize by direct synthesis. Diversifying the isomorphously substituted framework heteroatoms greatly broadens the applications of metallosilicates from traditional selective oxidation to Lewis acidic catalysis and photocatalysis. These new materials with extra-large pores or even mesopores, highly required by practical applications, should be useful in developing new green catalytic processes that enable processing bulkier substrates. Combining the intrinsic catalytic properties of metallosilicates with the second active site of noble metal will open new possibility to develop greener processes with intensification features, simplifying complex reaction steps and saving separation energy. In addition, spectroscopical techniques and theoretical calculations provide new information about the chemistry and coordination of framework heteroatom active sites and their catalysis mechanism involved in each reaction. Exciting achievements beyond our imagination are highly expectable by taking the advantage of the experimental and theoretical knowledge achieved so far. Last but not least, novel approaches in the synthesis, manipulation, and modification of metallosilicates clearly evidence a significant potential of these zeolites in new application areas.

It is our great pleasure and honor to work together with the excellent experts and researchers in heteroatom zeolite field to compile this thought-provoking book. We would like to express our sincere thanks to all of them for their great efforts to put together timely reviews on selected topics. This book, concentrated on the specific zeolites of metallosilicates, is expected to be helpful for the experienced academic researchers. We also wish the industrial researchers and the newcomers would find it as a useful reference book.

June 2023
Shanghai, China

Peng Wu
Hao Xu

1

Synthesis of Titanosilicates

Xinqing Lu

Zhejiang Normal University, Key Laboratory of the Ministry of Education for Advanced Catalysis Materials, Department of Chemistry, Yingbin Avenue, Jinhua, Zhejiang 321004, P.R. China

1.1 Introduction

Zeolites are microporous crystals that are constructed by tetrahedral SiO_4 and AlO_4 species interlinked by sharing O atoms, and they demonstrate remarkable application prospects in adsorption, separation, ion exchange, and heterogeneous solid-acid catalysis [1]. Typically, a part of the framework of Si and Al atoms can be replaced by heteroatoms, such as Ti, Sn, Ge, Zr, B, P, V, and Ga, via isomorphous substitution, resulting in heteroatomic zeolites or metallosilicates [2–4]. Among these heteroatomic zeolites, titanosilicate is the most representative one, and it can catalyze diverse selective oxidation reactions, such as alkene epoxidation, aldehyde or ketone ammoxidation, benzene or phenol hydroxylation, 1,4-dioxane oxidation, selective oxidation of pyridine derivatives, and oxidation desulfurization [5–9], as well as acid-catalyzed reactions, such as ring-opening reactions of epoxides [10–12], ethylenediamine condensation [13], and Beckmann rearrangement of oxime [14] (as shown in Figure 1.1). Moreover, the discovery of titanosilicates has expanded the application scope of zeolites, as heterogeneous catalysts, from acid catalysis to the redox field. Several reviews and monographs have proposed opportunities and challenges for titanosilicates in synthetic and catalytic applications [3–9, 15–18]. As depicted in Figure 1.2, the number of annual publications related to titanosilicates has rapidly increased from 1983 to 2023, and this number has remained at approximately 200–350 over the last decade.

Notably, titanosilicates can be divided into microporous, mesoporous, and hierarchical types based on their textural properties and pore sizes. Among these, microporous titanosilicates, with isolated tetrahedral Ti species, possess pores that are <2 nm in size, and these include small- and medium-pore titanosilicate zeolites with 8- or 10-membered ring (MR), 12-MR large-pore zeolites, and extra-large-pore zeolites with ≥14 MRs. Among the 255 ordered zeolite framework structures with three-letter codes and the partially disordered zeolite structures recognized by the International Zeolite Association Structure Commission (IZA), 28 structures

Micro-Mesoporous Metallosilicates: Synthesis, Characterization, and Catalytic Applications, First Edition. Edited by Peng Wu and Hao Xu.

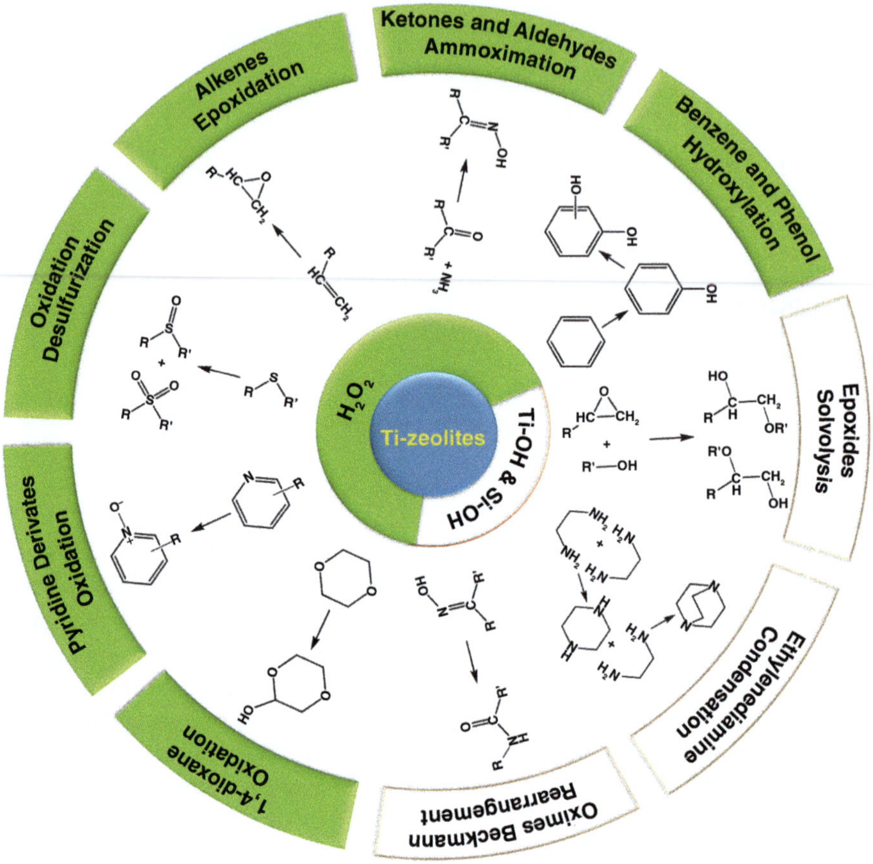

Figure 1.1 Reactions catalyzed by titanosilicates.

can be synthesized as microporous titanosilicates. Owing to their unique porosity and hydrophobicity, microporous titanosilicates can activate H_2O_2 molecules and catalyze selective oxidation reactions. Titanosilicalite-1 (TS-1), with MFI topology, was the first microporous titanosilicate to be employed as a commercial catalyst. For example, the application of TS-1 in the liquid-phase epoxidation of propylene to propylene oxide using H_2O_2 as the oxidant was first reported by EniChem in 1983 [19] and was implemented on a commercial scale by Evonik and SKC in South Korea in 2008. Mesoporous titanosilicates, such as Ti-MCM-41, Ti-MCM-48, Ti-KIT-5, Ti-SBA-15, and Ti-SBA-16, possess pores that are >2 nm in size and amorphous pore walls [5]. They are more active than microporous materials in the oxidation of bulky substrates with cumene hydroperoxide or *tert*-butyl hydroperoxide (TBHP) as the oxidant. However, they are much less active in oxidation reactions using hydrogen peroxide as the oxidant owing to their extremely high hydrophilicity derived from abundant surface silanols on their amorphous pore walls [20]. Hierarchical titanosilicates contain both micropores and mesopores and exhibit better catalytic properties than their microporous counterparts, particularly

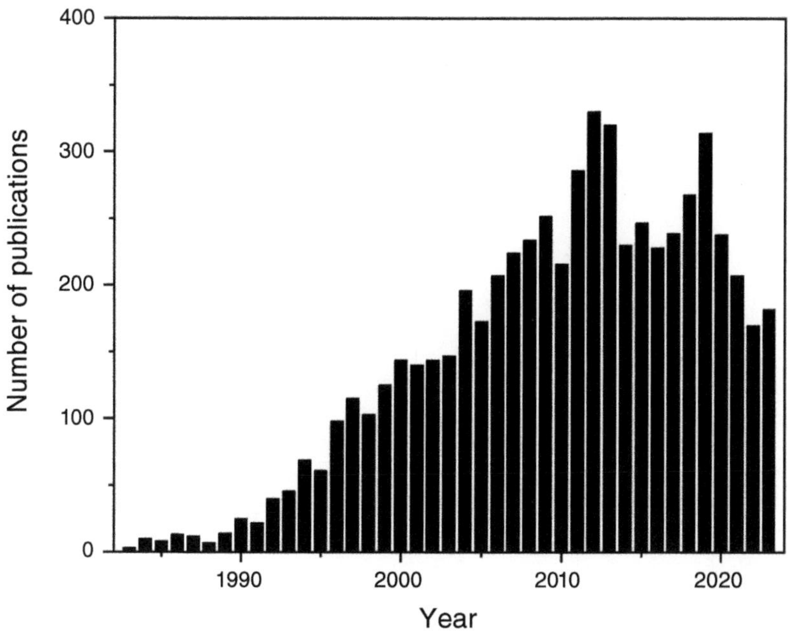

Figure 1.2 Change trend of annual publication number for titanosilicates. Source: SciFinder.

in catalytic reactions involving bulky substrates and/or organic hydroperoxide oxidants [5, 18].

Titanosilicates are primarily synthesized via hydrothermal synthesis (HTS), dry-gel conversion (DGC), fluoride-assisted synthesis, and post-synthesis methods (see Figure 1.3). Among these, HTS has been the most widely adopted approach for zeolite synthesis. This is because the contents and distributions of Ti species, crystal sizes, morphologies, and other physicochemical properties of titanosilicates can be tailored by adjusting the composition of synthetic gels and the crystallization conditions [15]. Particularly, the formation of anatase TiO_2 via the oligomerization of Ti monomers in a HTS process is generally easy owing to the faster hydrolysis rate of the Ti precursor compared to that of the Si precursor. However, the formation of anatase TiO_2 results in low activity and selectivity in catalytic reactions. Consequently, several strategies have been proposed to inhibit the generation of the anatase phase by using additional additives, such as H_2O_2, isopropanol, Triton X-100, Tween-20, and $(NH_4)_2CO_3$, as well as by accurately adjusting the feeding rate [21–27]. These methods generally slow the hydrolysis of Ti precursors to match that of the Si precursors, thereby lowering the anatase content. As an example, Lin et al. [28] proposed a reversed-oligomerization synthesis strategy to address the mismatched hydrolysis rates between Si and Ti precursors, which was implemented by the fast oligomerization of Ti monomers and subsequent de-oligomerization to Ti monomers with the aid of hydroxyl free radicals ($^{\bullet}OH$) generated in situ by ultraviolet (UV) irradiation.

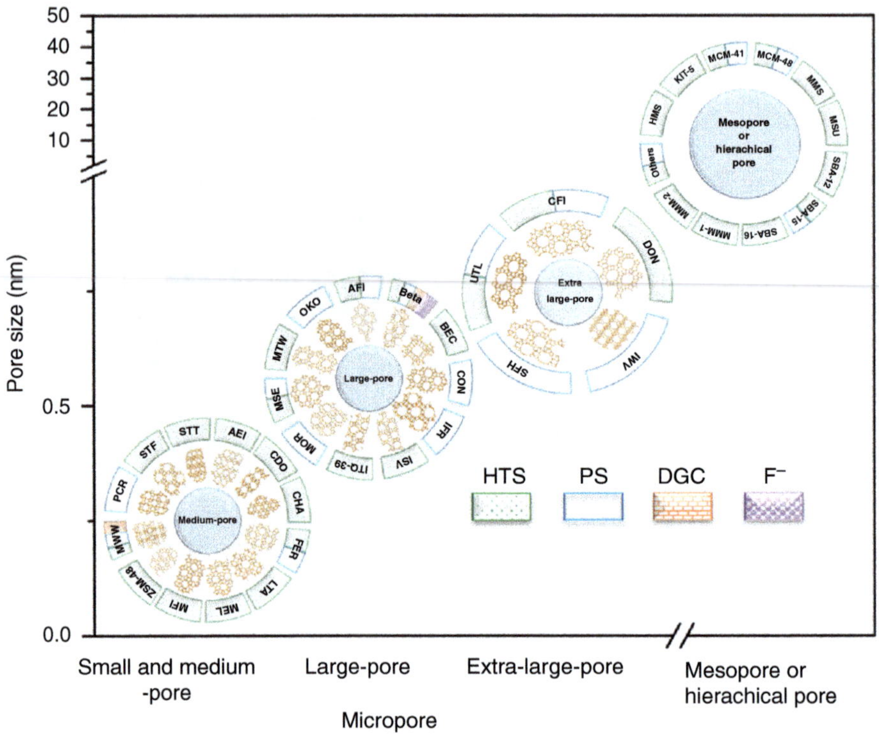

Figure 1.3 Overview of the synthesis methods of titanosilicates with different structures. HTS indicates hydrothermal synthesis, PS indicates post-synthesis, DGC indicates dry-gel conversion, F^- indicates fluoride-assisted method. Small and medium-pore stands for 8-MR and 10-MR titanosilicates, large-pore for 12-MR and extra-large pore for ≥14-MR.

The DGC method can be classified into vapor-phase transport (VPT) and steam-assisted crystallization (SAC) based on the volatility of structure-directing agents (SDAs) [9]. The VPT approach is applicable to volatile SDAs, where SDAs and water are not present in dried synthetic gels but are transferred to them via the vapor phase. The SAC approach can be realized by adding non-volatile SDAs to the dried synthetic gels with water placed below them; subsequently, the synthetic gels can be crystallized using steam. Compared with traditional HTS methods, the DGC method presents several advantages, such as lower SDA consumption, higher product yield, and shorter crystallization time. The crystal sizes obtained by DGC methods can differ from those obtained by direct HTS. For instance, for Ti-Beta synthesized through the DGC method, the crystal size is much smaller than that obtained by the HTS method [29], whereas the opposite is true for MWW-type titanosilicates [30]. In addition, DGC can be used to prepare hierarchical titanosilicate zeolites [31–33].

In fluoride-assisted synthesis, the presence of F^- can accelerate the crystallization process; however, titanosilicate crystals are usually larger than those obtained via HTS [34]. In contrast to aluminosilicates, the concentrations of alkali metal ions (such as Na^+ and K^+) in synthetic gels should be limited to obtain titanosilicates with high activities. Notably, MOR- and MSE-type zeolites barely crystallize in siliceous gels in the absence of Al^{3+} and alkali metal ions. Thus, the post-synthesis method is another available approach for titanosilicates. The post-synthesis method can proceed in the gas–solid [35], liquid–solid [36], or solid–solid phase [37], depending on the phase of the Ti source. In this chapter, we highlight the most remarkable achievements in the synthesis of titanosilicates with different pore topologies, including medium-pore, large-pore, extra-large-pore, mesopore, and Engelhard Ti silicates (ETS).

1.2 Synthesis of Medium-Pore Titanosilicates

1.2.1 TS-1 Synthesis

TS-1 (MFI topology), which possesses a three-dimensional (3D) medium-pore system (10-MR, ~0.55 nm), is one of the most studied titanosilicates and has been applied in many industrial processes, such as propylene epoxidation [19], phenol hydroxylation [38], and cyclohexanone ammoximation [39]. The first discovery of a TS-1 zeolite can be dated back to the patent disclosed by Taramasso et al. in 1983 [19], which was based on the matching hydrolysis of tetraethylorthosilicate (TEOS) and tetraethylorthotitanate (TEOT) using tetrapropylammonium hydroxide (TPAOH) as the SDA.

Extra-framework Ti species tend to form in TS-1 in the presence of Na^+ and K^+ from commercial aqueous TPAOH solutions as impurities [40, 41]. The diffraction peaks ascribed to the extra-framework anatase TiO_2 can even be detected in the X-ray diffraction patterns of TS-1 zeolites once the Na^+ concentration in the synthetic gel reaches a very high level. Moreover, residual alkali metal cations in the synthesized titanosilicates are detrimental to their reactivity in selective catalytic oxidation reactions [42].

In addition to alkali metal cations, the Ti content in synthetic gels and crystallization temperature were also found to be key factors for TS-1 synthesis [43]. The Ti content of the TS-1 samples is always lower than that of synthetic gels, implying that the Ti atoms in the synthetic gels could not be completely introduced into the MFI framework. Excess Ti in the synthetic gel increases the risk of forming extra-framework anatase TiO_2 as an impurity. Hence, several attempts have been made to maximize the amount of tetrahedrally incorporated Ti species in the framework to achieve a higher activity. It has been found that TPAOH should be slowly added into the synthetic gels at low temperatures under vigorous stirring to inhibit the formation of extra-framework Ti species [44]. In other words, it is necessary to match the hydrolysis rates between the Si and Ti precursors to obtain abundant silanol groups for condensation with the monomeric Ti species generated by hydrolysis. Fan et al. proposed that the Si/Ti molar ratio in the framework of the TS-1 zeolite can be decreased to 34 using $(NH_4)_2CO_3$ as a crystallization-mediating agent to match the hydrolysis rate of the Ti precursor with that of the Si precursor and the crystallization rate [23]. In contrast, the Si/Ti ratio was 58 in the absence of $(NH_4)_2CO_3$.

Recent research on the synthesis of TS-1 has mainly focused on (a) enhancing the accessibility of the framework Ti species by constructing hierarchical TS-1 zeolites by tuning the particle size and morphology and (b) developing alternative SDAs to replace expensive TPAOH for practical applications. Representative approaches for constructing hierarchical TS-1, such as mesoporogen-directed methods, mesoporogen-free methods (kinetic-regulated or DGC methods), and demetallization methods, have been reviewed previously [2, 5, 18]. On the other hand, the high cost of TPAOH has stimulated many researchers to concentrate on reducing its dosage or finding an alternative SDA. TS-1 can be prepared using a cheaper SDA of tetrapropylammonium bromide (TPABr); however, the large zeolite crystals (15 × 8 × 1.5 μm) obtained in earlier studies are undesirable [45]. Zuo et al. reported the synthesis of TS-1 with a crystal size of 200 nm using TPABr as the SDA, where seeding played a crucial role in reducing the crystal size [46].

TS-1 zeolites with novel morphologies have also been reported. In contrast to the raspberry morphology of traditional TS-1 zeolites, a lamellar TS-1 zeolite with oriented growth and sheet-like morphology (0.5–1.0 μm length and 20–50 nm thickness) was synthesized with a bifunctional surfactant as the SDA [47]. Due to its hierarchical structure, layered TS-1 showed a higher activity than conventional TS-1, Ti-MWW, Ti-Beta, or even mesoporous titanosilicate Ti-MCM-41 in cyclohexene epoxidation with TBHP as the oxidant. Wang et al. [48] reported the one-pot synthesis of TS-1 microspheres that were approximately 10 μm using a triblock copolymer (F127) as the supporting additive through a temperature-programed self-assembly strategy. The TS-1 microspheres were self-assembled from TS-1 nanoparticles (50–100 nm) and possessed more intercrystal mesopores than traditional TS-1 zeolite, resulting in higher activities during the oxidation of bulky substrates, such as 3-picoline and cyclohexene. Moreover, these TS-1 microspheres exhibited high mechanical stability under harsh thermal and hydrothermal conditions. Thus, because of their high reactivity and mechanical strength, TS-1 microspheres have potential applications in slurry reactors [49].

1.2.2 Ti-MWW Synthesis

Ti-MWW is another medium-pore titanosilicate with a two-dimensional (2D) pore system consisting of intralayer 10-MR sinusoidal pore channels and an interlayer 10-MR pore channel linked to intracrystalline 12-MR supercages [50]. Direct HTS of Ti-MWW is challenging [51]. A key breakthrough was achieved by introducing H_3BO_3 as a structure-supporting agent for the synthesis of Ti-MWW zeolites with high activity, inspired by the HTS of the borosilicate MWW-type zeolite ERB-1. Nevertheless, a higher B content in the synthetic gel is required for Ti-MWW zeolite than for ERB-1 because the introduction of Ti^{4+} ions considerably hinders the crystallization of the MWW framework [52]. The UV–visible spectra of the as-synthesized Ti-MWW had an adsorption band at ca. 220 nm, attributed to the tetrahedrally coordinated framework Ti species, together with another band at 260 nm, related to the extra-framework octahedral Ti species on the external surface of the Ti-MWW zeolites [52]. Irrespective of the Ti content in the synthetic gel, a band at approximately 330 nm was not observed in the UV–visible spectra of Ti-MWW, implying that the generation of the anatase TiO_2 phase was effectively inhibited.

Hexamethyleneimine (HMI) and piperidine (PI) can be employed as SDAs for Ti-MWW synthesis, and the obtained samples are denoted as Ti-MWW(HMI) and Ti-MWW(PI), respectively. The two Ti-MWW zeolites have similar specific surface areas but significantly different Ti^{4+} distributions owing to their different particle sizes. Irrespective of the Si/Ti molar ratio in the zeolites, the Ti-MWW(PI) zeolite showed the main adsorption band at approximately 260 nm, together with a band at approximately 220 nm, indicating more extra-framework Ti species (Figure 1.4). In contrast, the main absorption band in the UV–visible spectrum of Ti-MWW(HMI)

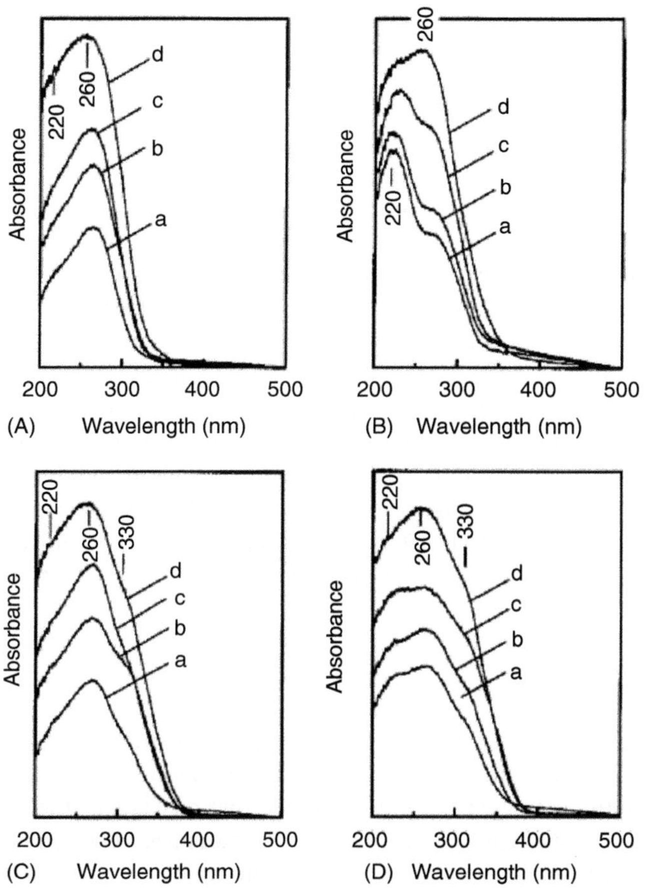

Figure 1.4 UV-visible spectra of as-synthesized (A) and calcined Ti-MWW(PI) (C), and as-synthesized (B) and calcined Ti-MWW(HMI) (D) with the Si/Ti molar ratios of 100 (a), 50 (b), 30 (c), and 10 (d). Source: Reprinted with permission from Wu et al. [52]. Copyright American Chemical Society (2001).

shifted from 260 to 220 nm with increasing Si/Ti molar ratio, indicating that the incorporation of Ti species preferentially occurred in the framework. After calcination of the as-synthesized Ti-MWW zeolites, a new band at approximately 330 nm, attributed to the anatase TiO_2 phase, emerged in the UV–visible spectra for both Ti-MWW(HMI) and Ti-MWW(PI), signifying that the neighboring external surface Ti species partially condensed and aggregated upon calcination. The formed anatase TiO_2 in the calcined Ti-MWW zeolite was barely removed by acid treatment. Therefore, acid treatment is usually performed on as-synthesized Ti-MWW zeolites rather than on calcined zeolites to remove the extra-framework Ti species on the surface, accompanied by a small amount of framework Ti species.

The B content in the Ti-MWW zeolite framework was far below that of the synthetic gel, implying that most of the B species were not involved in building the MWW structure. Additionally, owing to the weak acidity and increased electronegativity of the zeolite framework, the presence of residual framework boron is detrimental to selective oxidation reactions, although most of the framework boron atoms are extracted from the MWW framework by acid treatment. Hence, many other methods, such as the fluoride-assisted method [9] and DGC method [30], have been developed to synthesize Ti-MWW zeolites with reduced boron content in synthetic gels. Wu et al. developed a post-synthesis method to prepare boron-free Ti-MWW by incorporating Ti species into the framework of deboronated MWW via PI-assisted reversible 3D–2D–3D structure conversion [53]. However, boron-free Ti-MWW synthesized via the gas-phase atom planting strategy showed relatively low reactivity in oxidation reactions because of the presence of extra-framework octahedral Ti species and the anatase TiO_2 phase [54]. Lu et al. [36] recently reported the boron-free synthesis of Ti-MWW by introducing framework Ti species into an acidic medium, and the obtained Ti-MWW zeolites showed outstanding catalytic properties for 1-hexene epoxidation. In addition to post-synthesis methods, boron-free Ti-MWW can be one-pot hydrothermally synthesized using HMI and *N,N,N*-trimethyl-1-adamantammonium hydroxide (1-TMAdaOH) as dual-SDAs with a small amount of K^+ [55]. This method was inspired by the synthesis of a highly siliceous MWW zeolite (ITQ-1). After the detrimental K^+ was removed by mild acid treatment, the obtained Ti-MWW catalyst showed high activity in epoxidation reactions.

1.2.3 TS-2 Synthesis

TS-2 with its MEL topology of two intersecting 10-MR pore channels shows a catalytic performance similar to that of the TS-1 zeolite because of the similar secondary building units. In contrast to the abundant literature on TS-1 synthesis, studies on TS-2 synthesis are limited. The synthesis of TS-2 was first reported by Reddy et al. via the crystallization of synthetic gels containing TEOS, tetrabutyl orthotitanate (TBOT), and tetrabutylammonium hydroxide (TBAOH) [56, 57]. Moreover, the crystallization process can be accelerated by microwave-assisted heating [58]. Tuel et al. [59] found that tetrabutylphosphonium hydroxide was an alternative SDA for TS-2 synthesis. In addition to HTS, DGC [60] and F^--assisted methods [61] have also been applied to TS-2 synthesis. Moreover, TS-2 can be prepared by the solid–gas reaction of the borosilicate MEL-type zeolite B-ZSM-11 with saturated $TiCl_4$ vapor; however, it shows poor catalytic reactivity in epoxidation reactions owing to the presence of anatase TiO_2 [62]. Xu et al. recently reported an intergrowth-triggered TS-2 microsphere formed by adjusting the synthetic parameters [63]. The initially formed microspheres were well preserved during the crystallization process because of the special intergrowth stacking style occurred in the synthetic gels with high alkalinity.

1.2.4 Synthesis of Other Medium-Pore Titanosilicates

Ti-FER zeolites with 2D-intersecting 10 × 8-MR pore channels can be synthesized using HF/pyridine as the mineralizing agent with [64] or without seeds [65]. FER zeolites can also be synthesized as 2D-layered zeolites, on which structural modifications are applicable to construct Ti-FER zeolites with larger pore systems. The structural diversity of layered Ti-FER zeolites is available in Chapter 2. Ti-ZSM-48, with non-interpenetrating linear 10-MR pore channels (5.3 × 5.6 Å in diameter) and Si/Ti molar ratios ≥30, can be hydrothermally synthesized using diaminooctane, hexamethonium hydroxide, or trimethylpropylammonium hydroxide as SDAs [66–68]. However, Ti-ZSM-48, which has a partially disordered structure, is inactive for phenol hydroxylation with H_2O_2 because of diffusional limitations or subtle differences in the Ti microenvironment compared to TS-1 and TS-2 [68, 69]. In addition, small- and medium-pore titanosilicates with other topologies, such as STF [70], CHA [71–74], LTA [75, 76], CDO [77], AEI [78], STT [79], and PCR [80, 81], have also been successfully synthesized, but were inactive or showed less attractive activity in catalytic reactions because of mass transfer limitations.

1.3 Synthesis of Large-Pore Titanosilicates

1.3.1 Ti-Beta Synthesis

Titanoaluminosilicate Ti-Al-Beta, with 3D 12-MR pore channels, was first prepared by a HTS method via the isomorphous substitution of Si^{4+} by Ti^{4+} in the aluminosilicate Al-Beta. It showed higher activity than TS-1 in the selective oxidation of cycloalkanes with H_2O_2 as the oxidant [22], indicating that titanosilicates with relatively large pores are more efficient for oxidation reactions involving bulky substrates. Further study on the synthesis and characterization of Ti-Al-Beta was conducted to illustrate the influence of synthetic parameters on the Ti-coordinated states and catalytic properties in selective oxidation reactions [82, 83]. The Ti species introduced in Ti-Al-Beta tends to adopt extra-framework octahedral coordination rather than framework tetrahedral coordination. Moreover, the acidity derived from the framework tetrahedral Al in Ti-Al-Beta easily accelerates the ring opening of epoxides, decreasing epoxide selectivity in alkene epoxidation [83].

Thus, the amount of Al in Ti-Al-Beta should be decreased to eliminate the adverse effects of Al on catalytic performance. Al-free Ti-Beta titanosilicate has been successfully synthesized with tetraethylammonium hydroxide (TEAOH) and dealuminated Beta as the SDA and seed, respectively, as reported by Corma et al. [84]. However, Al-free Ti-Beta obtained by the above seeding methodology still showed very low-epoxide selectivity in alkene epoxidation. A similar phenomenon was observed for an Al-free Ti-Beta zeolite prepared by a solid–gas phase reaction between borosilicate B-Beta and saturated $TiCl_4$ vapor [85]. This phenomenon can be illustrated by the presence of numerous internal Si—OH groups (framework defects) in the Al-free Ti-Beta (Ti-Beta(OH)) obtained in OH^- media. Some Si—OH groups are formed after the elimination of TEA^+ and are balanced by SiO^- groups upon calcination [84]. Corma et al. found that TEA^+ can be neutralized by F^- instead of SiO^- for synthesis in an F^- medium, resulting in the formation of defect-less Ti-Beta zeolites (Ti-Beta(F)) with high crystallinity, hydrophobicity, and hydrothermal stability [34, 86]. Because of its hydrophobicity, Ti-Beta(F) is superior to Ti-Beta(OH) in unsaturated fatty oil epoxidation reactions in terms of reactivity and epoxide selectivity [34, 87]. Additionally, several other approaches, such as the DGC method, interzeolite transformation, and dissolution–recrystallization methods, have also been reported for the synthesis of Ti-Beta zeolites [29, 88–90]. Inspired by the DGC method for preparing high-silica Beta zeolite [91], large-pore Ti-Beta with high hydrophobicity has been prepared by the DGC method, even in the presence of alkali metal cations [29, 88]. The hydrophobic Ti-Beta obtained by the DGC method showed higher catalytic reactivity and selectivity to epoxides than the hydrophilic Ti-Beta(OH) in selective oxidation reactions [29, 92], which is in accordance with previous conclusions related to the effects of zeolite polarity on the catalytic properties in selective oxidation reactions [34, 86]. Zhu et al. [89] reported the interzeolite transformation of Ti-MWW zeolite into Ti-Beta zeolite (as shown in Figure 1.5), which showed excellent catalytic properties for cyclohexene epoxidation with H_2O_2 as the oxidant. The framework structure similarity between MWW and Beta zeolites and the addition of Beta seeds play crucial roles in the crystallization of Ti-Beta. The interzeolite transformation of Ti-MWW (10 × 10-MR pore channels) to large-pore Ti-Beta (12 × 12 × 12-MR pore channels) breaks the traditional interzeolite transformation concept (spontaneous transformation toward a denser zeolite framework) of zeolite synthesis [93–95]. The dissolution–recrystallization of dealuminated-Beta zeolite in a mixture of a tetraethylammonium aqueous solution and a Ti precursor produced Ti-Beta with nanosized crystals, high Ti content, and intercrystal mesoporosity, which showed outstanding catalytic properties in cyclohexene epoxidation with hydrogen peroxide or TBHP as the oxidant [90].

1.3.2 Ti-MOR Synthesis

Ti-MOR, which contains 12-MR pore channels, has shown unique catalytic properties both in the hydroxylation of toluene and ammoximation of ketones [96, 97]. Until now, MOR structures have not been crystallized in siliceous gels without

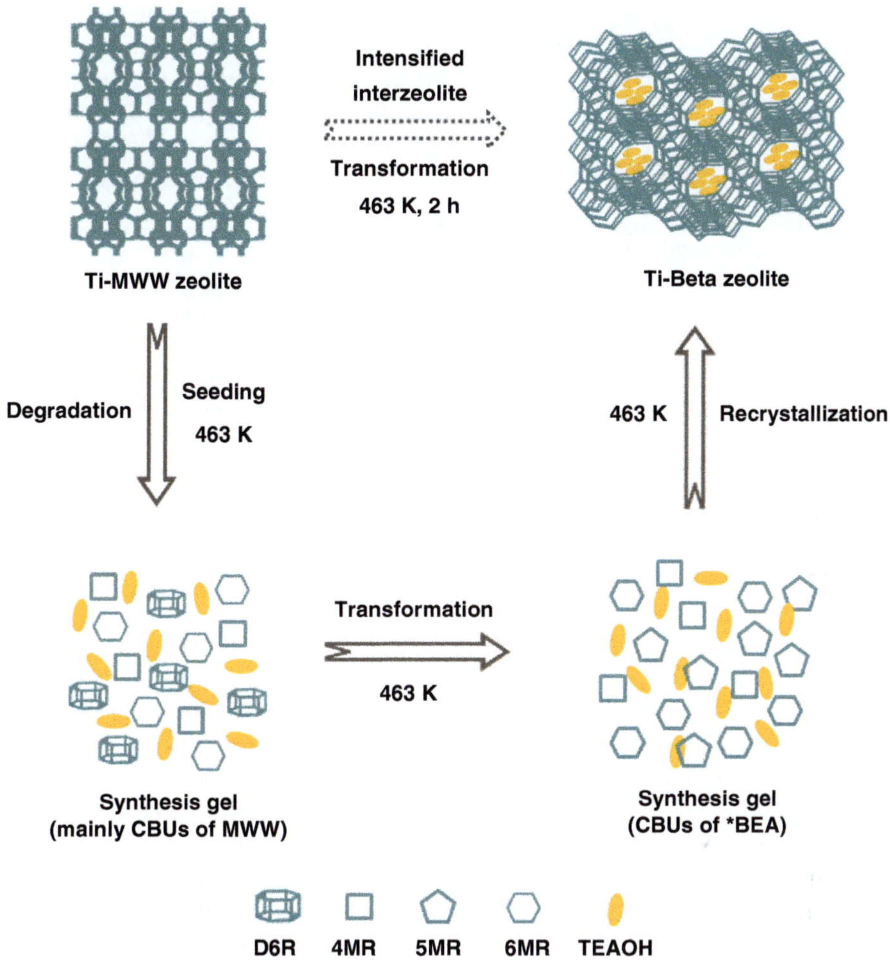

Figure 1.5 Flow Diagram for the interzeolite transformation of Ti-MWW zeolites into Ti-Beta zeolites. Source: Reprinted with permission from Zhu et al. [89]. Copyright Royal Society of Chemistry (2019).

the assistance of Al^{3+} [98]. Thus, a highly active Ti-MOR can only be prepared by dealumination of the aluminosilicate zeolite Al-MOR and the subsequent gas–solid reaction between $TiCl_4$ vapor and the dealuminated MOR at elevated temperature [99]. The transport of $TiCl_4$ molecules and reactants is strongly related to the length of the 12-MR pore channel along the *c*-axis. In this sense, Yang et al. investigated Ti-MOR zeolites with different crystal lengths along the *c*-axis to understand the role of the 12-MR pore channel length on the catalytic performance (Figure 1.6) [100]. The crystal length along the *c*-axis plays a critical role in the diffusion of framework Al species out of the 12-MR pore channels and the diffusion of Ti species into the 12-MR pore channels, which affects the amount of Ti inserted and the catalytic activity.

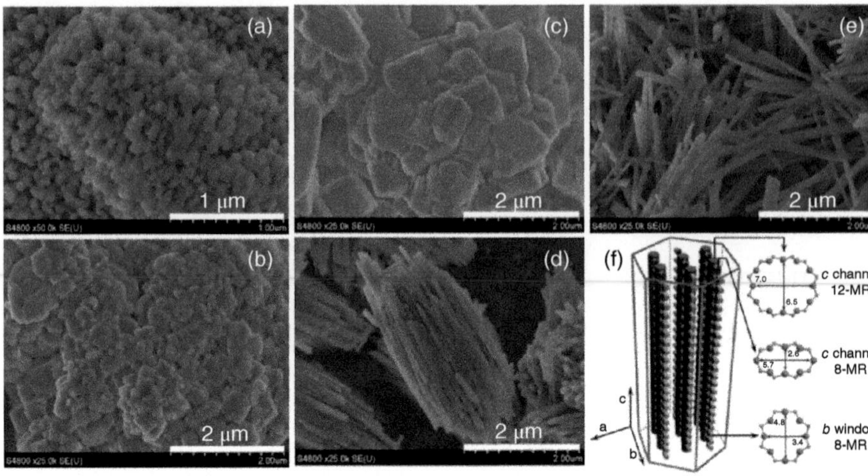

Figure 1.6 The SEM images of MOR zeolites with different crystal lengths of 110 nm (a), 230 nm (b), 630 nm (c), 2450 nm (d), 5160 nm (e) and the pore channels of MOR zeolite (f). Source: Yang et.al. [100]/Reproduced with permission from Elsevier.

1.3.3 Ti-MSE Synthesis

The Ti-containing MSE zeolite is another large-pore titanosilicate with 3D 12 × 10 × 10-MR pore channels and an 18 × 12-MR supercage linked to the 10-MR pore channel. Similar to the MOR, the MSE structure cannot be obtained via hydrothermal crystallization without the assistance of Al^{3+}. Kubota et al. attempted to synthesize Ti-MSE by introducing a Ti tetrabutoxide and H_2O_2 aqueous solution into an aluminosilicate gel with *N*,*N*,*N*′,*N*′-tetraethylbicyclo[2.2.2]oct-7-ene-2,3:5,6-dipyrrolidinium diiodide ($TEBOP^{2+}(I^-)_2$) as the SDA [101]. Most framework Al atoms were extracted from the zeolite framework via acid treatment. However, titanoaluminosilicates show low phenol hydroxylation activity after acid treatment, possibly because of their hydrophilic nature. Thus, highly active Ti-MSE zeolites can only be synthesized using the post-atom planting method. Ti-MSEs are known as Ti-MCM-68 [101–105], Ti-UZM-35 [106], and Ti-YNU-2 [107] according to the parent materials used for Ti planting. Ti-MCM-68 was first synthesized by dealumination of Al-MCM-68 (HTS using $TEBOP^{2+}(I^-)_2$ as the SDA) and a subsequent gas–solid reaction between $TiCl_4$ vapor and dealuminated MCM-68 [104]. The obtained Ti-MCM-68 showed superior catalytic performance compared to that of TS-1 for phenol hydroxylation with hydrogen peroxide in terms of reactivity and para-product selectivity. However, the HTS of Al-MCM-68 requires an extremely long crystallization period (>14 days) using $TEBOP^{2+}(I^-)_2$ as the SDA. Peng et al. [105] found that Al-MCM-68 could be rapidly crystallized in only one day via the inter-zeolite transformation method using siliceous Beta zeolite as the silica source. The corresponding Ti-MCM-68 showed high catalytic activity for anisole hydroxylation. In addition, another MSE titanosilicate, Ti-UZM-35, synthesized by the combination of the dealumination of Al-UZM-35 (synthesized using dimethyldipropylammonium hydroxide (DMDPAOH) as the SDA) and a subsequent gas–solid reaction [106] showed comparable reactivity and product selectivity to the Ti-MCM-68 zeolite in the hydroxylation of phenol. Ti-YNU-2 exhibited remarkably enhanced catalytic properties compared to Ti-MCM-68 although the preparation process was relatively complex [107]. To prepare Ti-YNU-2, an all-silica YNU-2P precursor with abundant internal defects was stabilized by steam treatment and subsequently treated with $TiCl_4$ vapor. Kubota et al. claimed that new five- or six-coordinated Ti species were present in Ti-YNU-2, which accounted for the enhanced catalytic reactivity and para-selectivity in phenol hydroxylation. In addition to the gas-phase Ti source, Ti-MCM-68, which has a high catalytic reactivity and para-selectivity for phenol hydroxylation, has been prepared using a liquid-phase source [102].

1.3.4 Synthesis of Other Large-Pore Titanosilicates

Other large-pore titanosilicates, such as Ti-ZSM-12 (unidimensional 12-MR pore channel) [108], Ti-ITQ-39 (3D 12 × 12 × 10-MR pore channel) [109], Ti-ITQ-7 (3D 12 × 12 × 12-MR pore channel) [110, 111], Ti-SSZ-33 (3D 12 × 12 × 10-MR pore channel) [112], Ti-IPC-2 (2D 12 × 10-MR pore channel) [80, 113], Ti-SSZ-42 (unidimensional 12-MR pore channel) [114], Ti-SAPO-5 (unidimensional 12-MR pore channel) [115], and Ti-ITQ-17 (3D 12 × 12 × 12-MR pore channel) [116], have also been successfully synthesized. Ti-ZSM-12 has been prepared using hexamethylene bis(diethylmethylammonium) hydroxide as the SDA via HTS [108]. However, the catalytic activities of Ti-ZSM-12 in cyclohexene epoxidation with H_2O_2 and TBHP were both much lower than those of Ti-Beta, possibly due to its smaller unidimensional micropore channels (5.6 × 6.0 Å) compared to Beta. Ti-ITQ-39, with its combination of medium- and large-pore channels, was synthesized using the HTS method [109]. Ti-ITQ-39 was highly active in linear and cyclic alkene epoxidation with H_2O_2, showing unique epoxide selectivities compared to TS-1 (medium-pore) and Ti-Beta (large-pore). Ti-SSZ-33 and Ti-SSZ-42, with topologies of CON [112] and IFR [114], respectively, were post-synthesized by the insertion of Ti species into the corresponding borosilicates and were active for cycloalkene epoxidation. Similar to the Beta zeolite, ITQ-7 possesses 3D 12-MR pore channels with one sinusoidal 12-MR pore channel along the *c*-axis and two straight 12-MR pore channels along the *a*- and *b*-axes [117]. Ti-ITQ-7 was synthesized using 1,3,3-trimethyl-6-azonium-tricyclo[3.2.1.46,6]dodecane as the SDA in F^- media [111] and showed catalytic properties comparable to those of Ti-Beta(F) owing to the similar secondary building units in the two zeolite structures. The crystallization time was reduced from 12 days to 12 hours, and the Ti incorporation efficiency was increased from 17.2% to 51% by introducing Ge atoms into the synthetic gels because of the enhanced structural stability of the Ge atoms located in the double-four ring cages [110, 118]. With a higher Ti content, Ti-Ge-ITQ-7 showed superior activity in the epoxidation of various alkenes compared to Ti-ITQ-7. ITQ-17, with the BEC topology, has a 3D 12-MR pore channel system (6.3 × 7.5 Å and 6.0 × 6.9 Å). However, the titanogermanium

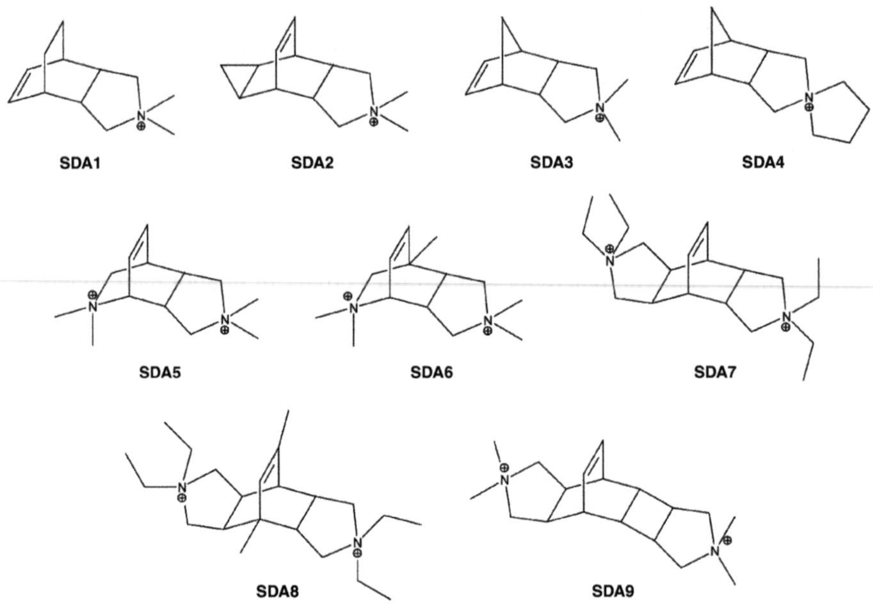

Figure 1.7 Different SDAs used for the synthesis of Ti-ITQ-17. Source: Reprinted with permission from Moliner et al. [116]. Copyright American Chemical Society (2008).

silicate ITQ-17 shows very low hydrothermal stability [116], and ITQ-17 with its stable zeolite structure can only be in the form of germanate [119] or silicogermanate [120]. Nevertheless, it was expected that Ge-free Ti-ITQ-17 should exhibit a superior catalytic behavior to Ti-Beta because the former has a larger pore opening compared to Ti-Beta (6.6 × 6.7 Å and 5.6 × 5.6 Å). All-silica ITQ-17 was first synthesized using 4,4-dimethyl-4-azonia-tricyclo[5.2.2.02,6]undec-8-ene Iodide as the SDA in buffered media containing hexafluorosilicate species and K^+ [121]. After the successful synthesis of all-silica ITQ-17, Ti-ITQ-17 was synthesized under similar conditions. However, the incorporation of Ti species into the zeolite framework was inhibited by K^+ [116]. Thus, further theoretical molecular modeling using different organic molecules as the SDA was performed to find a suitable SDA for Ti-ITQ-17 synthesis that would allow crystallization in the absence of K^+. Among the nine designed SDAs, SDA9 showed an optimum directing ability for the crystallization of Ti-ITQ-17 (see Figure 1.7) [116]. The obtained Ti-ITQ-17 exhibited higher reactivity and epoxide selectivity than Ti-Beta for the epoxidation of bulky cycloalkenes.

1.4 Synthesis of Extra-Large-Pore Titanosilicates

The synthesis of extra-large-pore Ti-zeolites with 14-MR or larger is desirable because of the combination of high hydrothermal stability and accessibility to bulky molecules [122]. In recent years, many extra-large-pore titanosilicates have been obtained by crystallization in highly concentrated gels in the presence of Ge atoms and F^- or by post-synthesis methods. Ti-DON [123] and Ti-CFI [124] with unidimensional 14-MR pore channels were synthesized using bis(pentamethylcyclopentadienyl) cobalt(III) hydroxide and *N*-methylsparteinium hydroxide as SDAs, respectively. However, the isomorphous incorporation of Ti atoms into the frameworks of DON and CFI is more difficult than that of TS-1. Moreover, extra-framework Ti species were found in both the Ti-DON and Ti-CFI samples. Recently, Ti-UTL, with a 2D intersecting 14-MR and 12-MR pore channels, was successfully synthesized in the form of titanogermanosilicate [80]. In addition, other extra-large-pore titanosilicates, such as Ti-ECNU-9 (2D 14×12-MR pore channels) [125] and Ti-IWV (2D 12×10-MR pore channels) [126], were successfully synthesized using post-synthesis methods, which will be further described in Chapter 2.

1.5 Synthesis of Mesoporous Titanosilicates

Ordered mesoporous molecular sieves with amorphous pore walls and uniform mesopore sizes up to 43 nm were first developed by Mobil in 1992 [127]. Corma et al. reported the first Ti-containing mesoporous molecular sieve, MCM-41 (Ti-MCM-41), which was active in selective epoxidation reactions involving bulky substrates [128]. Subsequently, a similar mesoporous titanosilicate Ti-HMS (hexagonal mesoporous silica) with higher activity than Ti-MCM-41 was developed by Tanev et al. [129]. Tatsumi et al. synthesized cubic Ti-MCM-48 with a 3D pore channel system [130] and Ti-SBA-15 with uniform tubular main channels varying from 5 to 30 nm and linked to micropores [131]. Ti-SBA-15 exhibited not only high hydrothermal stability but also excellent resistance to Ti leaching compared to Ti-MCM-41. Other mesoporous titanosilicates, such as hexagonal Ti-SBA-12 [132], cubic Ti-SBA-16 [132], and Ti-KIT-5 [133], have also been prepared. Generally, there are two strategies for synthesizing mesoporous titanosilicates: (i) direct HTS and (ii) the post-atom planting method [134]. In direct HTS, mesoporous titanosilicates are usually synthesized by adding Ti sources, such as TEOT, tetrapropyl orthotitanate, or TBOT, to the synthetic gels for the corresponding all-silica mesoporous materials and crystallization under similar conditions [129, 132]. However, some Ti atoms are trapped inside the amorphous pore walls and are inaccessible. Post-atom incorporation is a robust and universal method [134–136], which is more beneficial for introducing highly accessible Ti atoms than the direct HTS method. Benefiting from the high accessibility of the mesoporous structure, all of the above mesoporous titanosilicates are active in epoxidation reactions containing bulky substrates. The relationship between the synthesis conditions and formation of anatase TiO_2 is ambiguous although anatase TiO_2 has been observed in some mesoporous titanosilicates [132, 133]. Nevertheless, the anatase TiO_2 phase does not negatively affect epoxidation with organic hydroperoxides. In addition to the Ti coordination state, the hydrophilicity/hydrophobicity is another important property of mesoporous titanosilicates. The amorphous pore walls of mesoporous titanosilicates generate numerous silanol groups on the surface, which negatively affect epoxidation reactions in terms of activity and epoxide selectivity [20]. Silylation with trimethylsilyl chloride [137] or hexamethyldisilazane [135] is simple but very effective for enhancing the hydrophobicity of mesoporous titanosilicates, although the thermal stability of such silylated samples is limited. In contrast, mesoporous titanosilicates silylated with fluorosilane species can be calcined at an elevated temperature of 500 °C in air [138]. After the silylation of Ti-SBA-15 with triethoxyfluorosilane, the adsorption capacity of water decreased by 85% compared to that of the parent material [138].

Mesoporosity was introduced into microporous zeolites for the fabrication of hierarchical zeolites to enhance the accessibility of active sites inside the micropores. The construction methods for hierarchical zeolites can be divided into top-down and bottom-up approaches. In the top-down route, a secondary pore structure is formed by the selective extraction of framework atoms from microporous zeolites via post-treatments, such as acid or alkaline leaching, steaming, or irradiation [139–145]. For titanosilicate zeolites, alkaline-derived desilication is more effective than acid-derived demetallation for introducing mesoporosity because of the Si-rich framework composition of titanosilicate zeolites. The concentration of NaOH aqueous solution should be below 0.2 mol L^{-1} to avoid the collapse of zeolite structures when desilication of TS-1 is carried out in NaOH aqueous solution [146]. A hierarchical TS-1 zeolite with a hollow structure was synthesized using a mixture of TPAOH and NaOH. This TPAOH-assisted method is usually accompanied by a recrystallization process owing to the structure-directing nature of TPAOH, resulting in a well-preserved crystallinity and hierarchical structure. The removal of residual Na^+ is necessary after desilication because the presence of Na^+ is detrimental to the catalytic reactivity of the formed hierarchical titanosilicate zeolites [147]. In addition, hierarchical structures can be formed by treating microporous TS-1 zeolites with TBAOH and NaOH under appropriate treatment conditions [142]. Using bottom-up approaches, mesoporosity can be established during the synthesis of zeolites with or without the presence of a secondary template. Secondary templating methods can be further divided into hard- and soft-templating strategies. For hierarchical TS-1 zeolites obtained via hard-templating, the resultant shape and size of the mesopores are closely related to the template morphology. Nanocarbon particles, polymers, and biological materials can be used as hard templates for the synthesis of hierarchical titanosilicates [148–151]. This method can be used to successfully construct hierarchical zeolites with a controllable mesopore size distribution. The soft-templating route, using surfactants or organosilanes as secondary templates, is another option for constructing hierarchical titanosilicates. However, the relatively small SDAs for microporous structures and relatively large soft templates for mesoporous structures usually work in a competitive manner instead of cooperatively, leading to the formation of separated phases of microporous zeolites and mesoporous materials rather than hierarchical zeolite structures. Tuel et al. [152] reported the synthesis of a hierarchical TS-1 zeolite with organosilane as both a Si source and mesoporogenic agent that could be introduced into growing zeolite domains via the formation of covalent bonds with SiO_2 or TiO_2, thus avoiding the phase-separation phenomenon. Secondary soft templates, which are used in the synthesis of hierarchical titanosilicate zeolites, can also direct the synthesis of mesoporous titanosilicates [153–155]. Secondary template-free strategies, including seed assistance, kinetic regulation of crystallization, and DGC, are efficient synthetic approaches for the construction of hierarchical zeolites [156]. The seed-assisted method is used not only to accelerate the crystallization process, enhance zeolite purity, adjust the morphology and crystalline size, and lower the synthetic cost but also to construct hierarchical titanosilicates [157]. However, significant challenges exist in constructing hierarchical titanosilicates with regular mesopore sizes and elucidating the formation mechanism of mesopores. In addition to the seed-assisted method, the kinetically regulated crystallization method has also been employed to construct hierarchical TS-1 zeolites by adjusting the zeolite nucleation and crystal growth, as well as manipulating the coordination and distribution of Ti species [158, 159].

1.6 Synthesis of ETSs

ETS is a class of zeolite-type titanosilicate materials that share SiO_4 tetrahedra and TiO_6^{2-} octahedra and possess strong Lewis basicity derived from the TiO_6^{2-} octahedra. The most representative ETS type titanosilicates are ETS-10 and ETS-4. The microporous titanosilicate ETS-10 was developed for the first time by Kuznicki [160]. It possesses a 12-MR pore channel approximately 4.9 × 7.6 Å in diameter and is widely used in ion-exchange reactions, such as those involving rare-earth cations [161–164], dehydration [165], catalysis [166, 167], photocatalysis [168], and sensor materials [169]. ETS-10 can be hydrothermally synthesized in the presence of SDAs, including pyrrolidine, tetramethylammonium chloride, 1,2-diaminoethane, choline chloride, tetraethylammonium chloride, TPABr, tetrabutylammonium bromide, and hexaethyl diquat-5 bromide [170–177]. Several different Ti precursors, such as $TiCl_3$ [178], $TiCl_4$ [179], TiF_4 [178, 180], $(NH_4)_2TiF_6$ [178], and TiO_2 (anatase or rutile) [170, 178, 181], can be employed as Ti sources for the synthesis of ETS-10. The catalytic properties of ETS-10 are closely related to its morphology, which depends on the synthesis conditions and SDAs [182, 183]. ETS-10 can be synthesized as powders and pellets. In comparison with the powder form, ETS-10 in pellet form is easier to apply and recover. Moreover, self-bonded pellets avoid the negative effects of binders on the diffusion of reactants to the Ti species [184]. In addition, carbon nanotubes are also used for to prepare self-linked ETS-10 [185].

ETS-4, with two types of 8- and 12-MR pore channels, has a structure similar to that of mineral zorite [160, 186, 187]. Because the 12-MR pore channels are inaccessible owing to the pronounced stacking disorder along the *c*-direction, guest molecules can only enter the crystal interior via the 8-MR pore channels. The size of the 8-MR pore channels can be modulated by dehydration, i.e. the so-called molecular gate effect, which has been successfully applied in the commercial separation of nature gas [188]. The dehydration process is usually performed at 150–300 °C [188, 189]. The thermal stability of ETS-4 can be improved by using alkaline earth cations instead of Na^+ [188, 190]. The most representative example is Sr-exchanged ETS-4, which has a thermal stability up to 350 °C [188]. In addition to the thermal adjustment of the pore size, replacing the anions in ETS-4 for the adjustment of openings can be used for the purification of multicomponent gases [191]. ETS-4 has been employed in the adsorption–separation of CH_4/N_2, O_2/Ar, O_2/N_2, and H_2/CH_4 [192–194]. In particular, Sr-exchanged ETS-4 shows high stability for the adsorption–separation of CH_4 and N_2 in pressure swing adsorption processes [195, 196]. ETS-4 can adsorb and enrich CH_4 from natural gases to alleviate energy shortages.

1.7 Conclusions

Titanosilicates, which can be divided into microporous, mesoporous, and hierarchical titanosilicates based on their textural properties and pore sizes, possess isolated tetrahedral Ti species as active sites and have expanded the application scope of zeolites as heterogeneous catalysts from the acid catalysis to redox fields. Although the HTS, DGC, fluoride-assisted media, and post-synthesis methods have achieved great success in the synthesis of titanosilicates, several challenges with regard to increasing the content and accessibility of Ti active sites still remain. Therefore, efforts should be directed toward developing hierarchical titanosilicates with high Ti contents and manipulating the coordination and distribution of Ti species.

References

1 Cundy, C.S. and Cox, P.A. (2003). The hydrothermal synthesis of zeolites: history and development from the earliest days to the present time. *Chem. Rev.* 103: 663–702.

2 Bellussi, G. and Millini, R. (2018). *Stucture and Reactvity of Metals in Zeolite Materials*. Switzerland: Springer.

3 Xu, H. and Wu, P. (2019). Recent progress of metallosilicate catalysts in environment-friendly selective oxidation reactions. *Chin. J. Catal.* 40: S51–S56.

4 Wu, Q., Xu, C., Zhu, L. et al. (2022). Recent strategies for synthesis of metallosilicate zeolites. *Catal. Today* 410: 2–12.

5 Přech, J. (2018). Catalytic performance of advanced titanosilicate selective oxidation catalysts – a review. *Catal. Rev.* 60: 71–131.

6 Perera, A.S. and Coppens, M.-O. (2016). Titano-silicates: highlights on development, evolution and application in oxidative catalysis. *J. Catal.* 28: 119–143.

7 Xu, H. and Wu, P. (2017). Recent progresses in titanosilicates. *Chin. J. Chem.* 35: 836–844.

8 Smeets, V., Gaigneaux, E.M., and Debecker, D.P. (2022). Titanosilicate epoxidation catalysts: a review of challenges and opportunities. *ChemCatChem* 14: e202101132.

9 Wu, P., Xu, H., Xu, L. et al. (2013). *MWW-Type Titanosilicate, Synthesis, Structural Modification and Catalytic Applications to Green Oxidations*. New York: Springer.

10 Tang, B., Dai, W., Wu, G. et al. (2014). Improved postsynthesis strategy to Sn-Beta zeolites as Lewis acid catalysts for the ring-opening hydration of epoxides. *ACS Catal.* 4: 2801–2810.

11 Lu, X., Xu, H., Yan, J. et al. (2018). One-pot synthesis of ethylene glycol by oxidative hydration of ethylene with hydrogen peroxide over titanosilicate catalysts. *J. Catal.* 358: 89–99.

12 Xu, W., Wang, X., Hou, W. et al. (2022). Synergetic effects of Sn and Ti incorporated in MWW zeolites on promoting the oxidative hydration of ethylene with H_2O_2 to ethylene glycol. *J. Catal.* 413: 554–564.

13 Wang, Y., Liu, Y., Li, X. et al. (2009). Intermolecular condensation of ethylenediamine to 1,4-diazabicyclo[2,2,2]octane over TS-1 catalysts. *J. Catal.* 266: 258–267.

14 Jin, X., Peng, R., Tong, W. et al. (2022). Investigation of the active centers and structural modifications for TS-1 in catalyzing the Beckmann rearrangement. *Catal. Today* 405–406: 193–202.

15 Moliner, M. and Corma, A. (2014). Advances in the synthesis of titanosilicates: from the medium pore TS-1 zeolite to highly-accessible ordered materials. *Microporous Mesoporous Mater.* 189: 31–40.

16 Xia, C., Lin, M., Zhu, B. et al. (2016). *Hollow Titanium Silicalite Zeolite: From Fundamental Research to Commercial Application in Environmental-Friendly Catalytic Oxidation Processes*. Croatia: InTech.

17 Oleksiienko, O., Wolkersdorfer, C., and Sillanpää, M. (2017). Titanosilicates in cation adsorption and cation exchange – a review. *Chem. Eng. J.* 317: 570–585.

18 Bai, R., Song, Y., Bai, R., and Yu, J. (2021). Creation of hierarchical titanosilicate TS-1 zeolites. *Adv. Mater. Interfaces* 8: 2001095.

19 Marco, T., Perego, G. and Notari, B. 1983. *US Patent No. 4*,410,501.

20 Blasco, T., Corma, A., Navarro, M.T., and Pariente, J.P. (1995). Synthesis, characterization, and catalytic activity of Ti-MCM-41 structures. *J. Catal.* 156: 65–74.

21 Lamberti, C., Bordiga, S., Zecchina, A. et al. (1999). Structural characterization of Ti-silicalite-1: a synchrotron radiation X-ray powder diffraction study. *J. Catal.* 183: 222–231.

22 Camblor, M.A., Corma, A., Martínez, A., and Pérez-Pariente, J. (1992). Synthesis of a titaniumsilicoaluminate isomorphous to zeolite beta and its application as a catalyst for the selective oxidation of large organic molecules. *J. Chem. Soc. Chem. Commun.* 589–590.

23 Fan, W., Duan, R.-G., Yokoi, T. et al. (2008). Synthesis, crystallization mechanism, and catalytic properties of titanium-rich TS-1 free of extraframework titanium species. *J. Am. Chem. Soc.* 130: 10150–10164.

24 Thangaraj, A., Eapen, M.J., Sivasanker, S., and Ratnasamy, P. (1992). Studies on the synthesis of titanium silicalite, TS-1. *Zeolites* 12: 943–950.

25 Khomane, R.B., Kulkarni, B.D., Paraskar, A., and Sainkar, S.R. (2002). Synthesis, characterization and catalytic performance of titanium silicalite-1 prepared in micellar media. *Mater. Chem. Phys.* 76: 99–103.

26 Zhang, T., Chen, X., Chen, G. et al. (2018). Synthesis of anatase-free nano-sized hierarchical TS-1 zeolites and their excellent catalytic performance in alkene epoxidation. *J. Mater. Chem. A* 6: 9473–9479.

27 On, D.T., Kaliaguine, S., and Bonneviot, L. (1995). Titanium boralites with MFI structure characterized using XRD, XANES, IR, and UV-visible techniques: effect of hydrogen peroxide on the preparation. *J. Catal.* 157: 235–243.

28 Lin, D., Zhang, Q.D., Qin, Z.X. et al. (2021). Reversing titanium oligomer formation towards high-efficiency and green synthesis of titanium-containing molecular sieves. *Angew. Chem. Int. Ed.* 60: 3443–3448.

29 Tatsumi, T. and Jappar, N. (1998). Properties of Ti-Beta zeolites synthesized by dry-gel conversion and hydrothermal methods. *J. Phys. Chem. B* 102: 7126–7131.

30 Wu, P., Miyaji, T., Liu, Y. et al. (2005). Synthesis of Ti-MWW by a dry-gel conversion method. *Catal. Today* 99: 233–240.

31 Du, Q., Guo, Y., Wu, P., and Liu, H. (2018). Synthesis of hierarchically porous TS-1 zeolite with excellent deep desulfurization performance under mild conditions. *Microporous Mesoporous Mater.* 264: 272–280.

32 Luo, Y., Zhu, Y., Pan, J., and Chen, X. (2020). Fast synthesis of hierarchical Al-free Ti-BEA plate-like nanocrystals from low-templated dry gel via a steam-assisted conversion method. *Green Chem.* 22: 1681–1697.

33 Smeets, V., Gaigneaux, E.M., and Debecker, D.P. (2020). Hierarchical micro-/macroporous TS-1 zeolite epoxidation catalyst prepared by steam assisted crystallization. *Microporous Mesoporous Mater.* 293: 109801.

34 Blasco, T., Camblor, M.A., Corma, A. et al. (1996). Unseeded synthesis of Al-free Ti-β zeolite in fluoride medium: a hydrophobic selective oxidation catalyst. *Chem. Commun.* 2367–2368.

35 Wu, P., Komatsu, T., and Yashima, T. (1997). Preparation of titanosilicate with mordenite structure by atomplanting method and its catalytic properties for hydroxylation of aromatics. In: *Studies in Surface Science and Catalysis*, vol. 105 (ed. H. Chon, S.-K. Ihm, and Y.S. Uh), 663–670. Elsevier.

36 Huo, Y., Zhang, Y., Xu, W. et al. (2020). Acid-modulated synthesis of Ti-MWW zeolites with rich framework Ti species for efficient epoxidation. *Ind. Eng. Chem. Res.* 59: 19929–19937.

37 Tang, B., Dai, W., Sun, X. et al. (2014). A procedure for the preparation of Ti-Beta zeolites for catalytic epoxidation with hydrogen peroxide. *Green Chem.* 16: 2281–2291.

38 Klaewkla, R., Kulprathipanja, S., Rangsunvigit, P. et al. (2003). Phenol hydroxylation using Ti- and Sn-containing silicalites. *Chem. Commun.* 1500–1501.

39 Le Bars, J., Dakka, J., and Sheldon, R.A. (1996). Ammoximation of cyclohexanone and hydroxyaromatic ketones over titanium molecular sieves. *Appl. Catal. A Gen.* 136: 69–80.

40 Notari, B. (1988). Synthesis and catalytic properties of titanium containing zeolites. In: *Studies in Surface Science and Catalysis*, vol. 37 (ed. P.J. Grobet, W.J. Mortier, E.F. Vansant, and G. Schulz-Ekloff), 413–425. Elsevier.

41 Huybrechts, D.R.C., Vaesen, I., Li, H.X., and Jacobs, P.A. (1991). Factors influencing the catalytic activity of titanium silicalites in selective oxidations. *Catal. Lett.* 8: 237–244.

42 Khouw, C.B. and Davis, M.E. (1995). Catalytic activity of titanium silicates synthesized in the presence of alkali-metal and alkaline-earth ions. *J. Catal.* 151: 77–86.

43 Bellussi, G. and Fattore, V. (1991). Isomorphous substitution in zeolites: a route for the preparation of novel catalysts. In: *Studies in Surface Science and Catalysis*, vol. 69 (ed. P.A. Jacobs, N.I. Jaeger, L. Kubelková, and B. Wichterlov'), 79–92. Elsevier.

44 van der Pol, A.J.H.P. and van Hooff, J.H.C. (1992). Parameters affecting the synthesis of titanium silicalite 1. *Appl. Catal. A Gen.* 92: 93–111.

45 Li, G., Guo, X., Wang, X. et al. (1999). Synthesis of titanium silicalites in different template systems and their catalytic performance. *Appl. Catal. A Gen.* 185: 11–18.

46 Zuo, Y., Liu, M., Zhang, T. et al. (2015). Enhanced catalytic performance of titanium silicalite-1 in tuning the crystal size in the range 1200–200 nm in a tetrapropylammonium bromide system. *ChemCatChem* 7: 2660–2668.

47 Wang, J., Xu, L., Zhang, K. et al. (2012). Multilayer structured MFI-type titanosilicate: synthesis and catalytic properties in selective epoxidation of bulky molecules. *J. Catal.* 288: 16–23.

48 Wang, Z., Xu, L., Jiang, J.-G. et al. (2012). One-pot synthesis of catalytically active and mechanically robust mesoporous TS-1 microspheres with the aid of triblock copolymer. *Microporous Mesoporous Mater.* 156: 106–114.

49 Deng, Z., Yang, Y., Lu, X. et al. (2016). Studies on the epoxidation of methallyl chloride over TS-1 microsphere catalysts in a continuous slurry reactor. *Catal. Sci. Technol.* 6: 2605–2615.

50 Leonowicz, M.E., Lawton, J.A., Lawton, S.L., and Rubin, M.K. (1994). MCM-22: a molecular sieve with two independent multidimensional channel systems. *Science* 264: 1910–1913.

51 Corma, A., Díaz, U., Fornés, V. et al. (1999). Ti/ITQ-2, a new material highly active and selective for the epoxidation of olefins with organic hydroperoxides. *Chem. Commun.* 779–780.

52 Wu, P., Tatsumi, T., Komatsu, T., and Yashima, T. (2001). A novel titanosilicate with MWW structure. I. Hydrothermal synthesis, elimination of extraframework titanium, and characterizations. *J. Phys. Chem. B* 105: 2897–2905.

53 Wu, P. and Tatsumi, T. (2002). Preparation of B-free Ti-MWW through reversible structural conversion. *Chem. Commun.* 1026–1027.

54 Yan, M., Jin, F., Ding, Y. et al. (2019). Synthesis of titanium-incorporated MWW zeolite by sequential deboronation and atom-planting treatment of ERB-1 as an epoxidation catalyst. *Ind. Eng. Chem. Res.* 58: 4764–4773.

55 Liu, N., Liu, Y., Xie, W. et al. (2007). Hydrothermal synthesis of boron-free Ti-MWW with dual structure-directing agents. In: *Studies in Surface Science and Catalysis*, vol. 170 (ed. R. Xu, Z. Gao, J. Chen, and W. Yan), 464–469. Elsevier.

56 Reddy, J.S., Kumar, R., and Ratnasamy, P. (1990). Titanium silicalite-2: synthesis, characterization and catalytic properties. *Appl. Catal.* 58: L1–L4.

57 Reddy, J.S. and Sivasanker, S. (1991). Selective oxidation of cyclohexane over TS-2, a titanium silicate molecular sieve. *Catal. Lett.* 11: 241–244.

58 Serrano, D.P., Uguina, M.A., Sanz, R. et al. (2004). Synthesis and crystallization mechanism of zeolite TS-2 by microwave and conventional heating. *Microporous Mesoporous Mater.* 69: 197–208.

59 Tuel, A. and Ben Taârit, Y. (1993). Synthesis, characterization, and catalytic properties of titanium silicates prepared using phosphonium ions. *Zeolites* 13: 357–364.

60 Uguina, M.A., Serrano, D.P., Ovejero, G. et al. (1997). TS-2 synthesis from wetness-impregnated SiO_2-TiO_2 xerogels. *Zeolites* 18: 368–378.

61 Salehirad, F., Aghabozorg, H.R., Manoochehri, M., and Aghabozorg, H. (2004). Synthesis of titanium silicalite-2 (TS-2) from methylamine-tetrabutylammonium hydroxide media. *Catal. Commun.* 5: 359–365.

62 Zhang, F., Guo, X., Wang, X. et al. (2000). Preparation of titanium-containing zeolites with MEL structure from B-ZSM-11 and their characterization. *Appl. Catal. A Gen.* 192: 157–163.

63 Xu, H., Tian, W., Xu, L. et al. (2020). Crossed intergrowth triggered TS-2 microsphere: formation mechanism, modification and catalytic performance. *Chin. J. Catal.* 41: 1109–1117.

64 Kaur Ahedi, R. and Kotasthane, A.N. (1998). Synthesis of FER titanosilicates from a non-aqueous alkali-free seeded system. *J. Mater. Chem.* 8: 1685–1686.

65 Corma, A., Diaz, U., Domine, M.E., and Fornes, V. (2000). Ti-ferrierite and TiITQ-6: synthesis and catalytic activity for the epoxidation of olefins with H_2O_2. *Chem. Commun.* 137–138.

66 Serrano, D.P., Li, H.-X., and Davis, M.E. (1992). Synthesis of titanium-containing ZSM-48. *J. Chem. Soc. Chem. Commun.* 745–747.

67 Madhusudan Reddy, K., Kaliaguine, S., and Sayari, A. (1994). Synthesis of titanium containing silica ZSM-48 (TS-48) using hexamethonium hydroxide as template. *Catal. Lett.* 23: 169–173.

68 Tuel, A. and Ben Taârit, Y. (1995). A new template for the synthesis of titanium silicalites with the ZSM-48 structure. *Zeolites* 15: 164–170.

69 Madhusudan Reddy, K., Kaliaguine, S., Sayari, A. et al. (1994). Synthesis and characterization of TS-48, a titanium containing silica analog of ZSM-48. *Catal. Lett.* 23: 175–187.

70 Variani, Y.M., Rojas, A., and Pergher, S.B.C. (2018). Synthesis and characterization of Ti-STF zeolite using 1,2,3-triethyl-4-methylimidazolium as structure-directing agent. *Microporous Mesoporous Mater.* 262: 106–111.

71 Damin, A., Bordiga, S., Zecchina, A. et al. (2003). Ti-chabazite as a model system of Ti(IV) in Ti-zeolites: a periodic approach. *J. Chem. Phys.* 118: 10183–10194.

72 Eilertsen, E.A., Bordiga, S., Lamberti, C. et al. (2011). Synthesis of titanium chabazite: a new shape selective oxidation catalyst with small pore openings and application in the production of methyl formate from methanol. *ChemCatChem* 3: 1869–1871.

73 Imasaka, S., Ishii, H., Hayashi, J. et al. (2019). Synthesis of CHA-type titanosilicate zeolites using titanium oxide as Ti source and evaluation of their physicochemical properties. *Microporous Mesoporous Mater.* 273: 243–248.

74 Araki, S., Ishii, H., Imasaka, S., and Yamamoto, H. (2020). Synthesis and gas permeation properties of chabazite-type titanosilicate membranes synthesized using nano-sized seed crystals. *Microporous Mesoporous Mater.* 292: 109798.

75 Boal, B.W., Schmidt, J.E., Deimund, M.A. et al. (2015). Facile synthesis and catalysis of pure-silica and heteroatom LTA. *Chem. Mater.* 27: 7774–7779.

76 Hernandez-Tamargo, C.E., Roldan, A., Ngoepe, P.E., and Leeuw, N.H.d. (2017). Periodic modeling of zeolite Ti-LTA. *J. Chem. Phys.* 147: 074701.

77 Murase, T. and Komura, K. (2017). Hydrothermal synthesis of titanosilicate type zeolitic layered PLS-1 and CDS-1 molecular sieve with CDO topology. *J. Porous. Mater.* 24: 203–209.

78 Funase, N., Takata, T., Tsunoji, N. et al. (2016). Hydrothermal conversion of titanated FAU to AEI zeolite and its enhanced catalytic performance for NO_x reduction. *Adv. Porous Mater.* 4: 62–72.

79 Eilertsen, E.A., Giordanino, F., Lamberti, C. et al. (2011). Ti-STT: a new zeotype shape selective oxidation catalyst. *Chem. Commun.* 47: 11867–11869.

80 Přech, J. and Čejka, J. (2016). UTL titanosilicate: an extra-large pore epoxidation catalyst with tunable textural properties. *Catal. Today* 277: 2–8.

81 Přech, J., Morris, R.E., and Čejka, J. (2016). Selective oxidation of bulky organic sulphides over layered titanosilicate catalysts. *Catal. Sci. Technol.* 6: 2775–2786.

82 Blasco, T., Camblor, M.A., Corma, A., and Perez-Pariente, J. (1993). The state of Ti in titanoaluminosilicates isomorphous with zeolite .beta. *J. Am. Chem. Soc.* 115: 11806–11813.

83 Corma, A., Camblor, M.A., Esteve, P. et al. (1994). Activity of Ti-Beta catalyst for the selective oxidation of alkenes and alkanes. *J. Catal.* 145: 151–158.

84 Camblor, M.A., Costantini, M., Corma, A. et al. (1996). Synthesis and catalytic activity of aluminium-free zeolite Ti-β oxidation catalysts. *Chem. Commun.* 1339–1340.

85 Rigutto, M.S., de Ruiter, R., Niederer, J.P.M., and van Bekkum, H. (1994). Titanium-containing large pore molecular sieves from boron-Beta: preparation, characterization and catalysis. In: *Studies in Surface Science and Catalysis*, vol. 84 (ed. J. Weitkamp, H.G. Karge, H. Pfeifer, and W. Hölderich), 2245–2252. Elsevier.

86 Blasco, T., Camblor, M.A., Corma, A. et al. (1998). Direct synthesis and characterization of hydrophobic aluminum-free Ti–Beta zeolite. *J. Phys. Chem. B* 102: 75–88.

87 Camblor, M.A., Corma, A., Esteve, P. et al. (1997). Epoxidation of unsaturated fatty esters over large-pore Ti-containing molecular sieves as catalysts: important role of the hydrophobic–hydrophilic properties of the molecular sieve. *Chem. Commun.* 795–796.

88 Takashi, T., Qinghua, X., and Nizamidin, J. (1997). Synthesis of Ti-beta zeolite with high oxidation activity by a dry-gel conversion technique. *Chem. Lett.* 26: 677–678.

89 Zhu, Z., Xu, H., Wang, B. et al. (2019). Intensified interzeolite transformation: ultrafast synthesis of active and stable Ti-Beta zeolites without solvents. *Chem. Commun.* 55: 14279–14282.

90 Wang, B., Xu, H., Zhu, Z. et al. (2019). Ultrafast synthesis of nanosized Ti-Beta as an efficient oxidation catalyst via a structural reconstruction method. *Catal. Sci. Technol.* 9: 1857–1866.

91 Rao, P.R.H.P. and Matsukata, M. (1996). Dry-gel conversion technique for synthesis of zeolite BEA. *Chem. Commun.* 1441–1442.

92 Jappar, N., Xia, Q., and Tatsumi, T. (1998). Oxidation activity of Ti-Beta synthesized by a dry-gel conversion method. *J. Catal.* 180: 132–141.

93 Inoue, T., Itakura, M., Jon, H. et al. (2009). Synthesis of LEV zeolite by interzeolite conversion method and its catalytic performance in ethanol to olefins reaction. *Microporous Mesoporous Mater.* 122: 149–154.

94 Itabashi, K., Kamimura, Y., Iyoki, K. et al. (2012). A working hypothesis for broadening framework types of zeolites in seed-assisted synthesis without organic structure-directing agent. *J. Am. Chem. Soc.* 134: 11542–11549.

95 Goel, S., Zones, S.I., and Iglesia, E. (2014). Encapsulation of metal clusters within MFI via interzeolite transformations and direct hydrothermal syntheses and catalytic consequences of their confinement. *J. Am. Chem. Soc.* 136: 15280–15290.

96 Xu, H., Zhang, Y., Wu, H. et al. (2011). Postsynthesis of mesoporous MOR-type titanosilicate and its unique catalytic properties in liquid-phase oxidations. *J. Catal.* 281: 263–272.

97 Ding, J., Xu, L., Yu, Y. et al. (2013). Clean synthesis of acetaldehyde oxime through ammoximation on titanosilicate catalysts. *Catal. Sci. Technol.* 3: 2587–2595.

98 Fernández, R. and Cardoso, D. (2005). Synthesis and properties of Ti-MOR molecular sieve on the oxidation of cyclohexene: influence of the Ti source. *Catal. Today* 107–108: 844–848.

99 Wu, P., Komatsu, T., and Yashima, T. (1996). Characterization of titanium species incorporated into dealuminated mordenites by means of IR spectroscopy and ^{18}O-exchange technique. *J. Phys. Chem.* 100: 10316–10322.

100 Yang, Y., Ding, J., Xu, C. et al. (2015). An insight into crystal morphology-dependent catalytic properties of MOR-type titanosilicate in liquid-phase selective oxidation. *J. Catal.* 325: 101–110.

101 Ikehara, Y., Ohno, Y., Inagaki, S., and Kubota, Y. (2017). Preparation of MSE-type titanosilicate via crystallization of titanoaluminosilicate and its catalytic use for selective oxidation of phenol using H_2O_2. *Chem. Lett.* 46: 1842–1845.

102 Inagaki, S., Ishizuka, R., Ikehara, Y. et al. (2021). Highly efficient titanosilicate catalyst Ti-MCM-68 prepared using a liquid-phase titanium source for the phenol oxidation. *RSC Adv.* 11: 3681–3684.

103 Inagaki, S., Tsuboi, Y., Sasaki, M. et al. (2016). Enhancement of Para-selectivity in the phenol oxidation with H_2O_2 over Ti-MCM-68 zeolite catalyst. *Green Chem.* 18: 735–741.

104 Kubota, Y., Koyama, Y., Yamada, T. et al. (2008). Synthesis and catalytic performance of Ti-MCM-68 for effective oxidation reactions. *Chem. Commun.* 6224–6226.

105 Peng, R., Pan, H., Li, X. et al. (2022). Post-synthesis of MSE-type titanosilicates by interzeolite transformation for selective anisole hydroxylation. *Catal. Sci. Technol.* 12: 6098–6111.

106 Yin, J., Lu, X., Yan, J. et al. (2020). Postsynthesis of Ti-UZM-35 titanosilicate as efficient catalyst for phenol hydroxylation reaction. *Microporous Mesoporous Mater.* 305: 110321.

107 Sasaki, M., Sato, Y., Tsuboi, Y. et al. (2014). Ti-YNU-2: a microporous titanosilicate with enhanced catalytic performance for phenol oxidation. *ACS Catal.* 4: 2653–2657.

108 Tuel, A. (1995). Synthesis, characterization, and catalytic properties of the new TiZSM-12 zeolite. *Zeolites* 15: 236–242.

109 Moliner, M. and Corma, A. (2012). Direct synthesis of a titanosilicate molecular sieve containing large and medium pores in its structure. *Microporous Mesoporous Mater.* 164: 44–48.

110 Corma, A., Díaz-Cabañas, M.J., Domine, M.E., and Rey, F. (2000). Ultra fast and efficient synthesis of Ti-ITQ-7 and positive catalytic implications. *Chem. Commun.* 1725–1726.

111 Díaz-Cabañas, M.-J., Villaescusa, L.A., and Camblor, M.A. (2000). Synthesis and catalytic activity of Ti-ITQ-7: a new oxidation catalyst with a three-dimensional system of large pore channels. *Chem. Commun.* 761–762.

112 Přech, J., Vitvarová, D., Lupínková, L. et al. (2015). Titanium impregnated borosilicate zeolites for epoxidation catalysis. *Microporous Mesoporous Mater.* 212: 28–34.

113 Liu, X., Zhang, L., Xu, H. et al. (2018). Pore size-tunable titanosilicates post-synthesized from germanosilicate by structural reorganization and H_2TiF_6-assisted isomorphous substitution. *Appl. Catal. A Gen.* 550: 11–19.

114 Hamoudi, S., Larachi, F., and Sayari, A. (2001). Synthesis and characterization of titanium-substituted large pore SSZ-42 zeolite. *Catal. Lett.* 77: 227–231.

115 Tuel, A. (1995). Synthesis, characterization, and catalytic properties of titanium silicoaluminophosphate TAPSO-5. *Zeolites* 15: 228–235.

116 Moliner, M., Serna, P., Cantin, A. et al. (2008). Synthesis of the Ti-silicate form of BEC polymorph of beta-zeolite assisted by molecular modeling. *J. Phys. Chem. C* 112: 19547–19554.

117 Villaescusa, L.A., Barrett, P.A., and Camblor, M.A. (1999). ITQ-7: a new pure silica polymorph with a three-dimensional system of large pore channels. *Angew. Chem. Int. Ed.* 38: 1997–2000.

118 Zicovich-Wilson, C.M. and Corma, A. (2000). Modifying the catalytic activity of Ti-zeolites by isomorphic substitution of Si by Ge atoms. A periodic quantum-chemical study. *J. Phys. Chem. B* 104: 4134–4140.

119 Conradsson, T., Dadachov, M.S., and Zou, X.D. (2000). Synthesis and structure of $(Me_3N)_6[Ge_{32}O_{64}]\cdot(H_2O)_{4.5}$, a thermally stable novel zeotype with 3D interconnected 12-ring channels. *Microporous Mesoporous Mater.* 41: 183–191.

120 Corma, A., Navarro, M.T., Rey, F. et al. (2001). Pure polymorph C of zeolite Beta synthesized by using framework isomorphous substitution as a structure-directing mechanism. *Angew. Chem. Int. Ed.* 40: 2277–2280.

121 Cantín, Á., Corma, A., Díaz-Cabañas, M.J. et al. (2006). Synthesis and characterization of the all-silica pure polymorph C and an enriched polymorph B intergrowth of zeolite Beta. *Angew. Chem. Int. Ed.* 45: 8013–8015.

122 Corma, A. and Davis, M.E. (2004). Issues in the synthesis of crystalline molecular sieves: towards the crystallization of low framework-density structures. *ChemPhysChem* 5: 304–313.

123 Balkus, K.J., Gabrielov, A.G., and Zones, S.I. (1995). The synthesis of UTD-1, Ti-UTD-1 and Ti-UTD-8 using $CP^*{}_2CoOH$ as a structure directing agent. In: *Studies in Surface Science and Catalysis*, vol. 97 (ed. L. Bonneviot and S. Kaliaguine), 519–525. Elsevier.

124 Přech, J., Kubů, M., and Čejka, J. (2014). Synthesis and catalytic properties of titanium containing extra-large pore zeolite CIT-5. *Catal. Today* 227: 80–86.

125 Yang, B., Jiang, J.-G., Xu, H. et al. (2018). Synthesis of extra-large-pore zeolite ECNU-9 with intersecting 14*12-ring channels. *Angew. Chem. Int. Ed.* 57: 9515–9519.

126 Jiao, M., Zhao, Y., Jiang, J. et al. (2021). Extra-large pore titanosilicate synthesized via reversible 3D–2D–3D structural transformation as highly active catalyst for cycloalkene epoxidation. *ACS Catal.* 11: 2650–2662.

127 Kresge, C., Leonowicz, M., Roth, W.J. et al. (1992). Ordered mesoporous molecular sieves synthesized by a liquid-crystal template mechanism. *Nature* 359: 710–712.

128 Corma, A., Navarro, M.T., and Pariente, J.P. (1994). Synthesis of an ultralarge pore titanium silicate isomorphous to MCM-41 and its application as a catalyst for selective oxidation of hydrocarbons. *J. Chem. Soc. Chem. Commun.* 147–148.

129 Tanev, P.T., Chibwe, M., and Pinnavaia, T.J. (1994). Titanium-containing mesoporous molecular sieves for catalytic oxidation of aromatic compounds. *Nature* 368: 321–323.

130 Koyano, K.A. and Tatsumi, T. (1996). Synthesis of titanium-containing mesoporous molecular sieves with a cubic structure. *Chem. Commun.* 145–146.

131 Wu, P., Tatsumi, T., Komatsu, T., and Yashima, T. (2002). Postsynthesis, characterization, and catalytic properties in alkene epoxidation of hydrothermally stable mesoporous Ti-SBA-15. *Chem. Mater.* 14: 1657–1664.

132 Kumar, A. and Srinivas, D. (2012). Selective oxidation of cyclic olefins over framework Ti-substituted, three-dimensional, mesoporous Ti-SBA-12 and Ti-SBA-16 molecular sieves. *Catal. Today* 198: 59–68.

133 Anand, C., Srinivasu, P., Mane, G.P. et al. (2012). Preparation of mesoporous titanosilicate molecular sieves with a cage type 3D porous structure for cyclohexene epoxidation. *Microporous Mesoporous Mater.* 160: 159–166.

134 Maschmeyer, T., Rey, F., Sankar, G., and Thomas, J.M. (1995). Heterogeneous catalysts obtained by grafting metallocene complexes onto mesoporous silica. *Nature* 378: 159–162.

135 Guidotti, M., Psaro, R., Batonneau-Gener, I., and Gavrilova, E. (2011). Heterogeneous catalytic epoxidation: high limonene oxide yields by surface silylation of Ti-MCM-41. *Chem. Eng. Technol.* 34: 1924–1927.

136 Fraile, J.M., García, N., Mayoral, J.A. et al. (2015). Multifunctional catalysis promoted by solvent effects: Ti-MCM41 for a one-pot, four-step, epoxidation–rearrangement–oxidation–decarboxylation reaction sequence on stilbenes and styrenes. *ACS Catal.* 5: 3552–3561.

137 Tatsumi, T., Koyano, K.A., and Igarashi, N. (1998). Remarkable activity enhancement by trimethylsilylation in oxidation of alkenes and alkanes with H_2O_2 catalyzed by titanium-containing mesoporous molecular sieves. *Chem. Commun.* 325–326.

138 Kamegawa, T., Suzuki, N., Tsuji, K. et al. (2011). Preparation of hydrophobically modified single-site Ti-containing mesoporous silica (TiSBA-15) and their enhanced catalytic performances. *Catal. Today* 175: 393–397.

139 Qin, Z., Lakiss, L., Gilson, J.P. et al. (2013). Chemical equilibrium controlled etching of MFI-type zeolite and its influence on zeolite structure, acidity, and catalytic activity. *Chem. Mater.* 25: 2759–2766.

140 Svelle, S., Sommer, L., Barbera, K. et al. (2011). How defects and crystal morphology control the effects of desilication. *Catal. Today* 168: 38–47.

141 Werner, A., Bludovsky, P., Selzer, C. et al. (2017). Hierarchical Ti-Beta obtained by simultaneous desilication and titanation as an efficient catalyst for cyclooctene epoxidation. *ChemCatChem* 9: 3860–3869.

142 Tekla, J., Tarach, K.A., Olejniczak, Z. et al. (2016). Effective hierarchization of TS-1 and its catalytic performance in cyclohexene epoxidation. *Microporous Mesoporous Mater.* 233: 16–25.

143 Du, S., Chen, X., Sun, Q. et al. (2016). A non-chemically selective top-down approach towards the preparation of hierarchical TS-1 zeolites with improved oxidative desulfurization catalytic performance. *Chem. Commun.* 52: 3580–3583.

144 Lu, X., Guo, Y., Xu, C. et al. (2019). Preparation of mesoporous mordenite for the hydroisomerization of *n*-hexane. *Catal. Commun.* 125: 21–25.

145 Lu, X., Guo, Y., Zhang, Y. et al. (2020). Enhanced catalytic activity of Pt/H-ZSM-12 via alkaline post-treatment for the hydroisomerization of *n*-hexane. *Microporous Mesoporous Mater.* 306: 110459.

146 Silvestre-Albero, A., Grau-Atienza, A., Serrano, E. et al. (2014). Desilication of TS-1 zeolite for the oxidation of bulky molecules. *Catal. Commun.* 44: 35–39.

147 Jiao, W., He, Y., Li, J. et al. (2015). Ti-rich TS-1: a highly active catalyst for epoxidation of methallyl chloride to 2-methyl epichlorohydrin. *Appl. Catal. A Gen.* 491: 78–85.

148 Schmidt, I., Krogh, A., Wienberg, K. et al. (2000). Catalytic epoxidation of alkenes with hydrogen peroxide over first mesoporous titanium-containing zeolite. *Chem. Commun.* 2157–2158.

149 Xin, H., Zhao, J., Xu, S. et al. (2010). Enhanced catalytic oxidation by hierarchically structured TS-1 zeolite. *J. Phys. Chem. C* 114: 6553–6559.

150 Ok, D.-Y., Jiang, N., Prasetyanto, E.A. et al. (2011). Epoxidation of cyclic-olefins over carbon template mesoporous TS-1. *Microporous Mesoporous Mater.* 141: 2–7.

151 Jiang, Y., Li, X., Zhang, S. et al. (2014). Titanosilicate composite with zeolite-large-crystal core and mesoporous shell structures: one-step synthesis and application in catalytic oxidations. *Mater. Lett.* 132: 270–272.

152 Cheneviere, Y., Chieux, F., Caps, V., and Tuel, A. (2010). Synthesis and catalytic properties of TS-1 with mesoporous/microporous hierarchical structures obtained in the presence of amphiphilic organosilanes. *J. Catal.* 269: 161–168.

153 Kamegawa, T., Suzuki, N., Che, M., and Yamashita, H. (2011). Synthesis and unique catalytic performance of single-site Ti-containing hierarchical macroporous silica with mesoporous frameworks. *Langmuir* 27: 2873–2879.

154 Lin, K., Li, L., Sels, B.F. et al. (2011). Titanosilicate beads as versatile catalysts for the conversion of trioses to lactates and for the epoxidation of alkenes. *Catal. Today* 173: 89–94.

155 Zhang, S., Jiang, Y., Li, S. et al. (2015). Synthesis of bimodal mesoporous titanosilicate beads and their application as green epoxidation catalyst. *Appl. Catal. A Gen.* 490: 57–64.

156 Du, Q., Guo, Y., Wu, P. et al. (2019). Facile synthesis of hierarchical TS-1 zeolite without using mesopore templates and its application in deep oxidative desulfurization. *Microporous Mesoporous Mater.* 275: 61–68.

157 Bai, R., Sun, Q., Song, Y. et al. (2018). Intermediate-crystallization promoted catalytic activity of titanosilicate zeolites. *J. Mater. Chem. A* 6: 8757–8762.

158 Song, Y., Bai, R., Zou, Y. et al. (2020). Temperature-regulated construction of hierarchical titanosilicate zeolites. *Inorg. Chem. Front.* 7: 1872–1879.

159 Yang, G., Han, J., Qiu, Z. et al. (2020). An amino acid-assisted approach to fabricate nanosized hierarchical TS-1 zeolites for efficient oxidative desulfurization. *Inorg. Chem. Front.* 7: 1975–1980.

160 Kuznicki, S.M. 1989. *US Patent No. 4,853,202.*

161 Zinicovscaia, I., Yushin, N., Humelnicu, D. et al. (2021). Sorption of Ce(III) by silica SBA-15 and titanosilicate ETS-10 from aqueous solution. *Water* 13: 3263.

162 Thakkar, J., Wissler, B., Dudenas, N. et al. (2019). Recovery of critical rare-earth elements using ETS-10 titanosilicate. *Ind. Eng. Chem. Res.* 58: 11121–11126.

163 De Luca, P., Bernaudo, I., Elliani, R. et al. (2018). Industrial waste treatment by ETS-10 ion exchanger material. *Materials* 11: 2316.

164 Figueiredo, B.R., Cardoso, S.P., Portugal, I. et al. (2018). Inorganic ion exchangers for cesium removal from radioactive wastewater. *Sep. Purif. Rev.* 47: 306–336.

165 Paula, A.S., Possato, G., Ratero, D.R. et al. (2016). One-step oxidehydration of glycerol to acrylic acid using ETS-10-like vanadosilicates. *Microporous Mesoporous Mater.* 232: 151–160.

166 Xiang, M. and Wu, D. (2019). Facile preparation of pore- and morphology-controllable ETS-10 zeolite with enhanced biomass hydrogenation activity. *Chem. Eng. J.* 369: 180–194.

167 Xiang, M., Zhang, F., Tong, L. et al. (2020). Development of the hierarchical ETS-10 zeolite catalyst for improving the aqueous-phase biomass hydrodeoxygenation activity. *J. Mater. Sci.* 55: 10505–10521.

168 Nguang, S.Y., Liew, A.S.Y., Chin, W.C. et al. (2022). Effect of graphene oxide on the energy level alignment and photocatalytic performance of Engelhard Titanosilicate-10. *Mater. Chem. Phys.* 275: 125198.

169 Pina Iritia, M.P., Almazan, F., Eguizabal, A. et al. (2016). Explosives detection by array of Si mu-cantilevers coated with titanosilicate-type nanoporous materials. *IEEE Sensors J.* 16: 3435–3443.

170 Rocha, J., Ferreira, A., Lin, Z., and Anderson, M.W. (1998). Synthesis of microporous titanosilicate ETS-10 from $TiCl_3$ and TiO_2: a comprehensive study. *Microporous Mesoporous Mater.* 23: 253–263.

171 Valtchev, V. and Mintova, S. (1994). Synthesis of titanium silicate ETS-10: the effect of tetrametylammonium on the crystallization kinetics. *Zeolites* 14: 697–700.

172 Valtchev, V.P. (1994). Influence of different organic bases on the crystallization of titanium silicate ETS-10. *J. Chem. Soc. Chem. Commun.* 261–262.

173 Anderson, M., Philippou, A., Lin, Z. et al. (1995). Al, Ti avoidance in the microporous titanoaluminosilicate ETAS-10. *Angew. Chem. Int. Ed.* 34: 1003–1005.

174 Das, T.K., Chandwadkar, A.J., Budhkar, A.P., and Sivasanker, S. (1996). Studies on the synthesis of ETS-10 II. Use of organic templates. *Microporous Mater.* 5: 401–410.

175 Kim, W.J., Kim, S.D., Jung, H.S., and Hayhurst, D.T. (2002). Compositional and kinetic studies on the crystallization of ETS-10 in the presence of various organics. *Microporous Mesoporous Mater.* 56: 89–100.

176 Pavel, C., De Luca, P., Bilba, N. et al. (2005). On the crystallization mechanism of ETS-10 titanosilicate synthesized in gels containing TAABr. *Thermochim. Acta* 435: 213–221.

177 Pavel, C., Nagy, J.B., Bilba, N. et al. (2004). Influence of the TAABr salts on the crystallization of ETS-10. *Microporous Mesoporous Mater.* 71: 77–85.

178 Lv, L., Su, F., and Zhao, X. (2004). A reinforced study on the synthesis of microporous titanosilicate ETS-10. *Microporous Mesoporous Mater.* 76: 113–122.

179 Pavel, C., Vuono, D., Catanzaro, L. et al. (2002). Synthesis and characterization of the microporous titanosilicates ETS-4 and ETS-10. *Microporous Mesoporous Mater.* 56: 227–239.

180 Yang, X., Paillaud, J.-L., Van Breukelen, H. et al. (2001). Synthesis of microporous titanosilicate ETS-10 with TiF_4 or TiO_2. *Microporous Mesoporous Mater.* 46: 1–11.

181 Liu, X.S. and Thomas, J.K. (1996). Synthesis of microporous titanosilicates ETS-10 and ETS-4 using solid TiO_2 as the source of titanium. *Chem. Commun.* 1435–1436.

182 Kim, W.J., Lee, M.C., Yoo, J.C., and Hayhurst, D.T. (2000). Study on the rapid crystallization of ETS-4 and ETS-10. *Microporous Mesoporous Mater.* 41: 79–88.

183 Ji, Z., Warzywoda, J., and Sacco, A. Jr., (2008). Synthesis and morphological control of large titanosilicate ETS-10 crystals. *Microporous Mesoporous Mater.* 109: 1–11.

184 De Luca, P., Vuono, D., and Filice, M. (2009). Self-bonded ETS-10 pellets containing iron. *Environ. Eng. Manage. J.* 8: 1009–1015.

185 De Luca, P., Nappo, G., Siciliano, C. et al. (2018). The role of carbon nanotubes and cobalt in the synthesis of pellets of titanium silicates. *J. Porous. Mater.* 25: 283–296.

186 Nair, S., Jeong, H.-K., Chandrasekaran, A. et al. (2001). Synthesis and structure determination of ETS-4 single crystals. *Chem. Mater.* 13: 4247–4254.

187 Braunbarth, C., Hillhouse, H.W., Nair, S. et al. (2000). Structure of strontium ion-exchanged ETS-4 microporous molecular sieves. *Chem. Mater.* 12: 1857–1865.

188 Kuznicki, S.M., Bell, V.A., Nair, S. et al. (2001). A titanosilicate molecular sieve with adjustable pores for size-selective adsorption of molecules. *Nature* 412: 720–724.

189 Delgado, J.A., Uguina, M.A., Agueda, V.I., and García-Sanz, A. (2008). Adsorption and diffusion parameters of methane and nitrogen on microwave-synthesized ETS-4. *Langmuir* 24: 6107–6115.

190 Philippou, A. and Anderson, M.W. (1996). Structural investigation of ETS-4. *Zeolites* 16: 98–107.

191 Lin, C.C., Sawada, J.A., Wu, L. et al. (2009). Anion-controlled pore size of titanium silicate molecular sieves. *J. Am. Chem. Soc.* 131: 609–614.

192 Pillai, R.S., Peter, S.A., and Jasra, R.V. (2008). Adsorption of carbon dioxide, methane, nitrogen, oxygen and argon in NaETS-4. *Microporous Mesoporous Mater.* 113: 268–276.

193 Yu, M., Primera-Pedrozo, J.N., Marcano-Gonzalez, M.E., and Hernandez-Maldonado, A.J. (2014). Selective adsorption of N_2 over CH_4 in flexible Sr^{2+} and Ba^{2+} UPRM-5 (TEA) titanium silicates: effect of activation temperature. *Chem. Eng. J.* 252: 311–319.

194 Liu, Y., Xiao, S., Bai, P. et al. (2014). Adsorption separation performance of H_2/CH_4 on ETS-4 by concentration pulse chromatography. *J. Energy Chem.* 23: 213–220.

195 Uguina, M., Delgado, J., Águeda, V., and García-Sanz, A. (2008). Microwave synthesis of Sr-ETS-4 and evaluation of its molecular sieving properties in the nitrogen/methane separation. In: *Studies in Surface Science and Catalysis*, vol. 174, 591–594. Elsevier.

196 Javani, R., Maghsoudi, H., Darvishi Gilan, S., and Majidpour, M. (2021). Study on adsorption performance of different adsorbents in nitrogen/methane separation. *Sep. Sci. Technol.* 56: 2562–2577.

2

Layered Heteroatom-Containing Zeolites

Hao Xu[1,2] *and Peng Wu*[1,2]

[1] *Shanghai Key Laboratory of Green Chemistry and Chemical Processes, School of Chemistry and Molecular Engineering, East China Normal University, North Zhongshan Road 3663, Shanghai, 200062, China*
[2] *Institute of Eco-Chongming, 20 Cuiniao Rd., Shanghai 202162, China*

2.1 Introduction

Layered zeolites or zeolitic layered materials are composed of nanosheets containing limited number of unit cells. The continuous growth along one particular direction is prohibited, and the interlayer connection is flexibly hydrogen bonded, which is different from the rigid chemical bonding in the traditional 3-dimensional (3D) zeolites and endows the layered zeolites with modifiable structure property [1–3]. Layered zeolites are attractive because they can produce a variety of derivative materials with organic/inorganic species intercalating into the interlayer space or containing randomly stacked uni-lamellar zeolitic layers. Heterogeneous atoms are introduced into the frameworks of layered zeolites and their derivative zeolitic materials to produce corresponding heteroatom-containing zeolites still featured by layered structure. The layered structure favors the introduction of metal active sites in catalyst preparation processes and also enhances the accessibility of active sites in the catalytic reactions, benefiting from the flexible 2D structures and the improved diffusion property of derivative materials. Although the number of layered heteroatom-containing zeolites is limited, their extraordinary catalytic performance in certain reactions, such as cyclohexanone ammoximation and propene epoxidation, has aroused significant interests in both academic and industrial fields [4, 5]. This chapter focuses on the synthesis of layered heteroatom-containing zeolites and the corresponding derivative zeolitic materials obtained by structural modifications. The traditional layered heteroatom-containing zeolites, such as Ti-MWW and Ti-FER, and those novel ones synthesized using bifunctional Gemini-type structure-directing agent (SDA) or through the structural modification of germanosilicates as potential layered zeolites are all involved. In addition to the most studied heteroatom Ti, this chapter also involves the Sn-, Zr-, and Fe- containing layered zeolites. Moreover, the catalytic performance of layered heteroatom-containing zeolites is compared to the corresponding 3D ones to reveal the inherent relationship between pore structure and catalytic performance.

Micro-Mesoporous Metallosilicates: Synthesis, Characterization, and Catalytic Applications, First Edition. Edited by Peng Wu and Hao Xu.

2.2 Traditional Layered Heteroatom-Containing Zeolites

2.2.1 Heteroatom-Containing MWW-Type Layered Zeolites and Their Derivative Zeolitic Materials

Titanium-containing MWW-type zeolite is the most studied heteroatom-containing layered zeolite as it possesses the largest amount of derivative zeolitic materials through direct synthesis or post-modification and is highly active in catalyzing the liquid-phase selective oxidation reactions with either hydrogen peroxide or organic hydroperoxide as oxidant, including alkene epoxidation, ketone ammoximation, and oxidative desulfurization [6]. As has been discussed in **Chapter 1**, a series of synthetic methods have been developed for Ti-MWW layered zeolites, including the B-containing [7], B-less [8–10], and B-free hydrothermal synthesis [11, 12]. The as-synthesized Ti-MWW layered precursors contain organic SDA molecules in both the interlayer space and the intralayer 10-member ring (R) pores. The removal of organic SDA molecules by calcination at elevated temperature induces the interlayer condensation to produce a rigid 3D Ti-MWW zeolite with the interlayer hydrogen bonded silanols transformed into Si-O-Si linkages. The structural transformation can be easily characterized by the change in XRD patterns, that is, the shift of layer-related 001 and 002 reflections to higher-angle region, resulting in the disappearance of 002 reflection by overlapping with the 100 reflection. The SDA species can also be partially removed by a mild acid treatment, which breaks the original ordered layer-stacking, induces a partial structural delamination, and produces a derivative material of Ti-MCM-56 analogue (Figure 2.1) [13]. The disordered layer-stacking in Ti-MCM-56 analogue, characterized by the overlapping of 101 and 102 reflection peaks in the XRD pattern, is retained upon calcination and endows Ti-MCM-56 analogue with larger external surface area and more exposed Ti active sites compared to the corresponding 3D Ti-MWW. Benefiting from the enhanced accessibility of Ti active sites, Ti-MCM-56 analogue exhibited higher activity in the epoxidation of bulky cycloalkenes with a large-size oxidant *tert*-butyl hydroperoxide (TBHP) and the oxidative desulfurization of bulky dibenzothiophene than both 3D Ti-MWW and the Ti-Beta titanosilicate with 12R large-pores. This post-modification method was also applied to synthesize a Sn-MCM-56 analogue via the mild acid treatment of as-made MWW-type stannosilicate layered precursor [14]. Note that this partially delaminated structure is named as Ti-MCM-56 or Sn-MCM-56 analogue because its structure is analogous to the hydrothermally

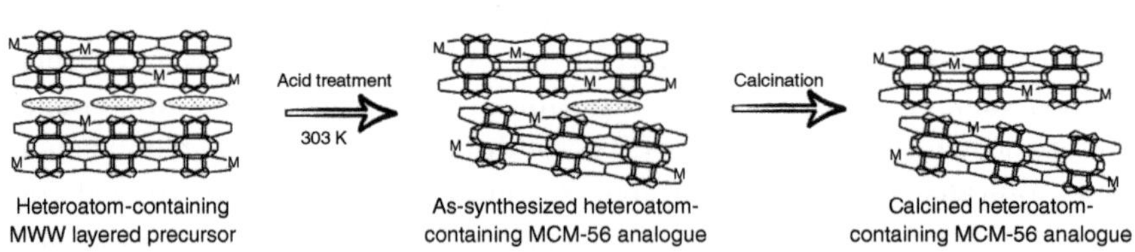

Figure 2.1 Schematic description of the preparation of heteroatom-containing MCM-56 zeolite from the corresponding MWW layered precursor via a mild acid treatment and subsequent calcination. Source: Reproduced from Ref. [13]/with permission of Elsevier.

synthesized MCM-56 aluminosilicate, which is crystallized from the same gel as MWW-layered precursor with shortened crystallization time [15]. However, it is not easy to control the crystallization extent and the construction of MCM-56 zeolite structure by direct hydrothermal synthesis is limited to aluminosilicate. Starting from a directly synthesized MCM-56 aluminosilicate, Ti-MCM-56 zeolite with pyramidal Ti-atoms in the framework was synthesized by selectively dealumination, Ti planting into vacancies and high-temperature calcination. Ti-MCM-56 zeolite was also active in the oxidative desulfurization of methyl phenyl sulfide and epoxidation of cyclooctene [16].

Contrarily, the full delamination of Ti-MWW layered precursor generally needs a more complex two-step process, involving swelling and ultrasonication treatments [17], which were first proposed to realize the phase-delamination of MWW-type aluminosilicate MCM-22 [18]. Amphiphilic surfactant molecules, such as cetyltrimethylammonium bromide (CTAB), are intercalated into the interlayer space of a Ti-MWW layered precursor in alkaline medium, which is provided by the organic base, tetrapropylammonium hydroxide (TPAOH). The organic base is used to break the some of the Si-O-Si bonds and deprotonate the Si-OH groups on layer surface, producing the Si—O^- moieties to interact with the electropositive CTA^+ cations, which favors the diffusion of bulky CTAB molecules into the interlayer space and construction of a swollen material. The appearance of a layer-related diffraction peak in the lower-angle region, corresponding to an enlarged interlayer space of ~3.9 nm, generally indicates the success of a swelling process. Since the alkaline TPAOH solution is corrosive to the layer structure with silica as main component, the amount of TPAOH should be delicately controlled to avoid the severe desilication and layer structure destroying. After the pre-swelling process, the layers are then forced apart through an ultrasonication treatment, producing a fully delaminated material Del-Ti-MWW, which is characterized by the decreased intensity of the diffraction peaks due to the loss of long-range order perpendicular to the layer stacking direction. Ultra-thin nanosheets with 1–3 layers in the HRTEM images can also be taken as a proof for the successful delamination of Ti-MWW layered precursor [17]. The full delamination significantly enhances the external surface area of Del-Ti-MWW, which was almost 2.6 times as large as that of the pristine 3D Ti-MWW [17]. In the epoxidation of bulky cyclic alkenes like cyclooctene and cyclododecane, the catalytic activity of Del-Ti-MWW was almost five times higher than that of 3D Ti-MWW. A remarkable decrease of turnover number (TON) upon the selective poisoning of Ti actives located on external surface indicated that the enhanced accessibility of active sites was critical for the improved catalytic performance of Del-Ti-MWW in the reactions involving bulky molecules. The delaminated MWW titanosilicate once has been synthesized by post-grafting a Ti complex on siliceous MWW (ITQ-2) delaminated in advance, benefiting from the abundant Si-OH groups on the layer surface of ITQ-2 [19]. The micro-environment of Ti active sites in the obtained Ti-ITQ-2 material was similar to those in mesoporous zeolites, such as Ti/MCM-41. Contrarily, the Ti active sites in Del-Ti-MWW resembled those in microporous zeolites as indicated by the comparable trans selectivity for Del-Ti-MWW and 3D Ti-MWW in catalyzing the epoxidation of 2-hexene isomers [17]. Thus, the hydrothermal stability of Ti active sites in Del-Ti-MWW was supposed to be higher than those in Ti-ITQ-2, although no experimental data have been provided.

In addition to the full structural delamination, swelling is also needed before the process of introducing inorganic silica or metal oxide species into the interlayer space to construct pillared or interlayer expanded structures because the swollen structure can provide enlarged interlayer space in comparison to the pristine layered precursor. The pillaring of a swollen structure is generally achieved by the hydrolysis of tetraethylorthosilicate (TEOS) in the interlayer space, resulting in amorphous silica pillars and mesopores between pillars. Metal alkoxide can be introduced to hydrolysis together with TEOS forming the pillars composed of silica and metal oxide. The characteristic diffraction peak for swollen structure in the low-angle range is well preserved in the pillaring and subsequent calcination process, indicative of a successful pillaring process. Two kinds of Ti-containing MWW-type pillared materials have been reported. The first one, named as Ti-MCM-36, was synthesized by the pillaring of a swollen Ti-MWW structure, and the Ti atoms were introduced in the hydrothermal synthesis of Ti-MWW layered precursor [20]. Benefiting from the interlayer mesopores, Ti-MCM-36 exhibited enhanced diffusion property compared to 3D Ti-MWW zeolite. In the epoxidation of both 1-hexene and propylene, Ti-MCM-36 showed higher TON values than 3D Ti-MWW and the TS-1 zeolite with 10R medium-pores. The second one, denoted as Si/Ti-MCM-36, was prepared by introducing the Ti active sites in the pillaring process of layered MCM-22 aluminosilicate or borosilicate [21, 22]. Tetrabutylorthotitanate (TBOT) was hydrolyzed together with TEOS, forming titanosilicate pillars. The introduced Ti atoms were proved to be well-dispersed in the titanosilicate pillars [21, 22]. The EXAFS analysis revealed that the micro-environment of Ti active sites in Si/Ti-MCM-36 resembles those in Ti-YNU-1 [21], which is an interlayer expanded structure with only micropores (see below). Additionally, Sn atoms can be introduced into the pillared MWW structure by the solid-state-exchange method, and the resultant catalyst was more active than both 3D Sn-MWW and large-pore Sn-Beta in the Baeyer–Villiger oxidation of 2-adamantanone and the isomerization of bulky sugar molecules [23].

Replacing the amorphous silica or titanosilicate pillars with the Si debris leached from zeolite framework or externally added silane agents can construct an interlayer expanded zeolite with interlayer ($\mathbf{n}+2$) or ($\mathbf{n}+4$) pore channels, where $\mathbf{n}$ represents the ring number of pores in the corresponding 3D zeolite obtained by direct calcination of a layered zeolite. The Si debris or guest silane agent molecules react with the Si-OH groups on the layer surface, linking the up-and-down layers and forming well-defined interlayer pores. The interlayer expanded structures resist to high-temperature calcination and are characterized by the shift of layer-related XRD diffraction peaks to lower-angle region, with the corresponding interlayer space increased by ~0.2 nm and ~0.5 nm for ($\mathbf{n}+2$) and ($\mathbf{n}+4$) pores, respectively. An interlayer expanded material, Ti-YNU-1 zeolite, was obtained unintentionally when refluxing the layered Ti-MWW zeolite precursor in an acid solution, during which the Si debris easily formed and diffused into the interlayer to connect the neighboring layers [24]. Compared to 3D Ti-MWW with interlayer 10R pores, Ti-YNU-1 possesses interlayer 12R pores, with the layer-related reflections of 001 and 002 shift to the lower-angle region and enlarged interlayer space revealed by the XRD and HRTEM techniques [25]. In addition to the Si debris, Ti debris was also proposed to attend the interlayer expansion process and existed in the pillar sites. However, the construction of Ti-YNU-1 zeolite was only restricted to the Ti-MWW layered precursor with higher Si/Ti ratio probably because more defect sites in high-silica Ti-MWW precursor can produce enough Si debris for interlayer expansion. Actually, the amount of leached Si debris under refluxing is uncontrollable. Thus, the introduction of silane agents with tailorable amount is more desirable. The silylation of a layered precursor involves the diffusion of silane agents into the interlayer space and the removal of SDA molecules, and the speed of two processes should be well-balanced by tailoring the concentration of acid aqueous solution and the temperature [26]. Moreover, the structure and amount of silane agents also affected the silylation process. The silane agents with two or three hydrolyzable substitute groups in acidic medium, such as methoxy or chloro group, are desirable, while the inert substitute groups prohibit intermolecular condensation [26]. The presence of four hydrolyzable substitute groups, e.g. TEOS, induces severe self-condensation and the formation of amorphous silica phase. As-made Ti-MWW layered precursor was silylated by diethoxydimethylsilane (DEDMS) to produce an interlayer expanded structure (IEZ-Ti-MWW) [27], which is structurally analogous to Ti-YNU-1. The acidic medium of silylation inevitably caused the leaching of Ti active sites. Nevertheless, both Ti-YNU-1 and IEZ-Ti-MWW exhibited higher activity compared to 3D Ti-MWW zeolite in the epoxidation of bulky cycloalkenes despite lower Ti content. This was mainly contributed by the expanded interlayer pores. Under identical reaction conditions, the activity of IEZ-Ti-MWW was inferior to that of Ti-YNU-1 probably because the Ti active sites in pillar sites of Ti-YNU-1 significantly enhanced the accessibility compared to those in the intralayer sites of IEZ-Ti-MWW [24, 27]. DEDMS is a monomeric silane agent, which can only introduce two additional Si atoms into each interlayer pore. Contrarily, a dimeric silane agent, such as $Cl(CH_3)_2Si–Si(CH_3)_2Cl$, can enlarge the interlayer 10R pores in MWW to 14R pores in the interlayer expanded one. However, the molecular size of dimeric silane agents encounter severe diffusion constraints in entering into the interlayer space of MWW layered precursor, and a pre-swelling process is needed to enlarge the interlayer space. The interlayer expansion of MWW layered precursor by the silylation with dimeric silane agent and subsequent liquid-phase Ti insertion with H_2TiF_6 aqueous solution produced IEZ-Ti-MWW(SiSi) zeolite, showing the 001 and 002 XRD reflections at lower 2θ degrees compared to the IEZ-Ti-MWW material silylated by monomeric silane agents. The interlayer distance of IEZ-Ti-MWW(SiSi) zeolite increased by 0.3 nm and 0.5 nm compared to IEZ-Ti-MWW and 3D Ti-MWW, respectively. Thus, the catalytic activity was further enhanced in the epoxidation of cycloalkenes, benefiting from the released intrapore diffusion constraints.

As described above, the swelling process is critical for the construction of delaminated, pillared, and interlayer-expanded structures. However, the corrosion of intralayer structure and the tedious processes are unavoidable. A hydrothermal synthesis of swollen Ti-MWW structure (named as Ti-ECNU-7P) was reported by Wu et al. [28]. The long-tail surfactant molecules, CTAB, were introduced into the interlayer space in the recrystallization of a pure-silicate MWW zeolite, ITQ-1, resulting in an alternative stacking of highly crystalline MWW-type inorganic layers and mesostructured organic CTAB layers. Although the delamination, pillaring, and silylation of Ti-ECNU-7 have not been performed, the successful pillaring and silylation of an analogous aluminosilicate have been reported [29, 30]. A hierarchical Ti-ECNU-7 zeolite, with disordered layered stacking, was obtained by the calcination of as-made Ti-ECNU-7P, where a mismatching condensation of the neighboring layers occurred due to the greatly enlarged interlayer spacing to mesoscale range. Compared to 3D Ti-MWW, Ti-ECNU-7 exhibited larger external surface area and mesopore volume. However, the superior activity of Ti-ECNU-7 to 3D Ti-MWW was not observed in the epoxidation of a linear alkene, 1-hexene, because the incomplete interlayer condensation of Ti-ECNU-7P left abundant free Si-OH groups on layer surface and it was thus more hydrophilic than 3D Ti-MWW [28]. The higher hydrophilicity favored the adsorption of H_2O and H_2O_2 rather than the organic substrates. Contrarily, in processing the epoxidation reaction of bulky cycloalkenes using H_2O_2 or TBHP as an oxidant, the superiority of Ti-ECNU-7 was observed in releasing the diffusion constraints and exhibited higher activity than 3D Ti-MWW. Additionally, the hierarchical Ti-ECNU-7 with mesopores is a potential catalyst for the industrial production of propylene oxide via the propylene epoxidation with cumene hydroperoxide (CHPO) [28].

The above structural modifications start from the 2D layered MWW precursor and involve the topotactic transformation to 3D derivative structure (Figure 2.2). The reverse structural transformation from 3D MWW to 2D layered one is also feasible, with the assistant of SDA molecules, such as piperidine (PI). The 3D–2D structural transformation can assist the introduction of heteroatoms into the structure of siliceous MWW, which has been described in **Chapter 1** as a unique boron-free synthetic method [11]. On the other side, the transformation of a 3D Ti-MWW to its corresponding 2D one with the hydrothermal treatment of PI aqueous solution can construct unique six-coordinated Ti species with PI molecules as one of the ligands [31]. The PI modification significantly enhanced the activity of MWW-type titanosilicate as both powder form and shaped form in the alkene epoxidation [32–35]. The detailed discussion on the PI modified Ti-MWW catalysts can be found in **Chapter 7**.

2.2.2 Heteroatom-Containing Layered Zeolites Built from *fer*-Layers

The structural modifications on heteroatom-containing MWW-type zeolite can serve as a model for other layered zeolites. However, not all the existing layered zeolites show potential activity with heteroatoms occupying the tetrahedral sites. Prior to heteroatom-containing MWW-type layered zeolites, FER-type layered zeolites, a collection of *fer* layers, have also been extensively studied. Both PREFER and PLS-3 zeolites, synthesized from different silica sources and SDAs, belong to FER-type layered zeolites, but they exhibit different morphologies [36, 37].

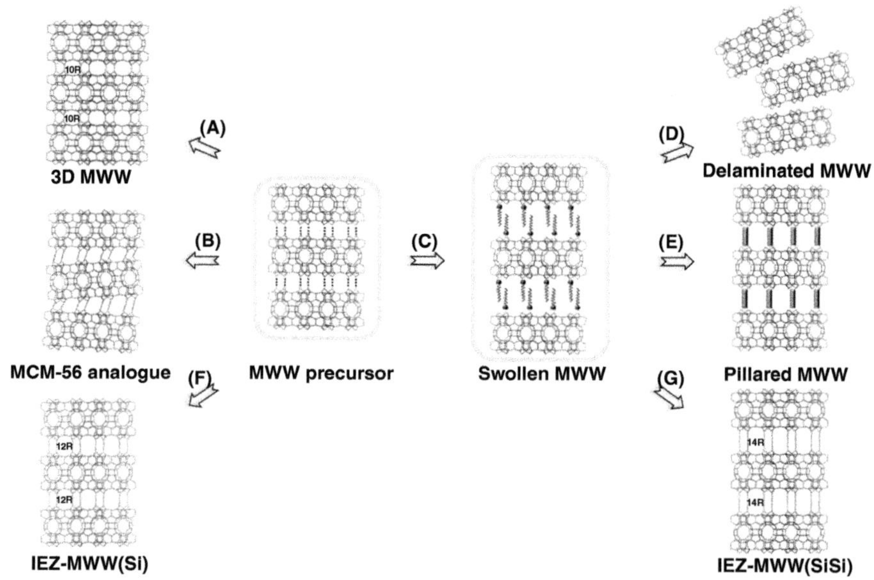

Figure 2.2 Summary of structural derivatives from MWW layered zeolites via different modifications: (A) calcination, (B) mild acid treatment, (C) swelling, (D) delamination, (E) pillaring, (F) silylation with monomeric silane agent, and (G) silylation with dimeric silane agent.

Developed earlier than PLS-3, PREFER was typically synthesized using 4-amino-2,2,6,6-tetramethylpiperidine (TEMP) as SDA with the assistant of F^- ions, showing micro-size plate-like morphology [36]. In contrast, PLS-3 was crystallized with the layered silicate of kanemite as silica source and a smaller-sized SDA of tetraethylammonium hydroxide (TEAOH), showing nano-size needle-like morphology. Recently, two novel FER-type layered materials, including ECNU-35 with a swollen FER structure [38] and delaminated ECNU-17 [39], have been directly synthesized by using special SDAs, which avoids the complex swelling and delamination processes and provides materials for further heteroatom-incorporation to construct potentially active catalysts. This has also expanded the FER-type zeolite family (Figure 2.3).

Inspired by the modifications on MWW-type layered precursors, the delamination of Ti-containing PREFER layered precursor was performed with the similar swelling-ultrasonication method, producing a fully delaminated material, Ti-ITQ-6. Although the delamination caused partial Ti leaching, Ti-ITQ-6 showed higher TON values compared to 3D Ti-FER zeolite in the epoxidation reactions of both linear 1-hexene and bulky norbornene, benefiting from enhanced accessibility of Ti active sites [40]. Moreover, the superior catalytic performance was also observed in the oxidative desulfurization reaction of Ph_2S [41]. The larger external surface area also makes Ti-ITQ-6 a better support compared to 3D Ti-FER. Ti-ITQ-6 was used as a support for Cu species to construct effective catalysts for low-temperature NH_3-SCR reaction, with framework Ti species as weaker acid sites, which favored

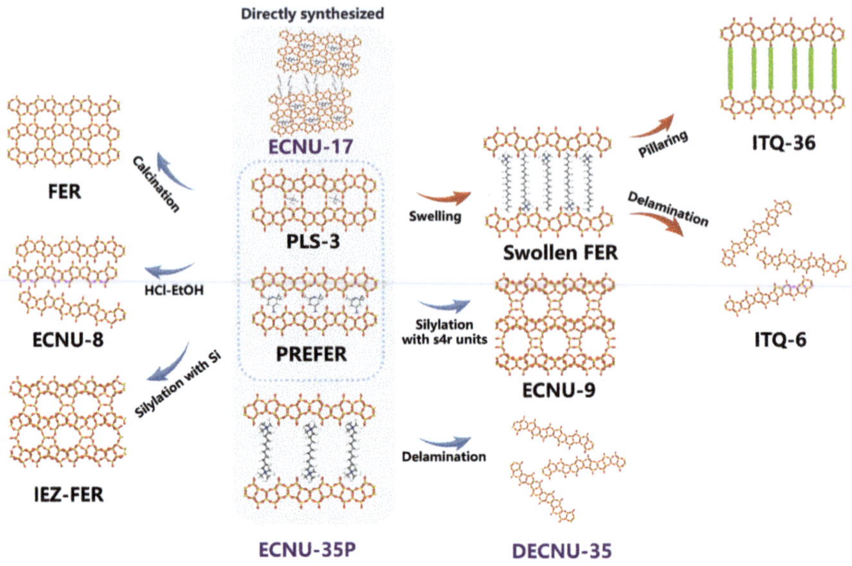

Figure 2.3 Summary of the FER-type related zeolites by both direct synthesis and post-modifications.

the desorption of ammonium compared to Al-related acid sites [42]. In the Cu deposition process, the significantly enhanced diffusion property of Ti-ITQ-6 resulted in the formation of mainly monomer copper cations, while aggregated copper oxide species formed over 3D Ti-FER. Since the monomer copper cations were more active in the NO oxidation to NO_2 than aggregated cooper oxide species, Cu-Ti-ITQ-6 was superior to Cu-Ti-FER in the low-temperature NH_3-SCR reaction.

The modifications on the layered zeolite already containing heteroatom in advance inevitably induce the leaching of the framework heteroatoms, especially for those involving acid medium. Introducing the heteroatoms during the modification process of a layered precursor can avoid this problem. Moreover, interlayer expansion with expensive silylating agents may count against practical applications. Treating PLS-3 layered precursor with the aqueous solution of a special Ti source, H_2TiF_6, realized the simultaneous interlayer expansion and Ti insertion (Figure 2.4), producing a Ti-containing interlayer expanded structure, IEZ-Ti-PLS-3 [43]. Partial desilication occurred during the treatment of H_2TiF_6 aqueous solution due to the existence of F^- ions, resulting in the formation of defects. Then, the dissolved Si debris diffused into the interlayer space to fabricate an interlayer expanded structure, while the Ti species provided by H_2TiF_6 were isomorphously incorporated into the framework defects. With intersecting 12×10R pore channels, IEZ-Ti-PLS-3 was inferior to Ti-MWW and TS-1 in the epoxidation of small-sized alkenes, such as 1-hexene and cyclopentene, from the perspective of TON value. Nevertheless, in processing the bulky cycloalkenes, IEZ-Ti-PLS-3 showed higher TON values than Ti-MWW, TS-1 and Ti-Beta for the cyclohexene epoxidation, and the superiority to Ti-MWW and TS-1 was kept when increasing the carbon

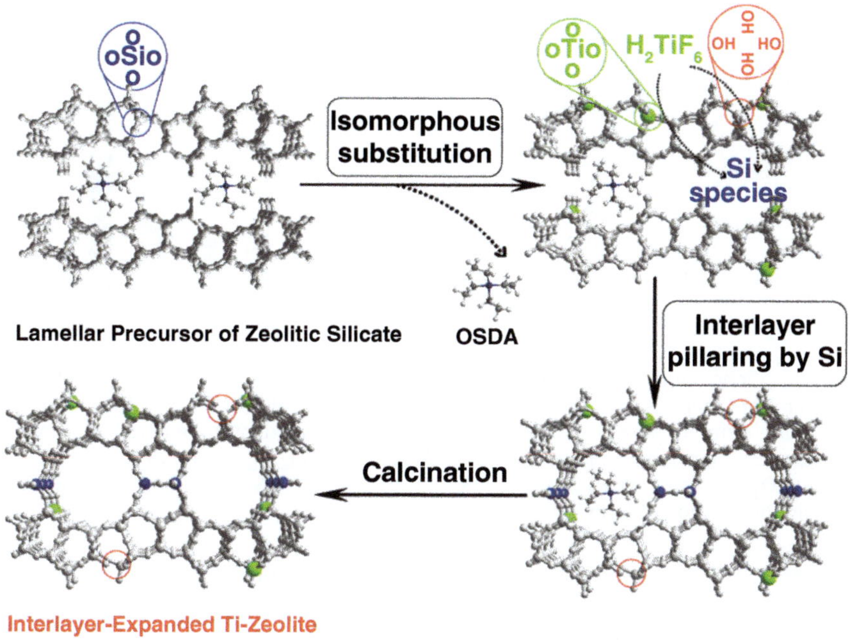

Figure 2.4 Schematic description of the IEZ-Ti-PLS-3 preparation via the treatment of H_2TiF_6 aqueous solution, simultaneously achieving the interlayer-expansion and Ti inserting. Source: Reprint from Ref. [43] with the permission from ACS publications.

number in cycloalkenes to 12. This one-pot method for heteroatom introduction and structural expansion was successfully extended to other layered precursors, including MCM-47, PLS-4, and RUB-39. Moreover, this strategy was not limited to the insertion of Ti atoms, and it can be used to prepare a Sn-containing interlayer expanded PLS-3 zeolite by the treatment of a mixture containing desired amount of $SnCl_4{\cdot}5H_2O$, HNO_3, and NH_4F [44]. The obtained IEZ-Sn-PLS-3 served as an active catalyst for the Baeyer–Villiger oxidation reaction of bulky 2-adamantanone.

Since the above H_2TiF_6 treatment can only produce interlayer expanded structure by introducing two addition Si atoms into the interlayer pores, this one-pot synthetic strategy is not general for all the structural modifications. The introduction of heteroatoms after structural modification is another alternative method to avoid the adverse effect of post-modifications on heteroatom-related active sites. A unique silane agent of 1,3,5,7-tetramethylcyclotetrasiloxane with single-4-ring (*s4r*) shape was used for the silylation of PLS-3 zeolite, introducing additional four atoms into the interlayer pores and constructing an extra-large 14 × 12R zeolite, ECNU-9 [45]. The *s4r*-shaped silane agent was too bulky to diffuse into the interlayer space. To solve this issue and avoid the structure-corrosive swelling process, a structural deconstruction-reconstruction strategy was performed to enhance the interlayer space. This strategy involved the first step of partial SDA (TEAOH) removal to disassemble the ordered PLS-3 structure, producing a sub-zeolite ECNU-8 with disordered layer stacking, and the second step of inserting the classic SDA

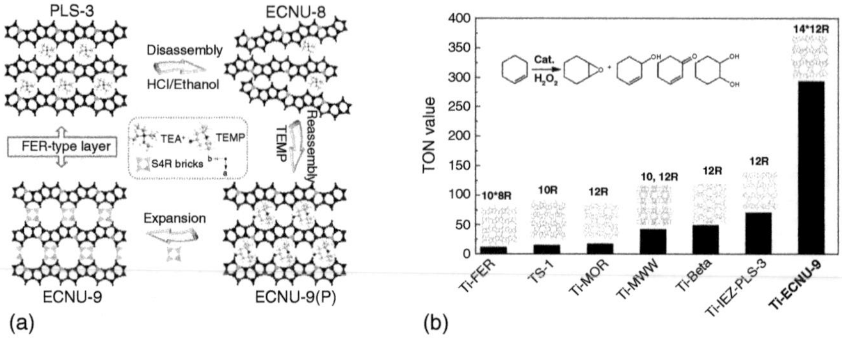

Figure 2.5 (a) Schematic description of the interlayer expansion of PLS-3 layered zeolite with *s4r* units. (b) Comparison of the catalytic performance of Ti-ECNU-9 with other Ti-zeolites with different pore structures in the cyclohexene epoxidation reaction in terms of TON value. Reaction condition: catalyst (0.05 g), acetonitrile (7.8 g), cyclohexene (10 mmol), H_2O_2 (10 mmol; 31 wt % aqueous solution), 333 K, 2 h. Source: Reproduced from Ref. [45]/with permission of John Wiley & Sons.

(TEMP) for PREFER synthesis into the interlayer space of ECNU-8 to construct a layered precursor of ECNU-9P with ordered layer stacking (Figure 2.5a) [45]. Both TEAOH and TEMP can direct the crystallization of FER-type layer structure. The reconstruction with larger-size TEMP resulted in enhanced interlayer space enough for the diffusion of bulky *s4r*-shaped silane without ruining the layer structure. Afterward, the bulky *s4r*-shaped silane was introduced to construct the interlayer expanded structure of ECNU-9, with intersecting 14 × 12R pores. The large pore size of ECNU-9 favored the diffusion of Ti species to access the defect sites during Ti inserting process by the treatment of H_2TiF_6 aqueous solution. In the epoxidation of bulky cyclohexene, Ti-ECNU-9 showed remarkably enhanced TON value, which was more than nine times higher than 3D Ti-FER zeolite and three times higher than IEZ-Ti-PLS-3, and also more active than the titanosilicates with other topologies like TS-1, Ti-MOR, Ti-Beta, and Ti-MWW (Figure 2.5b). This is considered to benefit from the greatly released diffusion constraints in 14R extra-large pores in Ti-ECNU-9.

CDO-type layered zeolites, including PLS-1 [46], PLS-4 [37], RUB-36 [47], ZSM-55 [48], and ECNU-57 [49], are also composed of *fer* layers. The different relative-positions of neighboring layers along the direction perpendicular to the layer-stacking one in the layered precursors induce the formation of different 3D zeolite structures (FER and COD-type) upon calcination. Despite both being composed of *fer* layers, FER and CDO-type zeolites possess different pore channels, with the former one having 10 × 8R pores and the latter one having 8 × 8R pores, and the structural transformation between them can be achieved by deswelling and reassembly or by ion-exchange [47, 48, 50]. The silylation of a Ti-containing RUB-36 zeolite (CDO-type) with dihydroxydimethylsilane (DHDMS) was once reported, producing the interlayer expanded structure of Ti-COE-3 and its calcined form of Ti-COE-4 [51]. The obtained Ti-COE-3 and Ti-COE-4 possess enlarged 10 × 10R pores and were active in the oxidation of hexane, 1-hexene, cyclohexene, and phenol. However, the main product of Ti-COE-3 and Ti-COE-4 was observed to be different from that of the traditional TS-1 zeolite, and the former two catalysts mainly produced direct oxidation products rather than the epoxidation products. This was probably caused by the different micro-environment of Ti species and the formation of a small amount of aggregated TiO_2 species in the silylation process. The presence of methyl groups in the as-made silylated structure of Ti-COE-3 endowed it with higher hydrophobicity than its calcined form of Ti-COE-4. Thus, Ti-COE-3 showed lower activity in the oxidation reaction of cyclohexene with H_2O_2 as oxidant, but comparable activity with TBHP as oxidant, compared to Ti-COE-4. In addition to the silylation with silica species, the RUB-36 zeolite was possibly silylated by other metal species, such as Fe, Sn, Zn, and Eu. In the acidic medium, the metal precursors diffused into the interlayer space, interacted with the surface Si-OH groups, and occupied the interlayer pillar T-sites, forming interlayer-expanded structure and catalytic functionality [52–54]. The introduction of Fe species into the pillar sites of Al-RUB-36 increased the Lewis acidity compared to that with Si occupying the pillar sites, and the obtained Al-COE-4/Fe showed higher activity in catalyzing reactions involving bulky substrates, such as the acylation of anisole with acetic anhydride and alkylation of toluene with benzyl chloride [54].

2.3 Novel Layered Heteroatom-Containing Zeolites

The traditional layered zeolites are spontaneously synthesized with an amine or a quaternary ammonium salt as the SDA, and the specific relationship between the SDA molecules and the unique layered structure is not clear. A breakthrough occurred with the development of a layered MFI-type zeolite, which was synthesized only as a 3D structure before, using an amphiphilic Gemini-type bifunctional SDA [55]. The hydrophilic head involving diammonium groups directs the crystallization of MFI-type structure, while the hydrophobic head involving a long-tail alky group interrupts the continuous growth of MFI structure along the *b* axis, leading to the crystal expansion limited to the *ac* plane. Shortly after this, another mile-stone progress in the field of layered zeolites occurred with the controllable degradation of germanosilicates into a series of novel layered zeolites by selective removal of Ge-rich double-4-ring (*d4r*) units [56–58]. The two remarkable progresses greatly prosper the layered-zeolite field and make a significant contribution to the construction of highly active heteroatom-containing zeolite catalysts.

2.3.1 Heteroatom-Containing MFI-Type Layered Zeolites

After reporting the first layered MFI in aluminosilicate form, Ryoo et al. introduced Ti species into the framework of layered MFI zeolite by direct hydrothermal synthesis using a Gemini-type surfactant of $C_{16}H_{33}$-$N^+(CH_3)_2$-C_6H_{12}-$N^+(CH_3)_2$-C_6H_{13} (abbreviated as $C_{16\text{-}6\text{-}6}$) [59]. The obtained layered TS-1 zeolite, with only (*h0k*) reflections in the XRD pattern, was composed of nanosheets, and the thickness along the *b* axis was around 2 nm (Figure 2.6a,b), corresponding to one unit-cell

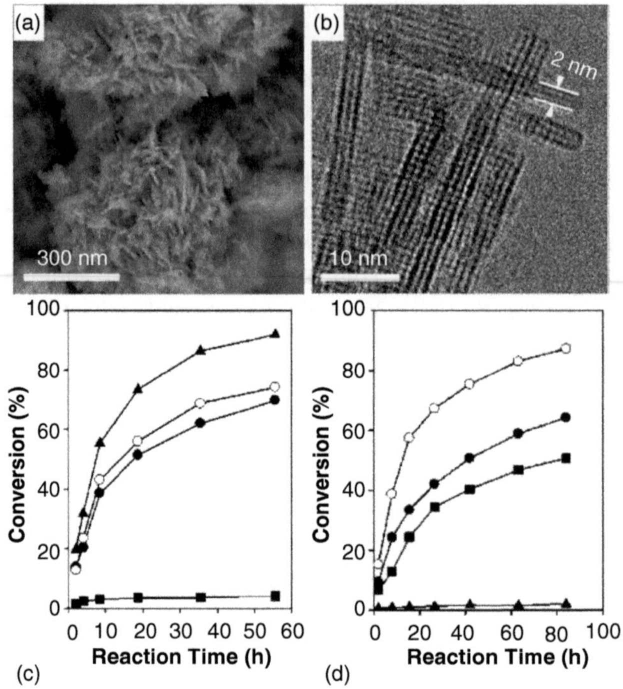

Figure 2.6 SEM (a) and TEM (b) images of nanosheet TS-1. The catalytic performance of bulk TS-1 (solid triangle), Ti-MCM-41 (solid square), nanosheet TS-1(solid circle), and nanosheet F-TS-1 (open circle) in the 1-hexene (c) and cyclooctene (d) epoxidation reaction. Reaction condition: 35 mg of catalyst, 12 mmol of 1-hexene or cyclooctene, 3.3 mmol of H_2O_2, 10 mL of acetonitrile, 60 °C. Source: Reproduced from Ref. [59]/with permission of American Chemical Society.

thickness of MFI topology. Despite having larger external surface area compared to bulk TS-1 catalyst, the nanosheet TS-1 was less active in 1-hexene epoxidation reaction [59]. Both the external and internal Ti sites were accessible to the small-size linear 1-hexene due to the absence of diffusion constraints in the micropores of TS-1 zeolite. The lower activity of external Ti sites compared to the internal ones in the epoxidation of 1-hexene was proposed to induce inferior catalytic performance of nanosheet TS-1 (Figure 2.6c). Contrarily, the nanosheet TS-1 was superior to bulk TS-1 in the epoxidation of bulky cyclohexene because the cyclohexene molecules hardly diffuse into the micropores of MFI zeolite and the external Ti species were the main active sites for the epoxidation of cyclohexene (Figure 2.6d). However, the abundant Si-OH groups with weak acidity on the highly exposed layer surface of nanosheet TS-1 caused ineffective H_2O_2 decomposition and lower epoxide selectivity due to the occurrence of ring-opening side reactions. These issues can be addressed by using TBHP as oxidant in H_2O-free reaction system or eliminating the Si-OH groups by 40% using NH_4F treatment. Similar catalytic performance of layered TS-1 zeolite has also been reported by Wu et al., although a different Gemini-type SDA with longer alky tail of C_{22} was used [60]. Since the external Ti sites with superior accessibility in layered TS-1 zeolite were able to catalyze the cyclohexene epoxidation reaction, a more active Ti-containing layered catalyst for this reaction was synthesized by constructing more external Ti sites via post-grafting method [61]. Wu et al. coated the nanosheet TS-1 on the traditional micro-sized TS-1 crystals via the desilication-recrystallization method with the assistance of a Gemini-type surfactant and constructed a highly active hierarchical MFI-type titanosilicates for the epoxidation of linear alkenes [62]. The strategy also achieved success in the desilication-recrystallization of a MFI-type aluminosilicate [63]. As the Gemini-type surfactants need complex synthesis, Wu et al. have tried to make full use of them. An acid–base bifunctional catalyst was constructed by a mild acid treatment and ion-exchange of the as-synthesized layered MFI [64]. With the Al atoms in the intralayer structure as acid sites and the OH^- counteranions of exposed quaternary ammonium cations and SiO^- groups as the base sites, this bifunctional MFI catalyst was highly active in the tandem deacetalization–Knoevenagel condensation reaction. Moreover, precise metals can also be introduced into this multilamellar MFI structure to provide another type of active site.

The as-synthesized multilamellar MFI-type zeolite shows the alternative stacking of inorganic MFI layer and organic surfactant layer, which resembles the swollen structure of a layered zeolite. Thus, the as-synthesized layered MFI was readily applied for a pillaring process. The pillaring of a Ti-containing, Al-containing, or siliceous MFI layered zeolite with the mixture of TEOS and TBOT not only introduced additional Ti active sites but also constructed a hierarchical MFI catalyst with intralayer micropores and interlayer mesopores [65–69]. Increasing the relative amount of TBOT had the risk of reducing the surface area although the Ti content could be enhanced. Thus, to construct a highly active pillared Ti-containing MFI catalyst, the ratio of TBOT/TEOS should be carefully controlled. The pillared Ti-containing MFI catalysts showed higher activity compared to both traditional bulk TS-1 and the directly calcined layered TS-1 zeolite in the reactions involving bulky substrates, such as the epoxidation of bulky alkenes [65, 68] and the selective oxidation of bulky sulfides [66, 69]. Specially, introducing titanosilicate pillars into an aluminosilicate layered MFI zeolite could result in an efficient photocatalyst for the degradation of bulky methyl orange molecules in water, with both Ti and Al active sites promoting the degradation and the pillared structure favoring the diffusion of bulky dye molecules [67].

The directly synthesized heteroatom-containing pillared MFI catalyst, which is also called self-pillared MFI zeolite, is more appealing compared to the one synthesized by post-modification of pillaring. A self-pillared Ti-containing MFI zeolite was synthesized by a dual-template method, involving the Bola-form surfactant of $[C_6H_{13}\text{–}N^+(CH_3)_2\text{–}C_6H_{12}\text{–}N^+(CH_3)_2(CH_2)_{12}\text{–O–(p–}C_6H_4)_2\text{–O–}(CH_2)_{12}\text{–}N^+(CH_3)_2\text{–}C_6H_{12}\text{–}N^+(CH_3)_2\text{–}C_6H_{13}]\ [Br^-]_4$ and TPAOH [70]. The addition of TPAOH not only favored the formation of self-pillared structure but also enhanced the amount of more active "TiO6" species, thus showing higher activity in the epoxidation of cyclooctene, compared to both conventional microporous TS-1 and mesoporous Ti-MCM-41 catalysts. A simple small-size quaternary ammonium, tetrabutylphosphonium hydroxide (TBPOH), can also direct the crystallization of a self-pillared

MFI structure (denoted as SPP), which is composed of single-unit-cell layers with rotational intergrowth interconnected by MEL structure [71]. Sn atoms were introduced into the self-pillared MFI structure by direct hydrothermal synthesis, and the Sn content should be controlled in a low level (Si/Sn ratio $\approx$ 200) to avoid the formation of SnO_2 and ensure the tetrahedral coordination state. The framework Sn derived Lewis acidity, the surface Si-OH groups related Brönsted acidity, and the unique diffusion property of layer-structure endowed Sn-SPP catalyst higher activity in the conversion of glucose to fructose and the conversion of even larger-size sugar of lactose to lactulose, compared to large-pore Sn-Beta and mesoporous Sn-MCM-41 catalysts [71].

2.3.2 Germanosilicate-Derived Heteroatom-Containing Zeolites

The meta-stability of Ge—O bonds endows germanosilicates with modifiable property and the potential being transformed into layered zeolites. The structural modification of germanosilicates is normally performed in acidic or alkaline medium, where the broken of Ge-O-Ge(Si) and the formation of new Si-O-Si bonding are favored [56, 72]. However, the operating condition is detrimental for the heteroatom-related active sites with risks of decreasing the content and altering coordination state of active sites. Thus, a preferred strategy is to simultaneously achieve the structural modification and insertion of heteroatoms. Sn atoms were introduced into the framework of UTL zeolite in the structural stabilization process of Ge-rich UTL zeolite, where a rapid structural deconstruction-reconstruction process occurred (Figure 2.7) [73]. The deconstruction, due to selective Ge removal, resulted in a layered intermediate, and making full use of this flexible layered structure the Sn species occupied the T-sites in UTL zeolite during the reconstruction process. The obtained Sn-containing UTL catalyst, with high hydrothermal stability and extra-large 14×12R pores, showed higher activity in the Baeyer–Villiger oxidation reaction of 2-adamantanone using TBHP as oxidant compared to large-pore Sn-Beta catalyst. Similarly, a Ti-containing IWV catalyst was synthesized taking advantages of the 3D–2D–3D structural transformation of IWV germanosilicate, which involved the complete removal of interlayer *d4r* units resulting in a layered intermediate (named as Hydro-IWV), and the subsequent reconstruction of Hydro-IWV to 3D IWV structure by the treatment of $(NH_4)_2TiF_6$ acidic solution [74]. With 12R large-pore channels and 14R double cavities in IWV structure, Ti-IWV exhibited higher activity compared to the traditional micropore Ti-MWW, TS-1, and Ti-Beta catalysts in the epoxidation of cycloalkenes. In the special case of cyclooctene epoxidation, the TON value of Ti-IWV catalyst was nearly 10 times higher than other titanosilicates due to the high concentration ability of cyclooctene in large pores and cavities.

Another two active Ti-containing catalysts (Ti-ECNU-14 and Ti-ECNU-15) with UTL-derived structures were synthesized by introducing Ti atoms after the structural transformation using liquid-phase treatment of H_2TiF_6 aqueous solution [75]. By controlling the acid-treatment time, the interlayer pores can be continuously tuned, resulting in the Ti-ECNU-14 with 12×10R pores and Ti-ECNU-15 with

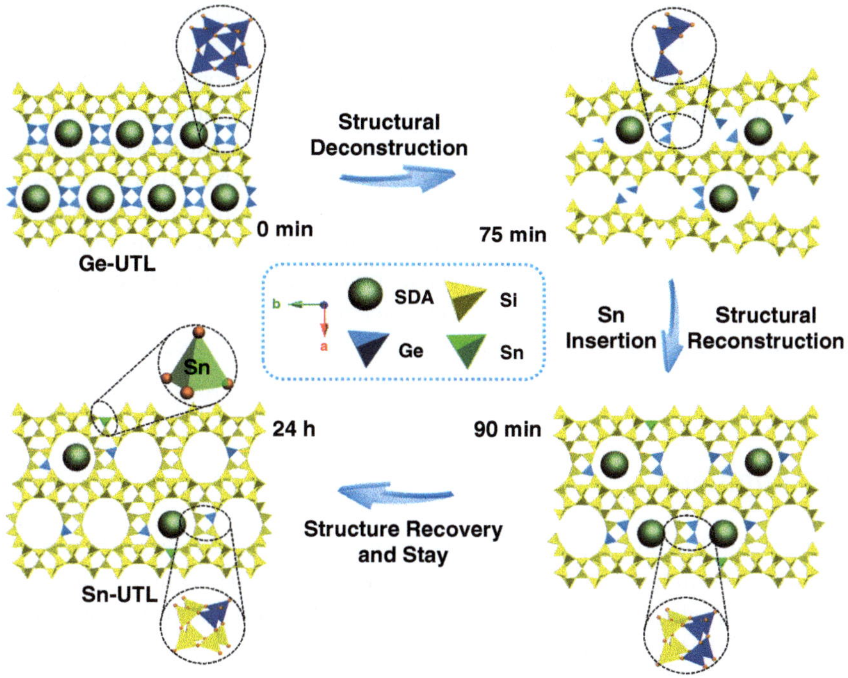

Figure 2.7 Schematic description of the synthesis of Sn-UTL zeolite via the structural deconstruction-reconstruction of Ge-rich UTL zeolite. Source: Reproduced from Ref. [73]/with permission of American Chemical Society.

(14–12) × (12–10)R pores. They showed pore-size dependent catalytic performance in the epoxidation of bulky cyclohexene and higher than traditional 3D Ti-zeolites, such as Ti-MWW and Ti-Beta. Following the classical structural modification strategies of layered zeolites, the pillaring and silylation of Ti-UTL-derived layered precursor, Ti-IPC-1P, were performed [66, 76]. The pillaring with pure silica and titanosilicate resulted in Ti-IPC-1-PI and Ti-IPC-1-PITi, respectively, while the silylation with monomeric silane agent produced Ti-IPC-2 (12 × 10R) zeolite. Through a delicate control of TBOT/TEOS ratio in pillaring Ti-IPC-1P, Ti-IPC-1-PITi exhibited much higher catalytic activity in the epoxidation of bulky alkenes than that of Ti-IPC-1-PI and was also superior to Ti-UTL and Ti-IPC-2 benefiting from the presence of interlayer mesopores and accessible Ti sites in titanosilicate pillars [76]. Similar catalytic performance was also observed in the selective oxidation of bulky sulfides [66].

2.4 Conclusions

Taking advantages of the modifiable property of layered structure, the accessibility of heteroatom-related active sites can be significantly enhanced by constructing larger pores or exposing them on external surface. The introduction process of heteroatoms into the layered zeolites and their derived structures should be carefully handled because the activity of the obtained heteroatom-containing catalysts is highly related to the content, micro-environment, and location of these active sites. Actually, the structural modification strategy has got remarkable progresses in the past decade for layered zeolites. However, the extension to heteroatom-containing layered zeolites are still restricted because the preservation of heteroatom-related active sites needs more cautiousness when performing the structural modifications. Additionally, the number of active layered heteroatom-containing zeolite catalysts is still limited. The exploration of heteroatom-containing zeolite with novel layered structure is highly desired. Last but not the least, the application of layered heteroatom-containing zeolites can be further extended to other fields, such as membrane catalysis, making full use of the unique 2D nanosheet structure.

Acknowledgments

The authors would like to acknowledge the financial supports from the National Natural Science Foundation of China (22222201 and 21972044), China Ministry of Science and Technology (2023YFB3810602, 2021YFA1501401), and the Research Funds of Happiness Flower ECNU (2020ST2203).

References

1 Díaz, U. and Corma, A. (2014). Layered zeolitic materials: an approach to designing versatile functional solids. *Dalton Trans.* 43: 10292–10316.

2 Roth, W.J., Nachtigall, P., Morris, R.E., and Čejka, J. (2014). Two-dimensional zeolites: current status and perspectives. *Chem. Rev.* 114: 4807–4837.

3 Xu, L. and Sun, J. (2016). Recent advances in the synthesis and application of two-dimensional zeolites. *Adv. Energy Mater.* 6: 1600441.

4 Přech, J. (2018). Catalytic performance of advanced titanosilicate selective oxidation catalysts – a review. *Catal. Rev.* 60: 71–131.

5 Xu, H. and Wu, P. (2021). Two-dimensional zeolites in catalysis: current state-of-the-art and perspectives. *Catal. Rev.* 63: 234–301.

6 Wu, P., Xu, H., Xu, L. et al. (2013). Structural modification of Ti-MWW: a door to diversity. In: *MWW-Type Titanosilicate: Synthesis, Structural Modification and Catalytic Applications to Green Oxidations*, 35–61.

7 Wu, P., Tatsumi, T., Komatsu, T., and Yashima, T. (2001). A novel titanosilicate with MWW structure. I. Hydrothermal synthesis, elimination of extraframework titanium, and characterizations. *J. Phys. Chem. B* 105: 2897–2905.

8 Liu, N., Liu, Y.M., Yue, C.C. et al. (2007). A new synthesis route for MWW analogues using octyltrimethylammonium cations as structure-directing agents under alkali-free conditions. *Chem. Lett.* 36: 916–917.

9 Yue, C.C., Xie, W., Liu, Y.M. et al. (2011). Hydrothermal synthesis of MWW-type analogues using linear-type quaternary alkylammonium hydroxides as structure-directing agents. *Micropor. Mesopor. Mater.* 142: 347–353.

10 Wu, P., Miyaji, T., Liu, Y.M. et al. (2005). Synthesis of Ti-MWW by a dry-gel conversion method. *Catal. Today* 99: 233–240.

11 Wu, P. and Tatsumi, T. (2002). Preparation of B-free Ti-MWW through reversible structural conversion. *Chem. Commun.* 1026–1027.

12 Liu, N., Liu, Y., Xie, W. et al. (2007). Hydrothermal synthesis of boron-free Ti-MWW with dual structure-directing agents. In: *Studies in Surface Science and Catalysis* (ed. R. Xu et al.), 464–469. Elsevier.

13 Wang, L., Wang, Y., Liu, Y. et al. (2008). Post-transformation of MWW-type lamellar precursors into MCM-56 analogues. *Microporous Mesoporous Mater.* 113: 435–444.

14 Liu, G., Jiang, J.-G., Yang, B. et al. (2013). Hydrothermal synthesis of MWW-type stannosilicate and its post-structural transformation to MCM-56 analogue. *Microporous Mesoporous Mater.* 165: 210–218.

15 Fung, A.S., Lawton, S.L., and Roth, J. (1994). Synthetic layered MCM-56, its synthesis and use. US Patent, US5362697A.

16 Korzeniowska, A., Grzybek, J., Roth, W.J. et al. (2019). Incorporation of Ti as a pyramidal framework site in the mono-layered MCM-56 zeolite and its oxidation activity. *ChemCatChem* 11: 520–527.

17 Wu, P., Nuntasri, D., Ruan, J. et al. (2004). Delamination of Ti-MWW and high efficiency in epoxidation of alkenes with various molecular sizes. *J. Phys. Chem. B* 108: 19126–19131.

18 Corma, A., Fornes, V., Pergher, S.B. et al. (1998). Delaminated zeolite precursors as selective acidic catalysts. *Nature* 396: 353–356.

19 Corma, A., Díaz, U., Fornés, V. et al. (1999). Ti/ITQ-2, a new material highly active and selective for the epoxidation of olefins with organic hydroperoxides. *Chem. Commun.* 779–780.

20 Kim, S.-Y., Ban, H.-J., and Ahn, W.-S. (2007). Ti-MCM-36: a new mesoporous epoxidation catalyst. *Catal. Lett.* 113: 160–164.

21 Jin, F., Chen, S.-Y., Jang, L.-Y. et al. (2014). New Ti-incorporated MCM-36 as an efficient epoxidation catalyst prepared by pillaring MCM-22 layers with titanosilicate. *J. Catal.* 319: 247–257.

22 Jin, F., Chang, C.-C., Yang, C.-W. et al. (2015). New mesoporous titanosilicate MCM-36 material synthesized by pillaring layered ERB-1 precursor. *J. Mater. Chem. A* 3: 8715–8724.

23 Ren, L., Guo, Q., Orazov, M. et al. (2016). Pillared Sn-MWW prepared by a solid-state-exchange method and its use as a Lewis acid catalyst. *ChemCatChem* 8: 1274–1278.

24 Fan, W., Wu, P., Namba, S., and Tatsumi, T. (2004). A Titanosilicate that is structurally analogous to an MWW-type lamellar precursor. *Angew. Chem. Int. Ed.* 43: 236–240.

25 Ruan, J., Wu, P., Slater, B., and Terasaki, O. (2005). Structure elucidation of the highly active Titanosilicate catalyst Ti-YNU-1. *Angew. Chem. Int. Ed.* 44: 6719–6723.

26 Wu, P., Ruan, J., Wang, L. et al. (2008). Methodology for synthesizing crystalline Metallosilicates with expanded pore windows through molecular Alkoxysilylation of Zeolitic lamellar precursors. *J. Am. Chem. Soc.* 130: 8178–8187.

27 Wang, L., Wang, Y., Liu, Y. et al. (2009). Alkoxysilylation of Ti-MWW lamellar precursors into interlayer pore-expanded titanosilicates. *J. Mater. Chem.* 19: 8594–8602.

28 Ji, X., Xu, L., Du, X. et al. (2017). Simple CTAB surfactant-assisted hierarchical lamellar MWW titanosilicate: a high-performance catalyst for selective oxidations involving bulky substrates. *Cat. Sci. Technol.* 7: 2874–2885.

29 Du, X., Yang, Y., Shi, D. et al. (2018). The construction of a series of hierarchical MWW-type zeolites and their catalytic performances for bulky aldol condensation. *Microporous Mesoporous Mater.* 268: 117–124.

30 Xu, L., Ji, X., Li, S. et al. (2016). Self-assembly of Cetyltrimethylammonium bromide and lamellar zeolite precursor for the preparation of hierarchical MWW zeolite. *Chem. Mater.* 28: 4512–4521.

31 Xu, L., Huang, D.-D., Li, C.-G. et al. (2015). Construction of unique six-coordinated titanium species with an organic amine ligand in titanosilicate and their unprecedented high efficiency for alkene epoxidation. *Chem. Commun.* 51: 9010–9013.

32 Ding, L., Yin, J., Tong, W. et al. (2021). Selective synthesis of epichlorohydrin via liquid-phase allyl chloride epoxidation over a modified Ti-MWW zeolite in a continuous slurry bed reactor. *New J. Chem.* 45: 331–342.

33 Tong, W., Yin, J., Ding, L. et al. (2020). Modified Ti-MWW zeolite as a highly efficient catalyst for the cyclopentene epoxidation reaction. *Front. Chem.* 8.

34 Yin, J., Xu, H., Wang, B. et al. (2020). Highly selective 1-pentene epoxidation over Ti-MWW with modified microenvironment of Ti active sites. *Cat. Sci. Technol.* 10: 6050–6064.

35 Yin, J., Jin, X., Xu, H. et al. (2021). Structured binder-free MWW-type titanosilicate with Si-rich shell for selective and durable propylene epoxidation. *Chin. J. Catal.* 42: 1561–1575.

36 Schreyeck, L., Caullet, P., Mougenel, J.C. et al. (1996). PREFER: a new layered (alumino) silicate precursor of FER-type zeolite. *Micropor. Mater.* 6: 259–271.

37 Ikeda, T., Kayamori, S., and Mizukami, F. (2009). Synthesis and crystal structure of layered silicate PLS-3 and PLS-4 as a topotactic zeolite precursor. *J. Mater. Chem.* 19: 5518–5525.

38 Wang, J., Fan, Y., Guo, X. et al. (2022). Direct synthesis and delamination of swollen layered Ferrierite for the reductive etherification of furfural. *ChemCatChem* 14: e202200535.

39 Peng, M., Wang, Z.-Q., Huang, J. et al. (2021). Two coexisting forms of simple molecules for directing Sesqui-unit-cell zeolite Nanosheets. *Chem. Mater.* 33: 6934–6941.

40 Corma, A., Diaz, U., Domine, M.E., and Fornés, V. (2000). New aluminosilicate and titanosilicate delaminated materials active for acid catalysis, and oxidation reactions using H2O2. *J. Am. Chem. Soc.* 122: 2804–2809.

41 Radko, M., Rutkowska, M., Kowalczyk, A. et al. (2020). Catalytic oxidation of organic sulfides by H2O2 in the presence of titanosilicate zeolites. *Microporous Mesoporous Mater.* 302: 110219.

42 Święs, A., Kowalczyk, A., Michalik, M. et al. (2021). Titanium–silicon ferrierites and their delaminated forms modified with copper as effective catalysts for low-temperature NH3-SCR. *RSC Adv.* 11: 10847–10859.

43 Yang, B., Jiang, J.-G., Zhang, K., and Wu, P. (2016). Synthesis of novel Titanosilicate catalysts by simultaneous Isomorphous substitution and interlayer expansion of Zeolitic layered silicates. *Chem. Mater.* 28: 5295–5303.

44 Yang, B. and Zou, Q. (2017). Post-synthesis and catalytic performance of novel interlayer expanded Stanosilicate IEZ-Sn-PLS-3. *Chem. Lett.* 46: 1781–1784.

45 Yang, B.T., Jiang, J.G., Xu, H. et al. (2018). Synthesis of extra-large-pore zeolite ECNU-9 with intersecting 14*12-ring channels. *Angew. Chem.-Int. Edit.* 57: 9515–9519.

46 Banno, S. and Komura, K. (2018). Large crystals synthesis of siliceous layered Zeolitic PLS-1 and CDS-1 zeolite by dry gel conversion method. *Cryst. Res. Technol.* 53: 1800036.

47 Zhao, Z., Zhang, W., Ren, P. et al. (2013). Insights into the Topotactic conversion process from layered silicate RUB-36 to FER-type zeolite by layer reassembly. *Chem. Mater.* 25: 840–847.

48 Roth, W.J., Gil, B., Makowski, W. et al. (2016). Interconversion of the CDO layered precursor ZSM-55 between FER and CDO frameworks by controlled deswelling and reassembly. *Chem. Mater.* 28: 3616–3619.

49 Yang, Q., Zhao, Y., Luo, F. et al. (2022). Stacking-faulted CDO zeolite nanosheets efficient for bulky molecular reactions. *Chem. Commun.* 58: 6008–6011.

50 Ahn, S.H. and Hong, S.B. (2020). Transformation control of a layered zeolite precursor by simple cation exchange. *Chem. Mater.* 32: 9740–9746.

51 Xiao, F.-S., Xie, B., Zhang, H. et al. (2011). Interlayer-expanded microporous Titanosilicate catalysts with functionalized hydroxyl groups. *ChemCatChem* 3: 1442–1446.

52 Gies, H., Feyen, M., De Baerdemaeker, T. et al. (2016). Interlayer expansion using metal-linker units: crystalline microporous silicate zeolites with metal centers on specific framework sites. *Microporous Mesoporous Mater.* 222: 235–240.

53 Bian, C., Wang, X., Yu, L. et al. (2020). Generalized methodology for inserting metal heteroatoms into the layered zeolite precursor RUB-36 by interlayer expansion. *Crystals* 10: 530.

54 De Baerdemaeker, T., Gies, H., Yilmaz, B. et al. (2014). A new class of solid Lewis acid catalysts based on interlayer expansion of layered silicates of the RUB-36 type with heteroatoms. *J. Mater. Chem. A* 2: 9709–9717.

55 Choi, M., Na, K., Kim, J. et al. (2009). Stable single-unit-cell nanosheets of zeolite MFI as active and long-lived catalysts. *Nature* 461: 246–249.

56 Roth, W.J., Shvets, O.V., Shamzhy, M. et al. (2011). Postsynthesis transformation of three-dimensional framework into a lamellar zeolite with modifiable architecture. *J. Am. Chem. Soc.* 133: 6130–6133.

57 Verheyen, E., Joos, L., Van Havenbergh, K. et al. (2012). Design of zeolite by inverse sigma transformation. *Nat. Mater.* 11: 1059–1064.

58 Roth, W.J., Nachtigall, P., Morris, R.E. et al. (2013). A family of zeolites with controlled pore size prepared using a top-down method. *Nat. Chem.* 5: 628–633.

59 Na, K., Jo, C., Kim, J. et al. (2011). MFI Titanosilicate Nanosheets with single-unit-cell thickness as an oxidation catalyst using peroxides. *ACS Catal.* 1: 901–907.

60 Wang, J., Xu, L., Zhang, K. et al. (2012). Multilayer structured MFI-type titanosilicate: synthesis and catalytic properties in selective epoxidation of bulky molecules. *J. Catal.* 288: 16–23.

61 Kim, J., Chun, J., and Ryoo, R. (2015). MFI zeolite nanosheets with post-synthetic Ti grafting for catalytic epoxidation of bulky olefins using H2O2. *Chem. Commun.* 51: 13102–13105.

62 Li, C.-G., Lu, Y., Wu, H. et al. (2015). A hierarchically core/shell-structured titanosilicate with multiple mesopore systems for highly efficient epoxidation of alkenes. *Chem. Commun.* 51: 14905–14908.

63 Luo, P., Guan, Y., Xu, H. et al. (2020). Postsynthesis of hierarchical core/shell ZSM-5 as an efficient catalyst in ketalation and acetalization reactions. *Front. Chem. Sci. Eng.* 14: 258–266.

64 Xu, L., Li, C.-G., Zhang, K., and Wu, P. (2014). Bifunctional tandem catalysis on multilamellar organic–inorganic hybrid zeolites. *ACS Catal.* 4: 2959–2968.

65 Přech, J., Eliášová, P., Aldhayan, D., and Kubů, M. (2015). Epoxidation of bulky organic molecules over pillared titanosilicates. *Catal. Today* 243: 134–140.

66 Přech, J., Morris, R.E., and Čejka, J. (2016). Selective oxidation of bulky organic sulphides over layered titanosilicate catalysts. *Cat. Sci. Technol.* 6: 2775–2786.

67 Emdadi, L., Tran, D.T., Zhang, J. et al. (2017). Synthesis of titanosilicate pillared MFI zeolite as an efficient photocatalyst. *RSC Adv.* 7: 3249–3256.

68 Wu, W., Tran, D.T., Wu, X. et al. (2019). Multilamellar and pillared titanium Silicalite-1 with long-range order of zeolite nanosheet layers: synthesis and catalysis. *Microporous Mesoporous Mater.* 278: 414–422.

69 Liu, B., Mu, Q., Huang, J. et al. (2020). Fabrication of titanosilicate pillared MFI zeolites with tailored catalytic activity. *Front. Chem. Sci. Eng.* 14: 772–782.

70 Wang, M., Wang, X., You, Q. et al. (2021). Dual-template synthesis of hierarchically layered titanosilicate-1 zeolites for catalytic epoxidation of cyclooctene. *Microporous Mesoporous Mater.* 323: 111207.

71 Ren, L., Guo, Q., Kumar, P. et al. (2015). Self-pillared, single-unit-cell Sn-MFI zeolite Nanosheets and their use for glucose and lactose isomerization. *Angew. Chem. Int. Ed.* 54: 10848–10851.

72 Liu, X., Luo, Y., Mao, W.T. et al. (2020). 3D Electron diffraction unravels the new zeolite ECNU-23 from the "pure" powder sample of ECNU-21. *Angew. Chem.-Int. Edit.* 59: 1166–1170.

73 Liu, X., Xu, H., Zhang, L. et al. (2016). Isomorphous incorporation of tin ions into Germanosilicate framework assisted by local structural rearrangement. *ACS Catal.* 6: 8420–8431.

74 Jiao, M., Zhao, Y., Jiang, J. et al. (2021). Extra-large pore Titanosilicate synthesized via reversible 3D–2D–3D structural transformation as highly active catalyst for Cycloalkene epoxidation. *ACS Catal.* 11: 2650–2662.

75 Liu, X., Zhang, L., Xu, H. et al. (2018). Pore size-tunable titanosilicates post-synthesized from germanosilicate by structural reorganization and H2TiF6-assisted isomorphous substitution. *Appl. Catal. A-Gen.* 550: 11–19.

76 Přech, J. and Čejka, J. (2016). UTL titanosilicate: an extra-large pore epoxidation catalyst with tunable textural properties. *Catal. Today* 277: 2–8.

3

Synthesis and Catalytic Applications of Sn- and Zr-Zeolites

Zhiguo Zhu

College of Chemistry and Chemical Engineering, Yantai University, 30 Qingquan Road, Yantai 264005, Shandong, China

3.1 Introduction

Isomorphous substitution of tetravalent heteroatoms such as Ti^{4+}, Sn^{4+}, and Zr^{4+} into silica matrices is an effective strategy for designing Lewis acid zeolites without the co-generation of Brønsted acid sites [1]. Titanium silicalite-1 (TS-1) with MFI topology was first developed in 1983 and was deemed as a milestone due to its unique feature of tetrahedral framework Ti^{4+} sites for activating peroxides [2]. Afterward, the incorporation of tetravalent metal, such as Sn^{4+} and Zr^{4+}, into siliceous zeolite framework sprang up like mushrooms, giving rise to Sn/Zr-zeolites. Representatively, Sn/Zr-Beta zeolites with high hydrophobicity and crystallinity are hydrothermally synthesized in a fluoride media [3]. Considering the water-tolerant property, the research studies on Sn/Zr-Beta zeolites have surged in recent decades in aqueous catalytic systems such as catalytic biomass conversions [3, 4].

Nonetheless, the fluoride-mediated method encounters the disadvantage of requiring long crystallization time (generally above 10 days), forming large-sized crystals at the micrometer level, generating limited metal contents in zeolite frameworks, and using environmentally unfriendly fluoride [3, 5, 6]. Remarkable endeavors have been made to address these issues, mainly by ameliorating synthesis methods and discovering other alternative metallosilicate analogues. The former comprised two strategies, that is, the bottom-up approaches, wherein a metal source is introduced into the initial synthesis gels, and top-down approaches, wherein a metal atom is inserted into an already-existing zeolite framework [7]. Each specific method possesses its own characteristics in terms of synthetic conditions and physicochemical property of targeted zeolites, likely producing different catalytic active sites, for example, open or closed, framework or extra-framework, and hydrated or dehydrated species. The latter approach concentrates on designing Sn/Zr-silicates with discrepant topology structures covering 8–12 MR zeolites and mesoporous silicates.

A Lewis acid is generally defined as an electron acceptor. In this regard, tetrahedrally coordinated metals in zeolite framework can serve as Lewis acid sites,

Micro-Mesoporous Metallosilicates: Synthesis, Characterization, and Catalytic Applications, First Edition. Edited by Peng Wu and Hao Xu.

which is derived from the coordinatively unsaturated metal ions when their valence electrons covalently link with adjacent framework oxygen atoms. The framework metal-based Lewis acid sites are capable of accepting electron pairs from organic guest molecules without forming charge imbalance in zeolite framework, favoring the adsorption and activation of reactants with electron-rich groups such as hydroxyl and carbonyl functional groups [3]. This chemical activation significantly relies on the heteroatoms, reaction substrates, and zeolite structures, which then impacts the stability, flexibility, and electronic structure of the resulting adducts [8].

The unique activation ability of Lewis acid in Sn/Zr-zeolites for hydroxyl and carbonyl groups has recently drawn increasing attentions in the field of catalytic conversions of biomass-derived substrates. Besides, metallosilicate zeolites can be utilized as supports to load metal particles or clusters [9]. These composite materials integrate the merits of Sn/Zr-silicates and metal species, which exhibit bifunctional feature or modulate the interface characteristics between metal and zeolite support [9, 10].

In this chapter, we systematically summarize the synthesis method of Sn- and Zr-containing silicate materials including microporous silicates (8 MR zeolite, CHA; 10 MR zeolites, MFI, MEL, MWW; 12 MR zeolites, *BEA, MSE, FAU, MTW, MOR, BEC; 14 MR zeolite, UTL) and mesoporous silicates (MCM-41, MCM-48, SBA-15, SBA-16) in recent decades. Two major methodologies are adopted to prepare stannosilicates and zirconosilicates, viz. bottom-up (hydrothermal synthesis, dry-gel conversion, interzeolite transformation, and structural reconstruction strategy) and top-down methods (interlayer-expanded method, atom-planting method, alkaline-assisted metalation, and demetallation–metalation). Then, the catalytic applications of Sn/Zr-silicates are reviewed in detail, where the catalytic performance will be correlated with the geometry structure of active site, topology structure, hydrophobicity, and accessibility. These catalytic transformations primarily involve redox catalysis (Baeyer–Villiger oxidation and Meerwein–Ponndorf–Verley–Oppenauer redox), Lewis acid catalysis (ring opening

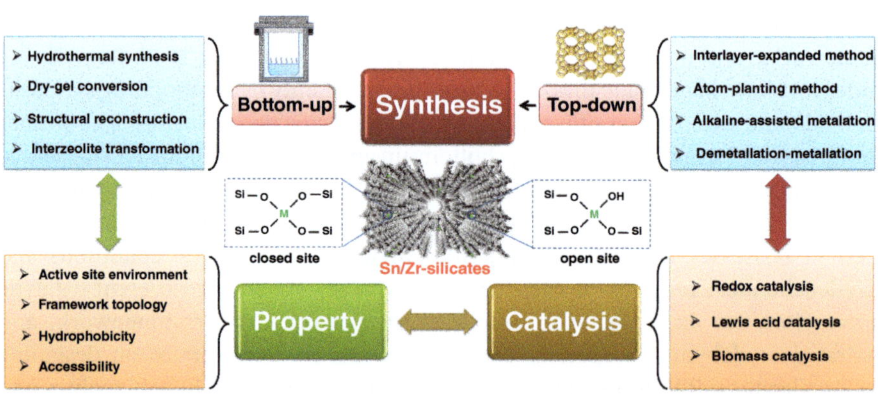

Figure 3.1 Illustration of the correlation between synthesis, property, and catalytic performance for Sn/Zr-zeolites. The green letter "M" in open and closed sites is the abbreviation of "metal," which represents Sn or Zr atom.

of epoxides, Aldol reaction, and so on) and biomass conversion (sugar isomerization, synthesis of lactic acid or alkyl lactates, and so on). The relationship between synthesis, properties, and performance is attempted to be established aiming to provide useful guidelines as shown in Figure 3.1.

3.2 Synthesis of Sn- and Zr-Zeolites

Regarding the synthesis of metallosilicate zeolites, it is imperative to form highly ordered framework structure during the process of crystallization. Apart from this, the heteroatoms are required to be isomorphously substituted into siliceous zeolite framework, acting as catalytically active sites with tetrahedrally coordinated state in framework to coordinate with electron-donating groups of the reaction substrate [11]. The formation of extra-framework metal species such as metal oxide is undesired because these species with higher coordination state are less active and even, in some cases, are detrimental for catalytic performances [3]. Hence, designing synthesis method is significantly important according to raw materials, process conditions, and practical demand. In general, metallosilicate zeolites can be prepared via two major strategies, i.e. bottom-up and top-down approaches, which are also suitable for the synthesis of Sn/Zr-containing zeolites and mesoporous silicates, and the detailed synthetic strategies are summarized in Figure 3.2.

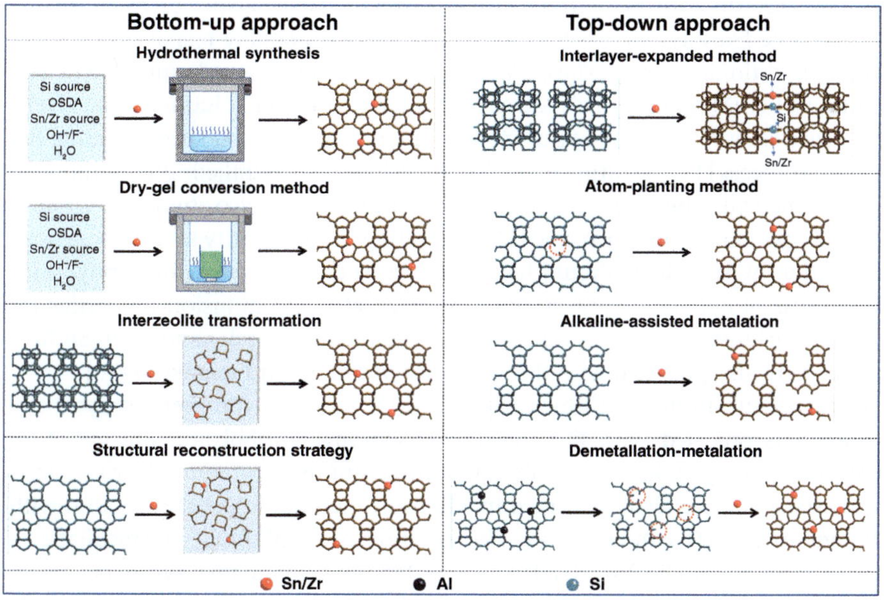

Figure 3.2 The main strategies for the synthesis of Sn/Zr-zeolites. For simplicity, Beta structure is taken as an example. Dark blue balls represent Al atoms in the initial structure for demetallation–metalation method. Aqua balls indicate Si atoms in interlayer-expanded method. The other balls are colored by red, which represent Sn/Zr atoms. The red dotted circle indicates silanol defects in zeolite.

3.2.1 Bottom-up Approaches

3.2.1.1 Hydrothermal Synthesis

The successful syntheses of TS-1 and all-silica zeolites under near neutral conditions with the mineralizer agent of fluoride open the door to prepare metallosilicate zeolites as some metal sources are poorly soluble or insoluble in conventional basic media [2]. Numerous heteroatoms such as Sn, Zr, W, Mo, and Hf are substituted into various zeolite topology structures [7]. Herein, this section mainly focuses on Sn/Zr-silicates with the topology structure of MFI, MWW, and *BEA.

3.2.1.1.1 MFI Structure Stannosilicate Sn-MFI was first hydrothermally synthesized by adding a tin source into a synthesis mixture of sodium hydroxide, colloidal silica, and tetrapropylammonium bromide (TPABr) [3]. Apart from this approach in a basic system, Sn-MFI can also be prepared in a fluoride medium with a pH value of 6.4 [12]. In this literature, authors also compared the influence of adding sequence of raw materials on zeolite crystallinity and catalytic activity in the hydroxylation of phenol. It was found that the procedure in which Sn source of $SnCl_4 \cdot 5H_2O$ was first dissolved in silica source of tetraethoxysilane followed by the addition of tetrapropylammonium hydroxide was the most favorable.

Subsequently, Vargas et al. reported that the presence of a residual tin/sodium surface phase was observed for the conventional hydroxide-mediated method with more than about 3 wt% Sn in the solid product [13]. Moreover, this tin/sodium overlayer was difficult to be removed by dissolving it in acidic or basic media. Incompatible hydrolysis rate of Sn and Si source was possibly responsible for the formation of tin/sodium overlayer. Yuan et al. utilized ethylenediaminetetraacetic acid tin salt (EDTA-Sn) and citrate as Sn source and buffer solution, respectively, to realize a slow-release rate of Sn^{4+} ions to interact with silicon species during the synthesis process [14], resulting in uniform distribution of tetrahedrally coordinated Sn ions. The hydrothermal synthesis of Sn-MFI zeolite was also attempted without the participation of sodium species [15], resulting in high purity and homogenous distribution of Sn, which should be of interest to the catalysis community. With the purpose of enhancing accessibility of active sites and diffusion properties for organic reactants in catalytic applications, hierarchical and nanosheet Sn-MFI zeolites were also prepared with the assistance of soft or hard templates [16].

As for Zr-substituted MFI-topology (Zr-MFI) zeolite, hydrothermal synthesis method was reported with the solvent of isopropyl alcohol and the OSDA of TPAOH at 180 °C for 48 h [17]. Subsequently, in situ incorporation of Zr into MFI zeolite structure was also performed in a fluoride medium to get MFI zeolites with different Al and Zr contents [18]. The addition of amphiphilic organosilane of $[(C_2H_5O)_3SiC_3H_6N(CH_3)_2C_{16}H_{33}]Cl$ in the hydroxide-mediated synthetic gel of traditional zirconosilicate afforded nanosized Zr-MFI crystals with tetrahedrally coordinated Zr^{4+} ions [19]. Zr-doped siliceous MFI zeolite with interconnected trimodal micro-meso-macroporous system was successfully made in the solvent of glycerin via a quasi-solid-state crystallization with the OSDA of TPAOH, where the addition of glycerin was of importance to obtain the porous hierarchy [20]. Soon afterward, the same group shed light on the effect of synthetic conditions such as scaffold templates, OSDA, crystallization time, and solvent types on the formation of targeted micro-meso-macroporous Zr-MFI zirconosilicate [21]. Besides, this methodology was also extended to the synthesis of aluminosilicate ZSM-5 and Beta zeolite with micro-meso-macroporous multiple systems [21].

3.2.1.1.2 MWW Structure The extraordinary catalytic performance of Ti-MWW zeolite in selective oxidations motivates the creation of other heteroatom-containing MWW zeolites such as Sn-MWW and Zr-MWW zeolites. Lv et al. firstly reported the hydrothermal synthesis of MWW-type zirconosilicate using hexamethyleneimine (HMI) or piperidine (PI) as OSDA with the help of crystallization-supporting agent of boric acid, where alkali metal ions were required for nucleation and crystallization [22]. Ti-MWW zeolite was successfully synthesized under alkali-free conditions [23], but only amorphous phase was obtained even at low Zr^{4+} contents (Si/Zr molar ratio = 500) in the synthesis gels without the aid of alkali metal cations [22]. Afterward, the same group also uses the similar approach to hydrothermally synthesize Sn-MWW zeolite [24]. Meanwhile, a mild acid treatment was carried out for the as-made Sn-MWW sample to generate Sn-MCM-56 analogue, which possessed high external surface area and more open reaction space, owing to partially delaminated MWW sheets. In the same year, Guo et al. hydrothermally prepared Sn-MWW zeolite by a reversible structural transformation from the highly deboronated MWW zeolite [25].

In order to overcome the drawbacks of multiple steps and co-existence of boron, Zhu et al. proposed a dual OSDA strategy for the hydrothermal synthesis of Sn-MWW and Zr-MWW zeolites using *N,N,N*-trimethyl-1-adamantammonium-hydroxide (TMAdaOH) and HMI as organic co-templates with the assistance of K^+ [26]. Yang et al. reported a hollow nest-structured Zr (Zr-HSZ) zeolite with intergrown single/double/multi-unit-cell MWW nanosheets by a one-step rotating hydrothermal method [27].

3.2.1.1.3 Beta Structure Unlike titanosilicate zeolites, it is difficult to introduce Sn atoms into siliceous zeolite frameworks because of the relatively larger ions size (0.71 Å) and Sn—O bond length (1.912 Å) than those (0.68 Å and 1.892 Å, respectively) of Ti^{4+}, as well as the incompatible hydrolysis rate of silica and Sn sources [3, 10].

The first hydrothermal synthesis of Sn-containing Beta (Sn-Beta) zeolite was reported by Mal et al. in 1997 [3]. When highly dealuminated Sn-Al-Beta was used as seeds, Al-free Sn-Beta was obtained, which demonstrated high hydrophilicity derived from numerous framework defects. Two years later, Corma et al. disclosed a patent about the synthesis of Sn-Beta zeolite, which is the most widely used method [28]. The highlight is the introduction of fluoride ions as effective mineralizing agents. Although nucleation became relatively convenient in the fluoride medium, the low supersaturation led to a long crystallization time (>40 days) and large crystal size at microsized level. The bulky Sn^{4+} ions caused a limited Sn content (generally Si/Sn molar ratio $\geq$ 100) in zeolite frameworks [5, 6]. Owning to the difficulty in the synthesis of Sn-Beta zeolite and its frustrated applications at that time, to our knowledge, the corresponding research studies are rarely reported in the following decade.

Until 2010, Davis et al. discovered that Sn-Beta zeolite was an unparalleled heterogeneous catalyst in the isomerization of glucose to fructose [29], and then researcher paid significant attention to the synthesis of Sn-Beta again. Adding all-silica Beta seeds can shorten the crystallization time to 11–12 days [7]. Fan et al. studied the influence of the morphology and dispersed state of Beta seeds in the synthetic gel on the crystallization kinetics of Sn-Beta zeolite [7]. A stable suspension with highly dispersed dealuminated zeolite seeds (200 nm) was adopted as a seeding solution, which enabled the synthesis time decrease from >20 to two days.

Moreover, the crystallization time of Sn-Beta zeolite was also closely related to Sn and H_2O amounts in the synthetic gels [30–33]. Meng et al. put forward an aerosol-seed-assisted method to synthesize Sn-Beta zeolite with the purpose of lowering the template and F^- contents and shortening the crystallization time [30]. Subsequently, to improve the Sn content and the diffusion property, aerosol-assisted hydrothermal method combined with further mild base treatment was conducted [31]. In 2020, the same group used aerosol-derived hollow Sn—Si mixed oxide as the precursor to produce Sn-Beta zeolite with high Sn content (Si/Sn = 38) and mesoporous structure [32]. Nonetheless, as revealed by UV–vis spectra, a large amount of extra-framework Sn species existed in these aerosol-assisted Sn-Beta zeolites. Analogously, Takayuki et al. successfully synthesized Sn-Beta zeolite with the Sn contents of 3.17 wt% (Si/Sn molar ratio of 60) with a Sn—Si mixed oxide precursor [33]. However, some atoms as contaminations such as Al and Y were inevitably introduced into zeolite.

As for the hydrothermal synthesis of all-silica or metallosilicate zeolite without the use of fluoride, the positive charge of organic OSDA in the as-synthesized sample is required to be compensated by the framework defect sites. It is well-established that fluoride is capable of facilitating the mineralization of silica source and compensating for positive OSDA ions [7]. Therefore, fluoride-mediated Sn-Beta zeolite exhibited less defect sites within zeolite crystals, resulting in high hydrophobicity [8]. However, the utilization of fluoride may cause environmental problems. Fluoride-free synthesis of Sn-Beta zeolite with high crystallinity was achieved using a complicated OSDA of *N*-cyclohexyl-*N,N*-dimethylcyclohexanaminium hydroxide (M_2Cy_2NOH) with the assistance of Na^+ ions and Beta seeds [34].

The hydrothermal synthesis of zirconosilicate zeolites was difficult due to the even larger ionic radius of Zr^{4+} than that of Sn^{4+}. Unlike Sn-Beta, the hydrothermal synthesis of Zr-substituted Beta (Zr-Beta) zeolites does not work in a traditional basic medium, which in general needs the assistance of Al^{3+} ions [35]. The Al-free Zr-Beta zeolite was first hydrothermally synthesized in the fluoride medium using dealuminated Beta zeolite as seeds [36, 37]. The longer crystallization time was required with the increase of Zr content in the synthesis gel. Soon afterward, many groups employed aforementioned fluoride-assisted approach to hydrothermally synthesize Zr-Beta zeolite [35].

Kots et al. investigated the mechanism of Zr incorporation during this fluoride-mediated synthesis of Zr-Beta zeolite [38]. In the first step, all-silica Beta was quickly crystallized through a solid–solid rearrangement mechanism. Zr species were then inserted into the Beta crystals, existing in the form of hexa-coordinated state and seating in square-bipyramidal positions within zirconium silicate fragments. In the second step, Zr-rich fragments began to interact with initially formed Si-Beta via a solid–solid rearrangement mechanism, resulting in the formation of four-coordinated and isolated Zr^{4+} ions in the zeolite framework. Further prolonging the crystallization time, the closed Zr^{4+} sites were converted into open sites [38]. Consequently, the optimization of crystallization time was essential for the catalytic performance of Zr-Beta zeolites, which was intrinsically determined by the contest of active sites (i.e. open and closed Zr^{4+} sites). Compared with Sn-Beta zeolite, so far the Zr-Beta zeolite material has received little attention. In most cases, it is deemed as a reference material to emphasize the merits of Sn-Beta zeolite in catalytic applications.

3.2.1.1.4 Other Structures Apart from zeolite carriers of MFI, MWW, and Beta structures, the substitution of Sn^{4+}/Zr^{4+} ions into other structures has also been attempted. A direct hydrothermal synthesis of all-silica chabazite zeolite (3D 8MR) substituted by framework Sn ions (Sn-CHA) was proposed by Harris et al [39]. Al-free Sn-ZSM-11 zeolite with MEL topology and 3D 10MR channels (Si/Sn molar ratio > 40) was hydrothermally synthesized using a similar protocol of all-silica ZSM-11 and tetrabutylammonium hydroxide (TBAOH) as the template [3]. Sn^{4+} ions can be also incorporated into all-silica ZSM-12 (MTW framework structure with 1D large-pore 12MR), which was achieved by hydrothermal method using a new OSDA of hexamethylene bis(benzyldimethyl ammonium hydroxide) [3]. Zr-Al-Y zeolite was prepared without organic template using $ZrSO_4$ as a Zr source at 98 °C for 18 h, resulting in high Al contents in the final products ($SiO_2/Al_2O_3 = 5.3$) [40].

Substitution of Sn^{4+}/Zr^{4+} ions into the framework of mesoporous silicates is an appealing and alternative method to improve diffusion performance and enhance the accessibility to the active sites. Compared with crystalline zeolites, the hydrothermal stability of mesoporous silicates was concerned due to the amorphous pore walls, whose structure was susceptible to be changed after suffering from hot water treatment [41]. Among these mesoporous materials, MCM-41 and SBA-15 were two main hosts for the preparation of Lewis acid catalysts. The hosts of Sn/Zr include SBA-16, MCM-48, HMS, TUD-1, KIT-5, KIT-6, MSU-3, and mesoporous silica nanospheres [42]. Sn/Zr ions in all these mesopore materials exhibited tetra-coordinated state in the framework structure. However, polymerized oxide clusters or particles appeared when the Sn contents were high in the gels. Moreover, the synthesis parameters and the type of raw materials, particularly the Sn/Zr source, greatly impacted the coordinated state of Sn/Zr species.

3.2.1.2 Dry-Gel Conversion Methods

Dry gel conversion (DGC) is an effective method for the synthesis of Sn/Zr containing zeolites. Niphadkar et al. found that DGC method resulted in smaller TPA^+

amount, shorter crystallization time, and less water waste compared to traditional hydrothermal method in the synthesis of Sn-MFI zeolite [43]. In 2013, Kang et al. adopted DGC method to synthesize Sn-Beta zeolite in fluoride medium [7]. The crystallized time was drastically reduced compared with conventional hydrothermal method. Subsequently, the same group noted that the interior structure of DGC autoclave greatly affected the crystallization behavior of Sn-Beta zeolite [44]. With the conventional autoclave, the condensed water from steam inside the autoclave generally dropped back into the dried gel, which in turn change the gels composition. To address this issue, an oval glass cover over the dried gel powers was constructed, guiding the condensed water back to the bottom, which was helpful for the formation of Sn-Beta zeolite [44]. In consideration of the environmental unfriendliness of fluoride, Chang et al. prepared Sn-Beta zeolite with the assistance of DGC method and dealuminated Beta seeds in fluoride-free medium [45].

3.2.1.3 Interzeolite Transformation

Interzeolite transformation (also called interzeolite conversion) is a zeolite synthesis strategy, which uses a parent zeolitic source to produce another daughter zeolite. The transformation is significantly affected by the nature and composition of the parent zeolite as well as synthetic conditions such as temperature and time [46]. Okubo et al. proposed a typical theory "composite building unit hypothesis" that at least one common CBU should be present between parent and daughter zeolites to conquer kinetic and thermodynamic hurdles [47].

Heteroatom-containing CHA-type zeolites (M-Al-CHA, M = Sn, Fe, Ga) were synthesized by interzeolite transformation of heteroatom-incorporated FAU-type zeolite (M-Al-FAU), which cannot be achieved using amorphous hydrogels as raw materials under the same hydrothermal conditions [51]. Interzeolite transformation method has also been used to synthesize Sn-Beta zeolite to solve the issues in conventional hydrothermal synthesis, such as low Sn content, time-consuming crystallization, and large crystal size. Zhu et al. synthesize Sn-Beta zeolite with a high Sn content of 3.03 wt% (corresponding to Si/Sn molar ratio of 63) using all-silica MWW zeolite (also termed as ITQ-1 zeolite) within a short time of three days (Figure 3.3a) [48]. This method was also suitable for the synthesis of pure silica Beta zeolite and Ti-Beta zeolite [52, 53]. The revolution process obeyed dissolution-recrystallization mechanism. Moreover, the addition of dealuminated Beta seeds and the structural similarity between the parent (MWW) and daughter (*BEA) zeolite were two crucial factors during the crystallization process. The obtained Sn-Beta zeolite demonstrated comparable hydrophobicity and smaller crystal size in comparison with traditional fluoride-assisted Sn-Beta zeolite [48]. Recently, Zhu et al. also put forward an interzeolite transformation of siliceous Y zeolite to Sn-Beta zeolite via a steam-assisted conversion in the absence of fluoride and alkali metals (Figure 3.3b) [49]. The crystal size of Sn-Beta (50–150 nm) was the minimum among all the present hydrothermal synthesized ones. Importantly, the Sn content of 2.39 wt% (Si/Sn = 81) in Sn-Beta zeolite exceeded that obtained by the classic fluoride-mediated method. However, its hydrophobicity was inferior, like Sn-Beta synthesized in alkaline medium [34]. In this work, the addition of

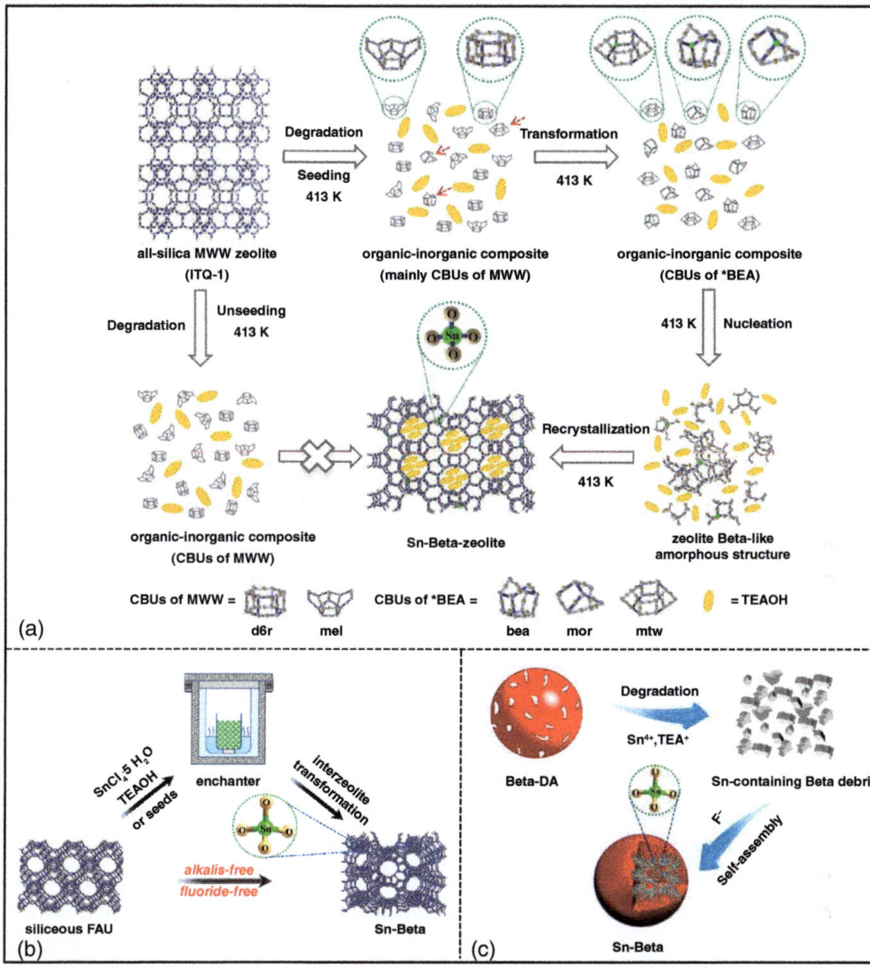

Figure 3.3 (a) Proposed crystallization mechanism for the interzeolite transformation of pristine ITQ-1 to Sn-Beta zeolite. Source: Reproduced from Ref. [48]/with permission of Elsevier. (b) The crystallization process for nanosized Sn-Beta zeolite synthesized from siliceous FAU zeolite without the participation of fluoride and alkali metal ions. Source: Reproduced with permission from Ref. [49]; © 2020 Elsevier. (c) Schematic illustration of the synthesis of Sn-Beta zeolite by structural reconstruction. Source: Reproduced from Ref. [50]/with permission of Royal Society of Chemistry.

Beta seeds was not prerequisite but just a crystallization accelerator, which was different to the aforementioned interzeolite conversion of ITQ-1 zeolite [48]. Without introducing Beta seeds into the synthesis gels, the fully crystallized time was prolonged from 3 to 13 days with the Si/Sn molar ratio of 120 [49].

Interzeolite transformation has shown substantial advantages for the synthesis of heteroatom-containing zeolites with short crystallization time, nanosized crystals, and high metal content. Besides, a higher supersaturation is proposed for this methodology [46], providing a great opportunity for the effective crystallization of a desired zeolite. Interzeolite transformation is an emerging method for the synthesis of metallosilicate zeolites, which will prosper the field of heteroatom-substituted zeolite synthesis and catalytic applications.

3.2.1.4 Structural Reconstruction Strategy

The hydrothermal crystallization process of zeolite contains several crucial steps as follows, depolymerization of silica source, polysilicate repolymerization, nucleation, nucleus growth, and crystal growth [46, 48]. It is generally recognized that the common building units (CBUs) are formed under the induction effect of organic template when using amorphous raw materials as silica sources, further establishing the zeolite framework via self-assembly. During the whole crystallization process, classic spontaneous nucleation is the core and rate-determining step [46]. For the conventional synthesis of metallosilicate zeolites, two phenomena are commonly noticed, (I) unsuccessful crystallization due to the difficulty in nucleation and (II) limited metal content and long crystallization time owing to the inhibiting effect of metal ions on nucleation and zeolite growth [6, 48].

From the above analysis, it is known that zeolite CBUs play important roles during the nucleation process. If these CBUs come from the degraded zeolite, rather than self-polymerization process of amorphous silica source with the assistance of template, obviously, the nucleation process will become easy, further forming more nuclei, accelerating zeolite growth, suppressing the formation of impurity phase, and promoting the generation of nano-sized crystals.

In 2017, Zhu et al. firstly utilized structural reconstruction strategy to synthesize Sn-Beta zeolite (Figure 3.3c) [50]. Deeply dealuminated Beta zeolite was degraded in alkaline TEAOH solution, and then Beta fragments and Sn species self-assembled into Sn-Beta zeolite in the presence of TEA^+ ions and fluoride. This method significantly reduced the crystallization time to 1 h, which was two orders of magnitude shorter than the traditional hydrothermal synthesis approach. Additionally, nano-sized Sn-Beta zeolite crystals (30–50 nm) and high isolated framework Sn content (Si/Sn molar ratio of 33) were obtained. UV-Raman characterization was conducted for the initial gel precursor, illustrating that primary structural units (four-, five-, and six-membered ring) of Beta zeolite was clearly observed. Contrarily, these signals were not shown for initial gel with amorphous TEOS as silica source. Probably, these degraded CBUs facilitated zeolite nucleation and crystal growth [50].

Subsequently, in 2020, Tang et al. employed this method to synthesize hierarchical Sn-Beta zeolite using polydiallyldimethylammonium chloride (PDADMAC) as mesoporous structure directing agent [54]. In 2021, this structural reconstruction approach was extended to the synthesis of MFI-type topology heteroatoms-containing zeolites, such as Sn-MFI, Al-MFI, TS-1, Fe-MFI, and Nb-MFI zeolites [55]. It was found that gripper-like silica species with rich Si-OH groups, mainly in Q^2 and Q^3 groups, played a significant role in heteroatom incorporation and uniform distribution. Metal ions can be locked by such gripper-like silica species, forming stable and isolated Si-O-metal units in the synthesis gels [55]. Reconstruction can also be performed over post-synthesized Sn-Beta obtained by solid-state incorporation to tune the hydrophobicity and morphology in the presence of ammonium fluoride (NH_4F) and tetraethylammonium bromide (TEABr) [56]. NH_4F acted as the dissolution agents of crystalline zeolite, while TEABr promoted the recrystallization of zeolite framework and then healed the silanol defects within zeolites.

3.2.2 Top-Down Approaches

Owning to the disadvantages of relatively long crystallization time, the utilization of toxic fluoride, limited framework metal content, and large crystal size for the direct synthesis of certain Sn/Zr-zeolites, top-down approaches were proposed, including direct metalation and demetallation–metalation.

3.2.2.1 Direct Metalation

3.2.2.1.1 Interlayer-Expanded Method Two-dimensional (2D) layered zeolite consists of lamellar sheets, which are interconnected by weak hydrogen bonding. This feature endows it great structural diversity such as interlayer expansion. In general, silylation agents or metal salts are employed to connect layers, giving rise to three-dimensional zeolite framework [57]. In the case of metal salts, heteroatom-containing interlayer-expanded zeolites (Me-IEZs) can be obtained. The introduction of metal ions not only expands the pore size between the layers but also makes siliceous zeolite functionalized (isolated metal ions serving as Lewis acid sites in zeolites).

Bian et al. used a Sn salt of bis(2,4-pentanedionate)-dichlorotin [$Sn(acac)_2Cl_2$] to connect the layers at 180 °C during the interlayer expansion of layered precursor COK-5 zeolite [57]. After interlayer expansion by Sn^{4+} ions, the interlayer distance was increased. Subsequently, interlayer expanded RUB-36 (Me-IEZ-RUB-36) zeolites with the linker sites containing Sn, Ti, Fe, Eu, and Al ions were performed by Gies et al. according to the similar procedure [58]. Rietveld analysis from powder XRD data proved Me-IEZ-RUB-36 framework structure was stable after calcination.

3.2.2.1.2 Atom-Planting Method In 1984, Mobil declared that direct incorporation of Al ions into zeolite framework was realized by treating ZSM-5 zeolites with $AlCl_3$ or $AlBr_3$ vapor at 375 °C [59]. Mechanism analysis indicated that the metal ions were integrated with defect sites (i.e. external and/or internal silanols) already existing in the zeolites. Beta zeolite with the random intergrowth of polymorphs A and B has a large quantity of silanol groups, which can be employed as anchoring sites to incorporate heteroatoms into Beta zeolite. Li et al. treated commercial Al-Beta zeolite with $Zr(NO_3)_4 \cdot 5H_2O$ in ethanol solution. This one-step strategy without pre-dealuminated process led to the formation of bifunctional Zr-Al-Beta zeolite with open Zr(IV) sites derived Lewis acid sites and framework Al derived Brønsted acid sites [60]. This method was also extended to prepare Sn/Ni/Fe-Beta and Zr-Al-MWW zeolites with bifunctional active sites [61]. Furthermore, metalation could proceed even for the starting zeolite without defect sites in fluoride-containing solution due to the etching effect of fluoride to silica. Motivated by this, in 1995, Skeels et al. directly treated Al—Y zeolite

(Si/Al = 2.5–7.0) with $(NH_4)_2SnF_6$ solution [62]. Analogously, treating Al-ZSM-5 (Si/Al = 19) in $(NH_4)_2ZrF_6$ solution enabled the successful synthesis of Zr-Al-ZSM-5 zeolite with tetrahedrally coordinated Zr^{4+} ions [63].

With the purpose of enhancing metal ions accessibility and decreasing the synthesis time, several post-synthesized methods are put forward to load metal ions on the surface of mesoporous silicates. Abundant silanol groups existed on the surface of mesoporous silicates due to the intrinsic property of amorphous pore wall, which are the perfect accommodation of metal ions without any pre-treatments [42]. Considering the existing state of metal source, chemical vapor deposition (CVD) and liquid grafting are distinguished. Herein, Sn-MCM-41 material is taken as an example to illustrate these synthesis methods. Treating MCM-41 sample with $SnCl_4$ vapor led to the production of Sn-MCM-41 [64]. In comparison with pristine MCM-41 and hydrothermal synthesized Sn-MCM-41, the hydrothermal stability of such post-synthesized Sn-MCM-41 material was significantly improved. This phenomenon may be resulted from the occupation of Sn ions in the silanol groups during the CVD process, affording higher hydrophobic surface. Liquid grafting is another technique for the incorporation of Sn ions into MCM-41 material, whose reproducibility is better than CVD method [3]. The grafting starts from siliceous MCM-41, which is conducted with a Sn source in solvents such as toluene and alcohol [3, 7]. Water was not popular during this treatment, and therefore, the parent MCM-41 pre-activating at 100 °C for 2 h was treated in anhydrous toluene [65]. Metal utilization efficiency was closely associated with the type and number of alkyl substituents in organic Sn source. This direct post-synthetic method was also suitable for incorporating Sn/Zr ions into other mesoporous silicates including SBA-15, MSU, and MCM-48 [42].

3.2.2.1.3 Alkaline-Assisted Metalation Alkaline treatment to zeolites has been extensively investigated in the past decades to introduce mesoporosity in zeolites for improving diffusion property [2, 4]. Simultaneously, partial desilication in the alkaline treatment led to the formation of silanol nests, which can act as the accommodation of metal species. Consequently, incorporating Sn ions into MFI zeolite was attempted via alkaline-assisted metalation [7]. A commercial high-silica MFI zeolite with a Si/Al molar ratio of 961 was treated by a Sn precursor in NaOH solution. Characterizations indicated that this method resulted in the insertion of Sn species on the crystal surface and the formation of extra-framework SnO_2 domains, which was different from the homogeneous distribution of Sn in the hydrothermally synthesized Sn-MFI zeolite material [66]. This one-step post-synthetic method was also adopted to Beta structure. In addition, the outstanding feature of this alkaline-assisted metalation was that the formed metallosilicate zeolites commonly had mesoporous structure with improved diffusion performance.

3.2.2.2 Demetallation–Metalation

Since direct metalation encounters the limitation of the relatively low metal content, the zeolite precursor with rich silanol nests is desired. Therefore, two step strategy, i.e. demetallation–metalation, is proposed. This method usually starts with an aluminosilicate zeolite and then a dealumination treatment that promotes the formation of silanol nests within zeolite before the further metalation. The silanol nests can accommodate metal ions, forming framework-substituted metallosilicate zeolites [7]. Through changing the dealuminated conditions such as acid concentration, treated time, and Si/Al molar ratio of parent zeolite, the dealuminated degree can be modulated. According to the phase state of the metalation process, three main approaches have been reported, including CVD, liquid grafting, and solid-state ion-exchange (SSIE).

As for CVD method, Li et al. treated highly dealuminated Beta zeolites (Si/Al molar ratio > 1700) with anhydrous $SnCl_4$ vapor, obtaining nanosized Sn-Beta zeolite with the crystal size of 60–80 nm [67]. The Sn loading was capable to be controlled by varying the contact time. The maximum Sn content can be achieved up to 0.5 mmol g^{-1} (Si/Sn molar ratio = 35), far outperforming that in conventional fluoride-mediated Sn-Beta zeolite. Jin et al. reported a highly dealuminated mesoporous Beta zeolite as the parent for Sn-Beta synthesis via CVD method [68] resulting in framework Sn content of 1.1 wt%.

Zhu et al. conducted the synthesis of Sn—Y from dealuminated ultra-stable Y (USY) zeolite and $SnCl_4$ vapor [69]. Note that the incorporated amount and coordinated state of Sn species significantly depended on the acid-treatment time of parent USY. A maximum framework Sn content of 2.6 wt% was achieved by precisely tailoring acid treatment time. Zhang et al. also reported the preparation of Sn-MSE zeolite (Sn-MCM-68) using dealuminated MCM-68 (Si/Al = 284) as the precursor [70].

With respect to liquid grafting, solvent was important for the quantity and quality of metal ions incorporated into zeolites [3, 7]. Impregnation of dealuminated Beta zeolite in anhydrous $SnCl_4$ liquid enabled the formation of Sn-Beta zeolite with maximum Sn contents of 5 wt% in isolated sites. A higher Sn amount led to the generation of Sn oxides in extra-framework position [71]. Similarly, high Sn loading up to 25 wt% can be achieved, but which exhibited very poor catalytic performance due to the formation of a large number of extra-framework SnO_2 species [72]. Sn-Beta prepared by grafting in water already exhibited partial extra-framework Sn species when the Sn content was only 2 wt%. The use of 2-propanol as grafting solvent brought about low Sn contents in isolated sites because the competitive adsorption of 2-propanol on the hydroxyl nests resulted in the low affinity of Sn species in solvent to silanol nests [73]. Solvents with low affinity for the silanol groups like dichloromethane were beneficial for the Sn ions grafting in hydroxyl nests of dealuminated Beta zeolite, affording high-quality Sn amount up to 6 wt% [73]. Zhu et al. reported liquid grafting of $(NH_4)_2SnCl_6$ in acidic solution over a partially dealuminated Y zeolite. The addition of HCl in solution promoted the further removal of Al species and the subsequent insertion of Sn in Y zeolite in tetrahedral state [74]. Owning to the difficulty in complete removal of Al ions from Y zeolite, the residual Al is inevitable irrespective of dealumination method. After metalation, bifunctional heteroatom-substituted Y zeolite containing metal and Al ions is naturally formed, which should be paid attention in catalysis. The combination of base and acid treatment for Al-containing zeolite precursors produces hierarchical dealuminated zeolites, which can be further treated by liquid grafting to obtain hierarchical heteroatom-containing zeolites such as Zr-Beta and Zr—Y [75]. Heteroatom-containing MWW zeolite was also synthesized by liquid grafting. MWW-type borosilicate zeolite precursor ERP-1P (Si/B = 11) was treated with $Zn(NO_3)_2$ in acidic solution, bearing the following merits, (I) removal of framework B, (II) formation of silanol nests, and (III) generation of delamination structure [76]. This delaminated MWW zeolite was named as DZ-1, and then heteroatoms (Sn, Zr, Ti, Hf, Nb, Ta) substituted DZ-1 were synthesized by reacting silanol nests with the organic metal precursor such as $Zr(OC_4H_9)_4$ in *n*-BuOH [76].

Regarding SSIE method, dealuminated zeolite is mechanically grounded with a metal source followed by a calcination procedure at elevated temperature, which is simple and bears the potential of large-scale applicability. The metal source is usually organometallic compound such as dimethyltin chloride and tin acetate. The minimum Si/Sn molar ratio can reach 16, corresponding to 10 wt% Sn [77]. This SSIE method has been used to synthesize Zr-Beta, Sn-Beta, Zr-SCM-1 (a new delaminated MWW-type material with a double-layered structure), bimetallic Pb-Sn-Beta [3, 7, 35]. SSIE method can be employed to synthesize hierarchical Sn/Zr-containing zeolites by the integration of acid and base treatment or by utilization of hierarchical pristine Al-Beta zeolite, which was similar to liquid grafting method [78].

3.3 General Remarks

Both bottom-up and top-down approaches can be used to successfully synthesize Sn/Zr-containing zeolites. Bottom-up approach comprised hydrothermal synthesis, dry-gel conversion method, structural reconstruction strategy, and interzeolite transformation. This approach favors the incorporation of Sn/Zr ions into zeolite framework, forming tetrahedrally coordinated metal sites. Sn/Zr-zeolites prepared by fluoride-mediated hydrothermal synthesis, structural reconstruction strategy, and interzeolite transformation usually possessed less defect sites within crystal and exhibited high hydrophobicity, which was beneficial for their catalytic application in aqueous media. Nonetheless, the utilization of fluoride for conventional hydrothermal synthesis not only causes environmental issues but also sometimes results in large crystal sizes in micrometer level due to the low supersaturation in the synthesis gels. Additionally, the inhibiting effect of metal ions for the crystallization limits the isolated framework metal and leads to a long crystallization time. Dry-gel conversion methods can shorten the synthesis period and increase the metal content to some extent, in some cases, even without the assistance of fluoride. According to the novel idea for efficient utilization of structural building units during the zeolite nucleation and crystal growth, structural reconstruction and interzeolite transformation strategies are proposed. When the topology structures for the starting and targeted zeolite are different, this strategy is called as interzeolite transformation. If they are the same, structural reconstruction is termed. These two methods are capable of decreasing zeolite crystal size, enhancing hydrophobicity, significantly increasing the amount of framework metal ions, and remarkably shortening crystallization time. However, fluoride is still needed as a mineralizing agent in some situations.

Top-down approach, also denoted as post-synthetic method, usually demonstrates short synthesis time, high metal loadings, and sometimes small crystal size in comparison with hydrothermal synthesis. First, hierarchical zeolites containing micropores and mesopores can be prepared by combining acid and base treatment or by employing common hierarchical aluminosilicate zeolites as a starting precursor. Second, polymerized Sn oxides were easily formed especially with extremely high metal content. However, this phenomenon can be minimized or even avoided through choosing approximate synthesis method (CVD, liquid grafting, and SSIE) and optimizing the preparation conditions (water content, the type of metal precursor, calcination atmosphere, and so on). Third, this approach is restricted to some special zeolites, not universal. For example, interlayer-expanded method was only suitable for the layered zeolite. Demetallation–metalation approach is limited to zeolites that can be easily dealuminated through acid treatment. Thus, this approach was not suitable for the starting zeolite with low Si/Al molar ratio such as LTL, MAZ, and EON zeolites because their frameworks can be easily collapsed once contacting with acid solution. If heteroatom-containing zeolite derived from low-silica precursor are needed, the more advanced dealuminated techniques are required to be developed in the future. Fourth, acid, alkaline, and even fluoride treatment are necessary for creating silanol nests to accommodate metal ions. However, metal ions do not occupy all the silanol nests in the metalation process, resulting in partially residual defect sites and therefore increasing surface hydrophilicity. Finally, although diffusion property of reaction substrates and accessibility of active sites are significantly enhanced for mesoporous silicates, these materials are not stable in aqueous media and commonly have extra-framework SnO_x/ZrO_x species regardless of synthesis techniques. In some cases, partial Sn/Zr ions are embedded into the amorphous pore wall, which are not accessible for substrates.

In conclusion, each method has its advantages and disadvantages, which is not universal for all Sn/Zr-zeolites with discrepant topology structures. Improving the present synthesis method and developing new methodology for the already existed zeolites like Sn-Beta or unfeasible ones such as Sn-MAZ are still on the way.

3.4 Catalytic Applications of Sn- and Zr-Zeolites

When Sn/Zr atoms are incorporated into a silicate framework, their electrons form covalent bonds with the oxygen atoms in the host. If the metal is coordinated to four framework oxygen atoms, this configuration is named as “closed” site. Whereas, if the metal is coordinated to three framework oxygen atoms and one hydroxyl oxygen atom, “open” site is generated. Due to the discrepant electronegativity, the covalent electrons are pulled toward oxygen atom. In fact, the tetrahedrally coordinated Sn/Zr atoms within zeolite framework are in coordinatively unsaturated state, which can serve as Lewis acid sites [35]. Correspondingly, the framework oxygen atoms display a weak basic character. Therefore, Si-O-metal groups in zeolite framework can also act as acid–base pairs in the catalytic applications. The framework-metal derived Lewis acid sites can accept electron pairs from electron-rich groups, leading to their adsorption and chemical activation. In the case of carbonyl groups, they are activated by the shift of an electron from the oxygen atom (mainly in p_y and p_z orbitals) of carbonyl group to the empty orbitals of metal atom (mainly the π[CO] in the Sn LUMO orbitals) and further a back-donation of the Sn HOMO orbitals to anti-bonding π^*(CO) orbitals in carbonyl group. This donation and further back-donation mechanism contribute to the increase of C—O bond length and positive charge density in carbonyl group [8]. The following part systematically summarized the catalytic applications of Sn/Zr-zeolites in redox catalysis, Lewis acid catalysis, and biomass conversion.

3.4.1 Redox Catalysis

3.4.1.1 Baeyer–Villiger Oxidation

The Baeyer–Villiger (B—V) oxidation reaction is first reported in 1899, which has been broadly applied in the transformation of ketones to esters and of cyclic ketones to lactones [42]. This reaction was initially proceeded in a homogenous system using peracids or persulfate salts as oxidants, which inevitably produced useless byproducts and had a low selectivity for desired product when other function groups like C=C bond coexisted in the substrates. As a consequence, heterogeneous catalytic system using hydrogen peroxide (H_2O_2) as the oxidant became a pursued goal. Numerous catalysts have been utilized containing hydrotalcites, acidic anionic resins, aluminosilicates, germanosilicate, and titanosilicates, through activating the oxidant of aqueous H_2O_2. In 1996, Bhaumik et al. found that TS-1 zeolite can catalyze B—V oxidation of cyclohexanone and acetophenone to corresponding esters. Peroxide Ti species were formed to drive this reaction, and meanwhile, its acidic nature was also responsible for the production of byproducts [79].

Five years later, Corma et al. utilized a stannosilicate Sn-Beta zeolite to proceed the B—V oxidation of unsaturated ketones with H_2O_2, which exhibited excellent activity and selectivity [6]. This reaction mechanism was different from that over titanosilicate zeolites, which comprised the following steps, (I) the coordination of ketone to Sn active sites, (II) attack of H_2O_2 toward carbon atom in carbonyl group, (III) rearrangement, and (IV) lactone desorption to perform a next cycle. Later, they found that partially hydrolyzed Sn sites (i.e. open Sn sites) exhibited higher catalytic activity for the conversion of cyclohexanone to lactone than closed Sn sites in Sn-Beta zeolite. Since then, a number of Sn-containing zeolite materials are developed as catalysts to catalyze the B—V oxidations, and Sn-Beta zeolite is usually deemed as a benchmark catalyst. Generally, 2-adamantanone is used as reaction substrate when evaluating a novel catalyst, which is too large to enter 10 MR channels but small enough to penetrate into 12 MR pores of Sn-Beta zeolites. Its polycyclic structure with high rigidity and stability results in outstanding selectivity in the B—V oxidation (generally above 98%). The substrates of cyclopentanone and cyclohexanone are not easily to be oxidized and corresponding selectivity is commonly low to 80% [80].

Li et al. proceeded B—V oxidation of 2-adamtanone with H_2O_2 over nanosized Sn-Beta post-synthesized by CVD method (Sn-Beta-CVD) [67]. It was found that chlorobenzene was the optimal solvent. Although Sn-Beta-CVD possessed less hydrophobicity than conventional Sn-Beta-F zeolite hydrothermally synthesized with the assistance of fluoride, nanosized crystal size beneficial for diffusion performance and high Sn content still facilitated the catalytic activity of the former zeolite. Sn-Beta zeolite synthesized by structural construction possessed small crystal size (30–50 nm), which was applied in the B—V oxidation of 2-adamantanone [50]. The catalytic performance in terms of space–time-yields (STY) and turnover frequency (TOF) of this Sn-Beta zeolite outperformed those of post-synthesized Sn-Beta zeolites prepared by CVD and SSIE methods. Interzeolite transformation mediated Sn-Beta also exhibited small crystal size and possessed extraordinary catalytic performance in the B—V oxidation of 2-adamantanone [48, 49].

Large pore and hierarchical Sn-zeolites were created to improve the diffusion property and enhance the accessibility to active sites. FAU-type topology structure zeolite has three-dimensional 12 MR channels and supercages, which is a promising candidate for the host of Sn [69, 74]. The catalytic activity was higher than Sn—Y zeolite in oxidation of 2-adamantanone and the conversion can achieve 97.5% within only 40 min, far outperforming those over hydrothermal synthesized materials, such as Sn-Beta-F, Sn-MFI, Sn-MWW, and Sn-MCM-41 [69]. For the sake of decreasing the residual Al amounts in Sn—Y zeolites (Si/Al = ca. 43), a novel liquid grafting method was also developed. The Si/Al molar ratio of Sn—Y was decreased to about 150 under optimal conditions [74]. This Sn—Y zeolite catalyst demonstrated 87.1% lactone selectivity in the B—V oxidation of cyclohexanone to ε-caprolactone with the oxidant of aqueous H_2O_2. The direct addition of alkali metal ions such as K_2CO_3 into reaction system was an effective method to improve the product selectivity [74]. Sn-UTL (intersecting 12 and 14 MR), Sn-MSE ($12\times10\times10$ MR), and UTL-derived pillaring zeolites IPC-1-SnPI also demonstrated extraordinary catalytic performance compared with other stannosilicate zeolites such as Sn-MFI, Sn-MWW, and even Sn-Beta zeolites in the 2-adamantanone oxidation, in particular, with the bulky oxidant of TBHP [42, 70]. Sn-containing mesoporous silicates were also applied in the B—V oxidation of ketones as heterogeneous catalysts, such as Sn-MCM-41, Sn-SBA-15, Sn-MCM-48, Sn-SBA-16, and Zr-SBA-16 [42].

Hierarchical Sn-Beta zeolite post-synthesized using a hierarchical Beta zeolite as precursor exhibited the mesopore size in the region of 20–50 nm and demonstrated excellent catalytic activity in the B—V oxidation of 2-adamantanone compared with traditional Sn-Beta-F zeolites hydrothermally synthesized in fluoride medium [68]. Sn-MFI nanosheets with about 2 nm thick in the [010] direction showed a prominent increase in activity in the B—V oxidation of cyclic ketones compared with bulk three-dimensional Sn-MFI [81]. Additionally, although the conversion and TOF of mesopore Sn-MCM-41 were slightly higher than those of Sn-MFI nanosheets, the latter presented high thermal and hydrothermal stability. Sn-MCM-56 zeolite, with partially delaminated MWW sheets, possessed more open reaction space and larger external surface area than three-dimensional Sn-MWW [24]. Consequently, this stannosilicate zeolite was more active than Sn-MWW and conventional Sn-Beta-F zeolite using bulky *tert*-butyl hydroperoxide (TBHP) as oxidant. Layered Sn-DZ-1, Sn-UCB-4, pillared Sn-MWW, and interlayer-expanded Sn-PLS-3 zeolites were also highly active in B—V oxidation [42].

3.4.1.2 Meerwein–Ponndorf–Verley Redox

The Meerwein–Ponndorf–Verley (MPV) reduction was individually discovered by Meerwein–Schmidt and Verley in 1925 and further developed by Ponndorf [35]. This hydrogen transfer reaction is complementary with Oppenauer oxidation. The common hydrogenation reaction is usually proceeded with the assistance of hydrogen under high temperature and pressure conditions, therefore bringing about dangerous operation, high energy consumption, low product selectivity, and high requirement for reaction equipment. For MPV reduction, alcohol is employed as hydrogen source to selectively reduce C=O bond in carbonyl compound even with C=C double bond under relatively mild liquid conditions via direct hydride transfer, which overcomes the problems of common hydrogenation reaction [35]. Homogeneous Lewis acid catalysts can trigger the MPV reduction such as aluminum isopropoxide and lanthanide complex, accompanying by the issues of difficult separation. Consequently, a series of heterogeneous catalysts are developed and the most prominent catalysts are the Lewis acid zeolites such as Sn- and Zr-zeolites.

Corma et al. used Sn-Beta to catalyze the MPV reduction of cyclohexanone with several alcohols and excellent catalytic performance was observed. In comparison with isolated framework Ti, Sn atoms seemed to be active and suitable for MPV reaction because of the stronger Lewis acidity of Sn-Beta zeolites [82]. Subsequently, the mechanism was investigated, and this reaction proceeded via the following steps, (I) adsorption of cyclohexanone and alcohol onto framework Sn sites, (II) deprotonation of alcohol, (III) hydride transfer through forming six-membered ring transition state, and (IV) transfer of proton from the catalyst to cyclohexanol.

The nature of silica host affects the catalytic performance of MPV reduction. Sn-Beta zeolites exhibited high activity with the bulky 4-*tert*-butylcyclohexanone as reactants in the absence of diffusion limitations [82]. However, relatively inferior catalytic activity was observed over mesoporous materials such as Sn-MCM-41 and Sn-SBA-15, indicating that the confinement of active sites in the micropore architecture has a positive effect on the activity [82, 83]. The creation of hierarchical Sn-Beta zeolite was necessary for large-size ketones such as cyclododecanone owing to significant diffusion limitation. As for cyclohexanone with small molecular size, microporous and hierarchical Sn-Beta demonstrated similar reaction rate, while bulky cyclooctanone and cyclododecanone were converted rapidly only over hierarchical Sn-Beta zeolite [3]. Additionally, ketone conversion was higher over hierarchical Sn-Beta zeolite than mesoporous Sn-MCM-41, supporting the conclusion of the positive confinement effect exerted by the zeolite structure [3].

The nature of Sn sites also affects the activity in the MPV reaction. Tetrahedrally coordinated Sn sites within zeolite framework were active in the MPV reduction of cyclohexanone, while octahedral SnO_2 was inactive [82]. Furthermore, open Sn sites were responsible for the catalytic reaction as proved by DFT calculation and experimental analysis [84]. As can be seen from the mechanism, the Lewis basicity of oxygen atom in Sn-OH indeed functioned the deprotonation for the alcohol. Lewis et al. proposed that the open and closed Sn sites can be in-situ transformed under reaction conditions [85]. Hence, the requirement of ex-situ open Sn sites within Sn-zeolite catalysts for MPV reduction reaction is still under debate.

Apart from Sn-Beta zeolite, Zr-Beta, Zr—Y, Hf-Beta, Hf—Y, and Zr-SBA-15 zeolites and Sn-grafted silica are also attempted as catalysts in the catalytic application of MPV reduction [3, 7, 35]. The combination of Sn/Zr atoms with Beta topology structure is the most common. The active site of Hf and the silica host of FAU-type structure are deserved to be paid attention in the future.

3.4.2 Lewis Acid Catalysis

3.4.2.1 Ring Opening of Epoxides

The ring opening of epoxides is one of the useful reactions for producing fine chemicals in chemical industry. Epoxides are significant electrophilic reagents and the C—O bond in three-member hetero-ring can be easily attacked by the nucleophilic reagent, therefore leading to the ring opening of epoxides. Typically, nucleophilic reagents are O- and N-containing molecules, for example, water, alcohol, ammonia, and amine [10, 35].

Ethylene glycol (EG) is broadly applied as solvents, automotive antifreeze, the raw material for polyesters, synthetic waxes, and so on, which can be produced by the direct hydration of ethylene oxide (EO) [86]. Sn-SSZ-13 zeolite post-synthesized by SSIE method with $(CH_3)_2SnCl_2$ was used as a catalyst in EO hydration [86]. The narrow pore opening of Sn-SSZ-13 (8 MR) allowed the free entrance of EO, H_2O, and EG molecules, but suppressed the diffusion of byproducts such as diethylene glycol and triethylene glycol generated via self-condensation of EG. It was found that the Lewis acidity from isolated framework Sn species was responsible for the formation of EG under mild conditions. Additionally, excellent recyclability was observed for this Sn-SSZ-13 catalyst. DFT calculation confirmed that the structure configuration of Sn sites greatly affected the activation energy of EO hydration, and the $Sn(OSi)(OH)_3$ species was the most favorable active sites [86]. Later, the same group developed a bifunctional TiSn-Beta catalyst to produce 1,2-diols using alkene as the starting reaction substrate [87]. The framework Ti and Sn Lewis acid sites functioned the alkene epoxidation and further epoxidation hydration, respectively, which integrated multistep reactions into a tandem relay catalysis over cooperative active sites within one catalyst.

The product from ring opening of epichlorohydrin with alcohol is widely applied in the field of pharmaceutical and solvents industry. Solid Lewis acid zeolite catalysts containing Sn-Beta, Zr-Beta, Hf-Beta, and Sn-SBA-15 were prepared to study the regioselective ring opening of epichlorohydrin with methanol [88]. Catalytic results revealed that the catalytic performance mainly depended on the types of host and Lewis acid sites. Sn-Beta-F zeolite was superior to other metallosilicate zeolites in terms of conversion and product regioselectivity. The combination of experimental results and DFT calculation further disclosed the reaction mechanism involving the activation of epichlorohydrin on an alcohol adsorbed on the Sn sites and further nucleophilic attack by another alcohol molecule. Besides, this Sn-Beta zeolite catalyst showed outstanding reusability without obvious decrease of conversion after the fourth run [88].

Since Ti-containing zeolite catalysts were discovered to trigger epoxides aminolysis under mild and solvent-free conditions, Ti/Zr-substituted MFI zeolites were also synthesized for the ring opening of epoxides with alcohol and amine [89]. Authors also found that the exceptional catalytic activity of Zr-MFI was derived from the isolated framework Zr species within MFI zeolite framework. Subsequently, hierarchical Zr-, Sn-, and Ti-Beta zeolites, post-synthesized by the integrating acid-alkaline treatment and SSIE method, were applied in the ring opening of styrene oxide with aniline [90]. Mesoporous Zr-Beta presented better catalytic performance in styrene oxide aminolysis than hierarchical Sn-Beta and Ti-Beta zeolites on account of the higher styrene oxide/aniline adsorption ratio over mesoporous Zr-Beta zeolite with moderate Lewis acidity. The presence of mesopore enables better accessibility of reactant molecules to active sites and higher diffusion performance, resulting in an enhanced catalytic activity for hierarchical Zr-Beta zeolites [90]. Zr-MOR zeolite also possessed high activity and regioselectivity in epoxide aminolysis reactions under ambient conditions [91]. The intrinsic nature of both reaction substrates (such as the basicity of amines) and Lewis acidity determines the catalytic performance in the ring opening of epoxide with amines.

3.4.2.2 Aldol Reaction

Stannosilicate zeolite is an important solid Lewis acid catalyst, which can catalyze aldol reaction to from new C—C bond. Representative Sn-Beta zeolite was found to catalyze the aldol reaction of glycolaldehyde with methanol aiming to produce lactates. α-Hydroxy-γ-butyrolactone (HBL) was also obtained from the coupling of dihydroxyacetone (DHA) and formaldehyde over different metal-containing zeolite catalysts (Ti-Beta, Sn-Beta, Zr-Beta, Hf-Beta, Al-Beta, Sn-MCM-41, and Sn-MFI) [92]. These catalysts exhibited comparable DHA conversion, but dramatically high yield of HBL was observed for Sn-Beta (60%), Sn-MFI (61%), and Sn-MCM-41 (64%) zeolites compared with other zeolites ($< 11\%$). Besides, it was confirmed that Sn oxide species were inactive for this reaction. Spectroscopic analysis, theory calculation, and isotopic labeling revealed the reaction mechanism consisted of three steps, including soft enolization of DHA, aldol addition of formaldehyde to the Sn-enolate, and the formation of erythrulose [92].

A series of stannosilicate zeolites (Sn-MFI, Sn-Beta, Sn-MCM-41, and Sn-SBA-15) were selected as catalysts to selectively transform glycolaldehyde to C_4 sugars in water under mild temperature ($< 100\,^\circ C$). High yield (74%) and selectivity (97%) of tetrose sugars were achieved over Sn-MFI zeolite with medium-size pores. However, other reference zeolite materials exhibited lower selectivity of C_4 sugars because of forming hexose sugars via consecutive aldol condensation of glycolaldehyde [3]. Lewis et al. reported an aldol reaction between aromatic aldehydes and acetone over Hf-, Zr-, and Sn-Beta zeolites [93]. Mechanism analysis indicated acid–base pairs in the Si-O-metal groups benefited for soft enolization via α-proton abstraction. Unlike conventional base catalysts, this Lewis acid catalysts can resist the harmful influence of water and acidic solution on catalytic activity. Herein, in most cases, Zr-Beta zeolite showed higher performance than Sn-Beta and Hf-Beta zeolites, which was greatly different with previous reports for the reactants of DHA and formaldehyde [92].

3.4.2.3 Propane Dehydrogenation

Light olefins are fundamental feedstock in the petrochemical industry with increasing needs, which are traditionally produced from catalytic cracking or methanol-to-olefins processes. The dehydrogenation of light alkanes gives an alternative route for alkenes production. A series of catalysts were employed into this catalytic conversion, such as Pt, CrO_x, Ga, V, and Sn based ones [10]. In 2016, Wang et al. found that metallic Sn can catalyze propane dehydrogenation as a highly active species, which breaks through the traditional concept that Sn is only a promoter [94]. Impregnated Sn/SiO_2 catalyst exhibited comparable catalytic performance to PtSn-based catalyst [94]. In order to effectively resist the Sn loss, Sn-containing hexagonal mesoporous silica (Sn-HMS) was synthesized to stabilize Sn species, which were located in the framework position [95]. The catalytic stability was significantly increased because of the confined effect of isolated Sn in HMS structure and the formation of Si-O-Sn bond, resisting to Sn loss and inhibiting the reduction of Sn ions to metallic Sn, respectively. It was speculated that isolated Sn species connecting to three framework Si—O groups adsorbed propane, forming intermediates by coordination of the coordinatively unsaturated $Sn^{\delta+}$ to isopropyl group and of lattice oxygen in Si-O-Sn group to α-H in propane. Finally, propene was formed via β-H cleavage and the released hydrogen was transferred to recover the Sn active sites, achieving one whole reaction cycle [95].

In 2021, Yue et al. discovered that Lewis acid sites in post-synthesized Sn-Beta zeolite can activate C—H bond in propane, accomplishing outstanding catalytic performance in dehydrogenation of propane to propene [96]. It was also found that Brønsted acid was the active site for cracking reaction. Consequently, high Lewis acid/Brønsted acid ratio favored the dehydrogenation reaction. When Sn-Beta zeolite was exchanged with Na^+ ions with the purpose of covering Brønsted acid sites, it possessed the best catalytic performance with the propane conversion and propene selectivity of 40 and 92%, respectively. Additionally, nearly no deactivation phenomenon was observed after the reaction time of 72 h. The open Sn sites seemed to be more active than closed Sn sites in the dehydrogenation reaction [96].

3.4.3 Biomass Conversion

The presence of isolated tetrahedral Sn^{4+}/Zr^{4+} ions substituted in zeolite framework results in the generation of abundant Lewis acid sites. This Lewis acid site from lower-coordinated metal species preferably interacts with electron-rich substituents such as carbonyl and alcohol groups, achieving an effect of adsorption and activation for reaction substrates [35]. Carbonyl and alcohol substituent groups are plentiful in bio-molecules. Thus, Sn/Zr-zeolites play a significant role in the biomass conversion into high-value fuels and other value-added chemicals.

3.4.3.1 Sugar Isomerization

The discovery about the isomerization of monosaccharides over Lewis acid catalysts is later than that over basic catalysts. The typical Lewis acid catalysts like $AlCl_3$ are commonly inactive in aqueous solution because water molecules preferentially coordinate to a metal cation, and further, it probably suffers from partial hydrolysis. Thus, Lewis acid catalyzed reactions are usually carried out in anhydrous systems. Nonetheless, using water as solvent has received great attention due to the increasing requirement of sustainable technologies and development [7]. As a result, the isomerization of carbohydrates in water over Lewis acid catalysts has prompted great research interest.

In 2010, Moliner et al. investigated the catalytic performance of glucose isomerization in aqueous system with the zeolite catalysts of Sn-MCM-41, Ti-MCM-41, Sn-Beta, and TS-1 [97]. Among these catalysts, Sn-Beta zeolite with high hydrophobicity exhibited best catalytic activity and fructose selectivity. This finding opens up a broad avenue for biomass conversions using Sn-Beta zeolite as heterogeneous Lewis acid catalyst. Sn-Beta zeolite was also applied in other isomerization reactions of carbohydrates containing dihydroxyacetone, lyxose, xylose, mannose, lactose, and galactose [3, 35].

The structure of the active sites in Sn-Beta zeolite played a significant role on its isomerization performance. It was predicted that the open sites of Sn active site were more active than closed Sn sites using theory calculation [98]. The open Sn sites and adjacent silanol group cooperatively stabilize the reaction transition state in the step of hydride transfer. Glucose in the open-chain form coordinated to Sn in a monodentate mode, whereas the –OH group linked to the silanol group via hydrogen bonding. This coordination state facilitated hydride transfer and further proton transfer, resulting in a lower activation energy. Authors also calculated the activation energy in the epimerization of glucose into mannose, which was predicted to carry out through the Bilik mechanism [98]. Subsequently, Bermejo-Deval et al. provided experimental evidence to exclude the participation of silanol groups in glucose–mannose epimerization via the cationic exchange of silanol groups with Na^+ ions [99]. The glucose–fructose isomerization occurred over pristine Sn-Beta zeolite, while Na^+-modified Sn-Beta triggered the glucose–mannose epimerization. When the parent Sn-Beta was modified with ammonia prior to catalytic application, a sharp decrease of catalytic activity in glucose–fructose isomerization was observed due to the deactivation of open Sn sites by NH_3 [99].

3.4.3.2 5-(Hydroxymethyl)Furfural (HMF) Synthesis

HMF, as a platform molecule, can be transformed into many chemical products and biofuels through a variety of upgrading routes such as hydrogenation, hydrogenolysis, and oxidation. HMF is usually obtained by the catalytic dehydration of glucose and fructose derived from abundant cellulosic biomass [97, 98]. The direct dehydration of fructose can produce HMF, while glucose-to-HMF needs two steps, that is, glucose to fructose and fructose to HMF [100]. Fructose-to-HMF was relatively easier than glucose-to-HMF and the former was commonly catalyzed by Brønsted acid sites. From section 3.3.1, the isomerization of glucose to fructose is assisted by Lewis acid sites [98]. Consequently, glucose-to-HMF reaction is proceeded with the aid of bifunctional active sites containing Brønsted and Lewis acid sites. Sn/Zr-zeolites are recognized as the promising catalysts for this transformation because Brønsted and Lewis acid sites can co-exist within one zeolite framework derived from isolated framework Al and heteroatoms of Sn or Zr, respectively.

Initially, the combination of a Lewis acid zeolite with a homogeneous Brønsted acid catalyst was performed to catalyze the glucose-to-HMF. It had been reported that Sn-Beta can function effectively for the glucose-to-fructose in aqueous solution even at extremely low pH of 1.0 [100]. This provided a possibility for achieving this task of one-pot synthesis of HMF from glucose in acidic media. Nikolla et al. combined Sn-Beta zeolite and an acid catalyst, in this case HCl, to convert glucose to HMF in the single and biphasic reaction systems, respectively [101]. However, the integration of Lewis and Brønsted acid sites within one heterogeneous catalyst to perform this one-pot reaction was increasingly desired. As for Sn-zeolite catalysts, the catalytic performance of glucose-to-HMF can be modulated by adjusting the Al or Sn amount within a specific zeolite and choosing different zeolite structures.

In 2015, Li et al. synthesized bifunctional Sn-Al-Beta zeolite with CVD method [102]. The Sn and Al amounts can be tuned by controlling dealuminated degree in the first step and the contact time with $SnCl_4$ vapor in the second step. The framework Al and tetrahedrally coordinated Sn endowed Sn-Al-Beta zeolite with Brønsted and Lewis acid sites, respectively. Under optimal reaction conditions, Sn-Al-Beta zeolite exhibited a good catalytic performance (glucose conversion, 60.0%; HMF selectivity, 62.1%) [102]. In 2020, a cyclic quaternary ammonium salt was applied in the hydrothermal synthesis of hierarchical Sn-Al-Beta zeolite at 170 °C within 24 h, which can robustly catalyze the synthesis of HMF from glucose in aqueous solution [103]. Other structures, like FAU with sodalite cavities and supercages, were also used to load Sn ions. Oozeerally et al. reported a post-synthesized method to prepare Sn-Al-Y zeolite with the Si/Sn and Si/Al molar ratio of 65 and 164, respectively [104]. This catalyst exhibited 17% HMF yield with the glucose conversion of 36% at 140 °C in aqueous system. However, this zeolite material deactivated quickly due to the formation of humin and/or the competitive adsorption of water molecules to framework Sn-derived Lewis acid sites. Once changing the solvent from water to dimethylsulfoxide (DMSO), stable catalytic performance was observed [104].

3.4.3.3 Synthesis of Lactic Acid or Alkyl Lactates

Lactic acid (LA) is used not only in the food industry but also as the building block of biopolymer (an eco-friendly substitute for petrochemical-derived plastics) and platform chemicals. Currently, LA is mainly synthesized through fermentation of sugars in large-scale production. It was also found that Lewis acid zeolite catalyst can effectively catalyze the formation of LA or alkyl lactates (AL) from sugars such as triose, pentose, and hexoses.

Dihydroxyacetone (DHA) and glyceraldehyde (GLA) are two kinds of triose sugars, which can be employed as reaction substrates to produce LA and AL over Lewis acid catalysts in water and alcohol, respectively [3]. The detailed mechanism was exhibited as follows: (I) dehydration of DHA/GLA to pyruvic aldehyde (rate-determining step, catalyzed by weak Brønsted acid or Lewis acid sites), (II) the transformation of pyruvic aldehyde to hydrate or hemiacetal in water or alcohol, respectively, and (III) the formation of LA/AL via intramolecular MPV or Cannizzaro reaction (catalyzed by Lewis acid sites). When employing DHA or GLA as reaction substrates, DHA can reversibly isomerize to GLA over Lewis acid catalyst by a hydride shift mechanism, which resulted in comparable LA/AL yields.

Among Sn-related zeolite materials containing Sn-Beta, Sn-MFI, Sn-MCM-41, and Sn-SBA-15, Sn-Beta zeolite with the strongest Lewis acidity possessed the highest catalytic activity for the production of AL from triose sugars in methanol [105]. However, 12 MR Sn-Beta and 10 MR Sn-MFI had similar catalytic activity in water. This indicated that diffusion limitation was not observed in the present reaction systems and the catalytic activity was related to other parameters such as Lewis acid strength, hydrophobicity, and even confinement effect. The rate-limiting step was performed faster over weak Brønsted acid sites than over Lewis acid sites [106]. So, adding weak Brønsted acid sites into catalysts was an effective method to enhance the catalytic performance. Other Sn-zeolite materials were also adopted into the conversions of trioses to LA/AL, which contained Sn-Beta, Sn-MCM-41, Sn-MFI, and Sn-MWW [3, 105].

Alternatively, pentose and hexose can be also exploited as reaction substrates for producing LA/AL over Sn/Zr-zeolites. The conversion of pentose to LA/AL consists of retro-aldol condensation and triose-to-LA/AL reactions. Similarly, the synthesis of LA/AL from hexose is also a tandem reaction, which requires the participation of Lewis acid sites or the combination of Lewis acid and weak Brønsted acid sites [107]. It has been reported that Zr-SBA-15 with both Lewis and Brønsted acid sites can catalyze the formation of methyl lactate (ML) from xylose and fructose in methanol, giving 41% and 44% ML yields, respectively, at 240 °C for 6 h [108]. The retro-aldol condensation was a vital step for the formation of ML. Post-synthesized Sn-MWW with Lewis acidity exhibited 44% ML yield from glucose at 160 °C for 20 h [25], which was comparable to that of Sn-MCM-41 [109]. Nanosized Sn-Beta was post-synthesized by grinding method using partially crystallized all-silica Beta zeolite as the precursor [110]. Forty-four percent ML yield was obtained over this Sn-Beta zeolite at 160 °C for 10 h, outperforming Sn-Beta-F and Sn-Beta-P prepared by classic dealumination–stannation approach. The excellent catalytic activity was ascribed to the enhancement of diffusion performance and the accessibility to active sites in the presence of mesoporosity. Bearing the merit of hierarchical Sn-zeolites in mind, Yang et al. reported a fluoride-free and low template content synthesis of hierarchical Sn-Beta zeolite, which contained three steps of dealumination, stannation, and TEAOH treatment [111]. The textural property of obtained Sn-Beta was closely related to the amount of TEAOH used. The optimal Sn-Beta zeolite gave a ML yield of 58% in the conversion of glucose in methanol at 160 °C for 10 h [111].

3.4.3.4 γ-Valerolactone Synthesis

Considering the versatile applications of γ-valerolactone (GVL) as fuel additive, solvent, and fine chemicals, its synthesis from furfural or sugars is attractive in the field of biomass upgrading. Using furfural as the starting substance, four successive cascade steps were required, that is, hydrogenation of furfural, ring opening of furfural ring, hydrogenation of levulinic acid, and cyclization [112]. The whole process requires the dual active sites of Brønsted and Lewis acid sites.

Initially, Lewis acid catalyst Zr-Beta and Brønsted acid catalyst Al-MFI with nanosheet morphology were combined to proceed the synthesis of GVL from furfural, giving rise to excellent targeted product yields of 78% in a 5 wt% furfural solution with the solvent and hydrogen source of 2-butanol [112]. In consideration of the proximity between Lewis and Brønsted acid sites, fixing these two active sites into one host is desired. Sn-Al-Beta zeolite with the combination of Lewis and Brønsted acidity was post-synthesized to perform the synthesis of GVL from furfural [113]. The amounts of residual Al and incorporated Sn were closely associated the Brønsted and Lewis acidity, respectively, then affecting the catalytic behaviors. An optimal GVL yield of about 60% was achieved over Sn-Al-Beta with Si/Al and Si/Sn of 473 and 63, respectively, at 180 °C in 2-butanol. The higher GVL yield over integrated Sn-Al-Beta than physical mixture of Sn-Beta and Al-Beta suggested that the close proximity between Lewis and Brønsted acid sites in Sn-Al-Beta zeolite was important for the formation of GVL from furfural [113].

Song et al. developed hierarchical meso-Zr-Al-Beta zeolites via multiple-step post-synthesis for the furfural-to-GVL reaction [78]. The presence of mesopores can enhance the accessibility to active sites. This catalyst demonstrated excellent catalytic activity, giving highest GVL yield of 95% and good recyclability in this cascade reaction [78]. Heteropolyacid (HPA) such as phosphotungstic acid (HPW) and silicotungstic acid (HSW) were anchored on Zr-Beta zeolite to construct an integrated HPA/Zr-Beta catalyst, where HPA and Zr-Beta served as Brønsted and Lewis acid sites, respectively. HPW/Zr-Beta exhibited a superior GVL yield of approximate 70% at 160 °C for 24 h, which was higher than that (about 40%) of conventional Sn-Al-Beta. This high activity was attributed to its high thermal stability and strong Brønsted acidity [114].

The levulinic acid was also employed as the reaction substrate to produce GVL only via a simple MPV reduction over Zr-Beta zeolites [115]. The GVL yields were up to 96% and 99% in a batch and continuous flow reactor, respectively. Employing xylose as a reaction substrate, an additional step (i.e. dehydration of xylose to furfuran, catalyzed by Brønsted acid sites) was observed in comparison with using furfuran as the starting reactant [116]. Bifunctional Zr-Al-Beta zeolite with Brønsted and Lewis acidity was used to catalyze the one-pot transformation of xylose to GVL in 2-propanol. The Brønsted and Lewis acidity can be regulated by changing the Al and Zr amounts, which were tightly related to the catalytic performance. Under optimal reaction conditions (temperature, 190 °C; catalyst loading, 15 g L^{-1}; xylose concentration, 30.5 g L^{-1}), GVL yield of about 34% was obtained within 10 h. Moreover, the structure remained stable in the reuse experiment by a regeneration method of direct calcination treatment at 550 °C [116]. Zr-Al-Beta zeolite was also applied in the catalytic application of glucose-to-GVL conversions [117]. Al/Zr molar ratio significantly impacted the product distribution. When the Al/Zr molar ratio reached 0.2, the GVL yield achieved the maximum value of 24% under the optimal reaction temperature (190 °C) and time (24 h) [117].

3.5 General Remarks

In summary, Sn/Zr-zeolites are effective in redox catalysis (MPV reduction, B—V oxidation), Lewis acid catalysis (ring opening of epoxide, aldol reaction, propane dehydrogenation, etc.), and biomass conversions (sugar isomerization, synthesis of LA or AL, γ-valerolactone synthesis, etc.). Five general remarks are drawn as follows:

(1) For a specific reaction, the synthesis method and parameter of a specific Sn/Zr-zeolite catalyst greatly affected the catalytic performance. The catalytic performance is highly related to the diffusion property (crystal size, channel systems), hydrophobicity, micro-environment of Sn sites (framework vs. extra-framework, open site vs. closed site), amount and strength of Lewis acid sites, distribution of Sn species, etc. In certain cases, even though they have the same structural topology, other features probably vary significantly for Sn/Zr-zeolites. Therefore, a perfect zeolite catalyst is difficult to be obtained, which urge us to focus on the synthesis procedure and parameters. Additionally, reaction conditions (reaction temperature, time, catalyst amount, type of solvent, pressure, agitating speed, etc.) should be considered to obtain optimal catalytic performance.

(2) Diffusion property, hydrophobicity, and confinement effect are three complementary parameters affecting catalytic performance. For bulky reaction substrates, diffusion constants generally exist for microporous Sn/Zr-zeolites, while ordered mesoporous silicates with large pore size benefit the diffusion of guest organic and accessibility to active site. Under diffusion constraints-free conditions, confinement effect can increase the catalytic activity on some occasions such as MPV reduction. The high hydrophobicity is advantageous for the stability and activity especially for aqueous catalytic systems. For certain reactions such as B—V oxidations, creating hierarchical Sn/Zr-zeolites is an effective methodology to compromise these three factors of diffusion property, hydrophobicity, and confinement effect.

(3) The optimal types of metal center and zeolite topology are diverse for different reactions. For instance, isolated Sn site is good for B—V oxidations, sugar isomerization, 5-(hydroxymethyl)furfural synthesis, and alkane dehydrogenation, while isolated Zr site is suitable for MPV reduction and the synthesis of γ-valerolactone. The topology structure probably impacted the micro-environment of Sn/Zr species, diffusion property, Lewis acid strengthen, etc. At present, even though numerous topology structures are used as the host of Sn/Zr atoms, *BEA topology with three-dimensional 12MR is still an appealing candidate.

(4) The tetrahedrally coordinated metal ions are generally recognized as the active sites for Sn/Zr-zeolite catalysts. Extra-framework metals are inactive or less active than framework ones. For the micro-environment of isolated metal, its configurations greatly affected the catalytic performance, which are simplified to open and closed sites. In certain reactions such as B—V oxidation, MPV reduction, and glucose isomerization, open metal sites are more active than

closed ones. However, the feature is revealed by ex-situ technique, which may not reflect the actual state under reaction conditions. The metal configuration may interconvert during the catalytic reaction. Moreover, the detailed configuration of metal ions is complex, and not limited to the typical open and closed two sites. Consequently, the extract nature of these species and their impact on catalytic behavior are still pending questions. The development of advanced in-situ characterizations such as FT-IR, MAS NMR, and EXAFS is highly desired to understand the real synthesis-structure-performance relationships, bridging the gap between laboratory investigation and large-scale production.

(5) The combination of Lewis and Brønsted acid makes a great contribution to multiple cascade reactions particularly in the field of biomass conversions. In most cases, Sn-zeolites seem to be more popular than Zr-zeolites in biomass conversions, due to the unique coordinated ability of carbonyl and hydroxyl groups (abundant in biomass-derived platform chemicals) to isolated Sn sites. Although all these tandem reaction processes require the nominal Lewis and Brønsted acid sites, the acid strength is different. For instance, using glucose as the reaction substrates, the integration of weak Brønsted and Lewis acid sites contributes to the formation of lactic acid or alkyl lactates in the solvent of water or alcohol, respectively. However, HMF is produced with the cooperation of strong Brønsted acid and Lewis acid sites with the same reactant of glucose in water.

Constructing new multifunctional catalyst materials and novel catalytic applications, particularly in the field of catalytic conversion of biomass, are two engines to drive the development of Sn/Zr-zeolites community.

References

1 Corma, A. (1995). Inorganic solid acids and their use in acid-catalyzed hydrocarbon reactions. *Chem. Rev.* 95: 559–614.

2 Přech, J. (2018). Catalytic performance of advanced titanosilicate selective oxidation catalysts – a review. *Catal. Rev.* 60: 71–131.

3 Ferrini, P., Dijkmans, J., Clercq, R.D. et al. (2017). Lewis acid catalysis on single site Sn centers incorporated into silica hosts. *Coord. Chem. Rev.* 343: 220–255.

4 Kubička, D., Kubičková, I., and Čejka, J. (2013). Application of molecular sieves in transformations of biomass and biomass-derived feedstocks. *Catal. Rev., Sci. Eng.* 55: 1–78.

5 Bates, J.S., Bukowski, B.C., Harris, J.W. et al. (2019). Distinct catalytic reactivity of Sn substituted in framework locations and at defect grain boundaries in Sn-zeolites. *ACS Catal.* 9: 6146–6168.

6 Corma, A., Nemeth, L.T., Renz, M., and Valencia, S. (2001). Sn-zeolite beta as a heterogeneous chemoselective catalyst for Baeyer–Villiger oxidations. *Nature* 412: 423–425.

7 Dapsens, P.Y., Mondelli, C., and Pérez-Ramírez, J. (2015). Design of Lewis-acid centres in zeolitic matrices for the conversion of renewables. *Chem. Soc. Rev.* 44: 7025–7043.

8 Román-Leshkov, Y. and Davis, M.E. (2011). Activation of carbonyl-containing molecules with solid Lewis acids in aqueous media. *ACS Catal.* 1: 1566–1580.

9 Zhang, M., Guan, X., Zhuang, J., and Yu, Y. (2022). Insights into the mechanism of ethanol conversion into 1,3-butadiene on Zr-β zeolite. *Appl. Surf. Sci.* 579: 152212.

10 Lei, Q., Wang, C., Dai, W. et al. (2021). Multifunctional heteroatom zeolites: construction and applications. *Front. Chem. Sci. Eng.* 15: 1462–1486.

11 Corma, A. and Garcia, H. (2002). Lewis acids as catalysts in oxidation reactions: from homogeneous to heterogeneous systems. *Chem. Rev.* 102: 3837–3892.

12 Mal, N.K., Ramaswamy, V., Rajamohanan, P.R., and Ramaswamy, A.V. (1997). Sn-MFI molecular sieves: synthesis methods, ^{29}Si liquid and solid MAS-NMR, ^{119}Sn static and MAS NMR studies. *Microporous Materials* 12: 331–340.

13 Vargas, N.G., Stevenson, S., and Shantz, D.F. (2012). Synthesis and characterization of tin(IV) MFI: sodium inhibits the synthesis of phase pure materials. *Micropor. Mesopor. Mater.* 152: 37–49.

14 Yuan, E., Dai, W., Wu, G. et al. (2018). Facile synthesis of Sn-containing MFI zeolites as versatile solid acid catalysts. *Micropor. Mesopor. Mater.* 270: 265–273.

15 Liu, Y., Xiao, Y., Xia, C. et al. (2020). Insight into the effects of acid characteristics on the catalytic performance of Sn-MFI zeolites in the transformation of dihydroxyacetone to methyl lactate. *J. Catal.* 391: 386–396.

16 Ren, L., Guo, Q., Kumar, P. et al. (2015). Self-pillared, single unit-cell Sn-MFI zeolite nanosheets and their use for glucose and lactose isomerization. *Angew. Chem.* 54: 10848–10851.

17 Dongare, M.K., Singh, P., Moghe, P.P., and Ratnasamy, P. (1991). Synthesis, characterization, and catalytic properties of [Zr]-ZSM-5. *Zeolites* 11: 690–693.

18 Shah, K.K., Saikia, J., Saikia, D., and Talukdar, A.K. (2012). Synthesis and characterization of isomorphously zirconium substituted Mobil five (MFI) zeolite. *Mater. Chem. Phys.* 134: 43–49.

19 Sarmah, B., Srivastava, R., and Satpati, B. (2016). Highly efficient silver nanoparticles supported nanocrystalline zirconosilicate catalyst for the epoxidation and hydration reactions. *ChemistrySelect* 5: 1047–1056.

20 Chen, L., Xu, S., Li, X. et al. (2012). Multimodal Zr-Silicalite-1 zeolite nanocrystal aggregates with interconnected hierarchically micro-meso-macroporous architecture and enhanced mass transport property. *J. Colloid Interface Sci.* 377: 368–374.

21 Sun, M., Chen, L., Li, X. et al. (2013). A comparative study of hierarchically micro-meso-macroporous solidacid catalysts constructed by zeolites nanocrystals synthesized via a quasi-solid-state crystallization process. *Micropor. Mesopor. Mater.* 182: 122–135.

22 Lv, A., Liu, Y., Liu, N. et al. (2008). Hydrothermal synthesis of MWW-type zirconosilicate. *Chem. Lett.* 37: 156–157.

23 Wu, P., Tatsumi, T., Komatsu, T., and Yashima, T. (2001). A novel titanosilicate with MWW structure. i. Hydrothermal synthesis, elimination of extraframework titanium, and characterizations. *J. Phys. Chem. B.* 15: 2897–2905.

24 Liu, G., Jiang, J., Yang, B. et al. (2013). Hydrothermal synthesis of MWW-type stannosilicate and its post-structural transformation to MCM-56 analogue. *Micropor. Mesopor. Mater.* 165: 210–218.

25 Guo, Q., Fan, F., Pidko, E.A. et al. (2013). Highly active and recyclable Sn-MWW zeolite catalyst for sugar conversion to methyl lactate and lactic acid. *ChemSusChem* 6: 1352–1356.

26 Zhu, Z., Guan, Y., Ma, H. et al. (2020). Hydrothermal synthesis of boron-free Zr-MWW and Sn-MWW zeolites as robust Lewis acid catalysts. *Chem. Commun.* 56: 4696–4699.

27 Yang, Y., Zhou, D., Zhang, H. et al. (2021). In-situ construction and catalytic property of highly exposed Lewis acidity on hierarchical Zr-zeolite assisted by K^+ cation. *Micropor. Mesopor. Mater.* 324: 110898.

28 Valencia, S.V. and Corma, A. (1999). Stannosilicate molecular sieves. US Patent No. 5968473A.

29 Bermejo-Devala, R., Assary, R.S., Nikolla, E. et al. (2012). Metalloenzyme-like catalyzed isomerizations of sugars by Lewis acid zeolites. *Proc. Natl. Acad. Sci.* 109: 9727–9732.

30 Meng, Q., Liu, J., Xiong, G. et al. (2018). Aerosol-seed-assisted hydrothermal synthesis of Sn-Beta zeolite and its catalytic performance in Baeyer–Villiger oxidation. *Micropor. Mesopor. Mater.* 266: 242–251.

31 Meng, Q., Liu, J., Xiong, G. et al. (2019). The synthesis of hierarchical Sn-Beta zeolite via aerosol-assisted hydrothermal method combined with a mild base treatment. *Micropor. Mesopor. Mater.* 287: 85–92.

32 Meng, Q., Liu, J., Xiong, G. et al. (2020). Fast hydrothermal synthesis of hierarchical Sn-Beta zeolite with high Sn content in fluoride media. *Micropor. Mesopor. Mater.* 294: 109915.

33 Iida, T., Takagaki, A., Kohara, S.J. et al. (2015). Sn-Beta zeolite catalysts with high Sn contents prepared from Sn–Si mixed oxide composites. *ChemNanoMat* 1: 155–158.

34 Xu, H., Wang, X., Ji, P. et al. (2018). Hydrothermal synthesis of Sn-Beta zeolites in F^--free medium. *Inorg. Chem. Front.* 5: 2763–2771.

35 Li, W., Sun, L., Xie, L. et al. (2019). Coordinatively unsaturated sites in zeolite matrix: construction and catalysis. *Chinese J. Catal.* 40: 1255–1281.

36 Zhu, Y., Chuah, G., and Jaenicke, S. (2003). Al-free Zr-zeolite beta as a regioselective catalyst in the Meerwein–Ponndorf–Verley reaction. *Chem. Commun.* 21: 2734–2735.

37 Zhu, Y., Chuah, G., and Jaenicke, S. (2006). Selective Meerwein–Ponndorf–Verley reduction of α,β-unsaturated aldehydes over Zr-zeolite Beta. *J. Catal.* 241: 25–33.

38 Pavel, A., Anna, V., Evgeny, V. et al. (2018). Mechanism of Zr incorporation in the course of hydrothermal synthesis of zeolite BEA. *Inorg. Chem.* 57: 11978–11985.

39 Harris, J., Liao, W., Iorio, J.R.D. et al. (2017). Molecular structure and confining environment of Sn sites in single site chabazite zeolites. *Chem. Mater.* 29: 8824–8837.

40 Meng, B., Ren, S., Li, Z. et al. (2020). Intra-crystalline mesoporous zeolite [Al,Zr]-Y for catalytic cracking. *ACS Appl. Nano Mater.* 9: 9293–9302.

41 Xiong, H., Pham, H., and Datye, A. (2014). Hydrothermally stable heterogeneous catalysts for conversion of biorenewables. *Green Chem.* 16: 4627–4643.

42 Cui, X.Z. and Shi, J.L. (2016). Sn-based catalysts for Baeyer-Villiger oxidations by using hydrogen peroxide as oxidant. *Sci. China Mater.* 59: 675–670.

43 Niphadkar, P.S., Kotwal, M.S., Deshpande, S.S. et al. (2009). Tin-silicalite-1: synthesis by dry gel con-version, characterization and catalytic performance in phenol hydroxylation reaction. *Mater. Chem. Phys.* 114: 344–349.

44 Kang, Z., Zhang, X., Liu, H. et al. (2013). Factors affecting the formation of Sn-Beta zeolites by steam-assisted conversion method. *Mater. Chem. Phys.* 141: 519–529.

45 Chang, C., Cho, J., Wang, Z. et al. (2015). Fluoride-free synthesis of a Sn-BEA catalyst by dry gel conversion. *Green Chem.* 17: 2943–2951.

46 Devos, J., Shah, M.A., and Dusselier, M. (2021). On the key role of aluminium and other heteroatoms during interzeolite conversion synthesis. *RSC Adv.* 11: 26188–26210.

47 Itabashi, K., Kamimura, Y., Iyoki, K. et al. (2012). A working hypothesis for broadening framework types of zeolites in seed-assisted synthesis without organic structure-directing agent. *J. Am. Chem. Soc.* 134: 11542–11549.

48 Zhu, Z., Xu, H., Jiang, J. et al. (2017). Sn-Beta zeolite hydrothermally synthesized via interzeolite transformation as efficient Lewis acid catalyst. *J. Catal.* 352: 1–12.

49 Zhu, Z., Ma, H., Li, M. et al. (2020). Fluoride-free hydrothermal synthesis of nanosized Sn-Beta zeolite via interzeolite transformation for Baeyer–Villiger oxidation. *Appl. Catal. A. Gen.* 590: 117370.

50 Zhu, Z., Xu, H., Jiang, J. et al. (2017). Structural reconstruction: a milestone in the hydrothermal synthesis of highly active Sn-Beta zeolites. *Chem. Commun.* 53: 12516–12519.

51 Takata, T., Tsunoji, N., Takamitsu, Y. et al. (2017). Incorporation of various heterometal atoms in CHA zeolites by hydrothermal conversion of FAU zeolite and their performance for selective catalytic reduction of NOx with ammonia. *Micropor. Mesopor. Mater.* 246: 89–101.

52 Zhu, Z., Xu, H., Jiang, J. et al. (2017). Hydrophobic nanosized all-silica Beta zeolite: efficient synthesis and adsorption application. *ACS Appl. Mater. Interfaces* 9: 27273–27283.

53 Zhu, Z., Xu, H., Wang, B. et al. (2019). Intensified interzeolite transformation: ultrafast synthesis of active and stable Ti-Beta zeolites without solvents. *Chem. Commun.* 55: 14279–14282.

54 Tang, B., Li, S., Song, W. et al. (2020). Fabrication of hierarchical Sn-Beta zeolite as efficient catalyst for conversion of cellulosic sugar to methyl lactate. *ACS Sustain. Chem. Eng.* 8: 3796–3808.

55 Chen, Z., Liu, X., Yu, Y. et al. (2021). Gripper-like silicon species for efficient synthesis of crystalline metallosilicates with spatially homogeneous heteroatoms in the framework. *Chem. Mater.* 33: 4988–5001.

56 Cho, H.J., Gould, N.S., Vattipalli, V. et al. (2019). Fabrication of hierarchical Lewis acid Sn-BEA with tunable hydrophobicity for cellulosic sugar isomerization. *Micropor. Mesopor. Mater.* 278: 387–396.

57 Bian, C., Wu, Q., Zhang, J. et al. (2015). Interlayer expansion of the layered zeolite precursor COK-5 with $Sn(acac)_2Cl_2$. *J. Energy Chem.* 24: 642–645.

58 Gies, H., Feyen, M., Baerdemaeker, T.D. et al. (2016). Interlayer expansion using metal-linker units: crystalline microporous silicate zeolites with metal centers on specific framework sites. *Micropor. Mesopor. Mater.* 222: 235–240.

59 Chang, C., Chu, C.T.W., Miale, J.N. et al. (1984). Aluminum insertion into high silica zeolite frameworks. 1. Reaction with aluminum halides. *J. Am. Chem. Soc.* 106: 8143–8146.

60 Li, G., Gao, L., Sheng, Z. et al. (2019). A Zr-Al-Beta zeolite with open Zr(IV) sites: an efficient bifunctional Lewis–Brønsted acid catalyst for a cascade reaction. *Cat. Sci. Technol.* 9: 4055–4065.

61 Shao, Y., Sun, K., Zhang, L. et al. (2019). Balanced distribution of Brønsted acidic sites and Lewis acidic sites for highly selective conversion of xylose into levulinic acid/ester over Zr-beta catalysts. *Green Chem.* 21: 6634–6645.

62 Skeels, G.W., Chapman, D.M., and Flanigen, E.M. (1995). *Subsittution of Sn in Place of Al in the Framework of Molecular Sieve Via Treatment with Fluoride Salts*. US5401488A. UOP.

63 Jin, L., Fang, Y., and Hu, H. (2005). Selective synthesis of 2,6-dimethylnaphthalene by methylation of 2-methylnaphthalene with methanol on Zr/(Al)ZSM-5. *Catal. Commun.* 7: 255–259.

64 Nishiyama, N., Yamaguchi, M., Nishiyama, Y. et al. (2005). Enhancement of structural stability of mesoporous silica via infiltration of $SnCl_4$ vapor. *J. Non Cryst. Solids* 351: 3218–3224.

65 de Aída, L.V.P., Alarcón, E., and de Montes, C.C. (2005). Nopol synthesis over Sn-MCM-41 and Sn-kenyaite catalysts. *Catal. Today* 107–108: 942–948.

66 Dapsens, P.Y., Mondelli, C., Kusema, B.T. et al. (2014). A continuous process for glyoxal valorisation using tailored Lewis-acid zeolite catalysts. *Green Chem.* 16: 1176–1186.

67 Li, P., Liu, G., Wu, H. et al. (2011). Postsynthesis and selective oxidation properties of nanosized Sn-Beta zeolite. *J. Phys. Chem. C* 115: 3663–3670.

68 Jin, J., Ye, X., Li, Y. et al. (2014). Synthesis of mesoporous Beta and Sn-Beta zeolites and their catalytic performances. *Dalton Trans.* 43: 8196–8204.

69 Zhu, Z., Xu, H., Jiang, J. et al. (2016). Postsynthesis of FAU-type stannosilicate as efficient heterogeneous catalyst for Baeyer-Villiger oxidation. *Appl. Catal. A: Gen.* 519: 155–164.

70 Zhang, Y., Huo, Y., Tang, K. et al. (2021). Role of the pore-opening structure and hydrophobicity of stannosilicate zeolites in Baeyer–Villiger oxidation. *J. Catal.* 394: 8–17.

71 van der Graaff, W.N.P., Li, G., Mezari, B. et al. (2015). Synthesis of Sn-Beta with exclusive and high framework Sn content. *ChemCatChem* 7: 1152–1160.

72 Dijkmans, J., Demol, J., Houthoofd, K. et al. (2015). Post-synthesis Snβ: an exploration of synthesis parameters and catalysis. *J. Catal.* 330: 545–557.

73 Vega-Vila, J.C., Harris, J.W., and Gounder, R. (2016). Controlled insertion of tin atoms into zeolite framework vacancies and consequences for glucose isomerization catalysis. *J. Catal.* 344: 108–120.

74 Zhu, Z., Xu, H., Jiang, J., and Wu, P. (2016). Postsynthesis and effective Baeyer–Villiger oxidation properties of hierarchical FAU-type stannosilicate. *J. Phys. Chem. C* 120: 23613–23624.

75 Zhang, H., Quek, Z.J., Jaenicke, S., and Chuah, G.K. (2021). Hydrophobicity and co-solvent effects on Meerwein-Ponndorf-Verley reduction/dehydration cascade reactions over Zr-zeolite catalysts. *J. Catal.* 400: 50–61.

76 Ouyang, X., Hwang, S., Xie, D. et al. (2015). Heteroatom-substituted delaminated zeolites as solid Lewis acid catalysts. *ACS Catal.* 5: 3108–3119.

77 Hammond, C., Conrad, S., and Hermans, I. (2012). Simple and scalable preparation of highly active Lewis acidic Sn-β. *Angew. Chem. Int. Ed.* 51: 11736–11739.

78 Song, S., Di, L., Wu, G. et al. (2017). Meso-Zr-Al-beta zeolite as a robust catalyst for cascade reactions in biomass valorization. *Appl. Catal. B: Environ.* 205: 393–403.

79 Bhaumik, A., Kumar, P., and Kumar, R. (1996). Baeyer-Villiger rearrangement catalysed by titanium silicate molecular sieve (TS-1)/H_2O_2 system. *Catal. Lett.* 40: 47–50.

80 Conrad, S., Wolf, P., Müller, P. et al. (2017). Influence of hydrophilicity on the Snβ-catalyzed Baeyer–Villiger oxidation of cyclohexanone with aqueous hydrogen peroxide. *ChemCatChem* 9: 175–182.

81 Luo, H., Bui, L., Gunther, W. et al. (2012). Synthesis and catalytic activity of Sn-MFI nanosheets for the Baeyer-Villiger oxidation of cyclic ketones. *ACS Catal.* 2: 2695–2699.

82 Corma, A., Domine, M.E., Nemeth, L., and Valencia, S. (2002). Al-free Sn-Beta zeolite as a catalyst for the selective reduction of carbonyl compounds (Meerwein–Ponndorf–Verley reaction). *J. Am. Chem. Soc.* 124: 3194–3195.

83 Samuel, P.P., Shylesh, S., and Singh, A.P. (2007). Catalytic properties of tin-containing mesoporous molecular sieves in the selective reduction of carbonyl compounds (Meerwein–Ponndorf–Verley (MPV) reaction). *J. Mol. Catal. A Chem.* 266: 11–20.

84 Boronat, M., Corma, A., and Renz, M. (2006). Mechanism of the Meerwein–Ponndorf–Verley–Oppenauer (MPVO) redox equilibrium on Sn– and Zr–Beta zeolite catalysts. *J. Phys. Chem. B.* 110: 21168–21174.

85 Lewis, J.D. and Van de Vyver, D.S. (2014). A continuous flow strategy for the coupled transfer hydrogenation and etherification of 5-(hydroxymethyl)furfural using Lewis acid zeolites. *ChemSusChem* 7: 2255–2265.

86 Dai, W., Wang, C., Tang, B. et al. (2016). Lewis acid catalysis confined in zeolite cages as a strategy for sustainable heterogeneous hydration of epoxides. *ACS Catal.* 6: 2955–2964.

87 Lei, Q., Wang, C., Dai, W. et al. (2021). Tandem Lewis acid catalysis for the conversion of alkenes to 1,2-diols in the confined space of bifunctional TiSn-Beta zeolite. *Chinese J. Catal.* 42: 1176–1184.

88 Deshpande, N., Parulkar, A., Joshi, R. et al. (2019). Epoxide ring opening with alcohols using heterogeneous Lewis acid catalysts: regioselectivity and mechanism. *J. Catal.* 370: 46–54.

89 Kore, R., Srivastava, R., and Satpati, B. (2013). Highly efficient nanocrystalline zirconosilicate catalysts for the aminolysis, alcoholysis, and hydroamination reactions. *ACS Catal.* 3: 2891–2904.

90 Tang, B., Dai, W., Sun, X. et al. (2012). Mesoporous Zr-Beta zeolite prepared by postsynthetic strategy as a robust Lewis acid catalyst for the ring-opening aminolysis of epoxides. *Green Chem.* 17: 1744–1755.

91 Tang, B., Song, W.C., Li, S.Y. et al. (2018). Post-synthesis of Zr-MOR as a robust solid acid catalyst for the ring-opening aminolysis of epoxides. *New J. Chem.* 42: 13503–13511.

92 Van de Vyver, S., Odermatt, C., Romero, K. et al. (2015). Solid Lewis acids catalyze the carbon–carbon coupling between carbohydrates and formaldehyde. *ACS Catal.* 5: 972–977.

93 Lewis, J.D., Van de Vyver, S., and Román-Leshkov, Y. (2015). Acid–base pairs in Lewis acidic zeolites promote direct aldol reactions by soft enolization. *Angew. Chem. Int. Ed.* 54: 9835–9838.

94 Wang, G., Zhang, H., Wang, H. et al. (2016). The role of metallic Sn species in catalytic dehydrogenation of propane: active component rather than only promoter. *J. Catal.* 344: 606–608.

95 Wang, G., Zhang, H., Zhu, Q. et al. (2017). Sn-containing hexagonal mesoporous silica (HMS) for catalytic dehydrogenation of propane: an efficient strategy to enhance stability. *J. Catal.* 351: 90–94.

96 Yue, Y., Fu, J., Wang, C. et al. (2020). Propane dehydrogenation catalyzed by single Lewis acid site in Sn-Beta zeolite. *J. Catal.* 395: 155–167.

97 Moliner, M., Román-Leshkov, Y., and Davis, M.E. (2010). Tin-containing zeolites are highly active catalysts for the isomerization of glucose in water. *Proc. Natl. Acad. Sci.* 107: 6164–6168.

98 Rai, N., Caratzoulas, S., and Vlachos, D.G. (2013). Role of silanol group in Sn-Beta zeolite for glucose isomerization and epimerization reactions. *ACS Catal.* 3: 2294–2298.

99 Bermejo-Deval, R., Orazov, M., Gounder, R. et al. (2014). Active sites in Sn-Beta for glucose isomerization to fructose and epimerization to mannose. *ACS Catal.* 4: 2288–2297.

100 Garces, D., Faba, L., Diaz, E., and Ordonez, S. (2019). Aqueous-phase transformation of glucose into hydroxymethylfurfural and levulinic acid by combining homogeneous and heterogeneous catalysis. *ChemSusChem* 12: 924–934.

101 Nikolla, E., Roman-Leshkov, Y., Moliner, M., and Davis, M.E. (2011). One-pot synthesis of 5-(hydroxymethyl)furfural from carbohydrates using tin-Beta zeolite. *ACS Catal.* 1: 408–410.

102 Li, L., Ding, J., Jiang, J. et al. (2015). One-pot synthesis of 5-hydroxymethylfurfural from glucose using bifunctional [Sn,Al]-Beta catalysts. *Chinese J. Catal.* 36: 820–828.

103 Saenluang, K., Thivasasith, A., Dugkhuntod, P. et al. (2020). In situ synthesis of Sn-Beta zeolite nanocrystals for glucose to hydroxymethylfurfural (HMF). *Catalyst* 10: 1249.

104 Oozeerally, R., Pillier, J., Kilic, E. et al. (2020). Gallium and tin exchanged Y zeolites for glucose isomerisation and 5-hydroxymethyl furfural production. *Appl. Catal. A: Gen.* 605: 117798.

105 Osmundsen, C.M., Holm, M.S., Dahl, S., and Taarning, E. (2012). Tin-containing silicates: structure–activity relations. *Proc. Roy. Soc. A: Math, Phys. Eng. Sci.* 468: 2000–2016.

106 Yang, X., Wu, L., Wang, Z. et al. (2016). Conversion of dihydroxyacetone to methyl lactate catalyzed by highly active hierarchical Sn-USY at room temperature. *Cat. Sci. Technol.* 6: 1757–1763.

107 Holm, M.S., Pagan-Torres, Y.J., Saravanamurugan, S. et al. (2012). Sn-Beta catalysed conversion of hemicellulosic sugars. *Green Chem.* 14: 702–706.

108 Yang, L., Yang, X., Tian, E. et al. (2016). Mechanistic insights into the production of methyl lactate by catalytic conversion of carbohydrates on mesoporous Zr-SBA-15. *J. Catal.* 333: 207–216.

109 Murillo, B., Sanchez, A., Sebastian, V. et al. (2014). Conversion of glucose to lactic acid derivatives with mesoporous Sn-MCM-41 and microporous titanosilicates. *J. Chem. Technol. Biotechnol.* 89: 1344–1350.

110 Yang, X., Liu, Y., Li, X. et al. (2018). Synthesis of Sn-containing nanosized beta zeolite as efficient catalyst for transformation of glucose to methyl lactate. *ACS Sustain. Chem. Eng.* 6: 8256–8265.

111 Yang, X., Bian, J., Huang, J. et al. (2017). Fluoride-free and low concentration template synthesis of hierarchical Sn-beta zeolites: efficient catalysts for conversion of glucose to alkyl lactate. *Green Chem.* 19: 692–701.

112 Bui, L., Luo, H., Gunther, W., and Romn-Leshkov, Y. (2013). Domino reaction catalyzed by zeolites with Brønsted and Lewis acid sites for the production of γ-valerolactone from furfural. *Angew. Chem. Int. Ed.* 52: 8022–8025.

113 Winoto, H.P., Ahn, B.S., and Jae, J. (2016). Production of γ-valerolactone from furfural by a single-step process using Sn-Al-Beta zeolites: optimizing the catalyst acid properties and process conditions. *J. Ind. Eng. Chem.* 40: 62–71.

114 Winoto, H.P., Fikri, Z.A., Ha, J.M. et al. (2019). Heteropolyacid supported on Zr-Beta zeolite as an active catalyst for one-pot transformation of furfural to γ-valerolactone. *Appl. Catal. B: Environ.* 241: 588–597.

115 Wang, J., Jaenicke, S., and Chuah, G. (2014). Zirconium–Beta zeolite as a robust catalyst for the transformation of levulinic acid to γ-valerolactone via Meerwein–Ponndorf–Verley reduction. *RSC Adv.* 4: 13481–13489.

116 Melero, J.A., Morales, G., Iglesias, J. et al. (2017). Efficient one-pot production of γ-valerolactone from xylose over Zr-Al-Beta zeolite: rational optimization of catalyst synthesis and reaction conditions. *Green Chem.* 19: 5114–5121.

117 Paniagua, M., Morales, G., Melero, J.A. et al. (2021). Understanding the role of Al/Zr ratio in Zr-Al-Beta zeolite: towards the one-pot production of GVL from glucose. *Catal. Today* 367: 228–238.

4

Synthesis of Germanosilicates

Jiuxing Jiang

Key Laboratory of Bioinorganicand Synthetic Chemistry of Ministry of Education, School of Chemistry, IGCME, Sun Yat-Sen University, No. 132, Waihuan East Road, Higher Education Mega Center, Panyu District, GuangZhou 510006, P.R. China

4.1 Introduction

4.1.1 General Property of Ge/Si Oxides

Germanium and silicon are elements of group IVA with two *s* electrons and two *p* electrons in their valance shell. Both are semiconductors in their simple substance. There are five commonly observed crystalline polymorphs of SiO_2 (dense SiO_2, except pure silica zeolite) that are good insulators. Among them, three phases occurred in ambient pressure, i.e. (α,β) quartz, (α,β_1,β_2)tridymite, and (α,β)cristobalite, where α is the low-temperature phase and β means high-temperature phase [1]. The three polymorphs share the same primary building unit, i.e. [SiO_4] tetrahedron. As we know, the secondary phase transition, e.g. α to β cristobalite, only changes the bond length/angles and symmetry of the structure, leaving the topology intact. In contrast, the phase transitions between quartz, tridymite, and cristobalite involve chemical bond breaking and reforming. High pressure leads to denser phases, i.e. coesite and stishovite. Whereas coesite prepared in 6.5 GPa still retains tetrahedron coordination [2], the pressure as high as 15 GPa leads to the formation of stishovite (isostructure of rutile) with octahedra coordination [3]. The densities of the polymorphs are 2.65, 2.26, 2.34, 2.92, and 4.35 g/cm^3 for quartz, tridymite, cristobalite, coesite, and stishovite, respectively. In zeolite science, another descriptor of density, i.e. framework density (FD), defined as framework atom (typically tetrahedra atom, T-atom) number in 1000 $Å^3$, is frequently used. Thus, the FD of the five polymorphs is 26.6, 22.6, 23.3, 29.3, and 42.9 Si atom/1000 $Å^3$, respectively. [4] The value of FD is generally proportional to the real density. Quartz, cristobalite, and tridymite are notoriously insoluble in all kinds of solvents, but they reacted with alkaline and HF. Stishovite shows the highest inner property that could resist the corrosion of HF. There are only two stable polymorphs of GeO_2 (except the germanates discussed in 4.2), i.e. tetragonal and hexagonal GeO_2. Hexagonal GeO_2 is structurally analog to β-quartz (tetrahedra

Micro-Mesoporous Metallosilicates: Synthesis, Characterization, and Catalytic Applications, First Edition. Edited by Peng Wu and Hao Xu.

coordination, chemical reagent of GeO_2) with a band gap of 6.6 eV. Generally speaking, it is an insulator. In contrast, the tetragonal version of GeO_2 with the isostructure of stishovite (rutile structure, octahedra coordination) shows a better conductivity, i.e. it is an ultra-wide-band-gap (4.68 eV) semiconductor [5].

Due to the larger covalent radii (Ge 122 p.m. vs. Si 111 p.m.), longer T-O bond length (c. 1.74 Å for Ge—O vs. c. 1.62 Å for Si—O), and smaller T-O-T angle (c. 130° for Ge-O-Ge vs. c. 145° for Si-O-Si), Ge exhibits rich coordination environments, i.e. $[GeO_4]$, $[GeO_5]$ even $[GeO_6]$ in the open-framework germanate [6]. Moreover, GeO_2 is much more reactive than SiO_2. It could be "solved" not only in acid and alkaline solutions but also in boiling water.

Germanosilicate is also called germanium silicate, silicogermanate, germanium silicon oxides, etc., and when the germanium amount is small, sometimes it is called Ge-doped SiO_2 or Ge-doped silica [7]. Owning a similar property, in some cases, Si and Ge could substitute each other in any ratio. Therefore, germanosilicate may be considered as a mixed oxide/oxyacid of Si and Ge with both element distribution homogeneously (macroscopic view) with Si/Ge ratio between 0 and ∞. If we define the Si/Ge ratio of germanate as zero, Si/Ge ratio of silicate as infinite. Then, the materials with Si/Ge < 1 could be considered silicogermanate, and the materials with Si/Ge > 1 could be named germanosilicate. Thus, crystalline and amorphous materials are within the scope of the definition.

4.1.2 Germanosilicate Glass

Historically, germanosilicates are firstly investigated as glass. Actually, germanosilicate glass is one of the functional glass materials with outstanding optical performance due to its stable chemical and mechanical properties. The doping of Ge is originally for increasing the refractive index of fused quartz glass that is wildly used in light-guide optical fiber [8]. However, two unexpected effects are also involved with the Ge incorporation. One is the optical loss. Second is the nonlinear effect. The initial investigation interest of the nonlinear effect is to control the optical loss [9]. Later research is mainly devoted to develop cost-effective glass to substitute the traditional nonlinear effect KDP (potassium dihydrogen phosphate), BBO (β-barium borate), crystals, etc. The second is the color center introduction due to the Ge defects [8]. Recently, amorphous Ge—Si mixed oxide has been found to show a tunable ultra-long phosphorescence and delayed fluorescence at high temperatures powered by the Ge defects [10]. Generally speaking, glass materials are isotropic due to their aperiodic structure. All secondary order nonlinearity (SON) component is 0 in glass-type materials. Therefore, traditionally glass materials are considered inert for the second-harmonic generation (SNG). However, the SON effect is confirmed in Ge-doped glass fiber under the induction of the Nd:YAG (Nd:$Y_3Al_5O_{12}$, solid-state laser emitting light of 1064 nm) laser irradiation (poling process) [11]. Soon, the SON effects were detected in silicate [7, 12], tellurates [13], borate [14], and sulfide glass [15] in poling processes such as thermal poling [16], corona poling [17], and electron/γ/laser beam irradiation [18]. Among the glass materials, quartz glass exhibits better chemical stability, mechanical property, and a wide transmitting window. The incorporation of Ge enables the glass to have a similar chemical property to quartz glass and a higher SON coefficient. Moreover, germanosilicate glass is photosensitive, i.e. the refractive index could be tuned by UV irradiation. Thus, the germanosilicate glass is widely used in tunable IR laser sources [19], grating [20], etc. Various techniques have been developed for the production of glass fiber, via preform (a piece of glass used to draw an optical fiber), including vapor deposition techniques [21], sol–gel [22], spatterings [7]; the rest part without involving preform, i.e. rod-in-tube [21], double-crucible methods [21].

4.2 Isomorphous Substitution in Germanosilicates

4.2.1 Isomorphous Substitution Si in Germanate

The crystalline silicogermanate material chemistry starts from pure germanates research, i.e. Si/Ge ratio = 0. Xu and coworkers performed the pioneering work in 1991 [23]. Following the method of zeolite synthesis, i.e. hydrothermal treatment of mixing GeO_2 and tetramethylammonium hydroxide, the first open framework porous germanate was obtained. Unfortunately, the porous germanate is not sufficiently robust to retain its structure after calcination. In the process, organic structure-directing agent (OSDA) molecules could be removed to evacuate the pore space like in zeolites. Despite that, this work initialized the research field of open framework germanate chemistry.

Table 4.1 lists the germanates reported to date. From the connectivity point of view, most materials are three-dimensional (27 of 37). At the same time, there are still zero-dimensional clusters, one-dimensional chains, and two-dimensional layers. Notably, the inorganic part of the zero-dimensional compound [DECDMP,F]-GeO_2-*D4R* [26] is independent Ge8-*d4r*-F (double four ring, Figure 4.1d,e). While in most of the structures, the *d4r* acts as only a secondary building unit (SBU). Besides Ge8-*d4r*, germanates exhibit rich chemistry of cluster SBU. Among them, Ge7 (Figure 4.1c) is most frequently observed (15 of 37). The cluster contains four GeO_4 tetrahedra, two GeO_5 trigonal bipyramidal, and one GeO_6 octahedron. The second most recorded is Ge9 (Figure 4.1f–h, 11 of 37), which comprises four GeO_4 tetrahedra, four GeO_5 trigonal bipyramidal, and one GeO_6 octahedron. However, according to the connectivity, Ge9 could be further classified into Ge9a (Figure 4.1f), Ge9b (Figure 4.1g), and Ge9c (Figure 4.1h). Ge11 (Figure 4.1i) is often reported as Ge10 due to one GeO_4 is shared by two clusters. The rest clusters, such as Ge3 (Figure 4.1a), Ge6 (Figure 4.1b), and Ge12 (Figure 4.1j,k), are only reported in some cases, or they could not be considered as clusters. Furthermore, the low $FD_{\text{Si or Ge}}$ value gives the structure a more chance of being an extra-large pore (extra-large pore in zeolite means the pore mouth is wider than 12 Si atoms), e.g. 30-ring-pore Su-M, JLG-12, 26-ring-pore SU-61, 24-ring-pore FDU-4, ASU-16 (SU-12), and 18-ring-pore SU-44.

From the synthesis point of view, 16 of 37 materials are synthesized in basic media (F^- free); 18 of 37 materials are obtained in F^- media; and 3 materials could be

Table 4.1 The statistics of germanate and silicogermanates reported to date.

Material name	Framework density Ge/1000Å^3	Connec-tivity	Ring opening	Si/Ge ratio	F^-/OH^-	OSDA	SBU	Reference
Ge_6O_{12}-NMe_4OH	–	3	–	0	OH^-	OSDA 24	–	[23]
$(C_2H_8N_2)(C_2H_{10}N_2)_2$ $[Ge_9O_{18}$-$(OH)_4]$	–	3	10	0	F^-	OSDA 15	Ge9a	[24]
$Ge_{10}O_{21}(OH),N_4C_6H_{21}$	15.3	3	–	0	OH^-	OSDA 11	Ge7 + 3	[25]
[DECDMP,F]-GeO_2-*D4R*	–	0	–	0	F^-	OSDA 52	Ge8-*d4r*	[26]
$[Co_2(en)_3][Ge_9O_{18}(OH)_4]$	–	3	–	0	F^-	OSDA 20	Ge9a	[27]
$(H_3NC_3H_7)_5(H_3O)$ $[Ge_{18}O_{36}(OH)_6],3H_2O$	11.7	3	16	0	OH^-	OSDA 5	Ge9b	[28]
ASU-14	13.4	3	8	0	F^-	OSDA 12	Ge9a	[29]
ASU-16	9.8	3	24	0	F^-	OSDA 7	Ge7	[30]
ASU-19	11.8	2	–	0	OH^-/F^-	OSDA 7	Ge7	[31]
ASU-20	11.2	2	–	0	OH^-/F^-	OSDA 15	Ge7	[31]
ICMM5	14.0	3	12	0	OH^-	OSDA 17	Ge8-*d4r*	[32]
ICMM6	14.9	3	–	0	OH^-	OSDA 5	Ge9c	[33]
ICMM7	14.0	3	14 × 12 × 12	0	OH^-	OSDA 16	Ge12a + 1	[34]
ICMM9	–	2	–	0	OH^-	OSDA 14	Ge3	[35]
FDU-4	11.1	3	24	0	OH^-	OSDA 11	Ge9b	[36]
FJ-1a,b	14.6	3	–	0	F^-	OSDA 19	Ge7	[37]
FJ-6	–	1	–	0	F^-	OSDA 19	Ge7	[38]
SU-M	7.1	3	30	0	F^-	OSDA 8	Ge11 (Ge10)	[39]

SU-8	11.5	3	16 × 10 × 8	0	F^-	OSDA 8	Ge7, Ge9a	[40]
SU-12 (ASU-16)	8.6	3	24 × 12 × 10	0.72 (0.09)	F^-	OSDA 3	Ge7	[41]
Su-21	–	2	–	0.9 (0.35)	F^-	OSDA 6	Ge3 + 1	[42]
SU-22 (ASU-20)	–	2	12 × 8	0	F^-	OSDA 10	Ge7	[43]
SU-23	–	2	10	0	F^-	OSDA 9	Ge7	[43]
SU-44	10.0	3	18 × 18 × 10	0	F^-	OSDA 8	Ge7, Ge9a	[40]
SU-61	10.2	3	26	0.1 (0.32)	OH^-	OSDA 8	Ge11 (Ge10)	[44]
SU-67	12.4	3	11 × 11 × 10	0	OH^- / F^-	OSDA 13	Ge11 (Ge10)	[45]
SU-75	11.4	3	12 × 11 × 10	0	OH^-	OSDA 10	Ge11 (Ge10)	[46]
SU-77	14.6	3	12 × 10 × 10	0	F^-	OSDA 5	Ge6	[47]
SU-79	12.6	3	11 × 10 3D	0	OH^-	OSDA 18	Ge12a + 1	[48]
JLG-4	9.7	1	–	0	F^-	OSDA 6	Ge7	[49]
JLG-5	9.2	1	–	0	F^-	OSDA 13	Ge7	[50]
JLG-12	9.3	3	30	0	F^-	OSDA 8	Ge7, Ge9a	[51]
IM-14	13.7	3	10 × 10 × 8	0	OH^-	OSDA 5	Ge11 (share 2Ge)	[52]
GeO-JU90	11.9	3	12 × 12 × 11	0	OH^-	OSDA 54	Ge7 + 1	[53]
HUT-2 (ICMM6)	–	3	11 × 8	0.1 (0.02)	OH^-	OSDA 5	Ge9c	[54]
PKU-15	14.4	3	12 × 10 × 7	0.11 (0.12)	OH^-	OSDA 25	Ge12b, *3r*,	[55]
PST-18	14.3	3	8, 7	1.5 (0.6)	OH^-	OSDA 24	Ge9a, *nat*	[56]

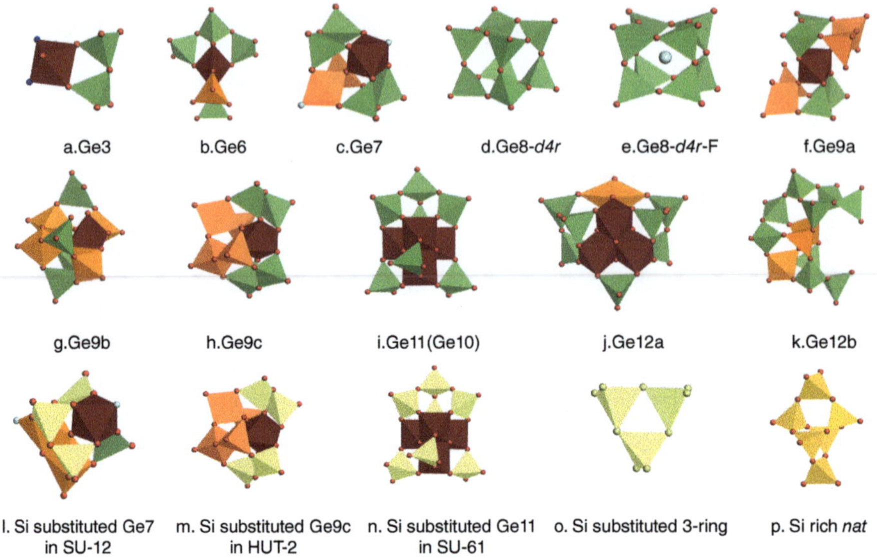

Figure 4.1 The germanate clusters. Tetrahedron (olive green) for GeO_4, tetrahedron (pale yellow) for Si substituted GeO_4, trigonal bipyramid (orange) for GeO_5, and octahedron (rust red) for GeO_6. Source: Jiuxing Jiang.

prepared with both OH^- and F^- as a mineralizer. Most OSDAs used for germanate are commercially available amines, with a few exceptions, Ge_6O_{12}-NMe_4OH, GeO-JU90, PKU-15, PST-18, prepared with simple quaternary ammonium.

The substitution of Si in germanate or silicogermanate is not random and vice versa. In a Si substitution as low as SU-12 (Si/Ge = 0.09), SU-61 (Si/Ge = 0.32), and HUT-2 (Si/Ge = 0.02), the cluster-based structure remains. Therefore, Si atoms can only go to GeO_4 in Ge7 (Figure 4.1l), Ge9c (Figure 4.1m), and Ge11 (Figure 4.1n) sites due to the four coordination habits.

Further increasing the Si/Ge ratio, like PKU-15 (Si/Ge = 0.12) and PST-18 (Si/Ge = 0.6), the Si amount is beyond the tolerance of the cluster GeO_4; therefore, some Si-rich SBU of zeolites such as *3r* (3-ring, Figure 4.1o) and *nat* (Figure 4.1p) are observed. Thus, the materials become "quasi-zeolite" [55]. Especially, PKU-15 develops thermal stability (stable up to 600 °C in a dry atmosphere) enough to remove the guest OSDA molecules. Therefore, PKU-15 gained substantial CO_2 and CH_4 adsorption ability thanks to the Si incorporation.

4.2.2 Isomorphous Substitution Ge in Silicates

On the germanosilicate side, initially, germanium was incorporated in ITQ-7 for introducing acid sites. ITQ-7 was first synthesized with OSDA 47 or OSDA 54 in pure silica form within 15 days under hydrothermal conditions [57]. However, Al could not be directly introduced in the zeolite synthesis. Thus, Ge and B were introduced into ITQ-7 for later Al replacement [58]. Surprisingly, the Ge incorporation significantly accelerated the crystallization process of ITQ-7 (15 days to

1 day) [59]. Furthermore, combining the ^{19}F and ^{29}Si NMR spectra of pure silicate and germanosilicate ITQ-7 zeolites shows that Ge atoms are preferentially located on the *d4r* site. Additionally, germanosilicate ITQ-7 emits luminescence upon excitation with three independent components (emission decays fit to variable proportions of three exponential kinetics). This evidence is in line with the three families of Ge atoms in the solid [60].

The isomorphous substitution of Ge has been investigated for ITQ-17 [61], ITQ-13 [62], and ITQ-21 [63]. Theoretical calculation explains that the smaller Ge-O-Si bond angle (<140°) relaxes the bending strain of the corner atom in *d4r* [63, 64]. As mentioned above, a zeolite case that enables the Si—Ge mutual substitution in any ratio is HPM-1 with **STW** structure [65]. With OSDA 80, which shows a strong structure-directing effect toward **STW** zeolite, the synthesis restriction of narrow Si/Ge ratio for SU-32 (originally synthesized with OSDA 4) is removed. Thus, it provides a rare opportunity to observe consecutive changes. From pure germanate (Si/Ge = 0) to pure silicate (Si/Ge = INF), the yield of zeolite first increases until Si/Ge is about 1.5 (Ge_f = Ge/[Ge + Si] = 0.4) to reach the maximum ca. 40% and then decreases. This trend aligns with free energy calculation (first drops, then raises). In other words, both Ge substitution and Si substitution are energy favorable for **STW** zeolite structure stabilization. Moreover, the unit cell expansion associated with Ge substitution in the most span of Si/Ge is also observed, except the 9 < Si/Ge < INF ($0 < Ge_f < 0.1$) due to above-mentioned strain relaxation effect of *d4r* structure. In contrast, the small amount of Ge incorporation ($0 < Ge_f < 0.13$) significantly increases the cell parameters of Ge-**MFI**, whose structure is without *d4r*. Therefore, within a Si/Ge period, the cell parameter expansion could potentially act as a "Si/Ge meter," like what we have done to evaluate the Si/Al ratio in Y zeolite. Zones et al. [66] summarized the SBU preference of the heteroatom substitution in silicate zeolite. Germanium exhibits a strong tendency to form *d4r*. Therefore, most germanosilicate zeolite possesses such SBU. In contrary, the *d4r*-free zeolite shows a low level of germanium incorporation tolerance due to the lack of preferential crystallographic site. However, the more complicated truth is that, the small amount of Ge in the original ITQ-47 (*d4r*-free) synthesis gel is essential [67].

4.3 Inorganic Structure-Directing Effects

4.3.1 Structure-Directing Effects of Ge

As mentioned above, Ge doping could not only stabilize the existing *d4r* in zeolite structures but also guide the zeolite synthesis. An interesting case is when OSDA 26, OSDA 33, OSDA 34, OSDA 38, OSDA 39, OSDA 40, OSDA 50, and OSDA 51 are used in pure silicate zeolite synthesis, ITQ-4, Beta, and ZSM-12 could be synthesized. However, when Ge is added, the products are formulated to *d4r*-containing zeolite Beta polymorph C [82]. We often call such a kind of commonly observed product as "default structure." We will discuss it later.

4.3.2 Structure-Directing Effects of F^-

The structure-directing effect of Ge toward *d4r* is shared with F^-. F^- route commonly used in the zeolite/open-framework structure synthesis. In the open-framework part, 16 of 37 materials could be synthesized in OH^- media (F^- free); 18 of 37 could only be prepared with the presence of F^-; the rest three compounds could be obtained in both media. In the zeolite part, 46 structures with the presence of F^- can be classified into two groups. The first type is bonded to the Ge and Si atom, e.g. in FJ-6, ASU-19, ASU-20, SU-12, SU-21, SU-22, SU-23, SU-77, and JLG-5. F atom bonds to the Ge atom as a terminal F. The second type is the residual ones in the center of *d4r* to form F^-@*d4r* SBU (Figure 4.1e). DFT and Kohn-Sham theory analyses show that the F^- forms highly delocalized hypervalent bonds to the *d4r* Si atoms with coordination number 8 and bond order 4.

More importantly, F^- in zeolite structure provides a probe to understand germanosilicate zeolites, especially the Si/Ge distribution in *d4r*. Among the isotope of Ge, i.e. ^{70}Ge, ^{72}Ge, ^{73}Ge, ^{74}Ge, and ^{76}Ge, only ^{73}Ge nuclei are NMR active. However, the ^{73}Ge NMR in solid exhibits a strong quadrupole interaction that prohibits its application in the study of germanosilicate materials. ^{29}Si-MAS-NMR Bloch-decay or ^{1}H-^{29}Si-MAS-NMR cross-polarization could provide clear evidence of the different types of Si coordination spheres and the presence of Si-OH. However, ^{29}Si-MAS-NMR convolutes the information of Si in the non-*d4r* crystallographic sites. Therefore, the F^- located in small cages such as *d4r* offers a unique opportunity to detect Si/Ge distribution in *d4r* via ^{19}F-MAS-NMR or ^{19}F-^{29}Si-MAS-NMR cross-polarization [62, 144–146]. Interestingly, F^- species in the non-*d4r* small cage could be exchanged by NH_4^+ or removed in a mild condition [147]. F^-@*d4r* previously proved to be removable only in germanosilicate ITQ-17 [148]. We recently demonstrated that, under a suitable temperature, i.e. 300 °C, the F^- could be reintroduced into calcined *d4r*-containing pure silica zeolite as F^-@*d4r*. Interestingly, such F^-@*d4r* exhibits mild acid property [149].

4.4 Organic Structure-Directing Agents in Germanosilicate Synthesis

4.4.1 Organic Structure-Directing Agent Types and Revolutions

The OSDA (sometimes called template when specific structure-directing effect is shown) was first introduced to zeolite synthesis to replace the alkaline metal ions, such as Na^+ and K^+, to increase the Si/Al ratio. Later, we witnessed the evolutions of zeolite synthesis, i.e. high silica (even pure silica) zeolites and the artificial creation of new zeolites due to OSDA usage. The very first OSDA introduced in zeolite synthesis is OSDA 24, i.e. tetramethyl ammonium [150]. Later, the substitutions on R1, R2, R3, and R4 enable a tremendous amount of quaternary ammonium to be synthesized and used. In 1982, the synthesis of aluminophosphate series AlPO-*n* first introduced amines as OSDA due to the weak alkalinity required. [151]

For the same reason, the high alkalinity is unnecessary for germanate (silicogermanate) synthesis. Therefore, many amines (second type of OSDA), herein OSDA 1–OSDA 23, were used as OSDA in germanates (silicogermanates). The amines may be protonated or remain unprotonated. N—H usually forms hydrogen bonding with the terminal Ge=O or Ge-OH to stabilize structures. Thus, it could merely be removed while retaining the host germanate (silicogermanate) structure intact, except for the quasi-zeolite PKU-15 [55]. The third type of OSDA is the transition-metal amines complex. The interesting feature of the amine as a ligand is it does not change the valance of the metal. Therefore, the Co^{2+} and Ni^{2+} metal complexes, i.e. OSDA 18–OSDA 20, still show the +2 valance. Furthermore, the chirality of Λ and Δ isomers could transfer to the host structures [6, 37]. Again, the weak thermal stability is the headache issue.

The fourth type of OSDA is P-derived OSDA, including phosphonium and phosphazene. Phosphonium shows a better substitution group tolerance than ammonium. P^+ could directly connect with aromatic groups, e.g. OSDA 109–OSDA 111. In contrast, N^+-aromatic is generally unstable (of course, there are exceptions, e.g. OSDA 103 for ITQ-29 [129]). The second feature of quaternary phosphonium hydroxide is Hoffman-decomposition free in contrast to the quaternary ammonium hydroxide. However, this does not mean that quaternary phosphonium hydroxides are more stable than quaternary ammonium due to the ligand-leaving reaction. The ligand-leaving reaction often occurs during the concentration of OSDA hydroxide in rotavapor, leading to the irreversible precipitate. Therefore, the concentration of phosphonium hydroxide solution should be controlled (for OSDA 98, 99, less than 0.1 mol/L). Phosphazene, e.g. OSDA 112, is a superbase molecule rather than a cation. Their alkalinity is much stronger than regular amine or amidine bases. The pk_a of protonated OSDA 112 is 42.7, the highest among the molecular base [152]. Therefore, it is sufficient to mobilize the Si species.

The fifth type of OSDA is sulfonium, i.e. OSDA 113 and OSDA 114. These organic cations are featured as unique tri-substitution molecular configurations rather than tetra-substitution of the quaternary ammonium/phosphonium. However, it is still a triangle pyramidal rather than a planar triangle due to the presence of the lone pair electrons of the center S atom.

4.4.2 Two Important Families of OSDA

There are two OSDA families we should take special notes on. The first one is the substituted isoindoline family, i.e. OSDA 60 and OSDA 65–OSDA 71. The OSDA family is synthesized with similar organic building blocks, i.e. 1,2-bis (bromomethyl)benzene, and a substituted piperidine/dicyclohexylamine. Like the usual alkylation of the secondary amine, a base, such as K_2CO_3 and Na_2CO_3, is necessary to adsorb the proton produced in the reaction. Later, the quaternary ammonium is separated with the salts using chloroform/CH_2Cl_2, etc. This series of OSDAs are designed to understand the size effect of OSDA in germanosilicate synthesis [73]. As shown in Figure 4.2, OSDA 65 and OSDA 66 (OSDA 1–OSDA 2 in the figure) can guide the synthesis of ITQ-7, ITQ-17, and beta zeolite. When the

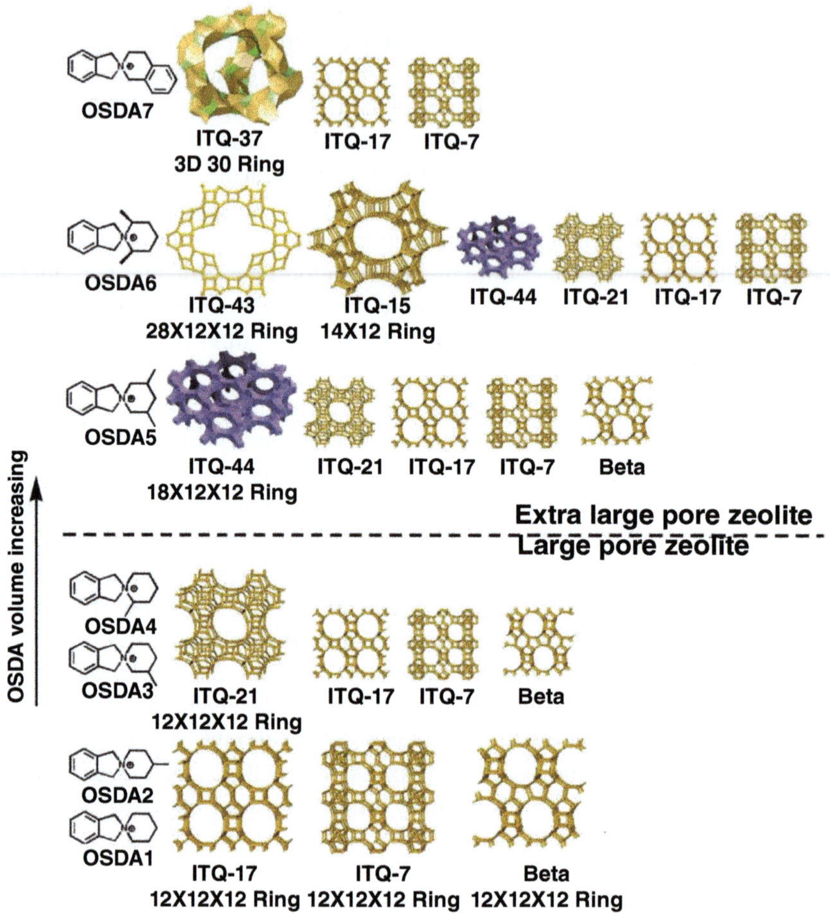

Figure 4.2 OSDA size effect on the germanosilicate zeolite synthesis. Reproduced from Ref. [73] with permission of ACS.

methyl group in the piperidine moves to two or three positions, i.e. OSDA 67 and OSDA 68 (OSDA 3 and OSDA 4 in the figure) lead to the formation of ITQ-21 in addition to the previous three zeolites. OSDA 69 (OSDA 5) with two methyl groups in positions 3 and 5 gives ITQ-44 in addition. The size of OSDA 70 (OSDA 6) enables the formation of hierarchical zeolite ITQ-43. The heaviest OSDA 71 could guide the construction of mesoporous ITQ-37. In this case, we observed a clear OSDA size boundary below that could form only large pore zeolite. If the size exceeds the limit, extra-large pore zeolite could be produced. Moreover, zeolites ITQ-7, ITQ-17, and ITQ-21 could be prepared with more than five OSDAs (Table 4.2). These zeolites are the most thermodynamically stable. We often address them as "default structures." ITQ-15 (**UTL** zeolite) is a notorious "default structure" that could be synthesized with more than 25 types of OSDAs [153].

Table 4.2 The statistics of germanate and germanosilicat zeolite reported to date.

Material name	IZA CODE	Isostruc-ture	Materials FD (T/1000 Å^3)	*d4r*(*d3r*) density N/1000 Å^3	Ring opening	Si/Ge ratio	F^-/ OH^-	OSDA	SBU	Reference
ASU-7	**ASV**	–	17.9	1.79	12	0	F^-	OSDA 1	*d4r, lau*	[68]
ASU-9	**AST**	–	15.8		–	0	F^-	OSDA 17	*d4r, ast*	[68]
UTM-3	**AST**	–	15.8		–	0	F^-	OSDA 24	*d4r*	[69]
FOS-5	**BEC**	ITQ-17	13.9	0.91	12×12×12	0	OH^-	OSDA 2	*d4r, mor, mtw*	[70]
SU-15	**SOF**	–	15.5	1.55	12×9	0.59	F^-	OSDA 4	*d4r*	[71]
SU-32	**STW**	HPM-1	15.2	1.52	10×8	0.9	F^-	OSDA 4	*d4r*	[71]
SU-78	–		14.8	0.46	12×12×12	1.5	OH^-	OSDA 44, 45, 46, 47,	*d4r*	[72]
ITQ-7	**ISV**		15.4	0.96	12×12×12	1-∞	F^-	OSDA 57, 58, 64, 65 66, 67, 68, 69, 70, 71	*d4r, mor, mtw*	[58, 59, 73]
ITQ-12	**ITW**		18.1	1.51	8×8	1-∞	F^-	OSDA 74, 77, 78, 79, 80	*d4r,*	[74–76]
ITQ-13	**ITH**	JLG-12	17.8	0.64	10×10×9	2-∞	F^-/OH^-	OSDA 31	*d4r, stf, lau, mel*	[62, 64, 77, 78]
ITQ-15	**UTL**	IM-12	15.2 (IM-12)	0.20	14×12×12	10	OH^-	OSDA 64	*d4r, mor, cas, fer, non, ton*	[79, 80]
ITQ-17	**BEC**	FOS-5	14.6	0.91	12×12×12	1-∞	F^-/ OH^-	OSDA 26, 33, 34, 35, 38, 39, 40, 50, 51, 58, 65, 66, 67, 68, 69, 70, 71	*d4r, mor, mtw*	[73, 79, 81–83]
ITQ-21	–		13.9	1.17	12×12×12	1–100	F^-/ OH^-	OSDA 61, 67, 68, 69, 70	*d4r, mor, mtw*	[79, 84, 85]
ITQ-22	**IWW**		16.1	0.58	12×10×8	3.2	OH^-/F^-	OSDA 31, 54	*d4r, mor, bre, stf, mel*	[64, 86]

(Continued)

Table 4.2 (Continued)

Material name	IZA CODE	Isostructure	Materials FD (T/1000 Å^3)	*d4r*(*d3r*) density N/1000 Å^3	Ring opening	Si/Ge ratio	F^-/ OH^-	OSDA	SBU	Reference
ITQ-24	**IWR**		15.5	0.55	12 × 10 × 10	5-∞	OH^-/F^-	OSDA 31, 41, 59	*d4r, bre, lau, mel*	[87–90]
ITQ-26	**IWS**		14.3	1.0.05	12 × 12	4	F^-	OSDA 88, 91, 98, 108	*d4r, mor, lau, mtw, mel*	[91–93]
ITQ-33	**ITT**		12.4	0.84	18 × 10 × 10	2	OH^-	OSDA 30, 31	*d4r, mel*	[94, 95]
ITQ-34	**ITR**	–	17.4	0.62	10 × 10 × 9	9	F^-	OSDA 106	*d4r, stf, lau*	[96]
ITQ-37	**-ITV**		10.3	1.07	30	1.4	F^-	OSDA 59, 71, 73, 100	*d4r, lau*	[97–99]
ITQ-38	**ITG**		16.1	0.58	12 × 10 × 10	4.5	F^-	OSDA 55, 56	*d4r, mel, stf, mtw, mor, bre*	[100]
ITQ-40	**-IRY**		10.1	1.06	16 × 16 × 15	1	F^-	OSDA 110, 111	*d3r, d4r, ddr*	[101]
ITQ-43	–		11.4	1.14	28 × 12 × 12	2	F^-	OSDA 70, OSDA 73	*d4r,*	[79, 98, 102]
ITQ-44	**IRR**		10.9	1.05	18 × 12 × 12	2	F^-	OSDA 26, 29, 37, 69, 70, 82, 114	*d3r, d4r, mel*	[95, 103–106]
ITQ-47	**BOG**		15.6	0	12 × 10		OH^-	OSDA 112	*-bre, bog, cas*	[67]
ITQ-53	**-IFT**		11.6	0.92	14 × 14 × 14	1	F^-	OSDA 105	*d3r, d4r*	[107]
ITQ-54	**-IFU**		11.1	0.70	20 × 14 × 12	2	F^-	OSDA 60	*d4r, ddr*	[108]
ITQ-56	–		12.4	0.70	22 × 12 × 12	2	F^-	OSDA 49	*d4r*	[109]
ITQ-62	**SOR**	SCM-14	17.5 (SCM-14)	0.71	12 × 8 × 8	3	F^-	OSDA 48	*d4r, mtw*	[110]

IM-10	**UOZ**		16.7	1.67	–	0	F^-	OSDA 31	*d4r, lau*	[111]
IM-12	**UTL**	ITQ-15	15.2	0.20	14 × 12 × 12	4.5	OH^-/F^-	OSDA 63	*d4r, mor, cas, fer, non, ton*	[112]
IM-16	**UOS**		17.9	1.49	10 × 8 × 8	1.4	F^-	OSDA 75	*d4r, mtw*	[113]
IM-17	**UOV**		16.0	0.55	12 × 10 × 8	1.5-	OH^-	OSDA 32	*d4r, mor, bre, lau, stf, mel, mtw*	[114]
IM-18	**UOE**		18.0	–	10 × 10 × 8		F^-	OSDA 21	*d4r, mtw*	[115]
IM-20	**UWY**		16.2	0.81	12 × 10 × 10	0.56–8.52	F^-	OSDA 75, 76	*d4r, stf, lau*	[116]
SSZ-77	**SVV**	–	17.6	0.31	–	4.4	OH^-	OSDA 36	*d4r, rte rut, ast*	[117]
PKU-12	**-CLO**		11.1 (-CLO)	1.39	20 × 20 × 20	0.53	F^-	OSDA 27	*d4r, lta, clo*	[118]
PKU-14	–	ITQ-21	13.9 (ITQ-21)	1.17	12 × 12 × 12	2	F^-	OSDA 45	*d4r*	[119]
PKU-16	**POS**		15.1	0.94	12 × 11 × 11		F^-	OSDA 21	*d4r, lau, mel, stf*	[120]
PKU-20	–		11.8	1.18	12 × 10	0.1	F^-	OSDA 28	*d4r, sti, asv*	[121]
PKU-23-Z	**UOS**	IM-16	17.9 (IM-16)	1.49	10 × 8 × 8	2	F^-	OSDA 21, 102	*Mtw, d4r,*$[4^2 5^4 8^2]$	[122]
PKU-27	–		–	0.73			F^-	OSDA 26	*d4r, bre, sti, pts.*	[123]
NUD-1	–		11.8	0.96	18 × 12 × 10	1–5	F^-	OSDA 29, 30, 85, 92	*d4r, d3r, mel*	[95, 124]
NUD-2	***CTH**[a]	CIT-13	16.5	0.51	14 × 10	5	F^-	OSDA 83, 84	*d4r non, cas*	[125]
NUD-3		ITQ-21, PKU-14	13.6	1.14	12 × 12 × 12	1	F^-	OSDA 97	*d4r*	[126]

(Continued)

Table 4.2 (Continued)

Material name	IZA CODE	Isostructure	Materials FD (T/1000 Å^3)	*d4r*(*d3r*) density N/1000 Å^3	Ring opening	Si/Ge ratio	F^-/ OH^-	OSDA	SBU	Reference
NUD-13		GeZA	13.2	1.01	15 × 12 × 12	1–3.3	F^-	OSDA 101	*d4r*	[127]
Germano-silicate LTA	**LTA**	ITQ-29	12.9 (Si—Al)	1.61	8 × 8 × 8	2	F^-	OSDA 90, 99, 103	*d4r, lta, sod,*	[128–131]
GeZA	–	–	13.2 (NUD-13)	1.01	15 × 12 × 12	1–5	F^-	OSDA 113	*d4r*	[104]
CIT-13	***CTH**[a]	NUD-2	16.4	0.51	14 × 10	2–8	F^-	OSDA 84, 86, 88, 89,	*d4r, non, cas*	[93, 132]
IWW-type germanosilicate	**IWW**	ITQ-22	16.1	0.58	12 × 10 × 8	4.9	OH^-	OSDA 62	*d4r, mor, bre, stf, mel*	[133]
SCM-14	**SOR**	ITQ-62	17.5	1.46	12 × 8 × 8	2.5	F^-	OSDA 23	*d4r, mtw*	[134]
SCM-15	**SOV**		15.2	0.95	12 × 12 × 8	5	F^-	OSDA 23	*d4r, stf, lau, mor, mtw*	[135]
SCM-25			14.7	0.86	12 × 12 × 10	2.4	F^-	OSDA 53	*d4r*	[136]
ECNU-16	**EOS**		17.6	1.49	10 × 8 × 8	1	F^-	93	*d4r, mtw*	[137]
ECNU-20	**IWR**	ITQ-24	15.5	0.55	12 × 10 × 10	3	OH^-	OSDA 22	*d4r, bre, lau, mel*	[138]
SYSU-3	**-SYT**		12.0	1.49	24 × 8 × 8	1	F^-	OSDA 72	*d4r, rpa*	[139]
HPM-1	**STW**	SU-32	15.2 (SU-32)	1.52	10 × 8	0-∞	F^-	OSDA 76, 80	*d4r*	[65, 106]
HPM-7	**POS**	PKU-16	15.1 (PKU-16)	0.94	12 × 11 × 11	3	F^-	OSDA 95, 96	*d4r, lau, mel, stf*	[140]

HPM-8			–	–		5.6–9	F^-	OSDA 95, 96	*d4r,*	[140]
HPM-12	***UOE**[a]	IM-18	18.0 (IM-18)	–	10×10×8	1–2.3	F^-	OSDA 94	*d4r, mtw*	[141]
HPM-14			14.1	0.86	16×9×8	0.42–2.33	F^-	OSDA 87, 100	*d4r*	[126]
HPM-16			14.7	0.87	12×12×12	2.33	F^-	OSDA 81	*d4r,*	[142]
GTM-3	**-ITV**	ITQ-37	10.3 (ITQ-27)	1.07	30	3	F^-	OSDA 42, 43	*d4r*	[143]

Note: The codes of the intergrowth structures (marked with prefix *) have been canceled by Structure Committee of International Zeolite Association. Considering the *Code commonly reported in the literatures, here we still listed them.

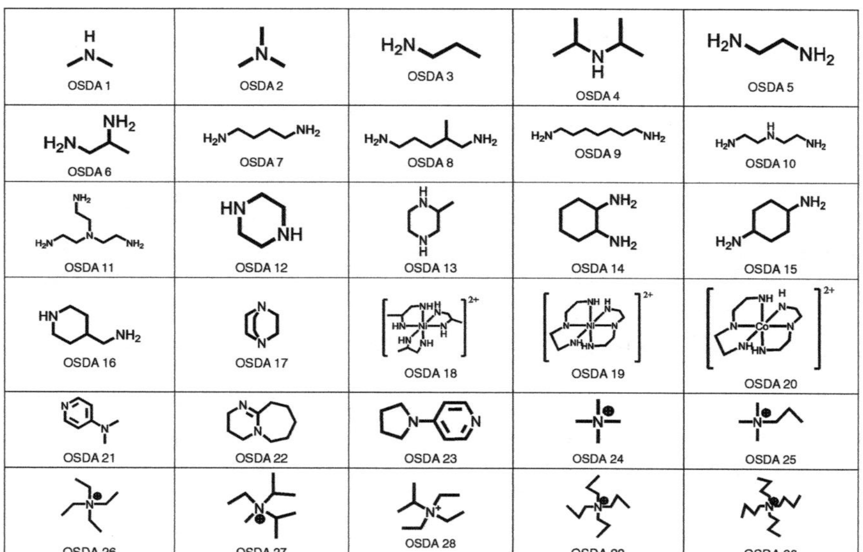

Table 4.2 (Continued)

Material name	IZA CODE	Isostructure	Materials FD (T/1000 Å^3)	*d4r*(*d3r*) density N/1000 Å^3	Ring opening	Si/Ge ratio	F^-/OH^-	OSDA	SBU	Reference

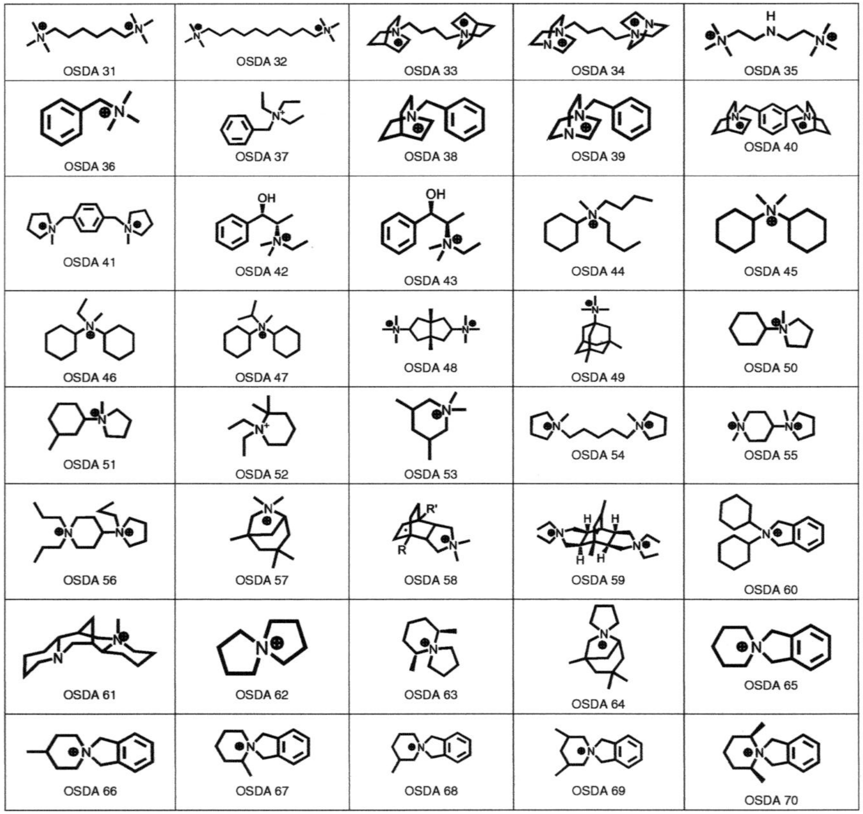

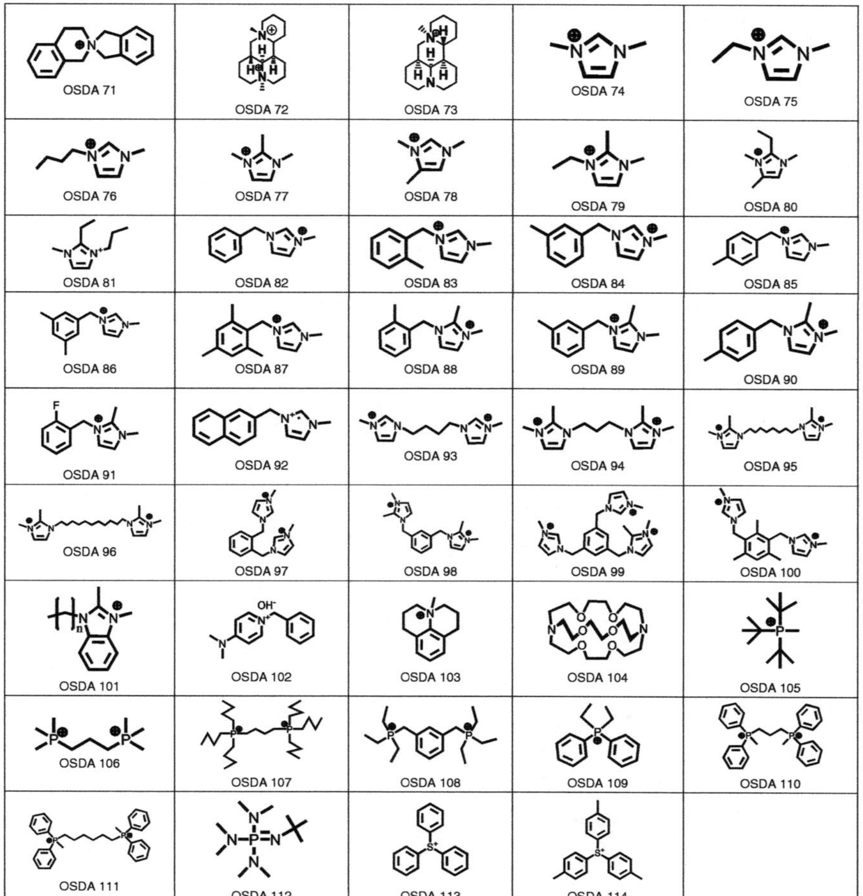
OSDA 71
OSDA 72
OSDA 73
OSDA 74
OSDA 75
OSDA 76
OSDA 77
OSDA 78
OSDA 79
OSDA 80
OSDA 81
OSDA 82
OSDA 83
OSDA 84
OSDA 85
OSDA 86
OSDA 87
OSDA 88
OSDA 89
OSDA 90
OSDA 91
OSDA 92
OSDA 93
OSDA 94
OSDA 95
OSDA 96
OSDA 97
OSDA 98
OSDA 99
OSDA 100
OSDA 101
OSDA 102
OSDA 103
OSDA 104
OSDA 105
OSDA 106
OSDA 107
OSDA 108
OSDA 109
OSDA 110
OSDA 111
OSDA 112
OSDA 113
OSDA 114

Although several OSDAs could guide the formation of the same zeolite, their structure-directing effects are different. Generally speaking, more powerful structure-directing effects render a wide crystallization zone, wide Si/Ge ratio, tolerance of F^- free, etc. For example, ITQ-24 was firstly synthesized as boro/aluminogermanosilicate using multiple-use hexamethonium, i.e. OSDA 31 with or without F^-. However, the synthesis was achieved with a narrow Si/Ge around 5. The bulky OSDA 59 that is used for ITQ-37 synthesis helps to obtain pure silica ITQ-24 [88]. Later, two independent works found that OSDA 41 could lead the synthesis of Ge-free aluminosilicate IWR zeolite [89, 90]. Unfortunately, none of them were successful with both Ge and F^- free to realize a commercial synthesis-friendly condition.

The second series of OSDA is the imidazolium-substituted family (Figure 4.3). Five imidazolium-based OSDAs can guide the synthesis of ITQ-12 germanosilicate (even pure silica), i.e. OSDA 74, OSDA 77, OSDA 78, OSDA 79, and OSDA 80. A noticeable cell parameter expansion is observed with the OSDA size growth [76]. Among them, the ideal fit of host and guest renders direct crystallization, and a wide crystallization zone of **ITW** is observed with OSDA 77 [154]. OSDA 78 first leads the **TON** formation and then changes to **ITW** (high framework density to low framework density) [154]. The second case of zeolite phase transition originated from OSDA 80. The OSDA 80 is the bulkiest OSDA that could synthesize **ITW**.

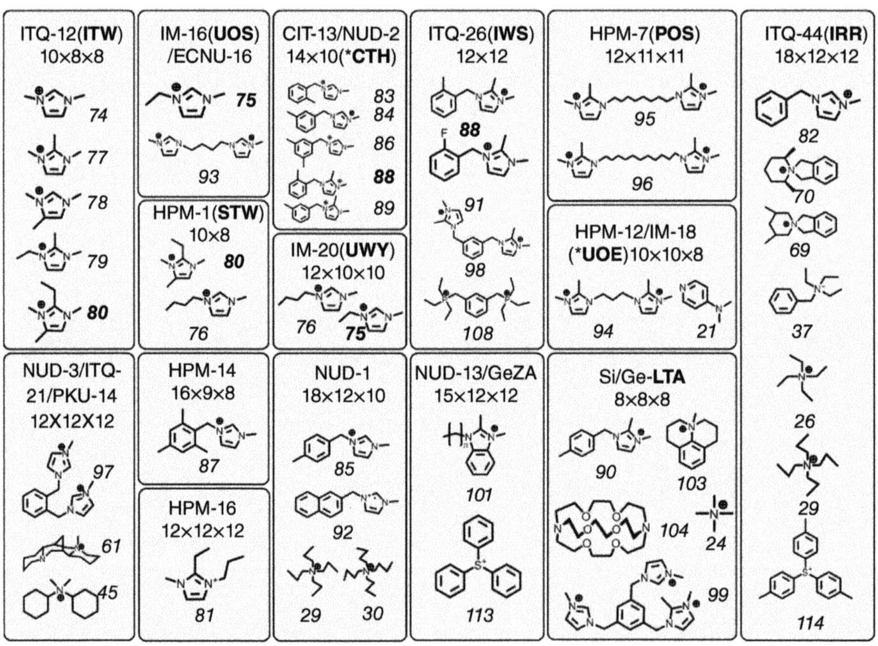

Figure 4.3 Substituted imidazolium OSDA family and corresponding zeolite produced. Prefix OSDA is omitted for clarity. Source: Jiuxing Jiang.

However, it kinetically favors the formation of HPM-1 (**STW**). With time elapsing, the product transforms into a thermodynamically stable **ITW**.

The benzyl-imidazolium series of OSDA shows great diversity due to the combination of the three factors: a) possibility of mono, di, tri, and even more imidazolium substitution; b) substitution of benzene ring; and c) substitution of imidazolium ring. This sub-group of OSDAs enables the discovery of CIT-13(***CTH**), HPM-14, NUD-1, and NUD-3. Noticeably, OSDA 88 could guide ITQ-26 (**IWS**) and CIT-13(***CTH**) with Si/Ge ratios 2 and 4, respectively [93]. Besides that, Si/Ge-**LTA** and ITQ-44 (**IRR**) are also witnessed to crystallize with the OSDA 90 and OSDA 82, respectively. OSDA 85 possesses only one more methyl group than OSDA 82, and the similar OSDAs could guide the synthesis of similar structures of NUD-1 (OSDA 85) and ITQ-44 (OSDA 82).

Before the discovery of OSDA 90, OSDA 99, and OSDA 103 for high silica zeolite **LTA** [129], zeolite Linde Type A could only be prepared with Si/Al ratio of 3 using TMAOH (OSDA 24). The increase of Si/Al ratio could significantly enhance the stability of zeolite and selective adsorption performance. Corma et al. designed *lta* cage space filler, i.e. a dimer of aromatic planer OSDA 103 via π–π interaction supramolecular assemblage. With the help of OSDA 103, Ge, and F^-, the high silica, even pure silica **LTA** zeolite, could be prepared. However, the reproducibility of the most interesting pure silica ITQ-29 is poor even though the seed is used. Furthermore, the complicated OSDA 103 synthesis is also a drawback. The crown ether Kryptofix 222, i.e. OSDA 104, was first applied in the synthesis of AlPO form **LTA** [130]. Later, it was also approved for germanosilicate **LTA** (membrane) synthesis. OSDA 99 was also helpful for germanosilicate **LTA** synthesis but in pure silica form [128]. In 2015, M. E. Davis' group [131] provided an elegant solution for pure silica **LTA** with OSDA 90. With the aid of TMA or seed (one of two), the pure silica LTA could be obtained within the acceptable synthesis time, c. 7 days. Nevertheless, the F^- route is the last gateway that prevents the commercial usage of high silica LTA.

Wu's group reported an interesting case of the imidazolium-based structure-directing effect. Using a dimer of OSDA 75 (OSDA for IM-16), i.e. OSDA 93, and a similar synthetic condition of IM-16, a similar structure of ECNU-16 (**EOS**) can be obtained (later approved as a pure phase of one phase component of intergrown IM-18 [155]). The experiment effect provides us with a new strategy of zeolite synthesis, i.e. halve or double the structure of existing OSDA molecules and try them in a similar synthetic condition with mother materials [137].

The third sub-group of imidazolium base OSDA is dication connected by a linear linker with varying lengths, i.e. OSDA 93, OSDA 95, and OSDA 96. With an increasing Si/Ge ratio, OSDA 95 and OSDA 96 give **MTW**, Beta, HPM-8, and HPM-7 (**POS**) zeolite products [141]. The density of the *d4r* unit in the zeolite structures follows the same trend. Thus, the phenomenon is in line with the strong structure-directing effect of Ge on the *d4r*. The structure-directing effect of OSDA relies on the distance matching between imidazolium positive charge and F^-@*d4r* negative charge [141, 156].

4.5 Structure Diversity of Germanosilicates/Silicogermanates

4.5.1 Relationship Between Composition and Structure

As summarized in Figure 4.4, on the silicogermanate side, clusters such as Ge7, Ge9, Ge10 (Ge11), and Ge12 are the main SBU of the structures. Due to the presence of terminal groups, the (silico)germanate is often categorized into open frameworks. The small amount of Si atoms acts as an isomorphous substitution. With the synthetic Si/Ge ratio increasing, quasi-zeolites, i.e. mixed structures of silicogermanate and zeolite, are produced. For example, with a synthetic Si/Ge ratio of 0.12, PKU-15 with Ge12b and typical zeolite SBU *3r* could be prepared. Another case is PST-18, with the structure composed of columns of Ge9a and *nat*(4^4). If the synthetic Si/Ge ratio is above 1, the products are generally zeolite or zeotype. However, the trend is not absolute, and there are various pure Ge zeolite reported, i.e. ASU-7 (**ASV**, displayed in zeolite column), FOS-5 (**BEC**, not shown in Figure 4.4 due to the isostructure of ITQ-17), and ASU-9/UTM-3 (**AST**, not shown due to the non-porous feature).

4.5.2 Pore Opening

Pore opening is another critical issue. The structures with pore mouth delimited 8-ring, 10-ring, 12-ring, and beyond 12-ring are classified as small, medium, large, and extra-large pore zeolites. Before 1988, all zeolites and synthetic open frameworks are equal or below 12-ring. M. E. Davis et al. reported a milestone work of 18-ring opening VPI-5, which led to an extra-large pore rush. Open framework (phosphate, phosphite, germanates) and zeolites with pore opening larger than 12-ring emerged, and for details, refer to the review articles that attribute to the extra-large pore zeolite [157]. Especially, researches of (silico)germanates open-framework and germanosilicate zeolites are highly active. Germanosilicates are the most fruitful for large/extra-large pore zeolite synthesis. Up to now, 15 extra-large pore zeolites were first reported as germanosilicates; more than 16 germanosilicates have been reported as large pore zeolites. Besides that, more than 10 types of medium/small pore germanosilicates zeolite are reported. The record of the upper limit of the pore mouth was continuously refreshed until the presence of SU-M (30-ring) [39], ITQ-37 (30-ring) [97], and ITQ-43 ($28 \times 12 \times 12$ ring) [102].

The structures of SU-M and ITQ-37 are characterized by the G(gyroid) minimal surface as the MCM-48. There are two sets of independent channels of SU-M. Like MCM-48, if the Ge cluster and the amines selectively block one channel of SU-M, the resulting structure SU-MB is a pair of enantiomers. As to the ITQ-37, the chirality naturally occurred. However, the overall product is racemic. SU-MB and ITQ-37 crystallized in the chiral space groups $I4_132$ and $P4_132$, respectively (or their counterparts). Although the structure of SU-M is not sufficiently robust to calcine, the mild condition ion exchange could completely remove organic amines due to the wide pore opening. Noticeably, as-synthesized SU-M shows a $214\,m^2/g$ BET surface area (type I of N_2 adsorption isotherm). ITQ-37 could remove

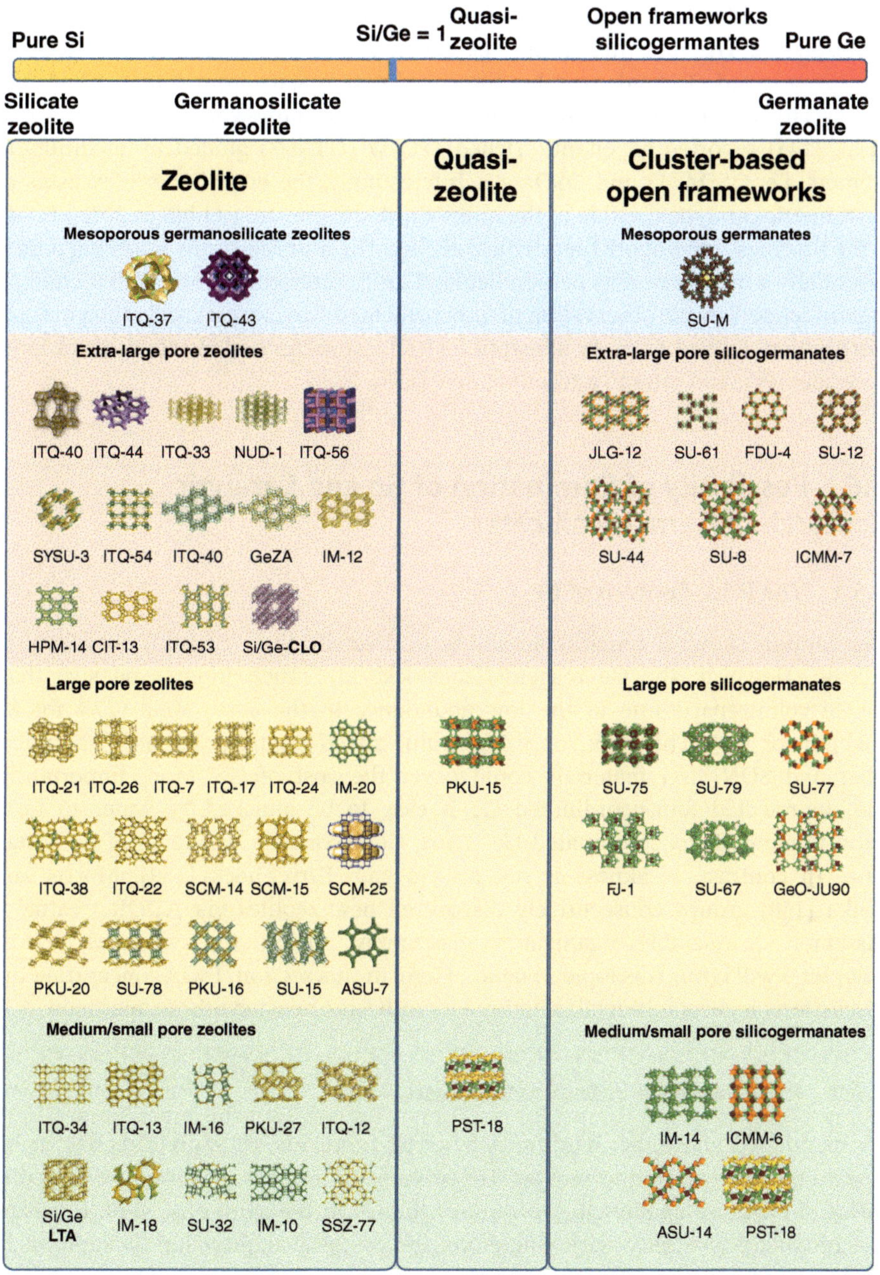

Figure 4.4 Structure diversity of germanosilicates/silicogermanates. Source: Jiuxing Jiang.

the OSDA and retain its structure upon the calcination up to 540 °C (in a dry atmosphere). The moisture harms the calcined germanosilicate zeolites because the unprotected Ge tends to relax the coordination state from 4 to 6. Then, the Ge goes to extra-framework, resulting in a collapsed framework. Recently, Ramon et al. [143] reported an enantio-riched ITQ-37 (GTM-3) guided by enantiopure isomers, i.e. OSDA 42 and OSDA 43. Interestingly, the enantiomeric excesses of asymmetric catalytic epoxide of the bulky *trans*-stilbene are as high as 30%. ITQ-43 is the first meso-micropore hierarchical zeolite. The four-leave clover 28-ring pore is essentially a mesopore plus perpendicular 12 × 12 intersect channels; that channel system renders us the observation of unprecedented two clear peaks in the pore size distribution derived from Ar adsorption [102]. More discussions about extra-large pore zeolites could refer to a recent review [158].

4.6 Possibility of Elimination of Ge and Catalytic Research of Germanosilicates

4.6.1 The Price Concern of Ge

Germanium shows a remarkable ability to diversify the structural chemistry of silicate. However, the Ge reagents/feedstocks are much more expensive than the Si counterparts due to the low abundance in the earth shell (27% for Si, 1.8 ppm for Ge). Therefore, only high-value-added applications such as optical fiber and SON/SNG materials could cover the cost of Ge. More importantly, the unprotected four-coordinated Ge is easy to be attacked by water to form extra-framework six-coordinated Ge. Thus, Ge becomes a weakness of the structure. In contrast, weakness is not always bad. Kirschhock [159], Morris and Cejka [160] groups consecutively discovered new zeolites via ADOR treatment (assembly–disassembly–organization–reassembly) using the weakness [161]. Chapter 5 will cover this topic in detail. Here, in this section, I will concentrate the discussion on germanium elimination by synthesis or post-synthesis methods.

4.6.2 Removal of Ge in Zeolite Synthesis

As mentioned above, germanium is a strong inorganic structure-directing agent toward the *d4r*-containing zeolites. However, as prominent the structure-directing effect of Ge is, as challenging to remove the Ge in the synthesis while retaining the product phase selectivity. Therefore, the essential request for Ge removal in zeolite synthesis is to look for other structure-directing agents, i.e. OSDA, F^-, and seed, toward target zeolite to compensate for the absence of Ge. ITQ-7, ITQ-12, and ITQ-13 were first synthesized as germanium free. The germanium was introduced to accelerate the crystallization process and to understand the structure-directing effect of Ge. So, these are not good cases. In the following attempt, ITQ-17 and ITQ-24 are the real cases of Ge removal. The first syntheses of ITQ-17 and ITQ-24 were performed with OSDA 26, 33, 34, 35, 38, 39, 40, 50, 51, and OSDA 31. However, none of them could realize the Ge-free **BEC** and **IWR** syntheses. The latter reports of OSDA 58 (three-step organic synthesis, OSDA also for ITQ-7) and OSDA 59 (four-step organic synthesis, OSDA also for ITQ-37) with strong structure-directing effects achieved the goal of pure silica. For ITQ-24, two groups from Hong [89] and Xiao [90] simultaneously found simple OSDA 41 (one-step organic synthesis) could also guide the pure silica ITQ-24. Furthermore, Hong's synthesis gets rid of seed, while Xiao's and original Corma's works need seed assistance. Interestingly, molecular mechanic-based theoretical calculation successfully predicts the OSDA 41 in Xiao's work, demonstrating the potential of computer simulation in target zeolite synthesis. However, like the above-mentioned case of **LTA**, the F^- route is the last crucial handicap to impede commercial usage. The last case is zeolite **STW**. **STW** zeolite was first synthesized by Zou's group as SU-32 [71] with Si/Ge = 0.9 using the simple amine OSDA 4. The usage of OSDA 76 by Camblor and coworkers achieved a Si—Ge free substitution in Si/Ge ratio from 0 to ∞ (refer to section 4.3 for a more detailed discussion).

4.6.3 Removal of Ge with Post-synthesis

As stated above, tetrahedrally coordinated Ge is easily hydrolysis in water vapor or moisture. Therefore, most calcinated germanosilicate zeolites are only stable in dry atmospheres. Moreover, Ge is much more expensive than Si and catalytic inert. So, at least three original motivations for post-synthesis are required: a) stabilizing the zeolite structure, b) introducing the active sites such as Al and Cu, and c) recycling Ge.

Valtchev's group [162] first tried to stabilize the germanosilicates by post-synthesis. Considering the poor stability of Ge-rich zeolite in an aqueous solution, they chose to treat the as-synthesized samples ITQ-17 with Al source (polyaluminum hydroxide chloride solution) and diluted HCl (0.01 mol/L). However, the Si/Ge ratio only increased from 3.6 to 6.7. Harsh treatment with 37% HCl (12 mol/L) could completely remove the germanium while maintaining the crystallinity [163]. Then, the researchers realized that the germanosilicate zeolites are not such fragile. The calcined samples had also been used due to the better mass transfer [164–166]. Other solutions are also used in post-treatment, such as ammonium hydroxide [167], HCl in ethanol [164], HCl in DMSO (dimethyl sulfoxide) [168], nitric acid [164, 169, 170], and $Al(NO_3)_3/Ga(NO_3)_3$ [165, 166]. Some dry routes are developed for germanium removal, such as plasma treatment [171], and H_2 reduction of Ge(IV) [172]. Most of the post-syntheses use the relatively stable framework, i.e. **UTL**, **UWY**, **BEC**, **IWR**, **ITH**, and **IWW**, as listed in Table 4.3. In contrast, most germanosilicates are not robust enough to be post-treated. Shamzhy and coworkers [180] claimed that the expensive Ge element could be recovered and reused in next round of synthesis. Wu and coworkers [179] suggested a crystallization–disassembly–dissolution–recrystallization (CDDR) strategy and prepared a high silica low defects analog of the parent zeolite. The Ge could also be recovered by filtration. However, the Ge recovery procedures are too complicated to achieve economic efficiency.

Table 4.3 Post-synthesis of germanosilicates and the catalytic applications.

Parent zeolite	Si/Ge synthesis	Treatment	Calcined/ Yes/ No	Si/Ge after removal	Catalytic sites introducing	Catalytic reaction	Ref.	Year
UTL	4.2	–	Yes	–	Al and B in synthesis	Decane hydroisomerization	[173]	2010
UTL	4.2				Ge, Pt	Decane hydroisomerization	[174]	2011
UTL	–	–	Yes	–	Al, B, Fe, Ga	Acylation Beckmann rearrangement	[175]	2013
UTL, UWY, BEC, IWR	4.7, 1.7, 2.2, 1.8, respectively	–	Yes	–	Ge	Baeyer–Villiger oxidation	[176]	2014
ITH	4–10	–	Yes	–	Al, B, Ga	Tetrahydropyranylation	[177]	2016
UTL **IWW**	4.2, 7.2	–	Yes	–	Ge	Baeyer–Villiger oxidation Etherification of levulinic acid	[170]	2020
ITR	10.9–60.5	–	Yes	–	Al	Catalytic cracking of *n*-decane	[178]	2022
BEC	3.6	Alumination/ 0.01 M HCl	No	6.7	Al	–	[162]	2009
IWW	5.6	0.01–12 M HCl	No	185	Al	–	[163]	2014
BEC	2.9	$TiCl_4$, $SiCl_4$, O_2 plasma	No	3.5–4.1	Ti	Photocatalysis of methanol to CO_2	[171]	2014
BEC	2.68	NH_4OH, 1 M/0.01 M HCl,	No	3.7	–	–	[167]	2014

UTL, UWY BEC, IWR	4.7, 1.7, 3.1, 1.8, respectively	1 m HCl ethanol 65% HNO_3	Yes	233, 100, 208, 15.1, respectively	–	–	[164]	2014
ITH	2.5–5.8	1 M $Al(NO_3)_3$	Yes	13.3–158	Al	Tetrahydropyranylation	[165]	2015
IWR	6.9	$Al(NO_3)_3$ $Ga(NO_3)_3$ solution 0.1–1.0 M	Yes	11–25.3	Al, Ga	Benzoylation of *p*-xylene	[166]	2015
UTL	4.7	1 M HNO_3 aqueous	No	59	Sn	MPV reduction Baeyer–Villiger	[169]	2016
BEC	2.4	$ZrOCl_2$ in 0.01 M HCl ZrOCl2 in DMSO	No	3.1–11.5	Zr	MPV reduction	[168]	2017
UWY UOS	3.4, 3.2, respectively	CDDR	Yes	18, 10.5, respectively	–	–	[179]	2018
ITH UTL IWW	3.1 3.2 5.3, respectively	HCl/ HNO_3 0.1, 1, 4 M	Yes	c. 5 c. 10 c. 26 respectively	Al	Tetrahydropyranylation	[180]	2020

4.6.4 Catalytic Research of Germanosilicates

Given the apparent drawbacks of germanosilicate as heterogeneous catalysts (expensive Ge, poor stability), people still found some exciting points. First, the Ge can catalyze! As early as the 2000s, many research works reported the acidic property of Ge-substituted **MFI** [181]. The conclusion is that the Brönsted acid strength of the Ge-site is roughly equal to pure silica MFI and weak than the B site. However, after the awareness of the direct incorporation of Al and B leads to catastrophic effects on the stability of UTL zeolite [173], Kirschhock's group proved that the pristine germanosilicate zeolite is not entirely inert in a Brönsted acid catalytic reaction, i.e. *n*-decane hydroisomerization/hydrocracking reaction [174]. Second, the Ge is better as the Lewis acid site than as a Bronsted acid site. The famous metal sites, i.e. Ti, and Sn, are well-known Lewis acid sites for the Baeyer–Villiger, Meerwein–Ponndorf–Verley (MPV) reduction, etc. Wu and their coworkers found an unprecedented conversion of a bulky ketone (2-adamantanone) and selectivity to the corresponding lactone over the series of germanosilicates (better than Sn-Beta) [176]. Furthermore, Ge sites are also demonstrated to be active for other reactions, such as ethylene oxide hydration [182] and etherification of levulinic acid [170].

Metal incorporation is the more common route to power germanosilicate. However, the direct introduction in synthesis gel may not always be successful. Furthermore, the incorporation of too many heteroatoms hinder the thermal stability of zeolite. Hence, the post-synthesis heteroatom-Ge exchange is a standard method to energize Ge-rich zeolites. Al, Ga, Sn, Ti, and Zr are reinserted as shown in Table 4.3. Here, I will not expand on this topic. For further discussions, please refer to recent review articles [183]. Finally, recent work on manipulating zeolite layers (naturally occurring or artificially made) dimmed synthesis and post-synthesis definitions, thus providing a new way to generate topological independent zeolite [161, 184].

4.7 Conclusions and Outlook

The tremendous structural diversity of germanosilicate aroused great research interest for two decades. Especially, the introduction of Ge into zeolite significantly enriches the International Structure Database. Moreover, the most newly code-assigned germanosilicate zeolites are large pore or extra-large pore zeolites that the zeolite community has long pursued. I believe that we will eventually find a suitable OSDA to get rid of Ge and F^- to realize the industrial-friendly zeolite synthesis.

References

1 Housecroft, C.E. and Sharpe, A.G. (2018). *Inorganic Chemistry*. Harlow: Pearson Education Limited.

2 Levien, L. and Prewitt, C.T. (1981). *Am. Mineral.* 66: 324–333.
3 Ross, N.L., Shu, J., and Hazen, R.M. (1990). *Am. Mineral.* 75: 739–747.
4 Baerlocher, C. and LB, M.C. Database of Zeolite Structures. In: . http://www.iza-structure.org/databases.
5 Chae, S., Lee, J., Mengle, K.A. et al. (2019). *Appl. Phys. Lett.* 114: 102104.
6 Lin, Z.-E. and Yang, G.-Y. (2010). *Eur. J. Inorg. Chem.* 2010: 2895–2902.
7 Khaled, J., Fujiwara, T., Ohama, M., and Ikushima, A.J. (2000). *J. Appl. Phys.* 87: 2137–2141.
8 Neustruev, V.B. (1994). *J. Phys. Condens. Matter* 6: 6901–6936.
9 Agrawal, G.P. (2021). *Fiber-Optic Communication Systems*. Wiley.
10 Chen, H., Wei, M., He, Y. et al. (2022). *Nat. Commun.* 13: 4438.
11 Österberg, U. and Margulis, W. (1986). *Opt. Lett.* 11: 516–518.
12 Garcia, F.C., Carvalho, I.C.S., Hering, E. et al. (1998). *Appl. Phys. Lett.* 72: 3252–3254.
13 Chen, H., Zhu, C., Lu, H. et al. (1999). *J. Mater. Sci. Lett.* 18: 1905–1906.
14 Xu, Z., Liu, L., Fei, Y. et al. (2000). *Appl. Phys. Lett.* 77: 70–72.
15 Liu, Q., Gan, F., Zhao, X. et al. (2001). *Opt. Lett.* 26: 1347–1349.
16 Dussauze, M., Cremoux, T., Adamietz, F. et al. (2012). *Int. J. Appl. Glass Sci.* 3: 309–320.
17 Okada, A., Ishii, K., Mito, K., and Sasaki, K. (1992). *Appl. Phys. Lett.* 60: 2853–2855.
18 Fleming, S.C. and An, H. (2008). *J. Cerma. Soc. Jpn.* 116: 1007–1023.
19 Anashkina, E.A., Andrianov, A.V., Koptev, M.Y. et al. (2012). *Opt. Express* 20: 27102–27107.
20 Meltz, G., Morey, W.W., and Glenn, W.H. (1989). *Opt. Lett.* 14: 823–825.
21 Chakrabarti, P. (2015). *Optical Fiber Communication*. New Delhi: McGraw Hill Education.
22 Pilz, S., Najafi, H., Ryser, M., and Romano, V. (2017). *Fibers* 5: 24.
23 Cheng, J., Xu, R., and Yang, G. (1991). *Dalton Trans.* 1537–1540.
24 Sun, K., Dadachov, M.S., Conradsson, T., and Zou, X.D. (2000). *Acta Crystallog. Sect. C: Cryst Struct. Commun.* 56: 1092–1094.
25 Beitone, L., Loiseau, T., and Ferey, G. (2002). *Inorg. Chem.* 41: 3962–3966.
26 Villaescusa, L.A., Lightfoot, P., and Morris, R.E. (2002). *Chem. Commun.* 2220–2221.
27 Julius, N.N., Choudhury, A., and Rao, C.N.R. (2003). *J. Solid State Chem.* 170: 124–129.
28 Attfield, M.P., Al-Ebini, Y., Pritchard, R.G. et al. (2007). *Chem. Mater.* 19: 316–322.
29 Li, H.L., Eddaoudi, M., and Yaghi, O.M. (1999). *Angew. Chem. Int. Ed.* 38: 653–655.
30 Plevert, J., Gentz, T.M., Laine, A. et al. (2001). *J. Am. Chem. Soc.* 123: 12706–12707.
31 Plevert, J., Gentz, T.M., Groy, T.L. et al. (2003). *Chem. Mater.* 15: 714–718.
32 Medina, M.E., Iglesias, M., Monge, M.A., and Gutierez-Puebla, E. (2001). *Chem. Commun.* 2548–2549.

33 Medina, M.E., Iglesias, M., Snejko, N. et al. (2004). *Chem. Mater.* 16: 594–599.

34 Medina, M.E., Gutierrez-Puebla, E., Monge, M.A., and Snejko, N. (2004). *Chem. Commun.* 2868–2869.

35 Gandara, F., Medina, M.E., Snejko, N. et al. (2008). *Inorg. Chem.* 47: 6791–6795.

36 Zhou, Y.M., Zhu, H.G., Chen, Z.X. et al. (2001). *Angew. Chem. Int. Ed.* 40: 2166–2168.

37 Liu, G.Z., Zhang, H.X., Lin, Z.E. et al. (2007). *Chem. Asian J.* 2: 1230–1239.

38 Zhang, H.X., Zhang, J., Zheng, S.T., and Yang, G.Y. (2003). *Inorg. Chem.* 42: 6595–6597.

39 Zou, X.D., Conradsson, T., Klingstedt, M. et al. (2005). *Nature* 437: 716–719.

40 Christensen, K.E., Shi, L., Conradsson, T. et al. (2006). *J. Am. Chem. Soc.* 128: 14238–14239.

41 Tang, L.Q., Dadachov, M.S., and Zou, X.D. (2005). *Chem. Mater.* 17: 2530–2536.

42 Tang, L.Q. and Zou, X.D. (2007). *Micropor. Mesopor. Mater.* 101: 24–29.

43 Shi, L., Bonneau, C., Li, Y.F. et al. (2008). *Cryst. Growth Des.* 8: 3695–3699.

44 Christensen, K.E., Bonneau, C., Gustafsson, M. et al. (2008). *J. Am. Chem. Soc.* 130: 3758.

45 Yue, H., Peskov, M.V., Sun, J., and Zou, X. (2012). *Inorg. Chem.* 51: 12260–12265.

46 Huang, S., Inge, A.K., Yang, S. et al. (2012). *Dalton Trans.* 41: 12358–12364.

47 Fang, L., Liu, L., Yun, Y. et al. (2014). *Cryst. Growth Des.* 14: 5072–5078.

48 Huang, S., Su, J., Christensen, K. et al. (2014). *Inorg Chem. Front.* 1: 278–283.

49 Pan, Q.H., Li, J.Y., Ren, X.Y. et al. (2008). *Chem. Mater.* 20: 370–372.

50 Pan, Q.H., Li, J.Y., Christensen, K.E. et al. (2008). *Angew. Chem. Int. Ed.* 47: 7868–7871.

51 Ren, X.Y., Li, Y., Pan, Q.H. et al. (2009). *J. Am. Chem. Soc.* 131: 14128.

52 Lorgouilloux, Y., Paillaud, J.L., Caullet, P., and Bats, N. (2008). *Solid State Sci.* 10: 12–19.

53 Xu, Y., Liu, L., Chevrier, D.M. et al. (2013). *Inorg. Chem.* 52: 10238–10244.

54 Ren, T., Xiao, P., Xing, X.-Y., and Christensen, K.E. (2010). *Cryst. Res. Technol.* 45: 1035–1040.

55 Liang, J., Xia, W., Sun, J. et al. (2016). *Chem. Sci.* 7: 3025–3030.

56 Jo, D., Choi, W., Shin, J., and Hong, S.B. (2018). *Dalton Trans.* 47: 17122–17126.

57 Villaescusa, L.A., Barrett, P.A., and Camblor, M.A. (1999). *Angew. Chem. Int. Ed.* 38: 1997–2000.

58 Corma, A., Díaz-Cabañas, M.J., and Fornés, V. (2000). *Angew. Chem. Int. Ed.* 39: 2346–2349.

59 Blasco, T., Corma, A., Díaz-Cabañas, M.J. et al. (2002). *J. Phys. Chem. B* 106: 2634–2642.

60 Corma, A., Díaz-Cabañas, M.J., García, H., and Palomares, E. (2001). *Chem. Commun.* 2148–2149.

61 Sastre, G., Vidal-Moya, J.A., Blasco, T. et al. (2002). *Angew. Chem. Int. Ed.* 41: 4722–4726.

62 Vidal-Moya, J.A., Blasco, T., Rey, F. et al. (2003). *Chem. Mater.* 15: 3961–3963.

63 Blasco, T., Corma, A., Díaz-Cabañas, M.J. et al. (2004). *J. Am. Chem. Soc.* 126: 13414–13423.

64 Sastre, G., Pulido, A., Castañeda, R., and Corma, A. (2004). *J. Phys. Chem. B* 108: 8830–8835.

65 Rigo, R.T., Balestra, S.R.G., Hamad, S. et al. (2018). *J. Mater. Chem. A* 6: 15110–15122.

66 Zones, S.I. (2011). *Micropor. Mesopor. Mater.* 144: 1–8.

67 Simancas, R., Dari, D., Velamazán, N. et al. (2010). *Science* 330: 1219–1222.

68 Li, H. and Yaghi, O.M. (1998). *J. Am. Chem. Soc.* 120: 10569–10570.

69 Xu, Y., Fan, W., Chino, N. et al. (2004). *Chem. Lett.* 33: 74–75.

70 Conradsson, T., Dadachov, M.S., and Zou, X.D. (2000). *Micropor. Mesopor. Mater.* 41: 183–191.

71 Tang, L.Q., Shi, L., Bonneau, C. et al. (2008). *Nat. Mater.* 7: 381–385.

72 Yu, Z.-B., Han, Y., Zhao, L. et al. (2012). *Chem. Mater.* 24: 3701–3706.

73 Jiang, J., Xu, Y., Cheng, P. et al. (2011). *Chem. Mater.* 23: 4709–4715.

74 Barrett, P.A., Boix, T., Puche, M. et al. (2003). *Chem. Commun.* 2114–2115.

75 Yang, X., Camblor, M.A., Lee, Y. et al. (2004). *J. Am. Chem. Soc.* 126: 10403–10409.

76 Rojas, A. and Camblor, M.A. (2014). *Dalton Trans.* 43: 10760–10766.

77 Corma, A., Puche, M., Rey, F. et al. (2003). *Angew. Chem. Int. Ed.* 42: 1156–1159.

78 Castañeda, R., Corma, A., Fornés, V. et al. (2006). *J. Catal.* 238: 79–87.

79 Smeets, S., Koch, L., Mascello, N. et al. (2015). *CrstEngComm* 17: 4865–4870.

80 Corma, A., Díaz-Cabañas, M.J., Rey, F. et al. (2004). *Chem. Commun.* 1356–1357.

81 Corma, A., Navarro, M.T., Rey, F., and Valencia, S. (2001). *Chem. Commun.* 1486–1487.

82 Corma, A., Navarro, M.T., Rey, F. et al. (2001). *Angew. Chem. Int. Ed.* 40: 2277–2280.

83 Cantin, A., Corma, A., Diaz-Cabanas, M.J. et al. (2006). *Angew. Chem. Int. Ed.* 45: 8013–8015.

84 Corma, A., Díaz-Cabañas, M.J., Martínez-Triguero, J. et al. (2002). *Nature* 418: 514–517.

85 Corma, A., Diaz-Cabanas, M.J., and Rey, F. (2003). *Chem. Commun.* 1050–1051.

86 Corma, A., Rey, F., Valencia, S. et al. (2003). *Nat. Mater.* 2: 493–497.

87 Castañeda, R., Corma, A., Fornés, V. et al. (2003). *J. Am. Chem. Soc.* 125: 7820–7821.

88 Cantín, A., Corma, A., Diaz-Cabanas, M.J. et al. (2006). *J. Am. Chem. Soc.* 128: 4216–4217.

89 Kemp, K.C., Seo, S., Ahn, S.H., and Hong, S.B. (2019). *Chem. Lett.* 48: 1445–1447.

90 Hong, X., Chen, W., Zhang, G. et al. (2019). *J. Am. Chem. Soc.* 141: 18318–18324.

91 Dorset, D.L., Strohmaier, K.G., Kliewer, C.E. et al. (2008). *Chem. Mater.* 20: 5325–5331.

92 Liu, D.-R., Gao, Z.-H., Zi, W.-W. et al. (2019). *Micropor. Mesopor. Mater.* 276: 232–238.

93 Boal, B.W., Deem, M.W., Xie, D. et al. (2016). *Chem. Mater.* 28: 2158–2164.

94 Corma, A., Díaz-Cabañas, M., Jordá, J. et al. (2006). *Nature* 443: 842–845.

95 Bai, R., Sun, Q., Wang, N. et al. (2016). *Chem. Mater.* 28: 6455–6458.

96 Corma, A., Diaz-Cabanas, M.J., Jorda, J.L. et al. (2008). *J. Am. Chem. Soc.* 130: 16482–16483.

97 Sun, J., Bonneau, C., Cantín, Á. et al. (2009). *Nature* 458: 1154–1157.

98 Zhang, C., Li, X., Li, X. et al. (2019). *Chem. Commun.* 55: 2753–2756.

99 Chen, F.-J., Gao, Z.-H., Liang, L.-L. et al. (2016). *CrstEngComm* 18: 2735–2741.

100 Moliner, M., Willhammar, T., Wan, W. et al. (2012). *J. Am. Chem. Soc.* 134: 6473–6478.

101 Corma, A., Díaz-Cabañas, M.J., Jiang, J. et al. (2010). *Proc. Natl. Acad. Sci.* 107: 13997–14002.

102 Jiang, J., Jorda, J.L., Yu, J. et al. (2011). *Science* 333: 1131–1134.

103 Jiang, J.X., Jorda, J.L., Diaz-Cabanas, M.J. et al. (2010). *Angew. Chem. Int. Ed.* 49: 4986–4988.

104 Jo, C., Lee, S., Cho, S.J., and Ryoo, R. (2015). *Angew. Chem. Int. Ed.* 54: 12805–12808.

105 Qian, K., Wang, Y., Liang, Z., and Li, J. (2015). *RSC Adv.* 5: 63209–63214.

106 Wang, M., Zhang, L., Guo, K. et al. (2019). *Chem. A Eur. J.* 14: 621–626.

107 Yun, Y., Hernández, M., Wan, W. et al. (2015). *Chem. Commun.* 51: 7602–7605.

108 Jiang, J., Yun, Y., Zou, X. et al. (2015). *Chem. Sci.* 6: 480–485.

109 Kapaca, E., Jiang, J., Cho, J. et al. (2021). *J. Am. Chem. Soc.* 143: 8713–8719.

110 Bieseki, L., Simancas, R., Jorda, J.L. et al. (2018). *Chem. Commun.* 54: 2122–2125.

111 Mathieu, Y., Paillaud, J.-L., Caullet, P., and Bats, N. (2004). *Micropor. Mesopor. Mater.* 75: 13–22.

112 Paillaud, J.-L., Harbuzaru, B., Patarin, J., and Bats, N. (2004). *Science* 304: 990–992.

113 Lorgouilloux, Y., Dodin, M., Paillaud, J.-L. et al. (2009). *J. Solid State Chem.* 182: 622–629.

114 Lorgouilloux, Y., Dodin, M., Mugnaioli, E. et al. (2014). *RSC Adv.* 4: 19440–19449.

115 Cichocka, M.O., Lorgouilloux, Y., Smeets, S. et al. (2018). *Cryst. Growth Des.* 18: 2441–2451.

116 Dodin, M., Paillaud, J.-L., Lorgouilloux, Y. et al. (2010). *J. Am. Chem. Soc.* 132: 10221–10223.

117 Earl, D.J., Burton, A.W., Rea, T. et al. (2008). *J. Phys. Chem. C* 112: 9099–9105.

118 Su, J., Wang, Y., Lin, J. et al. (2013). *Dalton Trans.* 42: 1360–1363.

119 Liang, J., Su, J., Wang, Y. et al. (2014). *Chem. A Eur. J.* 20: 16097–16101.

120 Hua, W., Chen, H., Yu, Z.-B. et al. (2014). *Angew. Chem. Int. Ed.* 53: 5868–5871.

121 Chen, Y., Su, J., Huang, S. et al. (2016). *Micropor. Mesopor. Mater.* 224: 384–391.

122 Wang, X., Firdous, A., Xu, L. et al. (2019). *Cryst. Growth Des.* 19: 2272–2278.

123 Chen, Y., Ma, N., Huang, S. et al. (2019). *Chem. A Eur. J.* 25: 3219–3223.

124 Chen, F.-J., Xu, Y., and Du, H.-B. (2014). *Angew. Chem. Int. Ed.* 53: 9592–9596.
125 Gao, Z.-H., Chen, F.-J., Xu, L. et al. (2016). *Chem. A Eur. J.* 22: 14367–14372.
126 Chen, F.-J., Gao, Z.R., Li, J. et al. (2021). *Chem. Commun.* 57: 191–194.
127 Cai, X., Zhao, Y., Zi, W. et al. (2022). *Chem. A Eur. J.* 28: e202200934.
128 Schmidt, J.E., Zones, S.I., Xie, D., and Davis, M.E. (2014). *Micropor. Mesopor. Mater.* 200: 132–139.
129 Corma, A., Rey, F., Rius, J. et al. (2004). *Nature* 431: 287–290.
130 Schreyeck, L., D'agosto, F., Stumbe, J. et al. (1997). *Chem. Commun.* 1241–1242.
131 Boal, B.W., Schmidt, J.E., Deimund, M.A. et al. (2015). *Chem. Mater.* 27: 7774–7779.
132 Kang, J.H., Xie, D., Zones, S.I. et al. (2016). *Chem. Mater.* 28: 6250–6259.
133 Yuan, R., Claes, N., Verheyen, E. et al. (2016). *New J. Chem.* 40: 4319–4324.
134 Luo, Y., Smeets, S., Peng, F. et al. (2017). *Chem. A Eur. J.* 23: 16829–16834.
135 Luo, Y., Smeets, S., Wang, Z. et al. (2019). *Chem. A Eur. J.* 25: 2184–2188.
136 Luo, Y., Fu, W., Wang, B. et al. (2022). *Inorg. Chem.* 61: 4371–4377.
137 Xu, L., Zhang, L., Li, J. et al. (2018). *Chem. A Eur. J.* 24: 9247–9253.
138 Zhang, L., Chen, Y., Jiang, J.-G. et al. (2017). *Dalton Trans.* 46: 15641–15645.
139 Zhang, C., Kapaca, E., Li, J. et al. (2018). *Angew. Chem. Int. Ed.* 57: 6486–6490.
140 Lu, P., Zhang, Y., Mayoral, A. et al. (2020). *Dalton Trans.* 49: 7037–7043.
141 Lu, P., Mayoral, A., Gómez-Hortigüela, L. et al. (2019). *Chem. Mater.* 31: 5484–5493.
142 Gao, Z.R., Balestra, S.R.G., Li, J., and Camblor, M.A. (2021). *Angew. Chem. Int. Ed.* 60: 20249–20252.
143 de la Serna, R., Nieto, D., Sainz, R. et al. (2022). *J. Am. Chem. Soc.* 144: 8249–8256.
144 Liu, X., Ravon, U., Bosselet, F. et al. (2012). *Chem. Mater.* 24: 3016–3022.
145 Kasian, N., Tuel, A., Verheyen, E. et al. (2014). *Chem. Mater.* 26: 5556–5565.
146 Liu, X., Chu, Y., Wang, Q. et al. (2017). *Solid State Nucl. Magn. Reson.* 87: 1–9.
147 Liu, X., Ravon, U., and Tuel, A. (2011). *Angew. Chem. Int. Ed.* 50: 5900–5903.
148 Liu, X., Ravon, U., and Tuel, A. (2011). *Chem. Mater.* 23: 5052–5057.
149 Zhao, Y.K., Liang, W.C., Wang, Y.H. et al. (2022). *Catalysts* 12.
150 Kerr, G.T. (1966). *Inorg. Chem.* 5: 1537–1539.
151 Wilson, S.T., Lok, B.M., Messina, C.A. et al. (1982). *J. Am. Chem. Soc.* 104: 1146–1147.
152 Saame, J., Rodima, T., Tshepelevitsh, S. et al. (2016). *J. Org. Chem.* 81: 7349–7361.
153 Shvets, O.V., Kasian, N., Zukal, A. et al. (2010). *Chem. Mater.* 22: 3482–3495.
154 Rojas, A., Martínez-Morales, E., Zicovich-Wilson, C.M., and Camblor, M.A. (2012). *J. Am. Chem. Soc.* 134: 2255–2263.
155 Hao X. (2022) Personal communication.
156 Lu, P., Gómez-Hortigüela, L., Gao, Z., and Camblor, M.A. (2019). *Dalton Trans.* 48: 17752–17762.
157 Jiang, J.X., Yu, J.H., and Corma, A. (2010). *Angew. Chem. Int. Ed.* 49: 3120–3145.
158 Wen, J., Zhang, J., and Jiang, J. (2021). *Chem. J. Chin. U.* 42: 101–116.

159 Verheyen, E., Joos, L., Van Havenbergh, K. et al. (2012). *Nat. Mater.* 11: 1059–1064.

160 Roth, W.J., Nachtigall, P., Morris, R.E. et al. (2013). *Nat. Chem.* 5: 628–633.

161 Eliasova, P., Opanasenko, M., Wheatley, P.S. et al. (2015). *Chem. Soc. Rev.* 44: 7177–7206.

162 Gao, F.F., Jaber, M., Bozhilov, K. et al. (2009). *J. Am. Chem. Soc.* 131: 16580–16586.

163 Burel, L., Kasian, N., and Tuel, A. (2014). *Angew. Chem. Int. Ed.* 53: 1360–1363.

164 Xu, H., Jiang, J.-g., Yang, B. et al. (2014). *Angew. Chem. Int. Ed.* 53: 1355–1359.

165 Shamzhy, M.V., Opanasenko, M.V., Ramos, F.S.O. et al. (2015). *Cat. Sci. Technol.* 5: 2973–2984.

166 Shamzhy, M. and Ramos, F.S.O. (2015). *Catal. Today* 243: 76–84.

167 Liu, X., Kasian, N., and Tuel, A. (2014). *Micropor. Mesopor. Mater.* 190: 171–180.

168 Kots, P.A., Sushkevich, V.L., Tyablikov, O.A., and Ivanova, I.I. (2017). *Micropor. Mesopor. Mater.* 243: 186–192.

169 Liu, X., Xu, H., Zhang, L. et al. (2016). *ACS Catal.* 6: 8420–8431.

170 Podolean, I., Zhang, J., Shamzhy, M. et al. (2020). *Cat. Sci. Technol.* 10: 8254–8264.

171 El-Roz, M., Lakiss, L., Vicente, A. et al. (2014). *Chem. Sci.* 5: 68–80.

172 Ma, Y., Xu, H., Liu, X. et al. (2019). *Chem. Commun.* 55: 1883–1886.

173 Kasian, N., Koranyi, T.I., Vanbutsele, G. et al. (2010). *Top. Catal.* 53: 1374–1380.

174 Kasian, N., Vanbutsele, G., Houthoofd, K. et al. (2011). *Cat. Sci. Technol.* 1: 246–254.

175 Shamzhy, M.V., Shvets, O.V., Opanasenko, M.V. et al. (2013). *ChemCatChem* 5: 1891–1898.

176 Xu, H., Jiang, J., Yang, B. et al. (2014). *Cat. Com.* 55: 83–86.

177 Shamzhy, M.V., Ochoa-Hernandez, C., Kasneryk, V.I. et al. (2016). *Catal. Today* 277: 37–47.

178 Cumplido, M.P. and Diaz-Cabañas, M.J. (2022). *Micropor. Mesopor. Mater.* 341: 112009.

179 Peng, M., Jiang, J., Liu, X. et al. (2018). *Chem. A Eur. J.* 24: 13297–13305.

180 Zhang, J., Yue, Q., Mazur, M. et al. (2020). *ACS Sustain. Chem. Eng.* 8: 8235–8246.

181 Tuan, V.A., Falconer, J.L., and Noble, R.D. (2000). *Micropor. Mesopor. Mater.* 41: 269–280.

182 Liu, X., Mao, W.T., Jiang, J.G. et al. (2019). *Chem. A Eur. J.* 25: 4520–4529.

183 Jiao, M., Jiang, J., Xu, H., and Wu, P. (2021). *Chem. J. Chin. U.* 42: 29–39.

184 Opanasenko, M., Shamzhy, M., Wang, Y. et al. (2020). *Angew. Chem. Int. Ed.* 59: 19380–19389.

5

Structural Modifications on Germanosilicates

Ondřej Veselý, Maksym Opanasenko, and Jiří Čejka

Charles University, Charles University Center of Advanced Materials, Faculty of Science, Department of Physical and Macromolecular Chemistry, Hlavova 8, 12843 Prague, Czech Republic

5.1 Introduction

The success of zeolites in diverse fields of industry originates from their favorable physicochemical properties (e.g. thermal stability, acidity, porosity) and great compositional and structural variability. To date, the International Zeolite Association recognized over 250 zeolite structures and others await the approval [1]. Yet, the research continues in pursuit of new zeolites with unique structures and topologies. Despite the unparalleled structural diversity of zeolites, certain structural features are rare. Notable examples of such features include odd-member channels [2] or extra-large pores [3]. The extra-large pore zeolites attract a particular interest for both industrial and laboratory applications. Processing of heavy-oil fractions, separation of large molecules or production of bulky chemicals would benefit from employing such materials [4, 5]. However, the preparation of extra-large pore zeolites commonly involves complex and costly structure-directing agents (SDAs) [6] or rare elements [7] such as gallium or germanium, while these zeolites themselves lack the stability and acidity.

Owing to several unique features, germanium aids the preparation of new zeolite frameworks. First and foremost, germanium increases the flexibility of T-O-T angles. For illustration, Si-O-Si bond angles commonly range from 137° to 180°, however, replacement of silicon with germanium increases the range from 137° to 180° up to 120°–180° [8, 9]. The increased angle flexibility enables the germanosilicate structural units to adopt new geometries, which rarely appear in purely siliceous zeolites. Such units include for example double-three rings (D3R) [10, 11] or double-four rings (D4R) [12]. The presence of such small-ring units is crucial in the formation of zeolites with low framework density, which explains, why germanium so commonly appears in the structure of large and extra-large pore zeolites (Figure 5.1) [11]. Notable examples include **UTL** (IM-12, ITQ-15) [13], CIT-13 (SAZ-1, NUD-2) [14], **BEC** (zeolite Beta polymorph C) [15], **IWW** (ITQ-22) [16], **ITH** (ITQ-13) [17], **UOV** (IM-17) [18], and **IWR** (ITQ-24) [19] germanosilicates.

Micro-Mesoporous Metallosilicates: Synthesis, Characterization, and Catalytic Applications,
First Edition. Edited by Peng Wu and Hao Xu.

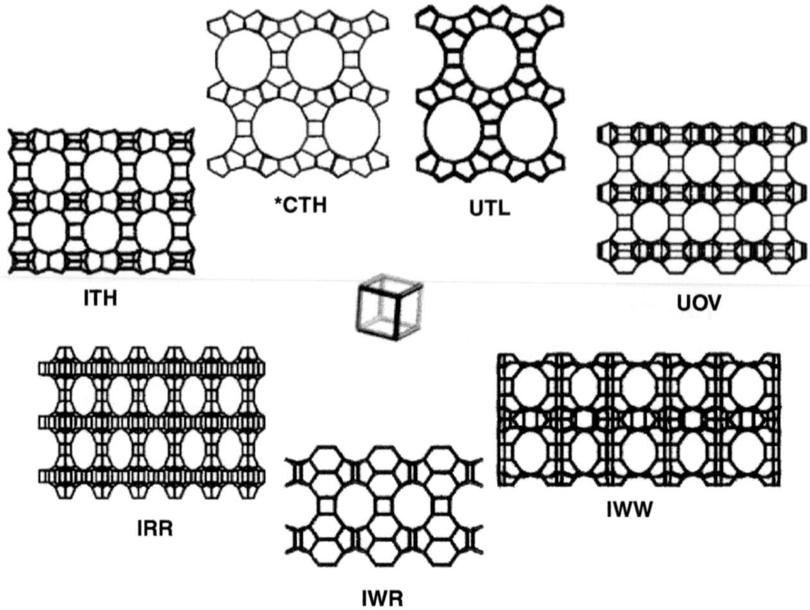

Figure 5.1 Selection of germanosilicate zeolites containing D4R units.

The germanium particularly bolsters the D4R formation well. In fact, octameric germanate ions, precursors to the D4R units, form in the synthesis gel in less than a minute upon adding the germanium [20, 21]. Presence of octameric ions in solution naturally promotes the formation of frameworks containing D4R units. This also illustrates why germanium preferentially occupies certain framework positions, particularly in D4Rs [22–24]. The preferential siting of germanium stems from its propensity to form cubical D4R units and its ability to accommodate to framework positions with smaller angles than silicon. Both of these factors play a critical role in the uneven distribution of germanium throughout the germanosilicate framework.

These effects manifest a unique structure directing ability of germanium. The addition of germanium can significantly alter the phase selectivity of a synthesis, particularly when one of the competing phases contains D4R or D3R units [25]. This structure directing ability can be demonstrated on the synthesis of the **BEC** (polymorph C of beta) zeolite [15]. The **BEC** can crystallize in the presence of special SDA (i.e. 4,4-dimethyl-4-azoniatricyclo[5.2.2]undec-8-ene) as a pure silicate [26] or even under the same synthesis conditions as regular Beta (combination of polymorphs A and B) if the synthesis mixture contains sufficient amount of germanium. However, in the absence of germanium the D4R units are not stable and hence the synthesis favors the disordered beta.

This chapter examines the role of germanium in zeolite chemistry with a particular focus on stability of germanosilicate zeolites, structural modifications of germanosilicates and their transformations to new materials. Controlled modifications of germanosilicates can produce layered or inorganic/organic-pillared hybrid materials or result in transformation to new zeolite frameworks with highly tunable pore-sizes, textural properties, or elemental compositions.

5.2 Germanosilicates to Layered Precursors

As explained in the previous section, germanium preferentially occupies specific T-positions within the framework, most commonly in the D4R units [24, 25]. This also determines that other T-positions are preferentially occupied by silicon atoms. The preference divides the zeolite into domains with different elemental composition. In other words, the germanosilicate framework contains structurally distinct silicon-rich and germanium-rich units.

Structures of numerous germanosilicate zeolites consist of silicon-rich layers, rods or cages connected by germanium-rich D4R units. For instance, the **UTL** zeolite consists of silicon-rich non-porous layers. Neighboring parallel layers are connected by germanium-rich D4R units, creating a two-dimensional system of intersecting pore channels in between the layers.

The specific silicon and germanium siting in germanosilicate zeolites significantly impacts their hydrolytic behavior. The Ge—O bonds are generally weaker compared to Si—O bonds, which negatively impacts the hydrothermal stability of germanosilicates [27–29]. Relative lability of the Ge—O bonds compared to the Si—O enables selective extraction of germanium from the framework, while keeping the silicon-rich domains mostly unaffected (discussed in detail in Section 5.3.5.1). Since germanium predominantly occupies the D4R units, its leaching removes or reconstruct the D4R units and thereby the connections between the silicon-rich segments. This process thus effectively transforms the three-dimensional germanosilicate zeolite into a two-dimensional (or lower) layered silicate material [30].

The concept of 3D–2D transformation is relatively common for producing layered materials. For example, preparation of MXenes from metal-aluminum carbides includes also selective etching of aluminum from between the layers [31]. However, in case of zeolites, only germanium exhibits preference for framework positions convenient for such transformations, making germanosilicates uniquely suitable for preparation of low-dimensional materials.

5.2.1 UTL to IPC-1P

The transformation of **UTL** to layered IPC-1P precursor (Figure 5.2) amply demonstrates the general principles of the 3D–2D transformation of germanosilicates. The relative weakness of Ge—O bonds enables the selective extraction of germanium-rich D4R units and thus interruption of the connections between the IPC-1P layers. The degermanation of the **UTL** commonly occurs in a presence of acids or water [28].

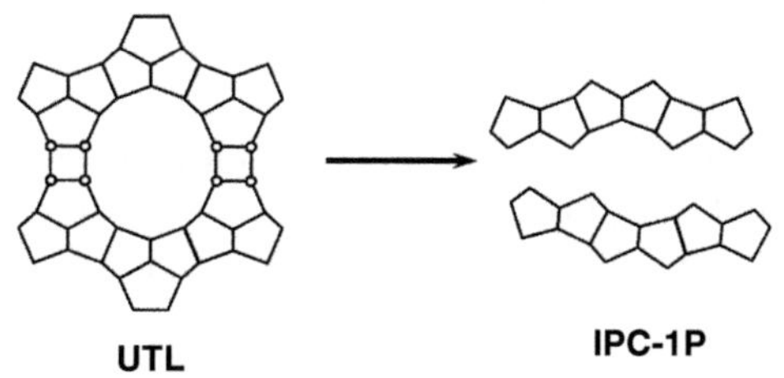

Figure 5.2 Transformation of **UTL** into layered zeolite.

Hydrolytic stability and hydrolysis. Water is a common denominator for structural decay of germanosilicate zeolites, including the **UTL**. Numerous zeolites with high germanium content (e.g. Si/Ge < 4; varies for individual zeolites) commonly undergo some form of structural changes or deterioration when treated with liquid water or even when exposed to air moisture [9, 14]. The decreased hydrolytic stability arises from a weaker character of Ge—O bonds and in turn their higher susceptibility to hydrolysis [32].

Both experimental studies and theoretical calculations show that when exposed to low water loadings (1 molecule per unit cell) the **UTL** germanosilicate can remain moderately stable and only after prolonged time shows structural changes [14]. The proposed mechanisms for bond breaking in D4R units at low water loadings require high activation energies to proceed and result in unstable products. This explains why certain germanosilicates, such as **UTL**, can be stored in desiccator for months without significant deterioration but undergo fast transformation when exposed to large volumes of water. Increased content of water in the micropores (18 molecules per unit cell) allows for an alternative hydrolysis mechanism. This mechanism involves proton transfer through a chain of water molecules and results in a more stable product. Moreover, the hydrolysis becomes more favorable with increasing germanium content. Above critical Ge threshold (e.g. Si/Ge lower than 8.5 for **UTL**), insertion of water molecule between Ge–O–Ge atoms becomes exergonic and nearly unrestricted [32]. Thus, the hydrolysis becomes more favorable and, consequently, more defects (silanols as the remnants of Si–O–Ge bonds) form. The newly formed defects increase the hydrophilicity of the zeolite, drawing more water into the channels and promoting further hydrolysis. This cycle continues until all the labile bonds are hydrolyzed [33–35].

In summary, water mediates the hydrolysis of labile Ge—O bonds in the **UTL**. However, water content determines not only the kinetics of the hydrolysis but also the energetics and feasibility of the process.

Intercalation. While water content bellow 18 water molecules per unit cell suffices for investigation of the mutual interactions and demonstrates the undergoing structural changes, it downplays the effects of the bulk liquid on the newly formed IPC-1P layers [32]. However, the hydrolysis in large volume of liquid water enables more water to intercalate from the bulk liquid into the interlayer space. Consequently, the volume in between the layers increases, leading to a rise in the interlayer distance. The increased interlayer distance facilitates the layer separation and, in fact, enables the formation of the IPC-1P [33, 35].

However, the intercalation affects more than just the separation of IPC-1P layers. The interlayer space may accommodate species other than water, both inorganic and organic in nature. The intercalated moieties control the relative arrangement of the layers or interlayer distance. Intercalation of, most commonly, organic agents may further change the ordering of the material or layers arrangement (see Section 5.3.5.1).

Moreover, intercalation enables increasing the interlayer distance of the IPC-1P, and thus, tuning the structure and pore size of the final material [36–38]. Zeolites commonly contain pores up to 1 nm in diameter, which makes them limited in

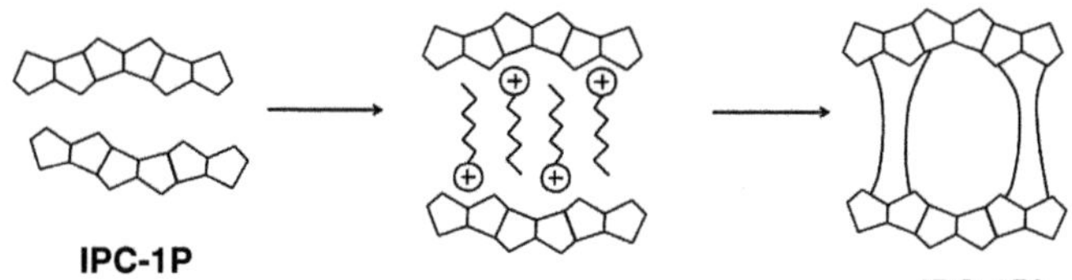

Figure 5.3 Intercalation of organic moieties to the layered precursor IPC-1P and subsequent pillaring.

regards to mass transfer and size of admissible molecules. On the other hand, the pore size of intercalated materials can exceed that of common 3D zeolites depending on the choice of the intercalant [36]. Highly common approach for increasing the interlayer distance utilizes amphiphilic surfactants as intercalants. The surfactant molecules interact with the silanol groups on the surface of the IPC-1P layers via their positively charged functional groups. However, the strong interaction between the silanols and the surfactant occurs only after deprotonation of the silanols under basic environment. The deprotonation induces negative charge on the surface of the layers. As a result, the positively charged groups of the surfactant molecules attach better to the surface. Under the basic pH and suitable concentration of the surfactant, the non-polar chains of surfactant molecules adopt vertical orientation with respect to the layers (Figure 5.3). This results in a formation of a surfactant bilayer in between the layers, which increases their respective spacing [39]. However, the basic environment causes partial dissolution of the layers, which may eventually result in complete amorphization of the zeolite or formation of mesoporous material such as M41S in the case of cetyltrimethylammonium commonly used as intercalant [40, 41]. Therefore, the intercalation conditions require delicate tuning to minimize the dissolution while successfully enlarging the interlayer spacing [39, 42].

The interlayer spacing of the intercalated IPC-1P depends mainly on the thickness of the surfactant bilayer. Ergo, the characteristics of the surfactant, more specifically the length of its hydrocarbon chain and the concentration, directly govern the interlayer distance. The distance shows near-linear dependency on the chain length with increase by 0.15–0.30 nm with prolongation of the hydrocarbon chain by one carbon. This enables a fine tuning of the interlayer distance and preparation of materials with controllable pore sizes up to 3.5 nm [39].

Pillaring. The intercalated "swollen" IPC-1P retains the enhanced interlayer distance due to the presence of the surfactant molecules. However, surfactant bilayer changes its structure based on external conditions such as pH or temperature. Moreover, the surfactant molecules de-intercalate from the interlayer space when treated with suitable solvent or decompose at high temperatures. With the surfactant removed the interlayer distance drops, accompanied by a partial loss of ordering of the material, resulting in a formation of disordered material such as IPC-1 [39]. In contrast, the layer orientation and distance can be preserved by pillaring. Pillaring introduces permanent rigid props that firstly intercalate in a form of low-molecular monomeric precursor into the interlayer space, then covalently bond to the layer surface and fix the layers in their relative distance after polymerization (Figure 5.3) induced by hydrolysis of the precursor and subsequent heating. On top of that, formation of the pillars may also introduce additional functionality by insertion of heteroelements (Al, Ti, ...) or organic functional groups [43–45]. The chemical nature of the pillars significantly impacts the pillar properties such as thermal or chemical stability.

Amorphous pillars. The most common pillaring method for layered zeolites consists of connecting the adjacent layers by amorphous silica. The silica pillars form when low-molecular silica source (e.g. tetraethoxysilane) intercalates in between the expanded IPC-1P layers and subsequently condenses after hydrolysis in a presence of water. After the condensation, the silica pillars provide a rigid support that preserves the interlayer distance imposed by the surfactant molecules and gives a constant mesoporosity to the material after removal of organic moiety (Figure 5.3) [39, 46].

Nevertheless, the amorphous silica provides support for the layers but lacks any source of catalytic activity. Commonly, the IPC-1P layers consist of silica and thus also lack any sources of catalytic activity. The IPC-1P contains acid sites only if the parent **UTL** zeolite already contained the active sites in a form of framework heteroelements (Al, Ga, Ti, ...) located in the layers. Nevertheless, pillaring of the layered material by amorphous silica "dilutes" the concentration of the active sites in the material as a whole. Introduction of heteroelements during the pillaring (generally in the form of metal oxides) introduces catalytic sites into the pillars. These additional sites may enhance the existing catalytic activity of the heteroelement-containing IPC-1P or provide a different type of centers to work in tandem with the existing active sites located in crystalline layers. As such, the mixed oxide pillars may contain acidic oxides such as alumina or titania or basic magnesium oxide [43–46].

Linkers. Another method of pillaring involves connecting the adjacent IPC-1P layers with more complex linkers for tuning the structure or functionality of pillared materials. The linker molecules are based on substituted silanes or siloxanes (e.g. diethoxy-dimethylsilane, diethoxy-tetramethyldisiloxane), and sometimes Si moieties are separated by organic fragments suitable for insertion of additional functional groups (e.g. 1,2-bis-(triethoxysilyl)ethane, 1,4-bis-(triethoxysilyl)benzene) [47, 48]. The chains of linker molecules end with alkoxysilane groups (or similar moieties) which enable covalent bonding of the linker to the IPC-1P layers via condensation of the surface silanols with the alkoxysilane groups [45, 47, 48].

The siloxane linkers may contain not only linear siloxane chains but also cycles or polyhedra, such as $(SiO)_4$ square in octamethylcyclotetrasiloxane or cubic $(SiO)_8$ unit in octamethylsilsesquioxane, respectively. The siloxane linkers play a particular role in preparation of "interlayer expanded zeolites" (IEZ). The IEZ form when layered zeolitic precursors (e.g. IPC-1P [48], MCM-22P [45], HUS-2 [49]) connect via silane or siloxane linkers into a 3D framework. However, the IEZs are not zeolites by definition, since the $(-O)_2Si(OH)_2$ interlayer bridges contain uncondensed hydroxyl groups that can be considered as structural defects. The presence of uncondensed groups results in a higher structural flexibility of the IEZs, which manifests as a lower degree of ordering compared to zeolites. On the other hand, the pore sizes of some IEZs exceed that of conventional zeolites, making them attractive for catalytic applications [45, 47].

The organic linkers provide even higher flexibility and larger structural variability than the siloxanes. The combination of the organic linkers with the zeolite layers creates a unique hybrid organic–inorganic system characteristic by high stability and mechanical durability of a zeolite and flexibility and functional variability of the organic linkers [47, 48]. The flexibility of the organic linkers depends considerably on their structure. Unsaturated or aromatic linkers (e.g. 1,2-bis(triethoxysilyl)ethene, 1,4-bis-(triethoxysilyl)benzene) provide increased rigidity, while fully saturated linkers (e.g. 1,2-bis(triethoxysilyl)ethane) show higher flexibility. High flexibility of the hydrocarbon chain reduces the degree of ordering of the material and commonly results as broad range of the interlayer distances and pore sizes of the material. More precisely interlayer distance can be maintained by providing additional rigid silica pillars. Also, rigid linkers show higher degree of ordering. Moreover, linkers with aromatic and highly hydrophobic functional groups (e.g. [pentafluorophenyl] silyl) show the ability to self-organize, providing additional means to stabilize the interlayer distance and ordering [47].

5.3 ADOR Strategy for Developing New Zeolite Structures

Relatively new concept for direct and targeted preparation of novel zeolites with high level of control over the synthesis products arises from assembly of pre-formed building blocks in an orderly manner [38, 50]. Such building blocks include two-dimensional zeolitic layers. The layers present an alternative route to preparing new zeolites simply by controlling the relative orientation and arrangement of the layers and the linkers that connect them. This protocol may be reminiscent of pillaring of the zeolite layers obtained by hydrolysis (see Section 5.2), but in fact the output of the new approach is the fully crystalline exclusively microporous material not the hybrid micro-mesoporous solid with only partially ordered domains. Replacing the linkers in between the layers leads to predictable frameworks with similar structural features and pore architecture, but different micropore-sizes. This includes zeolites with tunable pore sizes [38, 51] or with uncommon structural features (e.g. odd-member pore windows). Moreover, zeolites prepared through such means do not obey the same rules and limitations as the hydrothermal synthesis, which enables the preparation of structures otherwise considered “unfeasible” [37].

The synthesis approach is called ADOR (Assembly-Disassembly-Organization-Reassembly) transformation. The ADOR takes advantage of layered material prepared from a 3D zeolite (as described in Section 5.2) and allows to subsequently design new structures from the preformed layers. Since the layers originate from 3D structures with regular location of leaving units. The regular pattern of silanol groups formed after disassembly enables full re-condensation of layered precursor into a three-dimensional framework [38]. The preparation of new zeolite via the

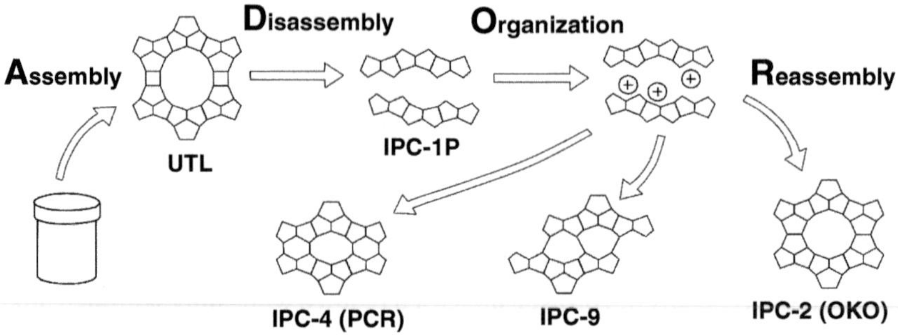

Figure 5.4 The ADOR transformation of the zeolite **UTL**.

ADOR includes four steps: (i) preparation of initial zeolite – Assembly; (ii) its selective degradation to two-dimensional building blocks (layers) – Disassembly; (iii) ordering of the blocks in suitable orientations – Organization; and (iv) their condensation to a three-dimensional structure – Reassembly (Figure 5.4).

5.3.1 Assembly

Assembly, the first step of the ADOR, refers to the hydrothermal synthesis of the parent germanosilicate. The germanosilicate synthesis as the conventional and well-established procedure seemingly plays a minor role in the whole process, however; careful choice and method of preparation of the parent material significantly impacts the subsequent transformations. The transformations necessitate the minimum of four germanium atoms in each D4R unit to enable complete hydrolysis of the unit and thus the removal of the connections between the layers. However, multiple arrangements of germanium within the D4R permit the existence of Si–O–Si bridges, which hinder the layer separation [50]. The number of such linkages decreases with increasing germanium content. Thereby, changing the germanium content in the parent zeolite controls the degree of the hydrolysis, its rate or the general stability of the zeolite in moist environment. Moreover, the stability depends not only on the content of germanium but also on its distribution throughout the framework. Zeolites prepared in hydroxide media generally contain more germanium atoms located in the D4R positions. On the other hand, germanium appears more randomly distributed in zeolites crystallized from fluoride-containing synthesis gel. The fluoride anion resides inside the D4R and stabilizes it. As a result, germanosilicates prepared in fluoride media are more resistant to hydrolysis and attempts at structural transformations than those prepared in hydroxide media, despite having the same crystal structure and germanium content [52].

Other synthesis conditions, such as the choice of SDA, play a crucial role in introducing other elements into the framework. Trivalent framework elements (e.g. B, Al) introduce negative charge into the framework and form ion pairs with the SDA. The stability of the ion pairs increases with the softness of the SDA cations, promoting the incorporation. The upper limit for the incorporation of heteroelements thus strongly depends on the properties of the SDA [53].

5.3.2 Disassembly

The Disassembly refers to a kinetically driven process of hydrolysis of the Ge-rich D4R units within the germanosilicate framework. Calcined germanosilicates undergo progressive structural degradation when exposed to liquid water or air moisture [14]. The water molecules can selectively cleave the labile Ge—O bonds which causes structural changes down the road (see Section 5.2.1) [32]. The hydrolysis causes a rapid increase in the Si/Ge molar ratio of the zeolite (e.g. for **UTL** the Si/Ge increases from 4 to around 30 within 5 min) due to selective leaching of Ge from the framework. The increasing number of germanium vacancies eventually results in complete removal of the Ge-rich D4R units and separation of the zeolite layers. The transformation of the zeolite to the layered form manifests as partial loss of ordering and decreased interlayer distance [50].

5.3.3 Organization

The layers formed immediately after the hydrolysis show a poor ordering and condense to disordered IPC-1 when calcined directly. To achieve full condensation into ordered 3D structure, the layers need to organize first, so that can easily form a crystalline material. The organization can either proceed spontaneously under sufficient conditions (pH, temperature, time) or with the assistance of organizing agent [50].

The spontaneous self-organization comprises a complex interplay of silicon dissolution, intercalation, and precipitation. Typically, it proceeds more rapidly with increasing acidity of the environment, temperature, and concentration of silica species in the solution. Highly acidic/basic environment favors breaking/formation of Si—O bonds and enables migration of Si species from the layers into the solution [38, 54]. Higher temperature increases the mobility of the Si species in the solution. The Si in solution then intercalates in between the layers where it participates into formation of linkages between silanol groups on the layers. The process is characteristic by decreasing number of defects, visible by methods such as ^{29}Si NMR. The intercalation of Si and subsequent restoration of interlayer linkages also causes an increase in the interlayer spacing, detectable by diffraction techniques [33–35].

Both interlayer distance (see Section 5.2.1) and layer ordering change upon intercalation of organic agents. The intercalated agents can interact with the surface silanol groups, break the hydrogen bonding between the layers or simply fill the interlayer space. This alters the respective orientation of the layers and enables more efficient re-condensation into a new 3D framework [36, 37].

5.3.4 Reassembly

The last step of the ADOR includes transformation of the layered precursor into a fully connected 3D zeolite framework via condensation of the surface silanol groups. The condensation occurs through heating of the layered material leading to formation of covalent interlayer bonds. The formation of the 3D framework via topotactic condensation of the layered material, which is irreversible under given conditions, differentiates the ADOR from the hydrothermal synthesis, where zeolite framework forms through a series of reversible steps. As a result, the zeolite structure prepared by ADOR does not obey the same energy-density criteria as the hydrothermal synthesis, which enables preparation of hydrothermally "unfeasible" zeolites [50, 55].

The preceding sections assumed an ideal state where the layers fully condense into the new material. However, a variety of factors may impact the degree of re-condensation and result in disorder or defects. The degree of the condensation depends on the distance between the silanol groups, presence of defects within the layers and the layer ordering. The unique arrangement of the silanol groups illustrates the difference between condensation of other layered zeolite precursors (e.g. PREFER [56, 57]) and the ADOR. The Disassembly produces a unique silanol quadruplet defect in place of the hydrolyzed D4R units. The relatively well-defined quadruplets make the layered material exceptionally suitable for subsequent re-condensation into a new 3D structure. Lastly, the respective layer arrangement plays a crucial role in the quality of the final material. The layer ordering, influenced by method of preparation (see Section 5.3.3) or by intercalation (see Section 5.2.1), often delimitates the difference between obtaining disordered assembly of layers and fully connected zeolite framework as the final material [28, 37, 50].

5.3.5 Liquid-phase ADOR

The ADOR transformation predominantly takes place in the liquid phase. Compared to the vapor-assisted (see Section 5.3.6) or solid-state (Sections 5.3.7 and 5.3.8) systems, the advantages of the liquid-phase setup are of simplicity and low demands on the equipment. The experimental procedure revolves immersing the parent germanosilicate in water or acidic solution for specific time period to obtain the layered precursor. The outcome of the transformation depends on the composition of the treating solution, temperature, liquid to solid ratio and treatment time, however, the basic concept remains the same. The outcome of the transformation results from the interplay of three main underlying processes: (i) hydrolysis of the D4R units; (ii) de/re-intercalation of the debris from/to the interlayer space; and (iii) structural rearrangement [50].

The rate of the hydrolysis of the germanium-rich D4R units depends on the concentration of water in the system. High water content makes the hydrolysis more favorable (discussed in detail in Sections 5.2.1 and 5.3.2) and thus it proceeds more rapidly in high volumes of water. Similarly, higher content of germanium in the D4R units makes the hydrolysis more favorable [32]. However, aside from the water and germanium content the hydrolysis seemingly proceeds regardless of the temperature or acidity of the system [35, 54].

The de/re-intercalation includes migration of the debris produced by the hydrolyzed D4R units from the interlayer space to the bulk solution or, oppositely, migration of leached framework species or external source of building blocks into the interlayer space. The rate of the migration impacts the concentration of building blocks in the interlayer space and thereby governs the structure and composition of the interlayer connecting units. Naturally, low concentration of building blocks favors layers directly connected by oxygen bridges, while higher concentration promotes formation of single-4-ring (S4R) units or restoration of the parent structure [35, 50, 58].

The formation of connections between the adjacent layers also depends on the rate of rearrangement. The rearrangement consists of a set of reactions that relate to Si—O bond breaking and formation. These reactions participate in leaching and re-insertion of Si (or other elements) from/into the framework and, consequently, in forming and healing defects and formation of structural units (such as S4R). The rate of the rearrangement increases with increasing mobility of the Si species. Typically, the ADOR is performed under acidic conditions where the mobility of Si grows with increasing acidity of the environment [50, 59].

The relative rate of these processes controls the outcome of the ADOR transformation. The liquid-phase ADOR stands out due to the facility of controlling these processes by experimental conditions. Higher concentration of the acid solution and higher temperature promote the rearrangement and facilitates the Si migration. On the other hand, higher water to solid ratio accelerates the hydrolysis, Si migration and enables layer separation by intercalation. Lastly, duration of the treatment determines the degree of hydrolysis or rearrangement and germanium content of the final product [35, 50, 54].

5.3.5.1 The UTL Case

The **UTL** germanosilicate is by far the most widely studied zeolite in the ADOR transformation. The **UTL** structure consists of silicon-rich layers connected via germanium-rich D4R units. The D4Rs are arranged in a periodically rectangular pattern, creating a 2D system of interconnected 14- and 12R channels in between the layers [13].

The sheer number of new structures produced by the ADOR transformation of **UTL** amply exemplifies the variability of the ADOR and also presents a great example for illustrating the processes involved in the ADOR [35, 50].

Continuously tunable porosity. The structure of zeolites produced by the ADOR transformation of **UTL** varies significantly depending on the experimental conditions, especially the acidity. The acidity impacts the rate of rearrangement (see Section 5.3.5) and thereby determines the structure and pore size of the final material. Mildly acidic to neutral solutions hydrolyze the **UTL** to the layered IPC-1P material. The IPC-1P with appropriate layer ordering (Section 5.3.3) can subsequently undergo full condensation into the IPC-4 (**PCR**) zeolite. The IPC-4 contains similar 2D interconnected pore system like the **UTL**, however, the layers in IPC-4 are directly connected via oxygen bridges instead of D4Rs. Replacing the D4R units with the oxygen bridges contracts the pore size from 14-12R (9.5×7.1 and 8.5×5.5 Å) to 10-8R (5.8×3.8 and 4.5×3.6 Å) pores for the **UTL** and IPC-4, respectively (Table 5.1) [38, 54].

Another structure, the IPC-2 (**OKO**) zeolite forms upon expanding the IPC-1P layers and connecting them by S4R units. The S4R connections provide the IPC-2 with 2D 12-10R (6.6×6.2 and 5.4×5.3 Å) pore system [38, 59]. The preparation of the

Table 5.1 Structural parameters of zeolites derived from the **UTL** germanosilicate via ADOR transformation.

Zeolite structure	Connecting units	Pore size (Å)								Refs.
		14R	12R	12R	10R	10R	9R	8R	7R	
UTL	D4R	9.5×7.1	8.5×5.5							[52]
IPC-7	D4R, S4R	9.5×7.1	8.5×5.5	6.6×6.2	5.4×5.3					[26]
OKO (IPC-2)	S4R			6.6×6.2	5.4×5.3					[21]
IPC-6	S4R, oxygen			6.6×6.2	5.4×5.3	5.8×3.8		4.5×3.6		[26]
PCR (IPC-4)	oxygen					5.8×3.8		4.5×3.6		[33]
IPC-10	S4R			8.6×5.9			5.5×4.9			[32]
IPC-9	Oxygen					6.3×4.9			4.6×3.5	[32]

IPC-2 can be achieved through several different approaches. First, the precursor to IPC-2, the IPC-2P, forms upon intercalation of the layered IPC-1P with a source of Si or other building units in 1 M acid solution (e.g. HCl). The acid promotes rearrangement of the building blocks, resulting in formation of –O–Si–O– bridges connecting the layers. The isolated bridges condense into S4R units upon calcination, resulting in fully condensed IPC-2 structure. The introduction of the external source of building blocks provides higher degree of control over the structure and the composition of the final material [60]. Second, the IPC-2P forms spontaneously under moderately acidic conditions (7 M HCl) even without any external Si source. Instead, the Si leached from the hydrolyzed D4Rs or from the layers participates in the rearrangement [59]. Lastly, the IPC-2 forms directly from the **UTL** upon hydrolysis under highly acidic conditions (12 M HCl). The high acidity enables an alternative transformation mechanism known as "inverse-sigma transformation." The inverse-sigma transformation results in direct formation of the final 3D structure while omitting the formation of the layered precursor [27]. Thus, the ADOR transformation of **UTL** under mildly acidic conditions produces the IPC-4, while moderately or highly acidic conditions result in the IPC-2.

However, aside from these three pure structures the ADOR can produce a whole spectrum of frameworks with continuously tuneable porosity. For instance, near-neutral conditions favor the formation of IPC-4 as the end product. However, gradually increasing the acidity enhances the rearrangement and promotes formation of interlayer –O–Si–O– linkages in a fraction of the material. With the increasing acidity the fraction of –O–Si–O– connections increases, and eventually results in the IPC-2P which contains exclusively the Si-bridges. Interestingly, carefully balancing the rate of hydrolysis and rearrangement by the acidity (i.e. 1.5 M HCl) produces a mixed structure. The structure, known as IPC-6, contains half interlayer connections resembling IPC-4 and half IPC-2 [54]. However, the IPC-6 is neither a physical mixture of IPC-2 and IPC-4 nor a random combination of both. Instead, the IPC-2 and IPC-4 interlayer connections alternate in an ordered fashion. The highly ordered nature results from the kinetics and mechanism of its formation. The IPC-6P forms from the IPC-1P by preferentially forming –O–Si–O– bridges in between pairs of adjacent layers. The formation of these IPC-6P bilayers is more favorable than random connections because the system made by bilayers possesses more degrees of freedom and higher flexibility. Ergo, the motif of two adjacent IPC-4 and IPC-2 layers repeats throughout the structure of IPC-6 [33].

Similar to IPC-6, another intermediate structure named IPC-7 forms upon treating the **UTL** germanosilicate by 5 M HCl. The acidic environment significantly promotes the rearrangement, resulting in a zeolite IPC-7 with half IPC-2 interlayer connections (S4Rs) and half **UTL** connections (D4Rs) [54].

Kinetics. The ADOR transformation of **UTL** involves the conversion of numerous intermediate structures. Therefore, the structure of the product depends also on the duration of the treatment and the rates of the underlying processes. First, high water content (e.g. 100 mL of solution per 1 g of the zeolite) favors rapid hydrolysis of the **UTL** to the IPC-1P. The hydrolysis involves fast degermanation and decrease in the interlayer distance (see Section 5.2.1). Subsequently, the hydrolyzed IPC-1P material undergoes a short induction period after which the interlayer distance starts increasing as the result of the progressing rearrangement. The structure gradually rearranges to IPC-6P (Section 5.3.5.1) and eventually to IPC-2P. The consequent transformation of **UTL** to IPC-1P and subsequently to IPC-2P evidences the importance of timing in the outcome of the transformation. Moreover, it shows that the IPC-1P forms as a kinetic product of the rapid hydrolysis, whereas the IPC-2P is the thermodynamic product [35].

Low water content in the system (8 $mL\,g^{-1}$) reduces the rate of hydrolysis but also the low volume of liquid in the pores inhibits the transport of leached species from the interlayer space and thereby augments the rearrangement. Ergo, the rearrangement sets in before the hydrolysis is complete. As a result, the IPC-1P ceases to be kinetically favorable and the **UTL** never fully hydrolyses to the layered form. Instead, the hydrolysis locally proceeds, "unzipping" the hydrolyzed layers and simultaneously forming new –O–Si–O– connections between them. The transformation under reduced water volume in the system eventually results in formation of IPC-2P while omitting the formation of the IPC-1P altogether [33, 34].

The slow disassembly of the **UTL** produces the IPC-1P only under decreased amount of water inhibiting the hydrolysis but large volume of inert liquid, which facilitates the transport of the leached species. Small alcohols (e.g. methanol, ethanol, iso-propanol) make a perfect example due to their high miscibility with water, high polarity, relatively high boiling points (64.7, 78.4 and 97 °C) and inert behavior toward the Ge—O bonds. The only downside is the limited solubility of the leached species, however; replacing water with acid (e.g. HCl) solves the obstacle.

The slow low-water high-volume hydrolysis proceeds via progressive degermanation of the **UTL** and gradual reduction of D4R units. The reduction steadily decreases the interlayer distance and continuously transforms the **UTL** to precursors of IPC-7, IPC-2, IPC-6 and eventually to IPC-1P [58].

Unfeasible zeolites. The IPC-7, IPC-2, IPC-6, and IPC-4 result from changing the respective layer position of the IPC-1P in the crystallographic a-direction, that is, the interlayer distance. Alternatively, the layers can shift in the *b*- or *c*-directions to provide different arrangement of the adjacent layers and form zeolites with new interlayer connectivities.

Notably, the IPC-9 zeolite forms upon intercalation of the IPC-1P with large amounts of choline chloride. The intercalated choline molecules take position in between the silanol quadruplets and by interaction with the adjacent layer, induce a shift by half unit cell in the *c*-direction. Condensation of the shifted precursor, IPC-9P, results in the formation of the IPC-9 zeolite with 10- and 7R pore windows and oxygen bridges connecting the layers. Introduction of additional silicon source forms –O–Si–O– linkages between the layers. The linkages condense to S4R upon calcination, providing the IPC-10 zeolite with 12- and 9R pores [37].

The IPC-9 and IPC-10 are particularly remarkable for two reasons. First, both frameworks possess odd-member pores. The odd rings scarcely appear among the known zeolites and could potentially offer unique selectivity and separation properties. Second, the IPC-9 and IPC-10 defy the LID (local atomic distances in four-cornered frameworks) criteria characteristic of zeolites prepared by the hydrothermal synthesis [55]. This illustrates the ability of ADOR is not only to synthesize new zeolites in a controlled and predictable manner but also to prepare new structures, which are unattainable by the conventional methods, such as the hydrothermal synthesis [37].

5.3.5.2 The CIT-13 Case

The CIT-13 (or SAZ-1) germanosilicate closely resembles the previously described **UTL**. The CIT-13 structure consists of non-porous silica-rich layers connected by germanium-rich D4R units in the *c*-direction [14]. Extraction of the germanium from the D4Rs produces a layered material CIT-13P (or SAZ-1P), much like hydrolysis of **UTL** produces the IPC-1P (Figure 5.5) [61, 62]. Similarly, the layers of CIT-13P can be connected by S4R and condensed to CIT-14 (or IPC-16) or by oxygen bridges to form the CIT-15 (IPC-15). Regarding the similarity to the **UTL**, the ADOR transformation of CIT-13 can likely produce "borderline structures" analogous to IPC-6 and IPC-7, however, such work has not been published to the best of our knowledge.

The CIT-13 closely resemble the **UTL** both structurally and in the way it transforms by ADOR. However, despite their many similarities, the two structures significantly differ in the geometrical arrangement of the D4R in the structure [14, 52].

The **UTL** contains intersecting 14- and 12R pore channels. The channels are separated by the D4R units arranged in a periodically rectangular pattern, when viewed from the *a*-direction. The CIT-13, on the other hand, contains intersecting 14- and 10R channels and the D4Rs between them follow a triangular pattern. The D4R units sharing the 10R window (*a*-direction) are closer (3.63 Å) than in the **UTL** (6.04 Å) [14]. This proximity allows for an additional unique rearrangement of the D4Rs in CIT-13 into a "zig-zag" chain, resulting in the **CFI** (CIT-5) structure (Figure 5.5) [14]. The **CFI** contains only one-dimensional 14R pores, while the 10R pores (originally present in the CIT-13) become completely closed by the zig-zag chain. On the contrary, the D4R in the **UTL** share 14- and 12R windows, which are too wide to enable the rearrangement of D4Rs into a zig-zag chain.

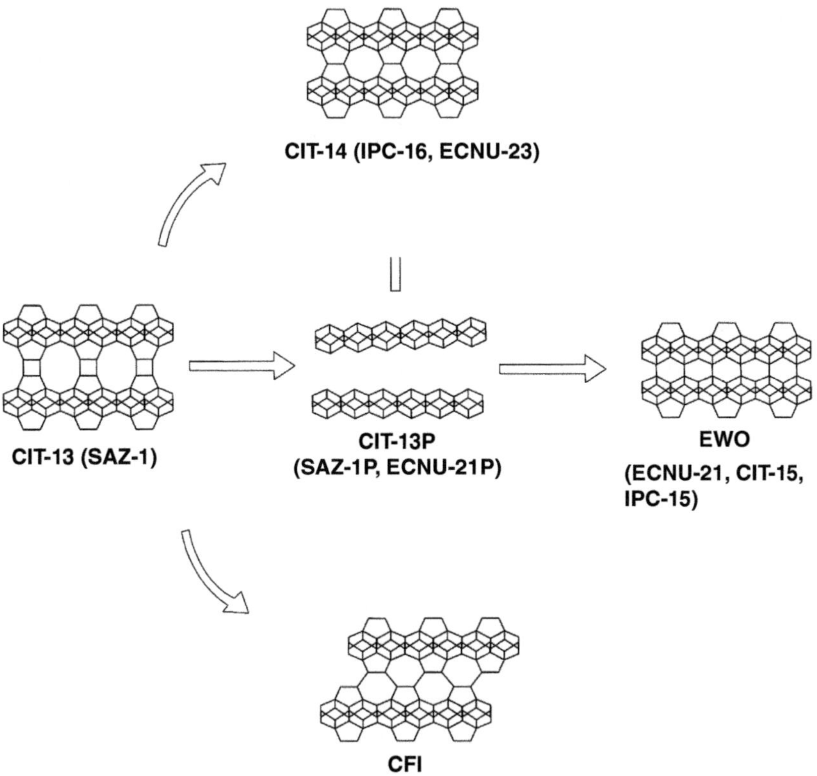

Figure 5.5 Transformations of the CIT-13 zeolite.

The germanium distribution in the framework is dependent on the synthesis method used to prepare the CIT-13. The CIT-13 zeolite is commonly prepared from a fluoride-containing gel. Only recently, Davis et al. developed a fluoride-free synthesis protocol for the CIT-13. However, despite having the CIT-13 structure, the two materials behaved rather differently. The fluoride-free CIT-13 was more susceptible to deterioration by air moisture and capable to undergo the inverse sigma transformation under highly acidic conditions. On the contrary, the fluoride-synthesized CIT-13 was more stable. It could not undergo the inverse sigma transformation and products of its hydrolysis in water or acidic solutions generally showed poor ordering.

The explanation for the discrepancy lies in the distribution of germanium throughout the two zeolites. While in the fluoride-free CIT-13 the germanium preferentially clustered in the D4R units, the fluoride-synthesized CIT-13 with equal Si/Ge showed more random distribution of germanium [52]. The lower occupancy of the D4R by germanium made Si–O–Si linkages within the units more common. The linkages between the neighboring layers prevented complete hydrolysis of the zeolite under the acidic conditions. Liu et al. addressed the issue by devising a base-assisted hydrolysis protocol. The mild basic conditions enabled hydrolysis of the Si–O–Si linkages within the D4R units, producing a layered material ECNU-21P [61]. The layered precursor prepared by the base-assisted treatment generally showed higher crystallinity compared to the SAZ-1P prepared via the acidic route. Reassembly of the layered ECNU-21P via S4R and oxygen bridges produced ECNU-23 and ECNU-21 zeolites isostructural to CIT-14 and CIT-15, respectively [61, 63].

5.3.5.3 The UOV Case

The **UOV** framework consists of Si-rich layers connected by the Ge-rich D4R units, much like the **UTL**. However, contrary to the **UTL**, the layers of **UOV** zeolite contain 12- and 8R pores penetrating them. System of parallel interconnected 10R pores separates the layers. The high porosity of the layers originally raised concerns about their stability in the ADOR transformation, however, the presence of wide pores in the layers does not compromise the stability of the **UOV** [64].

The **UOV** successfully undergoes ADOR transformation to a zeolite IPC-12, where direct oxygen linkages replace the original D4R (Figure 5.6). Replacing D4R units with direct linkages reduces the size of the 10R pores to 6R. In contrast, pores perpendicular to the layers retain their size during the transformation [65]. Although the ADOR transformation of **UOV** provided so far only one daughter structure, the IPC-12, the process also shows significant similarities to the transformation of **UTL**. The **UTL** transformation is a balance between the hydrolysis of **UTL** and the rearrangement of the hydrolyzed material. While mildly acidic conditions favor the hydrolysis to the layered precursor, high acidity promotes the rearrangement and re-intercalation of Si into the framework [54]. However, the re-intercalation of Si practically does not occur in the **UOV** case due to its open pore system. The large pores accelerate the Si deintercalation. Thus, the absence of Si in the interlayer space leads to recondensation of the layers of the hydrolyzed **UOV** under acidic conditions. As a result, the disassembly of **UOV** does not produce separate layers but instead the semi-connected IPC-12P precursor. Performing the ADOR transformation under increasing acidity leads to suppression of the hydrolysis, resulting in a smaller shift in the interlayer distance and eventually to decrease of crystallinity of the precursor. However, highly acidic conditions (e.g. 12 M HCl) enable an alternative mechanism for transformation, similar to the inverse sigma transformation of **UTL** to **OKO**. This transformation results in direct conversion of the **UOV** to highly crystalline IPC-12 zeolite [64, 66].

5.3.5.4 The IWW Case

The **IWW** structure consists of porous layers connected via D4R units into a 3D framework. The **IWW** framework possesses 8-, 10-, and 12R pores interconnected into a three-dimensional channel system [16, 67]. On the first look, the **IWW** closely resembles the **UOV**, which also contains porous layers interconnected by D4Rs. However, the **IWW** shows strong tendency to self-heal and higher resistance against moisture than the **UOV** [68]. The leeched Si can migrate quickly within the 3D channel system of **IWW** and rapidly heal the defects formed by the degermanation. The fast reinsertion of the Si species restores the original **IWW** zeolite and hinders its transformation to daughter structures [69, 70].

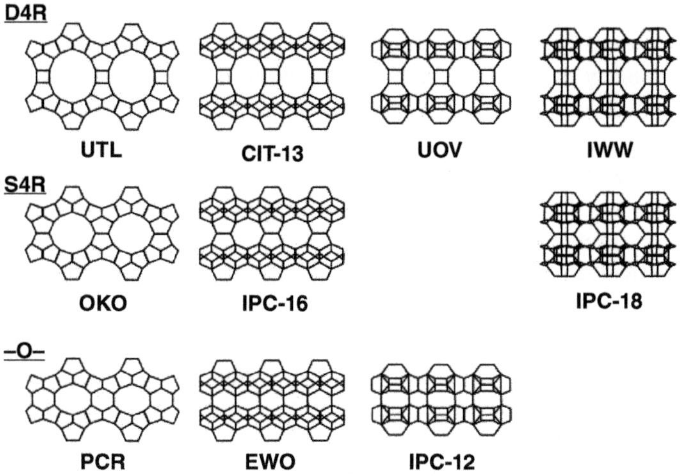

Figure 5.6 ADOR transformations of zeolites CIT-13, **UOV**, **IWW**, and **UTL**.

The successful ADOR transformation of **IWW** thus requires precise tuning of the conditions and the germanium content of the **IWW**. Ge-poor **IWW** (Si/Ge ~ 6,4) retains its structure during whole treatment and only undergoes degermanation. **IWW** with higher content of germanium (Si/Ge ~ 3.1) undergoes hydrolysis and transforms to a layered precursor IPC-5P. However, the layers of IPC-5P are not fully separated like in the case of IPC-1P. The IPC-1P can undergo swelling with surfactants, which increases the interlayer distance proportionally to the length of the surfactant (see Section 5.2.1). On the contrary, IPC-5P shows only a small increase in its interlayer distance upon swelling. The interlayer distance of "swollen" IPC-5P resembles that of **IWW**. This implies the horizontal orientation of the surfactant molecules and evidences the incomplete separation of IPC-5 layers. The structure of IPC-5P collapses upon direct calcination. However, when provided with additional source of silicon the IPC-5P transforms back to **IWW** with enhanced Si content [69].

Only **IWW** with high germanium content (Si/Ge < 2) undergoes full ADOR transformation to IPC-18 zeolite (Figure 5.6). However, the high content of germanium reduces the stability of **IWW**. The high-Ge **IWW** structure collapses when exposed to strongly alkaline or acidic conditions. Ergo, the ADOR transformation needs to take place under mild conditions, such as pure water. In water, the high-Ge **IWW** hydrolyses into IPC-18P precursor. Subsequent calcination of the precursor provides the IPC-18 zeolite with the same layers as **IWW** but connected via S4R units. The replacement of D4R by S4R results in decrease of size of the inter-layer pores from 10- to 8R and interlayer distance from 12.2 to 9.1 Å for **IWW** and IPC-18, respectively [70].

5.3.6 Vapor-phase ADOR

The liquid-phase ADOR transformation often fails on germanosilicate zeolites with low framework density. The low-density frameworks generally contain high amount of germanium which impairs their hydrothermal stability and the wide pores and cavities make the structure easily accessible and prone to attack by moisture or harsh reagents [14, 71]. The lability of these frameworks results in their degradation or amorphization upon contact with water. Moreover, the liquid water within the pores affects the transport of leeched Si and Ge species, which may also hinder the transformation, as illustrated by the **IWW** zeolite (see Section 5.3.5.4). The re-insertion of the leeched Si species heals the framework during hydrolysis and thereby hinders or prevents the structural transformation [3, 67, 69].

Since most of the obstacles arise from the presence of large volume of liquid water inside of the pores, they diminish when the ADOR transformation takes place at vapor-solid system, which uses acidic vapors to induce the transformation. The vapor-phase ADOR takes place in a vessel with two compartments, separated by a porous membrane. First compartment contains the source of the vapors, such as solution of hydrochloric acid or solid ammonium chloride. The second compartment contains dry germanosilicate zeolite. Upon heating, the acidic vapors pass through the porous membrane to the second compartment and diffuse into the sample. However, since the transport through the zeolite pores is not obstructed by liquid, the hydrolysis of the Si–O–Ge bonds proceeds rapidly. Additionally, the leeched germanium reacts with the acidic vapors and forms volatile germanium chloride, which easily diffuses out of the zeolite. On the other hand, the migration of Si is hindered which prevents its re-insertion into the framework and healing of the defects [68].

The suppression of Si migration and re-insertion during the vapor-phase ADOR most notably contributed to the first transformation of **IWW** zeolite into the IPC-18. The **IWW** mostly resisted the transformation attempts in the liquid phase (see Section 5.3.5.4) due to its interconnected 8-10-12R channel system, which facilitates the Si migration [67, 69]. However, the restricted movement of the Si species provided by the vapor-phase setup provided enough time for the extraction of germanium from the D4R units and their subsequent transformation to S4Rs. Replacing the D4R with S4R narrowed down the interlayer channels of **IWW** from 10- to 8R, providing the IPC-18 structure with 8-8-12R pore system [68].

On the other hand, the absence of liquid water imposes other limitations onto the vapor-phase ADOR. The vapor-phase transformation rapidly converts the parent germanosilicate to the daughter material, however; it prohibits the formation of layered material such as IPC-1P. The hydrolyzed material quickly forms new interlayer connections, thereby preventing the complete separation of individual layers, which occurs during hydrolysis in large volumes of liquid water (see Section 5.3.5.1) [68]. This excludes the preparation of layered, intercalated, or pillared materials and limits the vapor-phase ADOR to preparation of 3D zeolite structures.

5.3.7 Reductive Degermanation

Reduction in a gas phase offers an alternative to the acidic leeching of germanium in the liquid or vapor-phase ADOR. The reductive degermanation in hydrogen gas under high temperature obviates the need for strong acids or organic reagents and also enables immediate condensation of the precursor into the three-dimensional framework similarly to the vapor-phase transformation. The reduction selectively reduces and removes the germanium atoms that occupy the D4R positions. At the same time the germanium atoms located within the layers remain untouched due to a relatively large difference in their reduction temperature. While the germanium in D4R positions reduces at 673 °C the atoms in layer positions do not undergo reduction until 834 °C which implies that the germanium located within the layers is more stable. Similarly to the vapor-phase ADOR the gas-phase reduction also restricts Si migration which results in high selectivity and control over the transformation process. For example, the reductive transformation of **UTL** germanosilicate proceeds via progressive removal of the germanium from D4Rs accompanied by steady narrowing of the overall interlayer distance. The interlayer distance gradually narrows in time and thus enables controlling the final structure (i.e. IPC-2, IPC-6) by the experiment duration alone [72].

The framework germanium reduces to metallic form and forms separate particles within the zeolite channels. Unlike the germanium chloride formed during the vapor phase process (see Section 5.3.6) the metallic germanium is not volatile and thus remains within the channels. However, the precipitated metallic germanium obstructs the channels and needs to be removed by additional treatment. First, high temperature calcination in air converts the metallic germanium to germanium oxide. The germanium oxide can dissolve in water and ergo the pores of the transformed zeolite can be liberated by water washing [72].

5.3.8 Solid-state Transformations

The liquid and vapor-phase ADOR transformations as well as the reductive degermanation produce the final material via high temperature condensation of the interlayer silanol groups to form the 3D structure. Nevertheless, the condensation can also occur by imposing a high pressure onto the material and the pressure may even induce a structural transformation. Most attempts at zeolite transformation under high pressure result in amorphization of the material [50, 73]. However, some notable zeolites can transform under raised pressure and maintain their crystallinity. For instance, the ITQ-29 (**LTA**) undergoes pressure-induced transformation to ITQ-50 (**IFY**). The transformation carried out under pressure as high as 3.2 GPa results in irreversible conversion of ITQ-29 to ITQ-50. However, performing the transformation at lower pressure (i.e. 1.2 GPa) causes only temporary changes and the ITQ-50 reverts back to ITQ-29 once the pressure is released [74].

The high-pressure transformations also provide unique opportunities in combination with the ADOR protocol. Mazur et al. investigated the pressure-induced 2D–3D transformation of layered IPC-1P to IPC-2 (**OKO**) under 1 GPa and 200 °C. The high-pressure conditions prompt minimization of the volume of the system and induce condensation of the layers into a 3D framework. Interestingly, the less-dense IPC-2 forms during the pressure-induced condensation instead of the denser IPC-4 (**PCR**). The IPC-2 becomes more favorable because its wider pores can accommodate the water released during the condensation of terminal silanol groups on the IPC-1P layers. In contrast, the IPC-4 cannot sufficiently accommodate the released water which results in an increased volume of the system, making the transformation less favorable.

The formation of IPC-2 under raised pressure proceeds under lower temperature (200 °C) compared to the condensation by calcination (550 °C) and requires no solvent. Nevertheless, the conditions impose mechanical strain onto the material, which results in the formation of defects, less defined crystal shape and presence of trace amount of amorphous phase [73].

Mechanochemical treatments such as ball-milling generally produce less waste, reduce the need for heating and are easy to scale up. The ball-milling also reduces the need for a solvent and enables performing treatments under low liquid/solid conditions and reduced time, however; its application in preparation of catalysts and porous materials is still rather uncommon despite the compelling advantages [75, 76].

Interestingly, utilization of mechanochemical methods in the zeolite synthesis shows more than a reduced need for solvent. The mechanochemical synthesis proceeds under decreased amount of SDAs or seeds. On top of that, the crystallization time also decreased.

However, the mechanochemistry shows promising results also for ion exchange, crystal size control or post-synthetic modifications [77]. More recently, Rainer et al. developed a mechanochemically-assisted ADOR transformation of **UTL** to IPC-4. Similar to the direct synthesis, the mechanochemical route requires reduced time to achieve full transformation. The **UTL** transforms to IPC-4 within 30 minutes while the liquid-phase hydrolysis commonly takes 16 h. The method also requires only reduced liquid/solid ratio 50 $mL\,g^{-1}$. Nevertheless, the liquid phase, either water or acid, mediates the hydrolysis of the Ge–O–Si bonds and its presence is thus essential for the transformation. In contrast, mechanochemical treatment of the **UTL** in a non-hydrolytic solvent (e.g. ethanol) or without any solvent only reduces the crystal size but no structural transformation occurs [78].

Section 5.3.5.1 exhaustively describes how acidity of the solution affects the rate of hydrolysis and the outcome of the ADOR transformation [54]. Interestingly, the mechanochemically assisted ADOR transformations of **UTL** revealed no such dependency. Transformations in highly concentrated acids yield IPC-4 (–O–), 2 M to 9 M acidic solutions yield IPC-2 (S4R), 5 M solutions yield IPC-7 (D4R and S4R) and diluted solutions yield IPC-6 (S4R and –O–). The nature of the interlayer connections in the product structure appears to correlate with solubility of the leeched germanium species, which reaches a minimum in 5 M acid. The limited solubility along with the low solvent volume hinders the removal of germanium from the material. The germanium re-incorporates into the framework and stabilizes larger connecting units, such as S4Rs. The re-incorporation agrees with the higher germanium content observed in the mechanochemically prepared samples (Si/Ge = 10) compared to those prepared in aqueous phase (Si/Ge = 30) [78].

5.4 Structure Stabilization

The relative lability of Ge—O bonds compared to the Si—O enables the selective hydrolysis, germanium leaching and controlled transformation of numerous germanosilicates. Nevertheless, the hydrolytic lability of germanosilicates is also their greatest shortcoming. The reduced stability, which causes structural deterioration and eventually amorphization, along with the high cost of germanium prohibits their application on a larger scale. Decreasing the germanium content generally increases the hydrolytic stability of germanosilicates. However, the germanium often plays crucial role in crystallization of particular zeolites, namely zeolites containing D4R units, making its presence during the crystallization essential. Hence, considerable efforts have been aimed at developing methods for post-synthetic stabilization of germanosilicates by isomorphous substitution of germanium with more stable framework elements such as silicon or aluminum [3, 71, 79].

5.4.1 Degermanation

The hydrolytic stability of germanosilicates increases with the Si/Ge molar ratio and thus the zeolites are commonly stabilized by post-synthetic substitution of framework Ge atoms for Si. Some zeolites, such as the **IWW** (see Section 5.3.5.4) even show remarkable ability to self-heal and treatment in highly acidic solutions results in degermanation without a significant loss of crystallinity [67]. Similarly, zeolites with low germanium content undergo degermanation without losing their crystallinity due to the low number of defects introduced by the treatment [79]. On the other hand, degermanation and stabilization of zeolites with high germanium content becomes more intricate. Structural integrity of zeolites including **UTL**, **UWY**, or **BEC** quickly deteriorates in aqueous solutions due to the rapid leaching of framework germanium atoms. The rapid degermanation introduces a substantial number of defects, which eventually result in the collapse of the structure. In contrast, the reincorporation of silica proceeds relatively slowly and thus cannot fill the defects sufficiently fast to prevent the structural collapse [14, 32, 71, 79] (Figure 5.7).

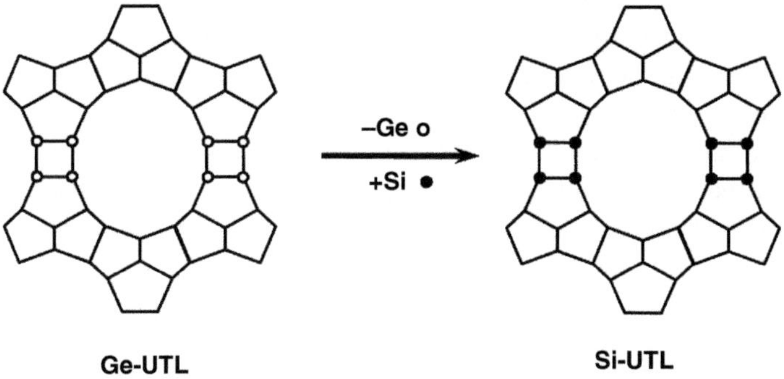

Figure 5.7 Stabilization of the **UTL** framework by the isomorphous substitution of Ge with Si.

Xu et al. developed an alternative method for post-synthetic stabilization of labile germanosilicates such as **UTL**, **UWY**, or **BEC** starting from the as-made zeolite. The zeolite micropores contain organic SDA molecules, which obstruct the pores, stabilize the framework and prevent its disassembly into layers or amorphization. The degermanation thus proceeds at lower rate. The acidic solution slowly removes the SDA molecules from the pores and progressively liberates the channel system. Simultaneously, silica debris from the layers re-incorporates into the framework and heal the vacancies after degermanation. Single treatment of as-made **UTL** zeolite increased its Si/Ge ratio from 4.7 to 59.4 while repeated degermanation yielded high-silica **UTL** with Si/Ge as high as 233. Similar results were reported for degermanated **UWY** and **BEC** with Si/Ge 100 and 208, respectively [71].

5.4.2 Functionalization With Catalytic Sites

The extraction of germanium creates silanol nests, vacancies, capable of re-incorporating Si or other framework heteroatoms. Incorporation of trivalent elements such as aluminum heals the defects and stabilizes the zeolite but also introduces a charge into the framework and generates ion exchange or Brønsted acid sites [80]. However, highly acidic conditions during the degermanation also promote extraction of aluminum from the zeolite. Thus, it is impossible to incorporate the aluminum into the framework under highly acidic conditions. Alternatively, milder acidic condition enables simultaneous degermanation and incorporation of aluminum or the degermanation and alumination can be carried out as separate steps [3, 79].

The applicability of both simultaneous and step-by-step degermanation and alumination varies for individual zeolites. **IWW** and **ITH** germanosilicates successfully undergo simultaneous degermanation-alumination in a mildly acidic solution of aluminum nitrate [79]. The acidic environment prompts the germanium leaching while the aluminum ions heal the defects and preserve the structure throughout the process. The rate of alumination varies depending on the diffusional properties of the channel system as evidenced by different degrees of alumination for large-pore **IWW** and medium pore **ITH** containing 7.1 and 1.5 mol% of aluminum after identical treatment, respectively [67, 79].

However, the diffusion limitations also prevent the stabilization of the **UTL** zeolite by the simultaneous degermanation-alumination. The **UTL** possesses two-dimensional channel system, which restricts the migration of the aluminum species. Thus, the rate of alumination decreases. In contrast, the degermanation proceeds rapidly, creating large number of structural defects and resulting in a loss of crystallinity. Ergo, the stabilization of **UTL** necessitates separate steps; firstly, degermanation and stabilization with Si (see Section 5.4.1) and subsequently incorporation of aluminum into the remaining vacant sites [71, 79].

Similarly, Lewis acid sites can be incorporated into the zeolite in a form of tetravalent ions such as Sn or Ti. Liu et al. described a two-step method for post-synthetic incorporation of tin into the **UTL** zeolite by degermanation and stabilization of the **UTL** in 1 M HNO_3 at 190 °C and subsequent incorporation of tin into the vacant framework sites. The tetrahedral Sn provides Lewis acidity to the zeolite framework. The combination of the Lewis acid sites and extra-large pores of the Sn-UTL make the zeolite suitable for Meerwein–Ponndorf–Verley reduction and Baeyer–Villiger oxidation of bulky molecules [81]. Jiao et al. applied an analogous method on the germanosilicate IWV and prepared Ti-IWV enriched with Lewis acid sites originating from the framework Ti. The Ti-IWV showed exceptional properties for epoxidation of bulky cycloalkenes such as cycloheptene or cyclooctene [82].

5.4.3 Slow Disassembly

Isomorphous substitution of germanium for silicon or aluminum stabilizes the zeolite and prevents its disassembly or decomposition. However, stable silicate or aluminosilicate zeolites can also form from labile germanosilicates through controlled deconstruction and subsequent reconstruction of the framework. Zhang et al. describes such exemplary procedure based on modified ADOR approach of the **UTL** zeolite [58]. The **UTL** undergoes decelerated hydrolysis under low-water high-liquid conditions (described in Section 5.3.5.1). The slow disassembly progressively transforms the **UTL** to IPC-2 and eventually to IPC-4. However, the transformation reverts in direction when framework-building elements, such as aluminum, are added into the mixture in high concentrations. The high concentration of aluminum species promotes the framework rearrangement and even enables spontaneous reconstruction of D4R units in the disordered material. The **UTL** thus undergoes a cyclic structural transformation involving leaching of Ge, followed by transformation to IPC-2P and subsequently restoration of the D4Rs with the aluminum species and recovery of the **UTL** with high content of framework aluminum [58].

The slow disassembly and the consecutive structural transformations are highlighted by characteristic changes in interlayer spacing, pore size distribution as well as elemental composition. The method incorporates large amount of aluminum into the **UTL** up to Si/Al = 20. Nevertheless, the slow nature of all involved processes painstakingly prolongs the duration of the treatment and contributes to the progressive decline in crystallinity of the material.

5.4.4 Reverse ADOR

The shortcoming of aluminum incorporation into the **UTL** zeolite using the slow disassembly approach lies in the lengthy treatment accompanied by progressive deterioration of the zeolite crystallinity. However, the drawbacks diminish by artificially accelerating the underlying processes. The interlayer distance expands more rapidly by intercalation of appropriate organic species (see Section 5.2.1) (e.g. 2,6-dimethyl-5-azoniaspiro[4,5]decane or tetrabutylammonium) compared to the spontaneous processes [36, 49, 83, 84]. Furthermore, D4R units form spontaneously and rapidly in the presence of germanium or fluoride anions, accelerating the structural reconstruction. This set of subsequent modifications, called "Reverse ADOR," includes the intercalation of the layered precursor IPC-1P with proper organic agent, reconstruction of the bridging D4R units under anhydrous conditions and subsequent condensation by calcination of the organics under high temperature [83, 84].

The reverse ADOR creates a cyclic **UTL**–IPC-1P–**UTL** transformation, which also enables incorporation of new heteroelements into the interlayer D4R positions. The reconstructed D4R units can consist of silicon or germanium as well as aluminum, titanium, or zirconium, providing large variability for the composition of the final zeolite. Furthermore, the reverse ADOR provides precise control over the elemental composition and often enables incorporation of larger quantities of the heteroelements than the direct hydrothermal synthesis. For example, the reconstructed **UTL** zeolite can accommodate aluminum up to Si/Al = 15 while the directly synthesized zeolite can reach only Si/Al = 44 [83, 85]. The incorporation of large amounts of each element relies on careful tuning of the synthesis conditions, especially the acid concentration during the reconstruction.

The reconstruction of the D4R units can occur due to stabilization by fluoride ions accommodating the center of the D4R cube or by germanium atoms introduced in the D4R positions. The fluoride-assisted reconstruction obviates the need for germanium in the structure and widens the range of elemental composition of the zeolite. However, the fluorides also partially dissolve the zeolite during the reconstruction, producing a large number of defects and decreasing the crystallinity [84]. Furthermore, the fluoride stabilization becomes less reliable under more-neutral conditions [83]. On the contrary, the germanium-assisted route is reliable under wider range of conditions and produces materials with higher level of crystallinity. However, the presence of germanium in the D4R units decreases the hydrolytic stability of the material. As a side note, the germanium also possesses Lewis acidity, which can impact the catalytic properties of the final material [83, 84].

5.5 Germanosilicate-Derived Catalysts

The previously described methods elaborated on the modification and transformation of germanosilicate zeolites with the aim to design and prepare new zeolite-based catalysts with pre-determined properties. The intercalation, pillaring or ADOR transformation alter the textural properties including pore size of the material, which determine its shape-selectivity and accessibility of the active sites. The changes in elemental composition (e.g. degermanation, alumination, pillaring with metal oxides) determine the nature and concentration of the catalytic active sites in the material. In case of germanosilicate zeolites described in this chapter, the acidity may originate either from the framework germanium or from additional heteroelement introduced during or after the hydrothermal synthesis [80]. The defects formed after degermanation can accommodate trivalent elements such as boron, aluminum or gallium to form Brønsted acid centers or tetravalent tin or titanium, which possess Lewis acidity [79, 86].

Framework germanium innately exhibits Lewis acidity due to its free orbitals [87]. However, weak Brønsted acid sites form upon interaction of water molecules with the germanium ions. Although germanium zeolite catalysts are scarcely studied, germanosilicates including **UTL**, **BEC**, **IWR**, **IWW**, or **ITH** succeeded in catalysis of Bayer–Villiger oxidation of bulky ketones (e.g. cyclohexanone, 2-adamantanone) to lactones with high selectivity (c. 90%) [87] or in esterification of levulinic acid [88]. Furthermore, Podolean et al. reported ketalization of polyols (e.g. ethylene glycol, glycerol and 1,4-butanediol) to dioxolanes over germanosilicate zeolites with exceptional selectivity (>99%) at room temperature. Nevertheless, the ketalization produces water as a side-product, which caused leeching of the germanium and decline of catalytic activity after repeated use, indicating the limitation for application of germanosilicates [89].

The Brønsted acid zeolites show high variability for tuning their properties to suit the needs of individual reactions. The concentration of active sites depends on the content of incorporated element, which can be controlled by pH and concentration of the treatment solution. Generally, increase in the pH (e.g. from 0.5 to 2.0) boosts the incorporation and results in higher concentration of active sites. The average strength of the acid sites originates from the particular element in respective order: B < Ga < Al. Strong active sites activate the reactant molecules more efficiently, however, they can also strongly adsorb products, leading to rapid deactivation by poisoning. Shamzhy et al. illustrated that on set of B, Ga, and Al-IWW zeolites in Friedel–Crafts acylation of p-xylene with benzoyl chloride. The acylation proceeded most rapidly over the moderately strong centers of Ga-IWR which enabled both sufficient activation of benzoyl chloride as well as easy desorption of the product. On the contrary, strong Brønsted sites of Al-IWR hydrolyzed the benzoylchloride to anhydride, severely halting the reaction [86].

Lewis acidic titanosilicate zeolites show particular ability to catalyze epoxidations of C=C double-bonds by peroxides. The titanium can be incorporated into the zeolites such as **UTL** by hydrothermal synthesis. The titano-germanosilicate **UTL** undergoes hydrolysis and ADOR transformations and produces Ti-IPC-n zeolites or pillared Ti-IPC-1P. Additionally, the amorphous pillars can contain added titanium oxide, which enhances the catalytic activity of the pillared material. The wide pores of the silica-titania pillared IPC-1P can catalyze reactions of bulky molecules and show performance, which exceeds that of conventional titanosilicate zeolites such as Ti-beta [46, 90].

The isoreticular IPC-n zeolites containing catalytic active sites (e.g. Ti, Al) provide a unique set of materials for studying the effects of pore size on the catalytic activity and selectivity. Both reactant conversion and product selectivity largely depend on the pore architecture of the zeolite [91]. However, while commonly used zeolites such as MFI, MOR or beta possess different size of pores, they also possess different structural features, which may affect their catalytic behavior. In contrast, the isoreticular zeolites produced via ADOR transformation of parent germanosilicate contain the same structural motif – layers – and differ only in the interlayer connections and channel size. Increasing size of the channels commonly correlates with increasing reactant conversion. However, understanding how the reaction mechanism or product selectivity depends on the channel size requires more elaborate analysis. By excluding the additional parameters, the isoreticular zeolites proved to

Figure 5.8 Encapsulation of Pt nanoparticles into IPC-2 and IPC-4 zeolites using the ADOR approach.

be essential in determining the role of pore size on selectivity in toluene alkylation and xylene isomerization [92, 93], on describing diffusion limitations and reaction rate constants in ethanol dehydration [51] or the role of confinement effect in epoxidation of cyclohexene [90].

Lastly, the 2D-to-3D zeolite transformation offers a unique and effective way of stabilizing metallic nanoparticles and controlling their size. The metallic nanoparticles, predominantly platinum and palladium catalyze hydrogenation reactions. However, the nanoparticles suffer from sintering at high temperature which decreases their surface area and thereby their catalytic activity. Zeolite frameworks can stabilize the nanoparticles by encapsulating them within the cavities inside the zeolite and thus preventing them from sintering and also providing the shape-selectivity of the zeolite framework. Corma et al. introduced a unique method for stabilizing the nanoparticles using the 2D-to-3D transformation of layered MCM-22P to 3D MWW framework. The MCM-22P layers intercalated by surfactants intake the nanoparticles into the interlayer space and trap them in the zeolite upon condensation of the layers [94]. The interlayer distance increases with the length of the surfactant chain (from 1.06 to 2.89 nm for C12 to C18, respectively). The mean size of the metallic nanoparticles also increases with the increasing interlayer distance (from 0.85 to 2.04 nm) and with metal loading [95, 96]. Following, Zhang et al. combined the method with the ADOR transformation, yielding IPC-2 and IPC-4 zeolites with encapsulated Pt nanoparticles (Figure 5.8) [97]. The materials showed promising results for shape-selective hydrogenation reactions of nitroarenes [96, 98].

5.5.1 Summary and Perspectives

Due to their unique chemistry, germanosilicates possess a great potential in preparation of new zeolites in virtue of structure-directing ability of germanium and its role in post-synthetic transformations. On one hand, introduction of germanium into a framework reduces hydrolytic stability of the zeolite. Hydrolysis of the relatively weak Ge—O bonds induces the formation of structural defects, which may even cause amorphization. On the other hand, the germanium also shows strong propensity to preferentially cluster in strained structural units such as double-four rings (D4R) or double-three rings (D3R). Numerous germanosilicate zeolites thus consist of such domains rich in germanium and separate domains rich in silicon, which differ in their hydrolytic stability. This enables selective hydrolysis of the germanium-rich units (e.g. D4R or D3R) and production of low-dimensional layered materials by hydrolysis of germanosilicates or preparation of new zeolite structures via the ADOR transformation.

The ADOR transformation represents a unique method for designing new zeolites by controlled and selective hydrolysis of D4R-containing germanosilicates into a layered form and subsequent re-connection of the layers into a new 3D framework. Meticulous tuning of the experimental conditions enables adjusting the interlayer distance and relative layer arrangement. Thereby, the ADOR process can provide zeolite materials with tunable porosity. Manipulation of the relative position of layers even produces odd-member ring zeolites considered "unfeasible" under hydrothermal conditions. Moreover, cyclic transformations such as the Reverse ADOR can reconstruct extra-large pore zeolites from layered precursors while changing their elemental compositions.

The promise of finding new zeolites and especially preparing porous materials with precisely tuned properties make the germanosilicate transformations, including ADOR, attractive for both deeper research on their fundamental principles and for finding new applications. For instance, the "ADORable" zeolites show promising results as new catalytic materials. Nevertheless, the research of the ADOR as well as other transformations of germanosilicates still calls for further expansion and investigation. To this day, only a handful of germanosilicate zeolites were studied by these methods; however, many more potential candidates remain unexplored. Notably zeolites with D4R located in multiple crystallographic dimensions appeal a great interest as starting materials for transformation into new zeolites; however, the attempts at ADOR transformations were hitherto unsuccessful due to their delicate properties and low stability. From a different perspective, the methods including interlayer expansion and Reverse ADOR open up the question of preparing new zeolites by increasing the interlayer distance and pore size instead of their decreasing during the transformation. The interlayer-expanded zeolites appeal especially as possible selective catalysts for reactions of bulky fine chemicals.

The cost of germanium, the key element in this chapter, complicates any eventual upscaling or applicability of the germanosilicates and their transformations. Although, studies on recovery of the spent germanium already proved that majority of the germanium can be reclaimed and recycled for the synthesis of new parent germanosilicates. Such cyclic process would significantly reduce the cost of production connected to the germanium usage. Another promising alternative is seeking other materials to replace the expensive germanosilicates by using a cheaper chemically labile element in its place. Elements such as boron or aluminum can selectively leach from the zeolite framework under specific conditions; however, unlike germanium, they generally disperse more randomly within the framework, which prevents the controlled disassembly of the zeolite into a low-dimensional material. Nevertheless, other groups of crystalline materials such as Metal-Aluminum carbides (MAXenes) or carbon materials can undergo 3D-to-2D disassembly analogously to the germanosilicates. Other materials including porous metallosilicates or Metal-Organic Frameworks (MOFs) possess similar structural motifs, chemically distinct units, which could be suitable for selective disassembly.

Anyway, the described field is in its infancy. The progress achieved up to now as well as nascent branching of germanosilicate zeolites area and its fast development ensure the bright future, and open up new approaches for targeted and controlled preparation of new advanced materials with physicochemical properties tailored to demand of particular applications.

Acknowledgements

The authors thank the Czech Science Foundation for funding this research through the ExPro project (19-27551X). M.O. thanks Ministry of Education, Youth and Sports of Czech Republic (ERC_CZ project LL 2104).

References

1 *Database of Zeolite Structures*. http://www.iza-structure.org/databases.
2 Li, X. and Deem, M.W. (2014). Why zeolites have so few seven-membered rings. *J. Phys. Chem. C* 118 (29): 15835–15839.
3 Burel, L., Kasian, N., and Tuel, A. (2014). Quasi all-silica zeolite obtained by isomorphous degermanation of an as-made germanium-containing precursor. *Angew. Chem. Int. Ed.* 53 (5): 1360–1363.
4 Du, H., Fairbridge, C., Yang, H., and Ring, Z. (2005). The chemistry of selective ring-opening catalysts. *Appl. Catal. Gen.* 294 (1): 1–21.
5 Corma, A., Iborra, S., and Velty, A. (2007). Chemical routes for the transformation of biomass into chemicals. *Chem. Rev.* 107 (6): 2411–2502.
6 Moliner, M., Rey, F., and Corma, A. (2013). Towards the rational design of efficient organic structure-directing agents for zeolite synthesis. *Angew. Chem. Int. Ed.* 52 (52): 13880–13889.
7 Zones, S.I. (2011). Translating new materials discoveries in zeolite research to commercial manufacture. *Microporous Mesoporous Mater.* 144 (1): 1–8.
8 Li, Y. and Yu, J. (2014). New stories of zeolite structures: their descriptions, determinations, predictions, and evaluations. *Chem. Rev.* 114 (14): 7268–7316.
9 Opanasenko, M., Shamzhy, M., Wang, Y. et al. (2020). Synthesis and post-synthesis transformation of germanosilicate zeolites. *Angew. Chem. Int. Ed.* 59 (44): 19380–19389.
10 Corma, A., Díaz-Cabañas, M.J., Jiang, J. et al. (2010). Extra-large pore zeolite (ITQ-40) with the lowest framework density containing double four- and double three-rings. *Proc. Natl. Acad. Sci.* 107 (32): 13997–14002.
11 Jiang, J., Jorda, J.L., Diaz-Cabanas, M.J. et al. (2010). The synthesis of an extra-large-pore zeolite with double three-ring building units and a low framework density. *Angew. Chem. Int. Ed.* 49 (29): 4986–4988.

12 Shvets, O.V., Zukal, A., Kasian, N. et al. (2008). The role of crystallization parameters for the synthesis of germanosilicate with UTL topology. *Chem. Eur. J.* 14 (32): 10134–10140.

13 Paillaud, J.-L., Harbuzaru, B., Patarin, J., and Bats, N. (2004). Extra-large-pore zeolites with two-dimensional channels formed by 14 and 12 rings. *Science* 304 (5673): 990.

14 Kang, J.H., Xie, D., Zones, S.I., and Davis, M.E. (2019). Transformation of extra-large pore germanosilicate CIT-13 molecular sieve into extra-large pore CIT-5 molecular sieve. *Chem. Mater.* 31 (23): 9777–9787.

15 Corma, A., Navarro, M.T., Rey, F. et al. (2001). Pure polymorph C of zeolite beta synthesized by using framework isomorphous substitution as a structure-directing mechanism. *Angew. Chem. Int. Ed. Engl.* 40: 2277–2280.

16 Corma, A., Rey, F., Valencia, S. et al. (2003). A zeolite with interconnected 8-, 10- and 12-ring pores and its unique catalytic selectivity. *Nat. Mater.* 2 (7): 493–497.

17 Corma, A., Puche, M., Rey, F. et al. (2003). A zeolite structure (ITQ-13) with three sets of medium-pore crossing channels formed by 9- and 10-rings. *Angew. Chem. Int. Ed.* 42 (10): 1156–1159.

18 Lorgouilloux, Y., Dodin, M., Mugnaioli, E. et al. (2014). IM-17: a new zeolitic material, synthesis and structure elucidation from electron diffraction ADT data and Rietveld analysis. *RSC Adv.* 4 (37): 19440–19449.

19 Castañeda, R., Corma, A., Fornés, V. et al. (2003). Synthesis of a new zeolite structure ITQ-24, with intersecting 10- and 12-membered ring pores. *J. Am. Chem. Soc.* 125 (26): 7820–7821.

20 Schaack, B.B., Schrader, W., and Schüth, F. (2009). How are heteroelements (Ga and Ge) incorporated in silicate oligomers? *Chem. Eur. J.* 15 (24): 5920–5925.

21 Schaack, B.B., Schrader, W., and Schueth, T. (2008). Detection of structural elements of different zeolites in nucleating solutions by electrospray ionization mass spectrometry. *Angew. Chem. Int. Ed.* 47 (47): 9092–9095.

22 Sastre, G., Pulido, A., and Corma, A. (2005). An attempt to predict and rationalize relative stabilities and preferential germanium location in Si/Ge zeolites. *Microporous Mesoporous Mater.* 82 (1–2): 159–163.

23 Blasco, T., Corma, A., Díaz-Cabañas, M.J. et al. (2002). Preferential location of Ge in the double four-membered ring units of ITQ-7 zeolite. *J. Phys. Chem. B.* 106 (10): 2634–2642.

24 Kasian, N., Tuel, A., Verheyen, E. et al. (2014). NMR evidence for specific germanium siting in IM-12 zeolite. *Chem. Mater.* 26 (19): 5556–5565.

25 Sastre, G. and Corma, A. (2010). Predicting structural feasibility of silica and Germania zeolites. *J. Phys. Chem. C* 114 (3): 1667–1673.

26 Cantín, Á., Corma, A., Díaz-Cabañas, M.J. et al. (2006). Synthesis and characterization of the all-silica pure polymorph C and an enriched polymorph B intergrowth of zeolite beta. *Angew. Chem. Int. Ed.* 45 (47): 8013–8015.

27 Verheyen, E., Joos, L., Van Havenbergh, K. et al. (2012). Design of zeolite by inverse sigma transformation. *Nat. Mater.* 11 (12): 1059–1064.

28 Roth, W.J., Shvets, O.V., Shamzhy, M. et al. (2011). Postsynthesis transformation of three-dimensional framework into a lamellar zeolite with modifiable architecture. *J. Am. Chem. Soc.* 133 (16): 6130–6133.

29 Villaescusa, L.A. and Camblor, M.A. (2016). Framework reduction of GeO2 zeolites during calcination. *Chem. Mater.* 28 (20): 7544–7550.

30 Chlubna, P., Roth, W.J., Greer, H.F. et al. (2013). 3D to 2D routes to ultrathin and expanded zeolitic materials. *Chem. Mater.* 25 (4): 542–547.

31 Lei, J.-C., Zhang, X., and Zhou, Z. (2015). Recent advances in MXene: preparation, properties, and applications. *Front. Phys.* 10 (3): 276–286.

32 Jin, M., Veselý, O., Heard, C.J. et al. (2021). The role of water loading and germanium content in germanosilicate hydrolysis. *J. Phys. Chem. C* 125 (43): 23744–23757.

33 Morris, S.A., Bignami, G.P.M., Tian, Y. et al. (2017). In situ solid-state NMR and XRD studies of the ADOR process and the unusual structure of zeolite IPC-6. *Nat. Chem.* 9 (10): 1012–1018.

34 Bignami, G.P.M., Dawson, D.M., Seymour, V.R. et al. (2017). Synthesis, isotopic enrichment, and solid-state NMR characterization of zeolites derived from the assembly, disassembly, organization, reassembly process. *J. Am. Chem. Soc.* 139 (14): 5140–5148.

35 Henkelis, S.E., Mazur, M., Rice, C.M. et al. (2019). Kinetics and mechanism of the hydrolysis and rearrangement processes within the assembly–disassembly–organization–reassembly synthesis of zeolites. *J. Am. Chem. Soc.* 141 (10): 4453–4459.

36 Mazur, M., Chlubná-Eliášová, P., Roth, W.J., and Čejka, J. (2014). Intercalation chemistry of layered zeolite precursor IPC-1P. *Catal. Today* 227: 37–44.

37 Mazur, M., Wheatley, P.S., Navarro, M. et al. (2016). Synthesis of 'unfeasible' zeolites. *Nat. Chem.* 8 (1): 58–62.

38 Roth, W.J., Nachtigall, P., Morris, R.E. et al. (2013). A family of zeolites with controlled pore size prepared using a top-down method. *Nat. Chem.* 5 (7): 628–633.

39 Shamzhy, M., Mazur, M., Opanasenko, M. et al. (2014). Swelling and pillaring of the layered precursor IPC-1P: tiny details determine everything. *Dalton Trans.* 43 (27): 10548–10557.

40 Smith, R.L., Eliášová, P., Mazur, M. et al. (2014). Atomic force microscopy of novel zeolitic materials prepared by top-down synthesis and ADOR mechanism. *Chem. Eur. J.* 20 (33): 10446–10450.

41 Kresge, C.T. and Roth, W.J. (2013). The discovery of mesoporous molecular sieves from the twenty year perspective. *Chem. Soc. Rev.* 42 (9): 3663–3670.

42 Chlubná, P., Roth, W.J., Zukal, A. et al. (2012). Pillared MWW zeolites MCM-36 prepared by swelling MCM-22P in concentrated surfactant solutions. *Catal. Today* 179 (1): 35–42.

43 Barth, J.-O., Jentys, A., Kornatowski, J., and Lercher, J.A. (2004). Control of acid–base properties of new nanocomposite derivatives of MCM-36 by mixed oxide pillaring. *Chem. Mater.* 16 (4): 724–730.

44 Přech, J., Eliášová, P., Aldhayan, D., and Kubů, M. (2015). Epoxidation of bulky organic molecules over pillared titanosilicates. *Catal. Today* 243: 134–140.

45 Roth, W.J., Makowski, W., Marszalek, B. et al. (2014). Activity enhancement of zeolite MCM-22 by interlayer expansion enabling higher Ce loading and room temperature CO oxidation. *J. Mater. Chem. A* 2 (38): 15722–15725.

46 Přech, J. and Čejka, J. (2016). UTL titanosilicate: an extra-large pore epoxidation catalyst with tunable textural properties. *Catal. Today* 277: 2–8.

47 Opanasenko, M., Shamzhy, M., Yu, F. et al. (2016). Zeolite-derived hybrid materials with adjustable organic pillars. *Chem. Sci.* 7 (6): 3589–3601.

48 Opanasenko, M., Parker, W.O.N., Shamzhy, M. et al. (2014). Hierarchical hybrid organic–inorganic materials with tunable textural properties obtained using zeolitic-layered precursor. *J. Am. Chem. Soc.* 136 (6): 2511–2519.

49 Yang, B., Jiang, J.-G., Xu, H. et al. (2018). Synthesis of large-pore ECNU-19 material (12 × 8-R) via interlayer-expansion of HUS-2 lamellar silicate. *Chin. J. Chem.* 36 (3): 227–232.

50 Eliášová, P., Opanasenko, M., Wheatley, P.S. et al. (2015). The ADOR mechanism for the synthesis of new zeolites. *Chem. Soc. Rev.* 44 (20): 7177–7206.

51 Zhou, Y., Kadam, S.A., Shamzhy, M. et al. (2019). Isoreticular UTL-derived zeolites as model materials for probing pore size–activity relationship. *ACS Catal.* 9 (6): 5136–5146.

52 Kang, J.H., Xie, D., Zones, S.I., and Davis, M.E. (2020). Fluoride-free synthesis of germanosilicate CIT-13 and its inverse sigma transformation to form CIT-14. *Chem. Mater.* 32 (5): 2014–2024.

53 Shvets, O.V., Shamzhy, M.V., Yaremov, P.S. et al. (2011). Isomorphous introduction of boron in germanosilicate zeolites with UTL topology. *Chem. Mater.* 23 (10): 2573–2585.

54 Wheatley, P.S., Chlubná-Eliášová, P., Greer, H. et al. (2014). Zeolites with continuously tuneable porosity. *Angew. Chem.* 126 (48): 13426–13430.

55 Li, Y., Yu, J.H., and Xu, R.R. (2013). Criteria for zeolite frameworks realizable for target synthesis. *Angew. Chem. Int. Ed.* 52 (6): 1673–1677.

56 Wu, P., Ruan, J., Wang, L. et al. (2008). Methodology for synthesizing crystalline metallosilicates with expanded pore windows through molecular alkoxysilylation of zeolitic lamellar precursors. *J. Am. Chem. Soc.* 130 (26): 8178–8187.

57 Schreyeck, L., Caullet, P., Mougenel, J.C. et al. (1996). PREFER: a new layered (alumino) silicate precursor of FER-type zeolite. *Microporous Mater.* 6 (5–6): 259–271.

58 Zhang, J., Veselý, O., Tošner, Z. et al. (2021). Toward controlling disassembly step within the ADOR process for the synthesis of zeolites. *Chem. Mater.* 33 (4): 1228–1237.

59 Mazur, M., Kubů, M., Wheatley, P.S., and Eliášová, P. (2015). Germanosilicate UTL and its rich chemistry of solid-state transformations towards IPC-2 (OKO) zeolite. *Catal. Today* 243: 23–31.

60 Mazur, M., Kasneryk, V., Přech, J. et al. (2018). Zeolite framework functionalisation by tuneable incorporation of various metals into the IPC-2 zeolite. *Inorg. Chem. Front.* 5 (11): 2746–2755.

61 Liu, X., Mao, W., Jiang, J. et al. (2019). Topotactic conversion of alkali-treated intergrown germanosilicate CIT-13 into single-crystalline ECNU-21 zeolite as shape-selective catalyst for ethylene oxide hydration. *Chem. Eur. J.* 25 (17): 4520–4529.

62 Firth, D.S., Morris, S.A., Wheatley, P.S. et al. (2017). Assembly–disassembly–organization–reassembly synthesis of zeolites based on cfi-type layers. *Chem. Mater.* 29 (13): 5605–5611.

63 Liu, X., Luo, Y., Mao, W. et al. (2020). 3D electron diffraction unravels the new zeolite ECNU-23 from the "pure" powder sample of ECNU-21. *Angew. Chem. Int. Ed.* 59 (3): 1166–1170.

64 Kasneryk, V., Shamzhy, M., Opanasenko, M. et al. (2018). Insight into the ADOR zeolite-to-zeolite transformation: the UOV case. *Dalton Trans.* 47 (9): 3084–3092.

65 Kasneryk, V., Opanasenko, M., Shamzhy, M. et al. (2017). Consecutive interlayer disassembly–reassembly during alumination of UOV zeolites: insight into the mechanism. *J. Mater. Chem. A* 5 (43): 22576–22587.

66 Kasneryk, V., Shamzhy, M., Opanasenko, M. et al. (2017). Expansion of the ADOR strategy for the synthesis of zeolites: the synthesis of IPC-12 from zeolite UOV. *Angew. Chem. Int. Ed.* 56 (15): 4324–4327.

67 Yuan, R., Claes, N., Verheyen, E. et al. (2016). Synthesis of an IWW-type germanosilicate zeolite using 5-azonia-spiro[4,4]nonane as a structure directing agent. *New J. Chem.* 40 (5): 4319–4324.

68 Kasneryk, V., Shamzhy, M., Zhou, J. et al. (2019). Vapour-phase-transport rearrangement technique for the synthesis of new zeolites. *Nat. Commun.* 10 (1): 5129.

69 Chlubná-Eliášová, P., Tian, Y., Pinar, A.B. et al. (2014). The assembly-disassembly-organization-reassembly mechanism for 3D-2D-3D transformation of germanosilicate IWW zeolite. *Angew. Chem. Int. Ed.* 53 (27): 7048–7052.

70 Lu, K., Huang, J., Jiao, M. et al. (2021). Topotactic conversion of Ge-rich IWW zeolite into IPC-18 under mild condition. *Microporous Mesoporous Mater.* 310: 110617.

71 Xu, H., Jiang, J.-g., Yang, B. et al. (2014). Post-synthesis treatment gives highly stable siliceous zeolites through the isomorphous substitution of silicon for germanium in germanosilicates. *Angew. Chem. Int. Ed.* 53 (5): 1355–1359.

72 Ma, Y., Xu, H., Liu, X. et al. (2019). Structural reconstruction of germanosilicate frameworks by controlled hydrogen reduction. *Chem. Commun.* 55 (13): 1883–1886.

73 Mazur, M., Arévalo-López Angel, M., Wheatley, P.S. et al. (2018). Pressure-induced chemistry for the 2D to 3D transformation of zeolites. *J. Mater. Chem. A* 6 (13): 5255–5259.

74 Jordá, J.L., Rey, F., Sastre, G. et al. (2013). Synthesis of a novel zeolite through a pressure-induced reconstructive phase transition process. *Angew. Chem. Int. Ed.* 52 (40): 10458–10462.

75 Amrute, A.P., De Bellis, J., Felderhoff, M., and Schüth, F. (2021). Mechanochemical synthesis of catalytic materials. *Chem. Eur. J.* 27 (23): 6819–6847.

76 Szczęśniak, B., Borysiuk, S., Choma, J., and Jaroniec, M. (2020). Mechanochemical synthesis of highly porous materials. *Mater. Horizons* 7 (6): 1457–1473.

77 Majano, G., Borchardt, L., Mitchell, S. et al. (2014). Rediscovering zeolite mechanochemistry – a pathway beyond current synthesis and modification boundaries. *Microporous Mesoporous Mater.* 194: 106–114.

78 Rainer, D.N., Rice, C.M., Warrender, S.J. et al. (2020). Mechanochemically assisted hydrolysis in the ADOR process. *Chem. Sci.* 11 (27): 7060–7069.

79 Shamzhy, M.V., Eliášová, P., Vitvarová, D. et al. (2016). Post-synthesis stabilization of germanosilicate zeolites ITH, IWW, and UTL by substitution of Ge for Al. *Chem. Eur. J.* 22 (48): 17377–17386.

80 Shamzhy, M., Opanasenko, M., Concepción, P., and Martínez, A. (2019). New trends in tailoring active sites in zeolite-based catalysts. *Chem. Soc. Rev.* 48: 1095–1149.

81 Liu, X., Xu, H., Zhang, L. et al. (2016). Isomorphous incorporation of tin ions into germanosilicate framework assisted by local structural rearrangement. *ACS Catal.* 6 (12): 8420–8431.

82 Jiao, M., Zhao, Y., Jiang, J. et al. (2021). Extra-large pore titanosilicate synthesized via reversible 3D–2D–3D structural transformation as highly active catalyst for cycloalkene epoxidation. *ACS Catal.* 11 (5): 2650–2662.

83 Veselý, O., Mazur, M., Přech, J., and Čejka, J. (2022). *Modified Reverse ADOR Assembles Al-Rich UTL Zeolite From IPC-1P Layers.* Inorganic Chemistry Frontiers.

84 Veselý, O., Eliášová, P., Morris, R.E., and Čejka, J. (2021). Reverse ADOR: reconstruction of UTL zeolite from layered IPC-1P. *Mater. Adv.* 2 (12): 3862–3870.

85 Shamzhy, M.V., Shvets, O.V., Opanasenko, M.V. et al. (2012). Synthesis of isomorphously substituted extra-large pore UTL zeolites. *J. Mater. Chem.* 22 (31): 15793–15803.

86 Shamzhy, M.V., Ochoa-Hernández, C., Kasneryk, V.I. et al. (2016). Direct incorporation of B, Al, and Ga into medium-pore ITH zeolite: synthesis, acidic, and catalytic properties. *Catal. Today* 277: 37–47.

87 Xu, H., Jiang, J., Yang, B. et al. (2014). Effective Baeyer–Villiger oxidation of ketones over germanosilicates. *Catal. Commun.* 55: 83–86.

88 Kasneryk, V.I., Shamzhy, M.V., Opanasenko, M.V., and Čejka, J. (2016). Tuning of textural properties of germanosilicate zeolites ITH and IWW by acidic leaching. *J. Energy Chem.* 25 (2): 318–326.

89 Podolean, I., Zhang, J., Shamzhy, M. et al. (2020). Solvent-free ketalization of polyols over germanosilicate zeolites: the role of the nature and strength of acid sites. *Cat. Sci. Technol.* 10 (24): 8254–8264.

90 Abdi, S., Kubů, M., Li, A. et al. (2022). Addressing confinement effect in alkenes epoxidation using 'isoreticular' titanosilicate zeolite catalysts. *Catal. Today* 390–391: 326–334.

91 Csicsery, S.M. (1984). Shape-selective catalysis in zeolites. *Zeolites* 4 (3): 202–213.

92 Žilková, N., Eliášová, P., Al-Khattaf, S. et al. (2016). The effect of UTL layer connectivity in isoreticular zeolites on the catalytic performance in toluene alkylation. *Catal. Today* 277: 55–60.

93 Remperová, N., Přech, J., Kubů, M. et al. (2022). Gas-phase isomerisation of m-xylene on isoreticular zeolites with tuneable porosity. *Catal. Today* 390–391: 78–91.

94 Liu, L., Díaz, U., Arenal, R. et al. (2017). Generation of subnanometric platinum with high stability during transformation of a 2D zeolite into 3D. *Nat. Mater.* 16 (1): 132–138.

95 Zhang, Y., Kubů, M., Mazur, M., and Čejka, J. (2019). Synthesis of Pt-MWW with controllable nanoparticle size. *Catal. Today* 324: 135–143.

96 Molitorisová, S., Zhang, Y., Kubů, M. et al. (2022). 2D-to-3D zeolite transformation for the preparation of Pd@MWW catalysts with tuneable acidity. *Catal. Today* 390–391: 109–116.

97 Zhang, Y., Kubů, M., Mazur, M., and Čejka, J. (2019). Encapsulation of Pt nanoparticles into IPC-2 and IPC-4 zeolites using the ADOR approach. *Microporous Mesoporous Mater.* 279: 364–370.

98 Zhang, Y., Fulajtárová, K., Kubů, M. et al. (2019). Controlling dispersion and accessibility of Pd nanoparticles via 2D-to-3D zeolite transformation for shape-selective catalysis: Pd@MWW case. *Mater. Today Nano* 8: 100056.

6

Heteroatom-Containing Dendritic Mesoporous Silica Nanoparticles

Bo Peng[1], Jia-Feng Zhou[1], Meng Ding[1], Laurent Bonneviot[2], and Kun Zhang[1,2,3]

[1] *East China Normal University, College of Chemistry and Molecular Engineering, Shanghai Key Laboratory of Green Chemistry and Chemical Processes, 3663 Zhongshan North Road, Shanghai 200062, China*
[2] *Université de Lyon, Institut de Chimie de Lyon, Laboratoire de Chimie, Ecole Normale Supérieure de Lyon, 46 Allée d'italie, Lyon 69364 CEDEX 07, France*
[3] *Institute of Eco-Chongming, 20 Cuiniao Road, Shanghai 202162, China*

6.1 Introduction

The pioneering synthesis of micelle templated mesoporous silicas by Mobil oil company in 1992 [1, 2] has triggered the imagination of researchers leading to an impressive number of variations. A spectacular control was developed on morphologies, topologies, and architectures in combination with tunable doping and even compositions notably in a range of rather high surface areas and large pore volumes opening a much wider field of applications than though at first. One of the first of them, MCM-41, remains the most widely documented and popular. It can be readily synthesized in the presence of cationic surfactant as a soft template in the alkaline aqueous solution. The assembly of MCM-41 silica with the templating micelles is driven by the electrostatic interaction between the cationic surfactant (S^+) and negatively charged silicates (I^-) named the $\{S^+I^-\}$ pathway [3]. It operates below the CMC of the surfactant according to cooperative self-assembly mechanism where mesoporous silica walls grow in the same time than the surfactants assemble into cylindrical micelles leading to a cetyltrimethyl-ammonium (CTA^+) silicate mesophase (silicatropic liquid crystal) with a hexagonal array of linear and parallel channels [4]. Although the principle of the preparation is quite simple, the scalable synthesis of mesoporous silica nanoparticles (MSNs) with larger pore size (>10.0 nm) and concurrent smaller grain size (<50 nm) is not easy [5].

More recently, different strategies to develop MSNs with both larger mesopore sizes and varied morphologies have appeared in the literature [6–14]. Among various investigated MSNs, relatively recently developed dendritic shaped nanoparticles (NPs) have attracted a particular interest [15, 16]. They present several advantages in comparison with more conventional MSNs counterparts: (i) large tunable porosity with pore sizes ranging from c. 10 to 20 nm, permitting the doping of heterometal

Micro-Mesoporous Metallosilicates: Synthesis, Characterization, and Catalytic Applications, First Edition. Edited by Peng Wu and Hao Xu.

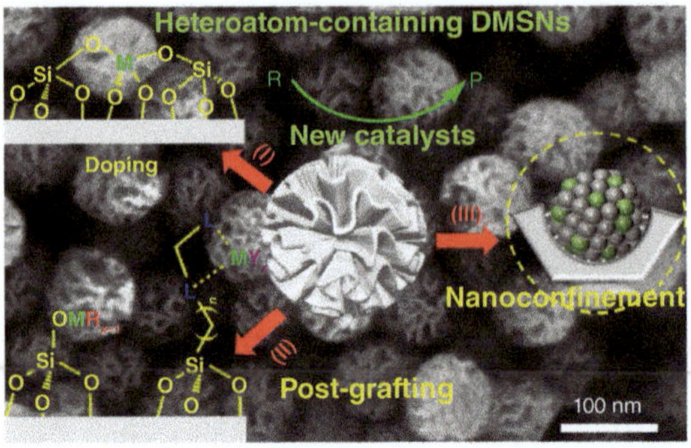

Figure 6.1 Heterometal containing dendritic mesoporous silica nanoparticles with unique center-radial wrinkled structures can be synthesized by three routes: (I) one-pot doping method, (II) post-grafting for surface metal complexes, and (III) loading of metal or metal oxide nanoparticles within the nanopores (nanoconfinement). M is heterometal atoms, such as Ti, V, Sn, Mn, Fe, Co, Al, Au, Ag, Cu, Pt, and Pd. R is alkyl complexes, and Y is counteranion or water complexes.

atoms and the loading of functional species with a size larger than 5 nm relevant for biomedical applications, catalysis, and others; (ii) tunable particle sizes in a very broad size range (from several tens of nanometers to several micrometers); and (iii) unique dendritic pore structures with a special center-radial wrinkled pore morphology ensuring better diffusion properties for the reactants inside the pores (Figure 6.1) [17–24].

Even though significant progresses have been made in the structural and morphological control over dendritic mesoporous silica nanoparticles (DMSNs) by various assembly strategies, mainly including the oil–water two-phase (OWTP) templating approach [15, 25–29], the microemulsion templating method [30–34], the organosilane coupling agent-assisted co-condensation method [35–39], the free-oil anion-assisted self-assembly (AAS) strategy [5, 16, 38, 40–42], and tetraalkoxysilane-assisted self-emulsification templating (TASET) strategy [43], the pure silica-based NPs are chemically inert and have no chemical reactivity and functionality, and the works devoted to the doping (or incorporation) of heterometal atoms and immobilization of metal species and/or oxides, and investigation of the chemical environment impact on the catalytic activity of incorporated metal species and/or oxides are relatively scarce.

In order to achieve more effective doping of heterometal atoms and loading of metal or metal oxide NPs, in this chapter, we first gave a brief introduction on the synthetic methods of pure silica-based DMSNs, and then based on the formation mechanism of DMSNs, we summarized the main synthetic strategies for the heteroatom-containing DMSNs and their typical application in catalysis. Finally, we put forward our personal outlook for the future research directions, including the main challenges faced by the current synthesis and how to maximize the advantages of the pore and structure of the heteroatom-doped DMSNs as heterogeneous catalysts, so as to improve their reaction activity and lifetimes.

6.2 Main Synthetic Methods and Formation Mechanism of Pure Silica-Based Dendritic Mesoporous Silica Nanoparticles (DMSNs)

6.2.1 Main Synthetic Methods of Dendritic Mesoporous Silica Nanoparticles (DMSNs)

Based on the main difference of gel compositions in the mother liquor, all the synthetic strategies of DMSNs were classified to four groups, including the OWTP templating approach (I), the organosilane-assisted co-condensation (OAC) method (II), the AAS strategy (III), and the TASET strategy (IV), which are summarized in detail in Table 6.1. The common feature of these strategies is using a cationic surfactant (CTA^+) as the soft-template, which follows the $\{S^+I^-\}$ synthetic route of classical MCM-41 silica. The differences of each method are as follows: approach (I) used a large number of organic solvents to stabilize the OWTP system for the synthesis of DMSNs, such as diethyl ether [27, 28], cyclohexane [31], toluene [48], 1-octadecene [29], benzyl acetate [44], chlorobenzene [45], and *p*-xylene [46], and the size of DMSNs obtained is usually greater than 200 nm, which may be due to the larger size of microemulsion droplets as the primary template for the assembly of silicate species and cationic surfactant micelles. Approach (II) is hardly applicable because the used co-condensed organosilane coupling agent is very special and commercially unavailable, such as tridecafluorooctyltriethoxysilane (Figure 6.2a) [35], phenyltriethoxysilane (Figure 6.2b,c), and bis[3-(triethoxysilyl)propyl]tetrasulfide (Figure 6.2d–j) [36, 37], and note that the DMSNs obtained by co-condensation method show a smaller particle size of less than 100 nm; without additional organic additives, approaches (III) and (IV) are the greenest and most attractive methods for preparing DMSNs. The two latter relies on simplicity, easy scale-up (up to Kg scale) and reproducibility. Interestingly, they both offer a large span of particle size from 30 to 300 nm and a variety of structure easily regulated by changing the type of anti-anion of the cationic surfactant (Table 6.2) and the pH value of the reaction system (Figure 6.3) [16, 43]. Both methods are also amenable to large-scale synthesis for DMSNs doped with heterometal atoms. This is indeed very useful as support or matrix for heterometal atoms, metals, and metal oxide NPs through post-treatment (Figure 6.1).

6.2.2 Unified Formation Mechanism of Dendritic Mesoporous Silica Nanoparticles

Although a variety of synthetic methods has been reported for DMSN, these particles exhibit very similar open 3D dendritic superstructures with a center-radial pore size gradually increasing from the center to the edge. A key question naturally arises whether their formation mechanism follows the same synthesis route. Lee and his coworker systematically investigated the formation mechanism of wrinkled silica nanoparticles (WSNs) (Figure 6.4), which have similar mesostructure and morphology than DMSNs. They showed that WSNs were generated in the

Table 6.1 Characteristics of the main synthetic strategies leading to a variety of dendritic mesoporous silica nanoparticles (DMSNs).

Synthetic strategy	Typical feature	Gel composition	Co-solvent	References
Oil–water two-phase (OWTP) templating approach (I)	Dynamic self-assembly at oil–water biphase interface	TEOS/CTAB/$NH_3 \cdot H_2O/H_2O$ (CTAB: cetyltrimethylammonium bromide)	Diethyl ether	[28, 30]
		TEOS/CTAB/L-lysine/Polystyrene/H_2O	Octane	[26]
		TEOS/CPB/Pentanol/Urea/H_2O (CPB: cetylpyridinium bromide)	Cyclohexane	[15]
		TEOS/CTAB/$NH_3 \cdot H_2O/H_2O$	Ethanol + ethyl ether	[27]
		TEOS/CPB/Butanol/Urea/H_2O	Cyclohexane/toluene	[31]
		TEOS/CTAC/$TEAH_3/H_2O$ (CTAC: cetyltrimethylammonium chloride; TEAH3: triethanolamine)	1-Octadecene	[29]
		TEOS/CTAB/$NH_3 \cdot H_2O/H_2O$	Acetone + benzyl acetate	[44]
		TEOS/CTAC/$TEAH_3/H_2O$	Chlorobenzene	[45]
		TEOS/CTAB/$NH_3 \cdot H_2O/H_2O$	Ethanol + *p*-xylene	[46]
Organosilane-assisted co-condensation (OAC) method (II)	Introduction of organic silane coupling agent with hydrophobic functions	TEOS/F13/CTAC/$TEAH_3/H_2O$ (F13: tridecafluorooctyltriethoxysilane)	Ethanol	[35]
		TEOS/PTES/CTAB /$NH_3 \cdot H_2O/H_2O$ (PTES: phenyltriethoxysilane)	Ethanol	[36]
		TEOS/BTES/CTAC/$TEAH_3/H_2O$ (BTES: bis[3-(triethoxysilyl)propyl]tetrasulfide)	Free organic solvent	[37]
Anion-assisted self-assembly (AAS) strategy (III)	Introduction of bulky counter anions with stronger affinity on the cationic surfactants (more details see Table 6.2)	TEOS/ CTATos/$TEAH_3/H_2O$ (CTATos: cetyltrimethylammonium tosylate)	Free organic solvent	[16]
Tetraalkoxysilane-assisted self-emulsification templating (TASET) strategy (IV)	Special use of tetraalkoxysilane with low hydrolysis rate, e.g. tetrabutoxysilane (TBOS)	TAOS/CTAB/$TEAH_3/H_2O$ (TAOS: tetraalkoxysilane)	Free organic solvent	[43]

Table 6.2 DMSN pore and particle sizes in function of the surfactant counterion (at low alkaline aqueous solution).

Year, ref.	Directing template/ anion	Anion structure	Catalyst	Pore size (nm)	Particle size (nm)
2013, [16]	CTA^+ Tos^-	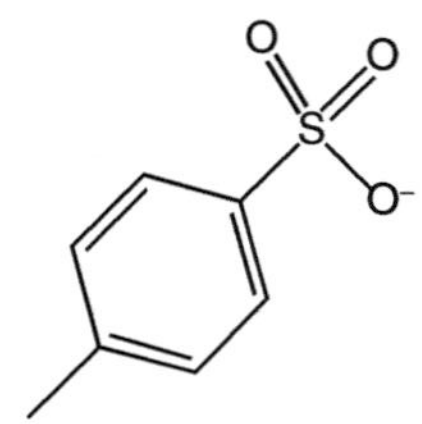	Triethanolamine	16, 17, 54	74, 115, 113
2014, [5]	CTA^+ Tos^-	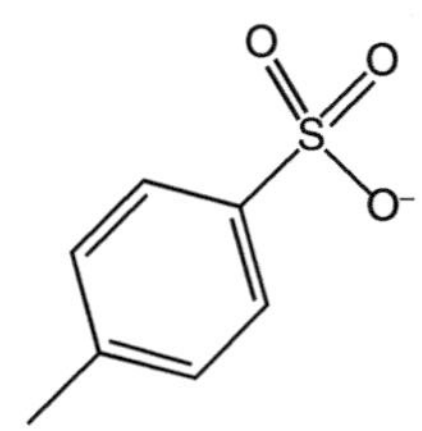	Triethanolamine	3.3, 10.6	40, 70, 85, 102, 115
2017, [41]	CTAB/ sodium dodecyl sulfate	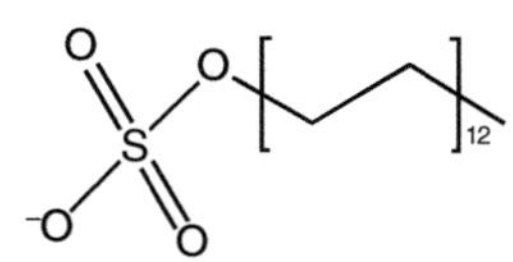	Triethanolamine	3.3, 10.6	105, 120, 150

Table 6.2 (Continued)

Year, ref.	Directing template/ anion	Anion structure	Catalyst	Pore size (nm)	Particle size (nm)
2016, [47]	CTAB/ sodium salicylate	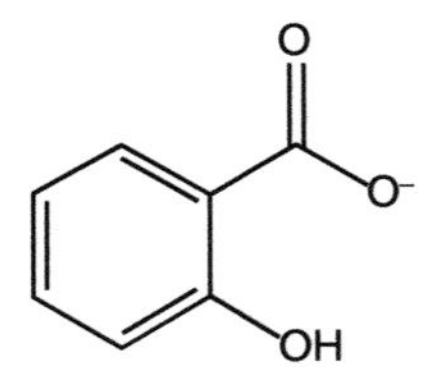	Triethanolamine	8.4, 17.5	140, 200
2016, [40]	CTAB/ fluorocarbon anion (F2 & F8)	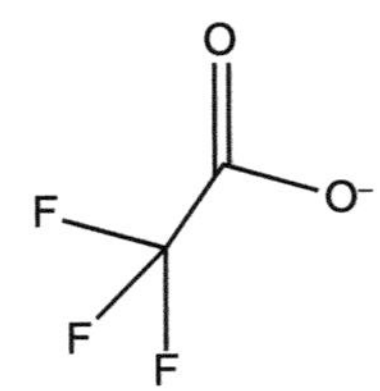	Triethanolamine	2.7, 22.4 2.4, 22.2	169, 86, 79
2018, [42]	CTAB/ fluorocarbon anion (F4)	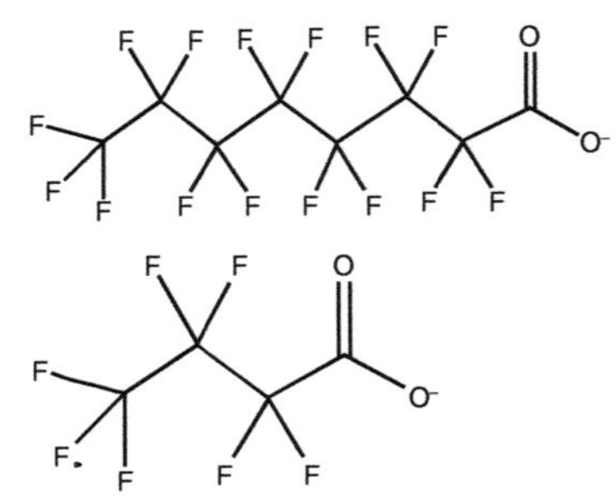	Triethanolamine	12–16	110

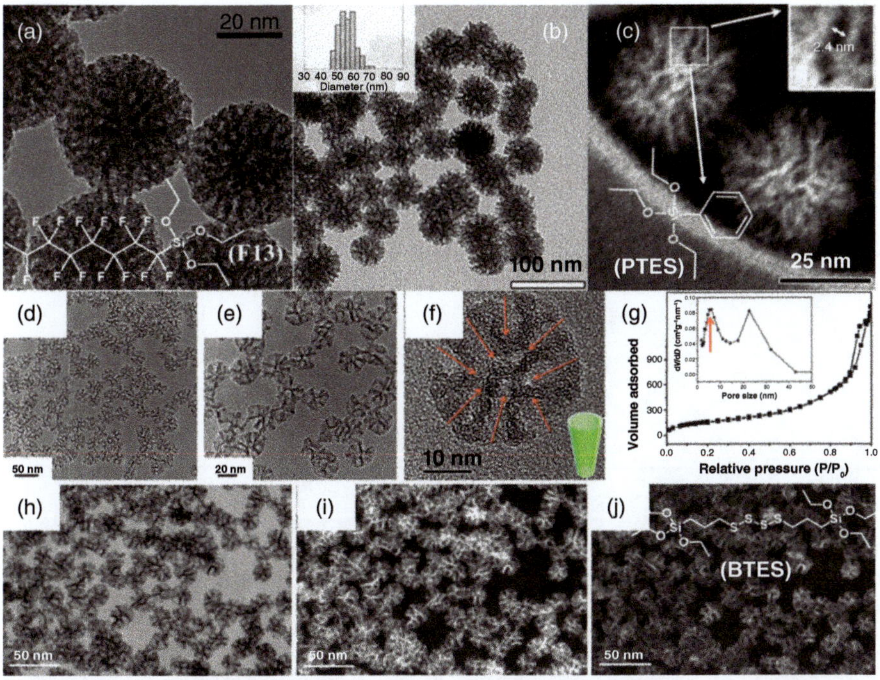

Figure 6.2 One-pot condensation strategy for dendritic mesoporous organosilica nanoparticles (DMONs) with varied organosilane coupling agents of tridecafluorooctyltriethoxysilane (a), Source: Pereira et.al. [35]/Reproduced with permission from American Chemical Society., The American Chemical Society; phenyltriethoxysilane (b, c), Source: Paula et.al. [36]/Reproduced with permission from Royal Society of Chemistry; and bis[3-(triethoxysilyl)propyl]tetrasulfide (d–j). Detailed structural analysis for DMONS with BTES as coupling agent (d–j): (d–f) TEM images of DMONs at different magnifications (inset of f: the pore shape of DMONs); (d) N_2 adsorption–desorption isotherm and (inset of g) the corresponding pore size distribution of DMONs; (h–j) bright-field (h) and dark-field (i) STEM images and their corresponding SEM image (g) of DMONs. Source: Wu et al. [37]/Reproduced with permission from John Wiley & Sons, Inc.

bicontinuous microemulsion phase of the Winsor III system [31]. They could tune both the structure of silica NPs and the inter-wrinkle distance of WSNs by simply adjusting the phase behavior of this system, which depends on the type of cosolvents (Figure 6.4a–c) and the water-surfactant-oil mixing ratio (Figure 6.4d) [31, 48]. Later, Polshettiwar and his coworkers further supported the idea that nanoreactor templates generated during the structural evolution of bicontinuous microemulsion droplets (BMDs) and their subsequent coalescence the conditions for the formation of these complex dendritic morphologies [34, 49].

The formation of BMDs in the oil–surfactant–water ternary system can be easily understood by changing the composition of ternary system. However, for the synthetic approaches (II), (III), and (IV), it is usually difficult to observe microemulsion droplets without directly introducing oil solvent, unless hydrophobic tetraalkoxysilane (TAOS) used as a silicon source is regarded as the oil phase. If this is true, the morphology and particle size of obtained MSNs will be strongly dependent on

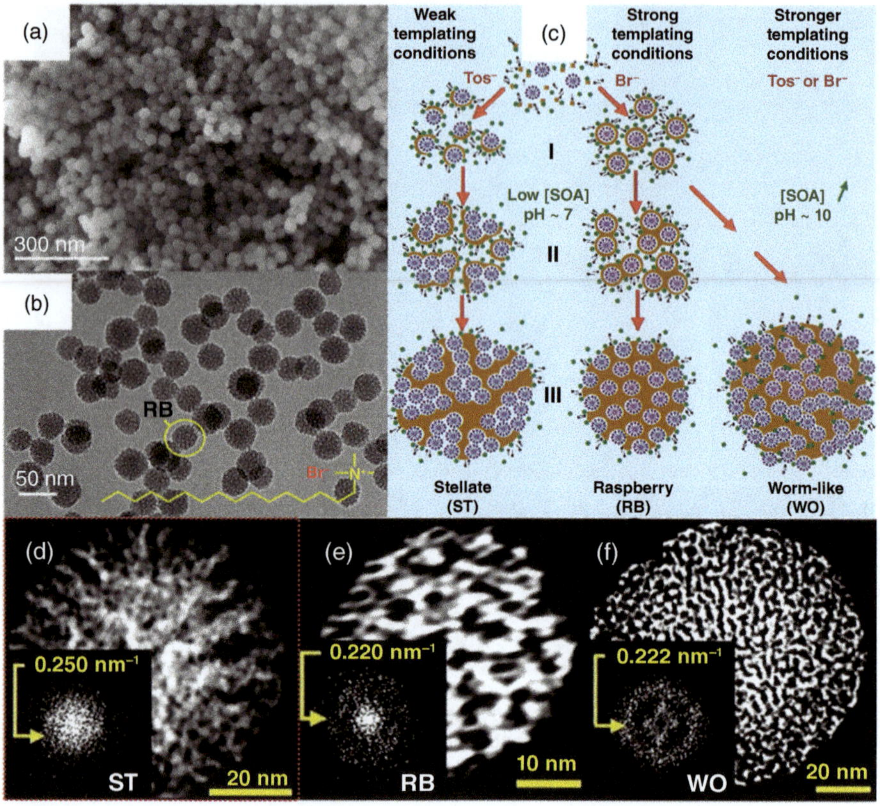

Figure 6.3 Anion-assisted self-assembly (AAS) strategy for mesoporous silica nanoparticles with varied mesostructures at different pH values. (a) SEM and (b) HRTEM images of MSNs with raspberry-like (RB) channel morphologies. (c) Schematic illustration of the synthesis mechanisms at low or high small organic amines (SOA, green dots) concentration (pH $\approx$ 7 or 10) and Tos^- (left) or Br^- (middle) (template counterion is not represented here): (I) partly silicated micelles, (II) block formation, and (III) block aggregation into stellate or raspberry nanoparticles. (d–f) Diffraction pattern of MSNs with (d) stellate (ST), (e) raspberry (RB), and (f) worm-like (WO) channel morphologies. Source: Zhang et al.[16]/Reproduced with permission from American Chemical Society.

the type of TAOS and pH value of reaction system. Obviously, these key reaction parameters will ultimately affect the loading, isolation state, and coordination environment of heterometal atoms in the silica framework by one-pot doping method.

The above hypothesis has been recently verified by Zhang et al. through delicately designed experiments [43]. When TMOS (Figure 6.5a,e) and TEOS (Figure 6.5b,f) with fast hydrolysis rate were used, the raspberry-shaped MSNs were obtained. However, if changing the silicon source to tetrapropoxysilane (TPOS) (Figure 6.5c,g) and tetrabutoxysilane (TBOS) (Figure 6.5d,h) with slower hydrolysis rates and stronger hydrophobicity, highly dispersed dendritic MSNs in a size of c. 60 nm could be easily prepared with CTAB as surfactant template in the aqueous solution. The characterization of N_2 physisorption showed that MSNs-M and NSNs-E only showed a narrow

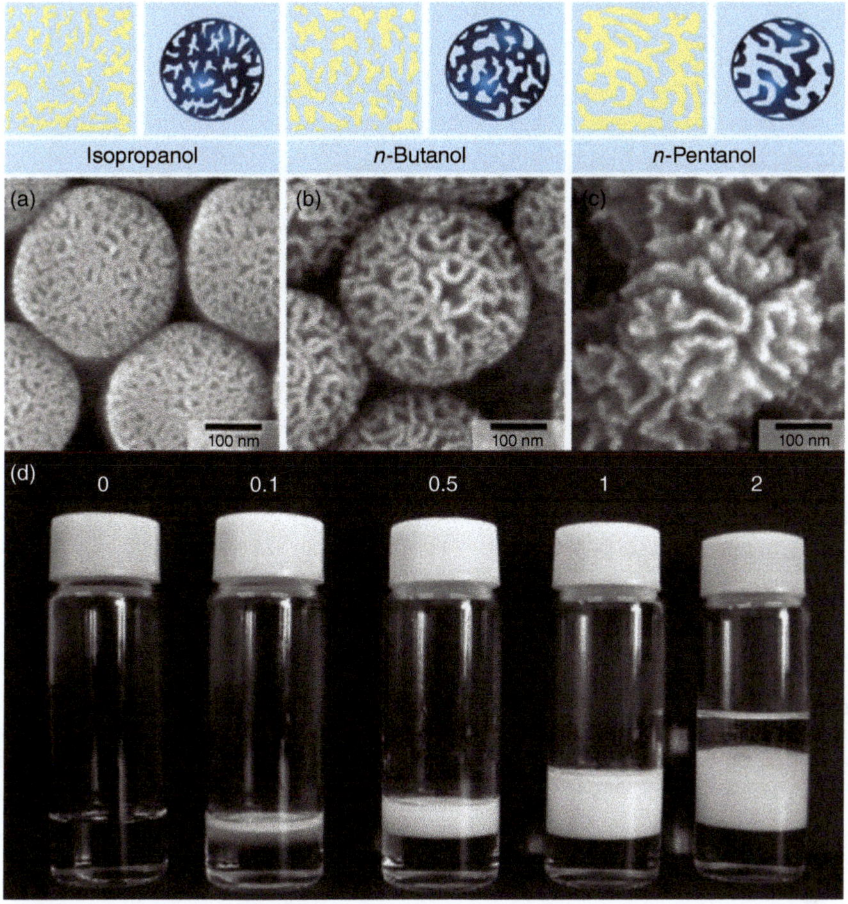

Figure 6.4 Schematic illustrations of microemulsion phases of reaction mixtures with identical molar amounts (6.0 mmol) of different cosolvents and SEM images of wrinkled silica nanoparticles (WSNs) generated from these mixtures: (a) *iso*-propanol, (b) *n*-butanol, and (c) *n*-pentanol. (d) Optical micrographs of separated phases of the Winsor III system at different volume ratios of cyclohexane to 15 mL of aqueous solution containing urea, cationic surfactant, and *iso*-propanol. Source: Moon and Lee[31]/Reproduced with permission from American Chemical Society.

size distribution at 1.9 and 2.5 nm (Figure 6.5i,j), similar to that of MCM-41, while MSNs-P and MSNs-B showed a bimodal pore size distributions centered at 5.5 and 6.6 nm with a shoulder at 3.0 nm, which is a typical structure feature of dendritic MSNs (Figure 6.5i,j).

The optical digital images provide the direct evidence that TAOS as silicon sources involves the formation of TAOS-surfactant-water ternary system and witness the evolution of phase behaviors with reaction time (Figure 6.6a–d). Optical images of equilibrium phases of each reaction stage were collected at precisely controlled time after stirring. At the initial stage, apart from TMOS which suffered rapid hydrolysis (Figure 6.6a), two clearly separated upper and lower phases were

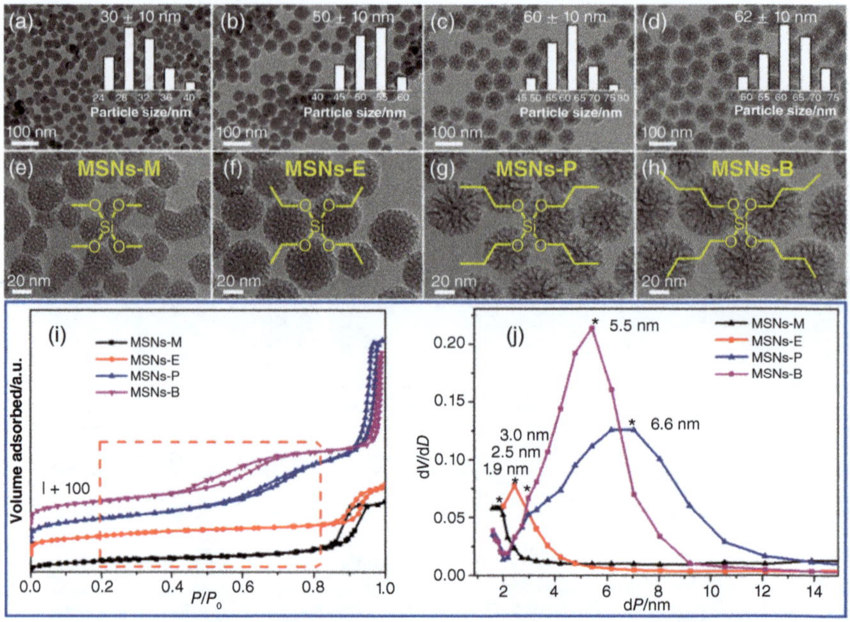

Figure 6.5 (a–h) TEM images of MSNs with varied silicon sources of tetramethoxysilane (TMOS) (a, e), tetraethoxysilane (TEOS) (b, f), tetrapropoxysilane (TPOS) (c, g), and tetrabutoxysilane (TBOS) (d, h). Particle size (PS) histograms obtained from the TEM pictures on N particles (N > 100 in each case) are provided as the insets of a, b, c, and d; N_2 adsorption–desorption isotherms (i) and pore size distribution (PSD) plots (j) of MSNs-R (the last capital letter R in the sample name of MSNs-R means the used silicon source of $Si(OR)_4$ with the abbreviation of alkyl groups, i.e. methyl, ethyl, propyl, and butyl). PSD calculated by the BJH method from adsorption branches [43]. Source: (a–h) Peng et al.[43]/Reproduced from ChemRxiv/CC BY-NC-ND 4.0. (i, j) [43]/American Chemical Society.

observed for other TAOS (TEOS, TPOS, and TBOS) at 0 min, which are composed of a white-milk emulsion and clear water phase, respectively, suggesting TAOS involves in the formation of microemulsion in the upper layer, and the stabilized time of upper-layer phase depends on the type of used TAOS with a trend of TEOS (less than 10 min) < TPOS (10 min) < TBOS (20 min) (Figure 6.6b–d). As the reaction proceeds further, due to the consumption of TAOS, the upper-layer phase gradually becomes clear and its volume fraction decreases, suggesting the breakage of microemulsion droplets and consecutive consumption of TAOS. Concomitantly, the bottom-layer phase becomes colloidal blue because of the formation of colloidal silica NPs (Figure 6.6b–d). Obviously, the longer retention time of emulsion droplets at the upper-layer is more conducive to the formation of DMSNs (Figure 6.5c,d). Thus, this synthetic methodology was called as a TASET for DMSNs. This also answers our previous observation that the final morphologies of obtained MSNs are dependent on the pH value (Figure 6.3c) and that under neutralized conditions (pH ≈ 6–7) with low hydrolysis rate of TAOS, DMSNs is favorably formed, while at high pH value, the raspberry-like MSNs were obtained [29].

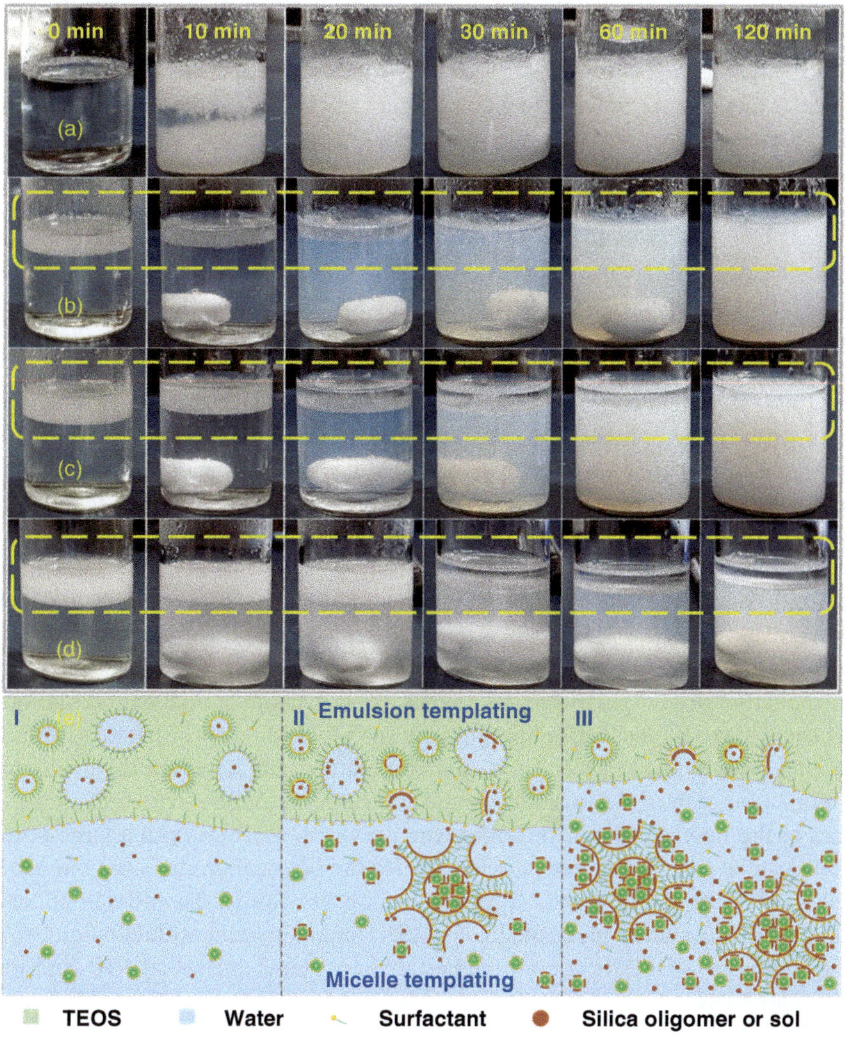

Figure 6.6 Equilibrium phases of reaction mixtures of (a) MSNs-M, (b) MSNs-E, (c) MSNs-P, and (d) MSNs-B at different reaction stage, 0, 10, 20, 30, 60, and 120 min. (e) Schematic illustration of formation mechanism of MSNs with controlled mesostructure using tetraalkoxysilane-assisted self-emulsification templating (TASET) strategy; here, tetraethoxysilane (TEOS) was used as a show example [43]. (Importantly note that, except for TMOS (a), the other three systems of TEOS (b), TPOS (c), and TBOS (d) are composed of two layers of different liquids at the early stage of reaction: the upper layer is milk emulsion, and the lower layer is water-based solution; with the time running, the milky upper-layer becomes clear, indicating the breakage of emulsion.) Source: Peng et al.[43]/Reproduced from ChemRxiv/CC BY-NC-ND 4.0.

Based on these observations, it can be concluded that all the synthesis approaches for DMSNs follow the same formation mechanism, i.e. a dynamic dual-templating synergetic co-assembly mechanism with microemulsion droplets as primary (or first level) template and cationic micelle as second template, which was illustrated by three main steps (Figure 6.6e): (I) at the initial stage of reaction, due to the presence of oily organic solvent or silica precursors (TAOS), immiscible oil/water biphase could be formed in the reaction system, and at the upper-layer (oil phase), water-in-oil microemulsion droplets were formed, while at bottom-layer, owing to the hydrophilicity of cationic surfactant head, the micelles are preferably formed in the water phase, and note that at two levels of templates, microemulsion droplets and micelles are both partially covered (or coated) by silica oligomers under alkaline condition, which favors the aggregation of microemulsion nanodroplets and silicate-micelles composites by condensation of surface silica to form the skeleton framework of MSNs. (II) With reaction proceeding under the vigorous stirring, two phases (oil and water) could be forcedly mixed together, and the microemulsion droplets with larger particle size collide, coalesce, or aggregate with each other to form a dendritic skeleton or framework of silica particles, which could act as nucleus for further growth with the filling of micelle-silica composites following the heterogeneous nucleation mechanism. (III) Finally, the solid dendritic DMSNs were collected.

The unified dynamic dual-templating self-assembly mechanism answers why the structure evolution of DMSNs depends on the thermodynamic stability of emulsion droplets and is kinetically controlled by a change of reaction parameters, such as the introduction of non-polar organic solvent, co-surfactant, special counter anion of surfactant and organosilane coupling agent, the way and rate of stirring, pH value of reaction system, and reaction time and temperature, etc. Based on the dynamic dual-templating synergetic co-assembly mechanism, the unique structural characteristics of the open 3D dendritic superstructures with a central radial large-pore channels can be clearly illustrated, especially for the physical origin of the bimodal or multimodal size distribution of N_2 physisorption. Obviously, the in-depth understanding on the formation mechanism of DMSNs and its structural elucidation benefit the synthesis of DMSNs containing heterometal atoms and metal NPs (and/or metal oxides NPs).

6.3 Synthesis of Heteroatom-Containing DMSNs and Their Catalytic Applications

Compared with traditional mesoporous silica (MCM-41 and SBA-15), dendritic MSNs with central radial pore structures are a new type of porous materials. Due to their unique structural advantages such as open three-dimensional dendritic super structure, large pore channels, highly accessible internal surface area, and variable particle sizes, they can be ideal nanoreactors for heterogeneous catalysis. Due to the chemical inertness of pure silica-based DMSNs, the key to realize the catalytic application of DMSNs containing heteroatoms is how to effectively introduce catalytic active components, such as heterometal atoms, surface metal complexes, metal NPs (or metal oxide NPs), and their composites.

6.3.1 One-Pot Doping Strategy for DMSNs Containing Heteroatoms (Al/Ti/V/Sn/Mn/Fe/Co)

Direct incorporation of heteroatoms into the silica framework with varied metal ion precursors is a simple way to synthesize mesoporous silica containing heteroatoms by one-pot strategy. The main group element aluminum (Al) and the transition metal group atoms (Ti/V/Sn/Mn/Fe/Co) are the most commonly used metals. The key of the direct synthesis method is how to realize the high dispersion of single metal atoms in the silica wall while maintaining the unique morphology of DMSNs. Zeng's work demonstrates that the conflict between morphological control and heteroatom incorporation can be resolved in a CTAB-stabilized toluene–water–ethanol microemulsion system [50]. For the first time, the monodisperse aluminosilicate spheres with macro–meso–micro-hierarchical porosity (denoted as MASS) are synthesized in one step at room temperature (Figure 6.7a). Simultaneous tuning of Al/Si ratio (0–0.35) and the hierarchical pore structure is realized by Hofmeister anion effects of the Al source itself, $[Al(OH)_4]^-$, which changes the geometry of CTAB micelles and giant vesicles, leading to the mesophase transition from p6mm to $Pm\overline{3}n$ (Figure 6.7a). The Al species in the as-made sample are in a tetrahedrally coordinated status evidenced by ^{27}Al magical-angle spinning NMR spectra (Figure 6.7b, left panel), and preliminary catalytic test in the model reaction of CO_2 hydrogenation shows that the improved acidity of MASS as a catalyst support benefits the production of dimethyl ether (DME) as a final product. Importantly note that even though Al atoms in as-made MASS are completely dispersed in the tetrahedral coordination state, after high temperature calcination, a small part of the four-coordination Al atoms transfer to the five-coordination Al (~30 ppm peak) and octahedral Al species (~0 ppm peak) (Figure 6.7b, right panel).

To further improve the dispersion of Al atoms in the silica framework, Yi et al. reported a convenient and effective method for the preparation of acid tunable nano-catalysts with a hierarchical and center-radial mesoporous structure, called pH-assisted delay addition [51]. In this method, the Al precursor was added to a pH adjusted solution, which contains preformed hierarchically and radially mesoporous (DMSNs) nanostructures. Through the second hydrothermal treatment, Al was incorporated into the silica framework in purely tetrahedral coordinated state, thus maintaining the unique structure of DMSNs with acidic properties [52]. Al-DMSNs synthesized with this method showed excellent catalytic performance for the acid catalyzed reactions, such as cracking of 1,3,5-triisopropylbenzene and hydrolysis of sucrose. In contrast, if a small amount of Al precursor was included in the initial solution, as in the case of the direct synthesis, the wrinkle center-radial structures of DMSNs cannot be maintained.

Similar "delay addition and second crystallization" strategy was used to synthesize DMSNs containing transition metals, such as Ti, Sn, and V. The Ti-DMSN catalysts exhibit a good dispersion of isolated tetrahedrally coordinated Ti atoms resulting

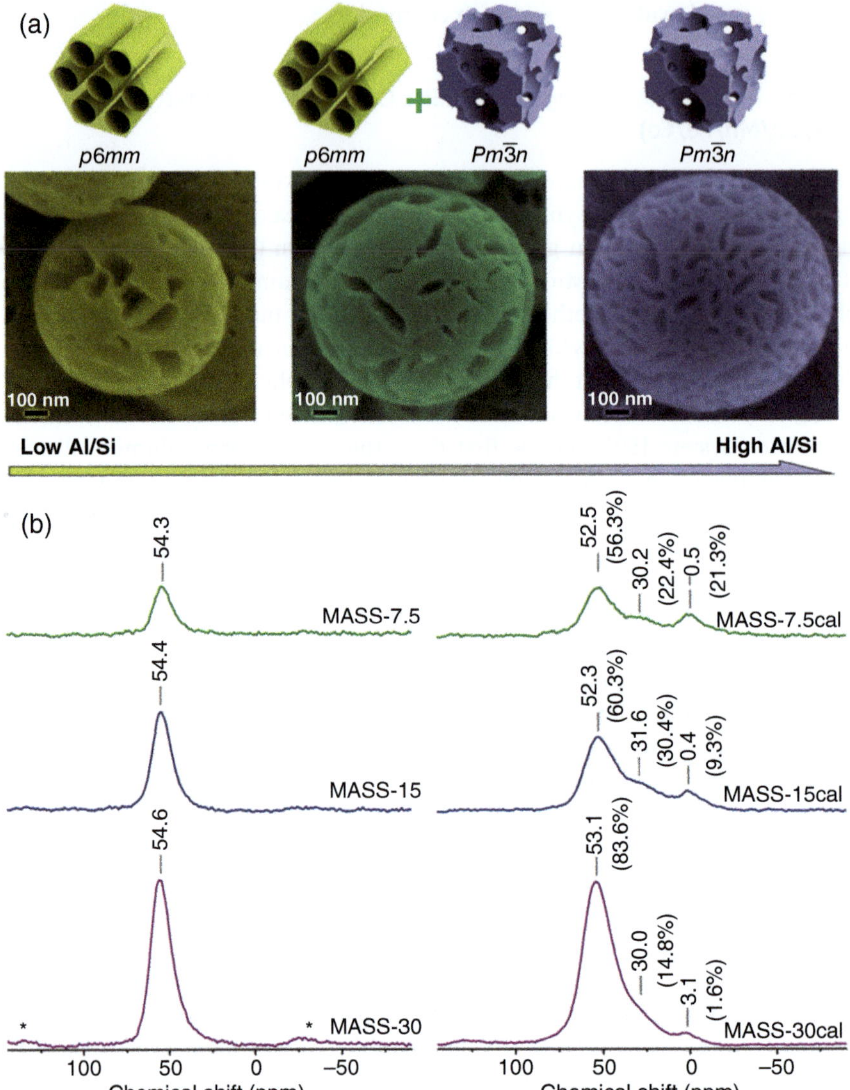

Figure 6.7 (a) Schematic illustration of mesophase transition of MASS from p6mm to Pm3̄n. (b) ^{27}Al magical-angle spinning NMR spectra of MASS samples before/after calcination. Numbers in parentheses are percentages of deconvoluted individual peak areas among total peak areas. Spinning sidebands are marked by asterisks. Source: Sheng and Zeng [50]/Reproduced with permission from American Chemical Society.

in higher catalytic activity than the traditional Ti-MCM-41 with one-dimensional pores for propylene oxidation using *tert*-butyl hydroperoxide as the oxidant [53, 54]. After post-treatment, the maximum turnover frequency of 1401 h^{-1} with an epoxide selectivity of 95% was obtained using the silylated Ti-DMSNs catalyst with a Si/Ti ratio of 209, suggesting that the delicate change of surrounding microenvironment of Ti active sites also significantly affects the catalytic performance of propylene oxidation [53]. Using tin chloride ($SnCl_4$) and ammonium metavanadate (NH_4VO_3) as metal precursors, Sn- and V-containing dendritic mesoporous silica nanoparticles (Sn-DMSNs and V-DMSNs) with highly dispersed SnO_x and VO_x species were also successfully prepared by the in-situ co-assembly method [55, 56]. Both catalysts exhibit excellent catalytic performance and stability for the non-oxidative dehydrogenation of propane to propylene.

In fact, heteroatom doped DMSNs not only have important applications in heterogeneous catalytic reactions but also show potential application prospects in the field of nano diagnosis and drug delivery. Very recently, manganese ions, iron ions, or cobalt ions and tetrasulfide bonds are co-incorporated into the framework of dendritic mesoporous organosilica (DMOS) to yield multifunctional NPs denoted as Mn-DMOS, Fe-DMOS, or Co-DMOS by directly doping metal ions during the one-pot synthetic process (Scheme 6.1a) [57]. Due to co-incorporation of metal ions and tetrasulfide bonds, these designed nanocarriers have multi-functions rather than only for cargo delivery. For example, as a proof of concept, the nanocomposite (NC) is established based on Mn-DMOS as an efficient nanocarrier for indocyanine green (ICG) delivery and modification with polyethylene glycol (Scheme 6.1b). In the tumor microenvironment, the generated hydrogen sulfide (H_2S) arising from the reaction between tetrasulfide bond and over-expressed glutathione (GSH) causes mitochondrial injury to reduce cellular respiration. The released Mn^{2+} from the rapidly decomposed NC catalyzes the endogenous hydrogen peroxide to produce oxygen (O_2). The photothermal effect from the released ICG initiated by the near-infrared (IR) light induces cancer cell apoptosis and simultaneously enhances the content of blood O_2 at tumor sites, consequently enhancing the photodynamic therapy efficiency of ICG.

6.3.2 Post-grafting for Surface Metal Complexes

The post-grating technique is an effective way to incorporate the heterometal complexes to the silica surface. Compared with the one-pot doping strategy, this method shows the following advantages: the mesostructure of DMSNs can be well maintained; the types of heterometal atoms are not limited; metal active sites can be easily accessible by reactants; the microenvironment of metal active sites can be precisely tuned, including surface hydrophilicity and hydrophobicity, the spatial distance between the metal active sites, and the density and type of adjacent functional groups [58–60]. In this technique, the anchoring point of metal complexes is usually the exposed surface silane group ($\equiv$ Si—OH) or pre-grafted organosilane, which has special coordination ability with the target metal.

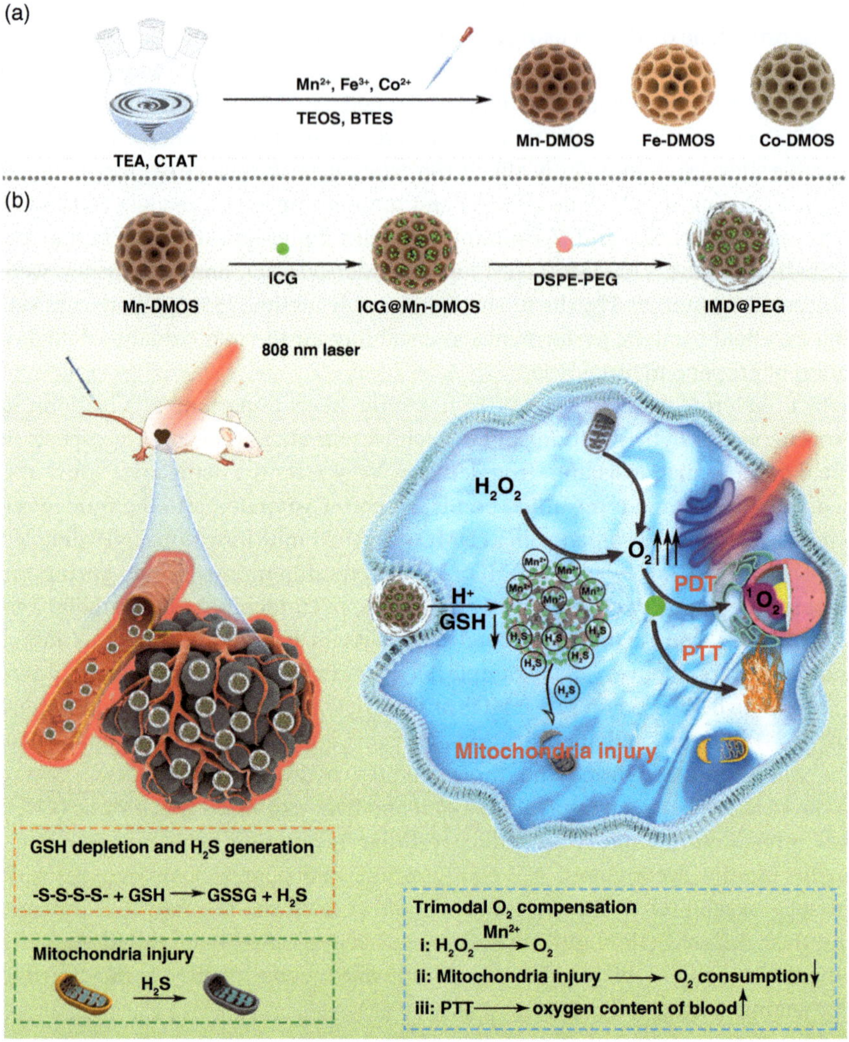

Scheme 6.1 Schematic illustration of (a) the synthetic process of Mn, Fe, and Co-DMOS and (b) the antitumor performance of the IMD@PEG NC. Reproduced with permission from Ref. [57] Copyright 2012, Wiley-VCH.

Surface organometallic chemistry has emerged as a powerful approach to generate well-defined surface sites with fully controlled ligand environments, and generally the bond formation between the metal center and the surface usually involves the protonolysis of an anionic ligand of a metal complex of general formula MR_x by surface OH groups (e.g. ≡SiOH for silica resulting in grafting through a ≡SiO-MR_{x-1} and the release of RH; Figure 6.1 and Scheme 6.2) [62, 63]. Recently, with the post-grating technique, the DMSN-supported tantalum hydride (TaH/DMSNs) was prepared by grafting the [Ta(=CH*t*Bu)(CH_2*t*Bu)$_3$] complex onto dendritic silica NPs, and then treatment under hydrogen atmosphere at 150 °C for 12 hours (Scheme 6.2) [64].

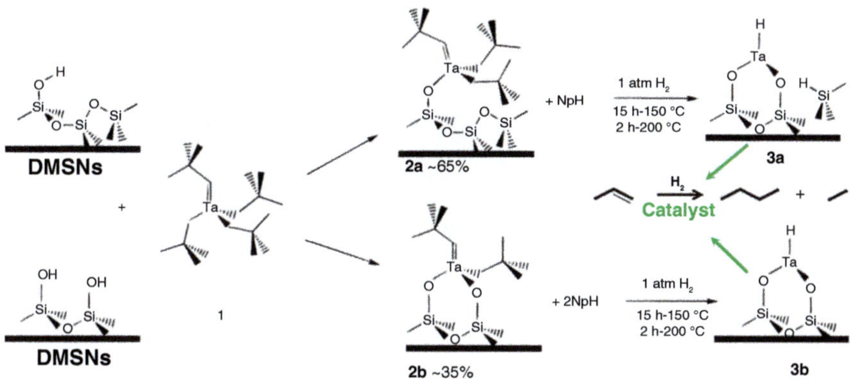

Scheme 6.2 Schematic illustration of the preparation of supported tantalum hydride by grafting of the [Ta(=CH*t*Bu)(CH_2*t*Bu)$_3$] complex onto DMSNs through protonolysis by surface silanols, which severed as a catalyst for hydrometathesis reaction of olefins. Reproduced with permission from Ref. [61] Copyright 1996, The American Chemical Society.

IR spectroscopy and solid-state NMR spectra (^{1}H MAS and ^{13}C-CP/MAS spectrum) confirmed that the Ta/hydrocarbyl complex was successfully grafted onto DMSNs. TaH/DMSN shows remarkable catalytic activity and stability, with excellent potential of regeneration, in the hydrometathesis reaction of olefins.

Another way to encapsulate heterometal complexes into nanostructured silica is to use the metal coördination of organic functional groups tethered on the silica surface by a dynamic linker of alkyl chain. The often-used organic functions mainly include amino group (—NH_2), thiol group (—SH), pyridine group, diamine group (—NH—CH_2—CH_2—NH_2), etc. Very recently, Guari and his coworkers showed that through a multi-step grafting process, a binuclear {Mn(bpy)(H_2O)}(μ-2-MeC_6H_4COO)$_2$(μ-O){Mn(bpy)(NO_3)} complex with catalase-mimicking activity can be readily immobilized onto the DMSNs (Scheme 6.3), which can be used as an efficient nanocatalyst for the dismutation reaction of hydrogen peroxide [22]. The high performance of nanocatalysts is contributed by the dual functionalization of DMSNs and their colloidal stability.

6.3.3 Loading of Metal and/or Metal Oxide Nanoparticles Within the Nanopores

The most significant feature of DMSNs is its bimodal pore size distribution where the relative fraction of dendritic large pores and small spherical mesopores can be rationally tailored by the sophisticated control of reaction parameters of the synthesis. In particular, the small open spherical mesopores in a size of less than 3.0 nm provide an ideal nanospace for the encapsulation of metal NPs and/or metal oxide NPs [65–68].

With DMSNs as an ideal confinement matrix, Ag NPs were trapped into the small mesopores and the size of Ag was limited to less than 3.0 nm due to the unique nanoconfinement (Figure 6.8a–c), which showed excellent catalytic performance in the selective reduction of 4-nitrophenol (4-NP) to 4-aminophenol

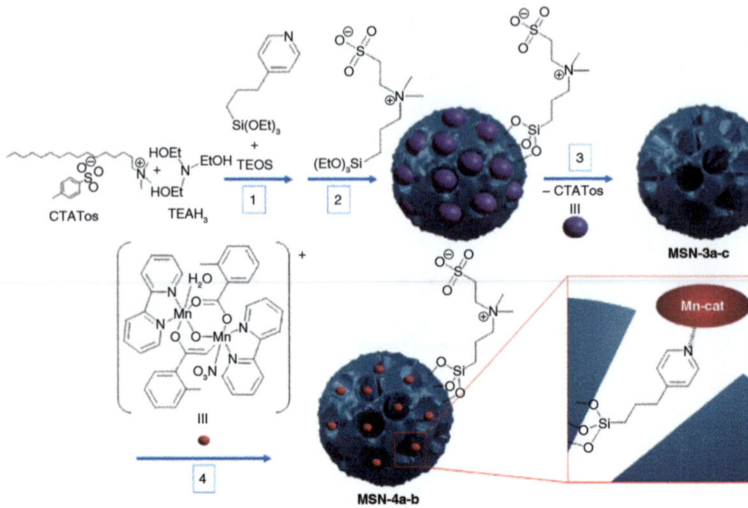

Scheme 6.3 Scheme of the strategy for inner and outer sequential *in situ* functionalization of DMSN nanoparticles with 1,4-pyridine and zwitterion followed by Mn-cat complex encapsulation [22]. Free access by the Creative Commons Attribution (CC BY) license, 2022, MDPI.

(4-AP) (Figure 6.8d) [69]. Combined with the characterizations of optical spectra, the authors confirm that the structural water molecules (SWs) adsorbed on Ag NPs in the form of $\{OH^- \cdot H_2O@Ag\ NPs\}$ are the real catalytic active site (not alone Ag NPs, Figure 6.8d), where SWs as a bridge connect $NaBH_4$ reducer and 4-NP reactant together, which accelerates the surface electron transfer by an interface non-covalent π conjugation due to the orbital overlap between the interacted multiple atoms, consequently enhancing the reaction kinetics of 4-NP reduction [70, 71]. This new mechanism also answers why the catalytic reduction of 4-NP does not follow the classic Langmuir Hinshelwood (L-H) mechanism, and the isotope labeling experiment proves that the hydrogen source of 4-AP product is water, not $NaBH_4$ reducer [70].

The similar strategy was used to synthesize the thiol-modified DMSNs supported Au NPs catalysts for the high-temperature gas-phase selective oxidation of benzyl alcohol with O_2 as oxidant. The chemical reactivity and stability of the obtained Au NPs@DMSNs catalysts were strongly dependent on the pore structure and the nature of organic functional groups grafted on the silica framework [39, 72]. The Au/DMSNs-4.4%SH catalyst with low loading weight of Au (~2%) and thiol ligand (~4.4%) achieved high conversion of benzyl alcohol (*Conv.* 91%) and selectivity for benzaldehyde (*Sel.* 98%) and unprecedented long lifetime up to 820 hours at 250 °C [72], which surpassed all the reported Au based nanocatalysts. The excellent catalytic performance and ultra-long lifetime of Au NPs@DMSNs catalyst were attributed to the nanoconfinement of DMSNs on the Au NPs and the unique packing of thiol groups on the surface of Au NPs, which limited the growth of Au NPs and concomitantly constructed a unique nanoconfined microenvironment at nanoscale interface for the synergistic adsorption and activation of reaction

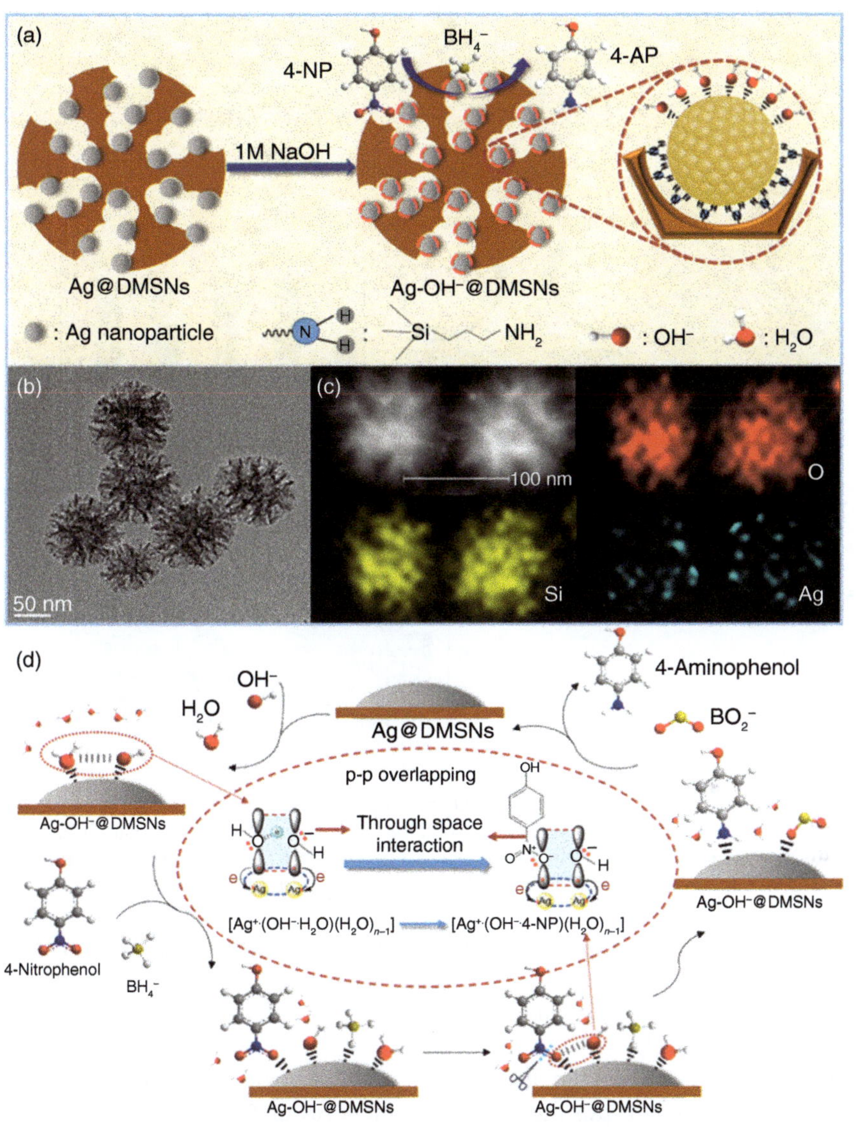

Figure 6.8 (a) Schematic illustration of the synthesis procedure of Ag@DMSNs. Bright-field transmission electron microscopy (b) and high-resolution HAADF-STEM (c) images of 2.0%Ag@DMSNs and the corresponding STEM-EDX mappings: Si (yellow), O (red), and Ag (blue), respectively. (d) Surface ligand delocalized *p* band intermediate state (PBIS) was proposed to understand the catalytic reduction mechanism of 4-nitrophenol (4-NP) where p band formed by space interaction between p orbitals of O atoms in the structural water and 4-NP provides the alternative channels for surface electron and proton transfer. Source: Hu et al. [69].

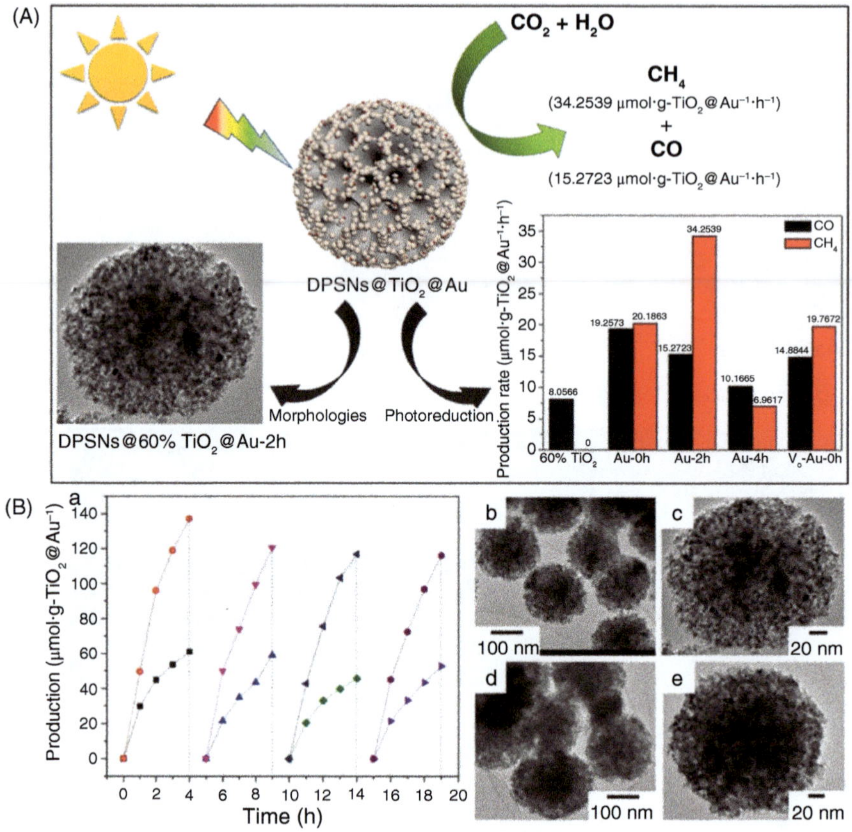

Figure 6.9 (A) Selective hydrogenation of CO_2 on DPSNs@TiO_2@Au nanocatalyst under simulated sunlight irradiation. (B) (a) Cycling runs for CO_2 reduction over DPSNs@60% TiO_2@Au-2h. TEM images of DPSNs@60% TiO_2@Au-2h before (b, c) and after (d, e) three cycling runs. Source: Wang et al. [75]/Reproduced with permission from Elsevier.

substrate and oxygen. Similar methodology was used to synthesize other noble metal nanocatalysts, such as Pd NPs and Ru NPs, which exhibit high catalytic performance in the Suzuki coupling reactions [73] and the hydrogenolysis of alkanes [74], respectively.

Not only single metal NPs but also the composite of metal NPs and metal oxide NPs can be incorporated into the confined nanopores of DMSNs, and they also show ultra-high catalytic performance. With a multi-step synthetic strategy, Du and his coworkers reported the preparation of a series of DPSNs@TiO_2@Au NCs with different sizes of Au NPs, which can be used as highly active catalysts for the selective hydrogenation of CO_2 under simulated sunlight irradiation with 300 W Xe lamp (Figure 6.9A) [75, 76]. Due to the introduction of Au with the local surface plasmon resonance (LSPR) effect, DPSNs@TiO_2@Au with appropriate size of Au NPs (3–5 nm) exhibited superior CH_4 production rate of 34.25 µmol·g-TiO_2@Au^{-1}·h^{-1}, medium CO production rate of 15.27 µmol·g-TiO_2@Au^{-1}·h^{-1}, and high cycle stability (Figure 6.9B).

6.4 Summary and Perspectives

In this chapter, we summarize the main synthetic strategies for the synthesis of DMSNs and conclude that their synthesis follows a unified dynamic dual template collaborative self-assembly mechanism with microemulsion droplets as primary template and micelles as secondary template. Subsequently, the main synthesis protocol and several typical catalytic applications of DMSNs containing heterometal atoms are reviewed. Due to the inherent structural characteristics of DMSNs, it has a center-radial shaped pores from the core to outer surface, allowing fast diffusion/mass transport inside the pores. Therefore, DMSN catalysts containing heteroatoms show high catalytic performance. In terms of the future development of DMSNs containing heterometal atoms, in my opinion, we must pay attention to two research directions: one is how to realize green and large-scale synthesis of DMSNs at low cost and move toward industrial practical application and another one is how to understand the nature of nanoconfined catalysis at nanometer scale from the perspective of catalytic mechanism research using DMSNs as a model system. For example, the surface state theory involved by SWs model provides a new insight into the synergetic electron and proton transfer at nanoscale interface [67, 69–71, 77–81], beyond the conventional d-band center theory for heterogeneous catalysis [82].

Acknowledgments

This research was funded by the National Key R&D Program of China (2021YFA1501401), the National Science Foundation of China (22172051, 21872053, and 21573074), the Science and Technology Commission of Shanghai Municipality (19520711400), the Research Funds of Happiness Flower ECNU (2020ST2203), the Open Project Program of Academician and Expert Workstation, the Shanghai Curui Low-Carbon Energy Technology Co., Ltd., and the JORISS program. K.Z. thanks ENS de Lyon for a temporary position as an invited professor in France.

References

1 Beck, J.S., Vartuli, J.C., Roth, W.J. et al. (1992). A new family of mesoporous molecular sieves prepared with liquid crystal templates. *J. Am. Chem. Soc.* 114: 10834–10843.

2 Kresge, C.T., Leonowicz, M.E., Roth, W.J. et al. (1992). Ordered mesoporous molecular sieves synthesized by a liquid-crystal template mechanism. *Nature* 359: 710–712.

3 Huo, Q.S., Margolese, D.I., Ciesla, U. et al. (1994). Generalized synthesis of periodic surfactant/inorganic composite materials. *Nature* 368: 317–321.

4 Firouzi, A., Kumar, D., Bull, L.M. et al. (1995). Cooperative organization of inorganic-surfactant and biomimetic assemblies. *Science* 267: 1138–1143.

5 Yu, Y.J., Xing, J.L., Pang, J.L. et al. (2014). Facile synthesis of size controllable dendritic mesoporous silica nanoparticles. *ACS Appl. Mater. Interfaces* 6: 22655–22665.

6 Trewyn, B.G., Slowing, I.I., Giri, S. et al. (2007). Synthesis and functionalization of a mesoporous silica nanoparticle based on the sol-gel process and applications in controlled release. *Acc. Chem. Res.* 40: 846–853.

7 Rosenholm, J.M., Sahlgren, C., and Lindén, M. (2010). Towards multifunctional, targeted drug delivery systems using mesoporous silica nanoparticles-opportunities & challenges. *Nanoscale* 2: 1870–1883.

8 Valtchev, V. and Tosheva, L. (2013). Porous nanosized particles: preparation, properties, and applications. *Chem. Rev.* 113: 6734–6760.

9 Wu, S.H., Mou, C.Y., and Lin, H.P. (2013). Synthesis of mesoporous silica nanoparticles. *Chem. Soc. Rev.* 42: 3862–3875.

10 Argyo, C., Weiss, V., Bräuchle, C., and Bein, T. (2014). Multifunctional mesoporous silica nanoparticles as a universal platform for drug delivery. *Chem. Mater.* 26: 435–451.

11 Yamamoto, E. and Kuroda, K. (2016). Colloidal mesoporous silica nanoparticles. *Bull. Chem. Soc. Jpn.* 89: 501–539.

12 Möller, K. and Bein, T. (2017). Talented mesoporous silica nanoparticles. *Chem. Mater.* 29: 371–388.

13 Lin, H., Chen, Y., and Shi, J.L. (2018). Nanoparticle-triggered *in situ* catalytic chemical reactions for tumour-specific therapy. *Chem. Soc. Rev.* 47: 1938–1958.

14 Huang, J.X. (2020). Welcome to accounts of materials research. *Acc. Chem. Res.* 1: 1–2.

15 Polshettiwar, V., Cha, D., Zhang, X.X., and Basset, J.M. (2010). High-surface-area silica nanospheres (KCC-1) with a fibrous morphology. *Angew. Chem. Int. Ed.* 49: 9652–9656.

16 Zhang, K., Xu, L.L., Jiang, J.G. et al. (2013). Facile large-scale synthesis of monodisperse mesoporous silica nanospheres with tunable pore structure. *J. Am. Chem. Soc.* 135: 2427–2430.

17 Du, X. and Qiao, S.Z. (2015). Dendritic silica particles with center-radial pore channels: promising platforms for catalysis and biomedical applications. *Small* 11: 392–413.

18 Wang, Y.B., Du, X., Liu, Z. et al. (2019). Dendritic fibrous nano-particles (DFNPs): rising stars of mesoporous materials. *J. Mater. Chem. A* 7: 5111–5152.

19 Hao, P., Peng, B., Shan, B.Q. et al. (2020). Comprehensive understanding of the synthesis and formation mechanism of dendritic mesoporous silica nanospheres. *Nanoscale Adv.* 2: 1792–1810.

20 Wang, Y.B., Zhang, B.L., Ding, X.P., and Du, X. (2021). Dendritic mesoporous organosilica nanoparticles (DMONs): chemical composition, structural architecture, and promising applications. *Nano Today* 39: 101231.

21 Malekmohammadi, S., Mohammed, R.U.R., Samadian, H. et al. (2022). Nonordered dendritic mesoporous silica nanoparticles as promising platforms for advanced methods of diagnosis and therapies. *Mater. Today Chem.* 26: 101144.

22 Pelluau, T., Sene, S., Garcia Cirera, B. et al. (2022). Multifunctionalized mesostructured silica nanoparticles containing Mn2 complex for improved catalase-mimicking activity in water. *Nanomaterials* 12: 1136.

23 Polshettiwar, V. (2022). Dendritic fibrous nanosilica: discovery, synthesis, formation mechanism, catalysis, and CO_2 capture–conversion. *Acc. Chem. Res.* 55: 1395–1410.

24 Xu, C., Lei, C., Wang, Y., and Yu, C.Z. (2022). Dendritic mesoporous nanoparticles: structure, synthesis and properties. *Angew. Chem. Int. Ed.* 134: e202112752.

25 Lin, H.X., Cui, K., Yao, Y.W. et al. (2005). A simple route for preparing radiolarian-like mesoporous silica from water–diethyl ether binary solvent system. *Chem. Lett.* 34: 918–919.

26 Nandiyanto, A.B.D., Kim, S., Iskandar, F., and Okuyama, K. (2009). Synthesis of spherical mesoporous silica nanoparticles with nanometer-size controllable pores and outer diameters. *Microporous Mesoporous Mater.* 120: 447–453.

27 Du, X. and He, J.H. (2010). Fine-tuning of silica nanosphere structure by simple regulation of the volume ratio of cosolvents. *Langmuir* 26: 10057–10062.

28 Zhang, H.J., Li, Z.Y., Xu, P.P. et al. (2010). A facile two step synthesis of novel chrysanthemum-like mesoporous silica nanoparticles for controlled pyrene release. *Chem. Commun.* 46: 6783–6785.

29 Shen, D.K., Yang, J.P., Li, X.M. et al. (2014). Biphase stratification approach to three-dimensional dendritic biodegradable mesoporous silica nanospheres. *Nano Lett.* 14: 923–932.

30 Cai, Q., Geng, Y., Zhao, X. et al. (2008). Morphological classification of mesoporous silicas synthesized in a binary water–ether solvent system. *Microporous Mesoporous Mater.* 108: 123–135.

31 Moon, D. and Lee, J. (2012). Tunable synthesis of hierarchical mesoporous silica nanoparticles with radial wrinkle structure. *Langmuir* 28: 12341–12347.

32 Gustafsson, H., Isaksson, S., Altskar, A., and Holmberg, K. (2016). Mesoporous silica nanoparticles with controllable morphology prepared from oil-in-water emulsions. *J. Colloid Interface Sci.* 467: 253–260.

33 Ernawati, L., Balgis, R., Ogi, T., and Okuyama, K. (2017). Tunable synthesis of mesoporous silica particles with unique radially oriented pore structures from tetramethyl orthosilicate via oil–water emulsion process. *Langmuir* 33: 783–790.

34 Maity, A., Das, A., Sen, D. et al. (2017). Unraveling the formation mechanism of dendritic fibrous nanosilica. *Langmuir* 33: 13774–13782.

35 Pereira, C., Alves, C., Monteiro, A. et al. (2011). Designing novel hybrid materials by one-pot co-condensation: from hydrophobic mesoporous silica nanoparticles to superamphiphobic cotton textiles. *ACS Appl. Mater. Interfaces* 3: 2289–2299.

36 Paula, A.J., Montoro, L.A., Filho, A.G.S., and Alves, O.L. (2012). Towards long-term colloidal stability of silica-based nanocarriers for hydrophobic molecules: beyond the Stöber method. *Chem. Commun.* 48: 591–593.

37 Wu, M.Y., Meng, Q.S., Chen, Y. et al. (2015). Large-pore ultrasmall mesoporous organosilica nanoparticles: micelle/precursor co-templating assembly and nuclear-targeted gene delivery. *Adv. Mater.* 27: 215–222.

38 Kalantari, M., Liu, Y., Strounina, E. et al. (2018). Superhydrophobic dendritic mesoporous organosilica nano-particles with ultrahigh-content of gradient organic moieties. *J. Mater. Chem. A* 6: 17579–17586.

39 Shan, B.Q., Xing, J.L., Yang, T.Q. et al. (2019). One-pot co-condensation strategy for dendritic mesoporous organosilica nanospheres with fine size and morphology control. *CrystEngComm* 21: 4030–4035.

40 Wang, Y., Nor, Y.A., Song, H. et al. (2016). Small-sized and large-pore dendritic mesoporous silica nanoparticles enhance antimicrobial enzyme delivery. *J. Mater. Chem. B* 4: 2646–2653.

41 Liu, P.C., Yu, Y.J., Peng, B. et al. (2017). A dual-templating strategy for the scale-up synthesis of dendritic mesoporous silica nanospheres. *Green Chem.* 19: 5575–5581.

42 Wang, Y., Song, H., Yu, M.H. et al. (2018). Room temperature synthesis of dendritic mesoporous silica nanoparticles with small sizes and enhanced mRNA delivery performance. *J. Mater. Chem. B* 6: 4089–4095.

43 Peng, B., Zhou, J.F., Chen, H. et al. (2022). Tetraalkoxysilane assisted self-emulsification templating for mesoporous silica nanoparticles with controlled mesostructures. *ChemRxiv. Cambridge: Cambridge Open Engage* https://doi.org/10.26434/chemrxiv-2022-tz5ss.

44 Lim, S.W., Jang, H.G., Sim, H. et al. (2014). Preparation of dandelion-type silica spheres and their application as catalyst supports. *J. Porous. Mater.* 21: 797–809.

45 Xu, C., Yu, M.H., Noonan, O. et al. (2015). Core-cone structured monodispersed mesoporous silica nanoparticles with ultra-large cavity for protein delivery. *Small* 11: 5949–5955.

46 Si, Y.S., Huang, T., Xie, H.Q., and Chen, M. (2022). Synthesis of biomimetic brochosome-shaped microspheres via droplets assembly strategy. *Chem. Mater.* 34: 7271–7279.

47 Yang, Y.N., Bernardi, S., Song, H. et al. (2016). Anion assisted synthesis of large pore hollow dendritic mesoporous organosilica nanoparticles: understanding the composition gradient. *Chem. Mater.* 28: 704–707.

48 Moon, D. and Lee, J. (2014). Formation of wrinkled silica mesostructures based on the phase behavior of pseudoternary systems. *Langmuir* 30: 15574–15580.

49 Bahadur, J., Maity, A., Sen, D. et al. (2021). Origin of the hierarchical structure of dendritic fibrous nanosilica: a small-angle X-ray scattering perspective. *Langmuir* 37: 6423–6434.

50 Sheng, Y. and Zeng, H.C. (2015). Monodisperse aluminosilicate spheres with tunable Al/Si ratio and hierarchical macro-meso-microporous structure. *ACS Appl. Mater. Interfaces* 7: 13578–13589.

51 Choi, Y., Yun, Y.S., Park, H. et al. (2014). A facile approach for the preparation of tunable acid nano-catalysts with a hierarchically mesoporous structure. *Chem. Commun.* 50: 7652–7655.

52 Jiao, J.Q., Fu, J.Y., Wei, Y.C. et al. (2017). Al-modified dendritic mesoporous silica nanospheres-supported NiMo catalysts for the hydrodesulfurization of dibenzothiophene: efficient accessibility of active sites and suitable metal–support interaction. *J. Catal.* 356: 269–282.

53 Wang, Z.J. and Balkus, K.J. (2017). Synthesis and modification of titanium containing wrinkled mesoporous silica for cyclohexene epoxidation. *Microporous Mesoporous Mater.* 243: 76–84.

54 Wang, Z.J. and Balkus, K.J. (2017). Liquid phase propylene oxidation with tert-butyl hydroperoxide over titanium containing wrinkled mesoporous silica. *Catal. Commun.* 96: 15–18.

55 Liu, Q.L., Yang, Z., Luo, M.S. et al. (2019). Vanadium-containing dendritic mesoporous silica nanoparticles: multifunctional catalysts for the oxidative and non-oxidative dehydrogenation of propane to propylene. *Microporous Mesoporous Mater.* 282: 133–145.

56 Liu, Q.L., Luo, M.S., Zhao, Z., and Zhao, Q.N. (2020). K-modified Sn-containing dendritic mesoporous silica nanoparticles with tunable size and SnOx-silica interaction for the dehydrogenation of propane to propylene. *Chem. Eng. J.* 380: 122423.

57 Liu, B., Feng, L.L., Bian, Y.L. et al. (2022). $Mn^{2+}/Fe^{3+}/Co^{2+}$ and tetrasulfide bond co-incorporated dendritic mesoporous organosilica as multifunctional nanocarriers: one-step synthesis and applications for cancer therapy. *Adv. Healthc. Mater.* 11: 2200665.

58 Zhang, K., Albela, B., He, M.Y. et al. (2009). Tetramethyl ammonium as masking agent for molecular stencil patterning in the confined space of the nano-channels of 2D hexagonal-templated porous silicas. *Phys. Chem. Chem. Phys.* 11: 2912–2921.

59 Zhang, K., Lam, K.F., Albela, B. et al. (2011). Mononuclear-dinuclear equilibrium of grafted copper complexes confined in the nanochannels of MCM-41 silica. *Chem. Eur. J.* 17: 14258–14266.

60 Albela, B. and Bonneviot, L. (2016). Surface molecular engineering in the confined space of templated porous silica. *New J. Chem.* 40: 4115–4131.

61 Vidal, V., Théolier, A., Thivolle Cazat, J. et al. (1996). Synthesis, characterization, and reactivity, in the C–H bond activation of cycloalkanes, of a silica-supported tantalum(III) monohydride complex: $(\vdots SiO)_2Ta^{III}-H$. *J. Am. Chem. Soc.* 118: 4595–4602.

62 Coperet, C., Comas Vives, A., Conley, M.P. et al. (2016). Surface organometallic and coordination chemistry toward single-site heterogeneous catalysts: strategies, methods, structures, and activities. *Chem. Rev.* 116: 323–421.

63 Gioffrè, D., Rochlitz, L., Payard, P. et al. (2022). Grafting of Group-10 organometallic complexes on silicas: differences and similarities, surprises and rationale. *Helv. Chim. Acta* 105: e202200073.

64 Polshettiwar, V., Thivolle Cazat, J., Taoufik, M. et al. (2011). "Hydro-metathesis" of olefins: a catalytic reaction using a bifunctional single-site tantalum hydride catalyst supported on fibrous silica (KCC-1) nanospheres. *Angew. Chem. Int. Ed.* 50: 2747–2751.

65 Chen, Y.T., Yang, T.Q., Pan, H.F. et al. (2014). Photoemission mechanism of water-soluble silver nanoclusters: ligand-to-metal–metal charge transfer vs strong coupling between surface plasmon and emitters. *J. Am. Chem. Soc.* 136: 1686–1689.

66 Yang, T.Q., Dai, S., Yang, S.Q. et al. (2017). Interfacial clustering-triggered fluorescence–phosphorescence dual solvoluminescence of metal nanoclusters. *J. Phys. Chem. Lett.* 8: 3980–3985.

67 Yang, T.Q., Shan, B.Q., Huang, F. et al. (2019). P band intermediate state (PBIS) tailors photoluminescence emission at confined nanoscale interface. *Commun. Chem.* 2: 1–11.

68 Yang, T.Q., Peng, B., Shan, B.Q. et al. (2020). Origin of the photoluminescence of metal nanoclusters: from metal-centered emission to ligand-centered emission. *Nanomaterials* 10: 261.

69 Hu, X.D., Shan, B.Q., Tao, R. et al. (2021). Interfacial hydroxyl promotes the reduction of 4-nitrophenol by Ag-based catalysts confined in dendritic mesoporous silica nanospheres. *J. Phys. Chem.* 125: 2446–2453.

70 Shan, B.Q., Zhou, J.F., Ding, M. et al. (2021). Surface electronic states mediate concerted electron and proton transfer at metal nanoscale interfaces for catalytic hydride reduction of $-NO_2$ to $-NH_2$. *Phys. Chem. Chem. Phys.* 23: 12950–12957.

71 Ding, M., Shan, B.Q., Peng, B. et al. (2022). Dynamic $Pt-OH^{-}\cdot H_2O-Ag$ species mediate coupled electron and proton transfer for catalytic hydride reduction of 4-nitrophenol at the confined nanoscale interface. *Phys. Chem. Chem. Phys.* 24: 7923–7936.

72 Zheng, L.X., Peng, B., Zhou, J.F. et al. (2022). High efficient and stable thiol-modified dendritic mesoporous silica nanospheres supported gold catalysts for gas-phase selective oxidation of benzyl alcohol with ultra-long lifetime. *Microporous Mesoporous Mater.* 342: 112140.

73 Fihri, A., Cha, D., Bouhrara, M. et al. (2012). Fibrous nano-silica (KCC-1)-supported palladium catalyst: Suzuki coupling reactions under sustainable conditions. *ChemSusChem* 5: 85–89.

74 Fihri, A., Bouhrara, M., Patil, U. et al. (2012). Fibrous nano-silica supported ruthenium (KCC-1/Ru): a sustainable catalyst for the hydrogenolysis of alkanes with good catalytic activity and lifetime. *ACS Catal.* 2: 1425–1431.

75 Wang, X., Xuan, X.N., Wang, Y.B. et al. (2021). Nano-Au-modified TiO_2 grown on dendritic porous silica particles for enhanced CO_2 photoreduction. *Microporous Mesoporous Mater.* 310: 110635.

76 Xuan, X.N., Tu, S.C., Yu, H.J. et al. (2019). Size-dependent selectivity and activity of CO_2 photoreduction over black nano-titanias grown on dendritic porous silica particles. *Appl. Catal. B Environ.* 255: 117768.

77 Hoffmann, R. (1971). Interaction of orbitals through space and through bonds. *Acc. Chem. Res.* 4: 1–9.

78 Tao, R., Shan, B.Q., Sun, H.D. et al. (2021). Surface molecule manipulated Pt/TiO_2 catalysts for selective hydrogenation of cinnamaldehyde. *J. Phys. Chem. C* 125: 13304–13312.

79 Yang, T.Q., Hu, X.D., Shan, B.Q. et al. (2021). Caged structural water molecules emit tunable brighter colors by topological excitation. *Nanoscale* 13: 15058–15066.

80 Zhou, J.F., Yang, T.Q., Peng, B. et al. (2021). Structural water molecules confined in soft and hard nanocavities as bright color emitters. *ACS Phys. Chem. Au* 2: 47–58.

81 Wang, P.Y., Zhou, J.F., Chen, H. et al. (2022). Activation of H_2O tailored by interfacial electronic states at a nanoscale interface for enhanced electrocatalytic hydrogen evolution. *JACS Au* 2: 1457–1471.

82 Medford, A.J., Vojvodic, A., Hummelshøj, J.S. et al. (2015). From the Sabatier principle to a predictive theory of transition-metal heterogeneous catalysis. *J. Catal.* 328: 36–42.

7

Chemical Post-Modifications of Titanosilicates

Fang Nan, Liu Dongxu, Yu Yunkai, and Liu Yueming

East China Normal University, Shanghai Key Laboratory of Green Chemistry and Chemical Processes, School of Chemistry and Molecular Engineering, North Zhongshan Rd. 3663, Shanghai 200062, P. R. China

7.1 Introduction

Titanosilicates are crystalline heteroatomic zeolites with Ti atoms implanted in the framework through direct synthesis (bottom-up strategy) or post-synthesis (top-down strategy). TS-1, the first titanosilicate developed by Enichem in 1983 [1], has been widely studied as efficient oxidation catalyst with the green oxidant H_2O_2 in various reactions, such as epoxidation of alkenes, hydroxylation of benzene or phenol, liquid-phase ammoximation of ketones or aldehydes, selective oxidation of alcohols, and amines. Its excellent catalytic activity, selectivity, and stability have opened the door to the rapid development of other crystalline titanosilicates with different zeolite topologies, mesoporous titanosilicates, crystalline/amorphous composite titanosilicates, and organic-inorganic hybrid titanosilicates [2–6].

The oxidation reaction catalyzed by titanosilicates/H_2O_2 system (TS/H_2O_2 system) follows a typical mechanism, involving the initial formation of active intermediate Ti—O—O—H by the activation of H_2O_2 over Ti active sites, then the transfer of active "O" atom to the reactant molecules, and finally the yield of the target product [7, 8]. Thus, the relationship between the structure of Ti active sites and their catalytic performance is a fundamental chemical issue in the study of titanosilicates. Additionally, increasing the content of Ti active sites and improving the intrinsic catalytic ability per single Ti site are another two important aspects. Meanwhile, the reaction of TS/H_2O_2 system is a typical multiphase catalytic process, in which the solvent plays a critical role to show a significant solvent effect, affecting the diffusion, adsorption/desorption, and surface reactions.

This chapter focuses on the post-modifications for enhancing the catalytic activity, selectivity, and stability of titanosilicates by controlling the diffusion, adsorption/desorption, and surface reactions, as well as the role of solvents. The acid/base modification, fluorination, and metal and non-metal ion modification tailor the surface properties, pore structure, and microenvironment of Ti active sites and then affect their catalytic performance. With the purpose to clarify the commonality and

Micro-Mesoporous Metallosilicates: Synthesis, Characterization, and Catalytic Applications, First Edition. Edited by Peng Wu and Hao Xu.

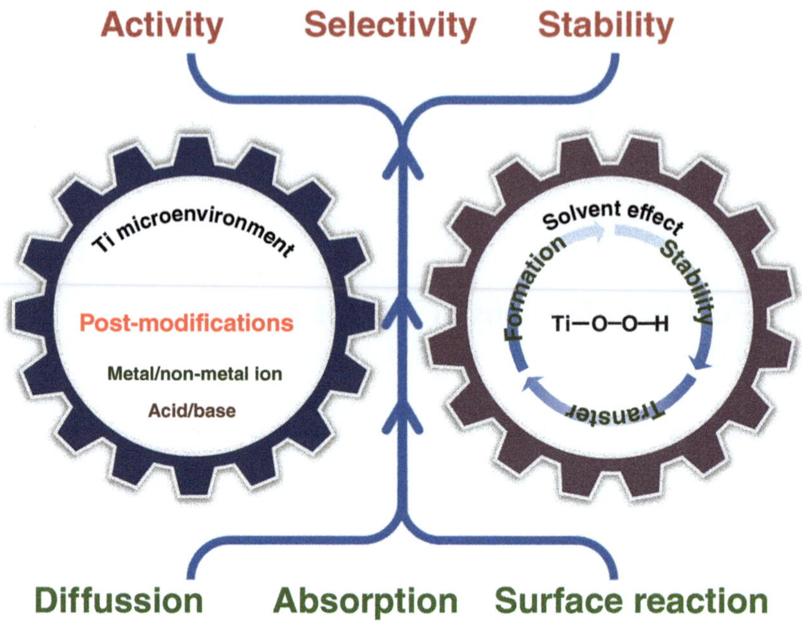

Scheme 7.1 The coordination and microenvironment of Ti synergistically cooperate with solvent effect to give full play to the catalytic oxidation of titanosilicates.

individuality of each modification method, we divide them into three categories (diffusion, adsorption/desorption, and surface reaction). In general, fine adjusting Ti microenvironment is the core of post-modification. The coordination between the structure of Ti active sites and solvent effect is the key to the formation, stabilization, and transfer of active intermediates Ti—O—O—H, which is the fundamental to give full play to the catalytic oxidation functionality of titanosilicates (Scheme 7.1). Finally, an outlook is provided for the future development in this field.

7.2 Diffusion and Adsorption/Desorption

Diffusion, adsorption of substrates, and desorption of products are essential steps in heterogeneous catalysis. Enhancement of the accessibility of Ti active sites and the appropriate control of hydrophobicity/hydrophilicity of titanosilicate surface allow the catalytic cycle to continuously going on.

7.2.1 Hierarchical Titanosilicates

Hierarchical structures can enhance accessibility and thereby the catalytic activity of titanosilicates in the reactions involving steric hindrance/diffusion constraints. There are at least two kinds of pore structures in hierarchical titanosilicates with mesopores and/or macropores additionally introduced to the microporous zeolites. In recent years, researchers have proposed various methods to construct hierarchical titanosilicates. During the bottom-up crystallization process, the formation of hierarchical structures is achieved by hard-templating, soft-templating, or non-templating method [9–17]. The top-down post-treatments are also useful for constructing hierarchical structures within microporous titanosilicates, such as desilication. Compared with the direct synthesis, the post-treatments can effectively avoid the uncertain influences on the nucleation and crystallization of titanosilicates.

When hierarchical titanosilicates are prepared by desilication under alkaline conditions, the silicon species are preferentially removed from the framework to create intracrystal mesopores. Chao et al. treated TS-1 with different inorganic bases (LiOH, NaOH, KOH, CsOH) for desilication and found that the rate of silicon dissolution increased with increasing alkalinity: LiOH < NaOH < KOH [18]. Meanwhile, Du et al. applied NH_4F-HF etching on TS-1 to obtain mesopores and macropores and found that NH_4F-HF could remove Si and Ti atoms without selectivity so that the obtained hierarchical pore TS-1 still maintained a similar Si/Ti ratio to the pristine one, effectively avoiding the change in framework composition [19]. The desilication inevitably made the TS-1 zeolite partially dissolved in the alkali solution, leading to the formation of amorphous phases, and a significant loss of product yield. In contrast, the dissolution-recrystallization post-treatment can avoid this problem. The organic quaternary ammonium cations with templating effect, such as tetrapropylammonium hydroxide (TPAOH) and tetraethyl ammonium hydroxide (TEAOH), are generally used to recrystallize the dissolved silicon species onto the surface of titanosilicates, forming a hollow structure. Lin et al. reported the preparation of TS-1 with hollow structure where the mesopores were created by an acid treatment followed by an alkali treatment or by a direct alkali treatment [20]. Wang et al. performed a hydrothermal treatment of TS-1 using TPAOH solution and found that mesopores were formed in TS-1, while some of the titanium and silicon species were removed and then recrystallized [21]. The additional introduction of ammonium salts can intensify the dissolution-recrystallization process [22]. Wang et al. systematically studied the post-treatment of TS-1 with several organic bases and found that TPAOH and tetrabutylammonium hydroxide (TBAOH) have an oriented crystallization growth effect on the recrystallization of dissolved silicon species [23].

7.2.2 Surface Hydrophilicity and Hydrophobicity

The hydrophilicity and hydrophobicity of titanosilicates are the important factors affecting the adsorption/desorption of H_2O_2 and organic substrates in reaction systems, which in turn affect the catalytic performance.

The increased hydrophilicity facilitates the enrichment of H_2O_2 and adsorption of hydrophilic substrates. Corma et al. assumed in 1996 that Ti-Beta could catalyze the epoxidation of 1-hexene with higher reaction activity in high polarity non-protonic solvent due to the higher hydrophilicity than TS-1 [24]. The hydrophilicity of titanosilicates can be enhanced by using silylation reagents like dimethyl dichlorosilicane (DCDMS) or by doping with heteroatoms (Fe^{3+}, B^{3+}) [25, 26]. Dialkyl-substituted silica species are the most used silylation reagents, which not only expand the layered titanosilicates but also introduce additional hydroxyl groups upon calcination [25, 27]. The doping of cationic heteroatoms may generate new hydroxyls in the framework of the titanosilicates, which enhances the surface hydrophilicity.

Improving hydrophobicity facilitates the adsorption of hydrophobic substrates. The commonly used method is the silylation modification with trialkyl-substituted silicon species, such as chlorotrimethylsilane (TMCS) and hexamethyldisiloxane (HMDS), which reduces the number of hydrophilic hydroxyl groups [28–34]. In 1998, Corma et al. used HMDS to modify Ti-MCM-41 and enhanced the hydrophobicity, successfully prohibiting the formation of diols produced by ring-opening side reaction in the epoxidation reaction of cyclohexene with tert-Butyl hydroperoxide (TBHP) [35]. In 2017, Lv et al. treated TS-1 with TPAOH for desilication and subsequently performed silylation with HMDS. The obtained hydrophobic hollow TS-1 with significantly increased contact angle exhibited good hydrophobicity in the Pickering interfacial catalytic reaction [32]. Besides, the fluorination treatment can also eliminate the surface hydroxyl groups of titanosilicates [30, 36, 37]. Alcohol modification and ion-exchange are also efficient ways to improve hydrophobicity [38–45]. In 2019, Wei et al. used ethanol to functionalize hierarchical TS-1 (HTS-1) and significantly improved the reaction activity and selectivity in methyl oleate (MO) epoxidation reaction [39]. The MO conversion could reach 92.7%, which exceeded the pristine HTS-1 (only 56.9%). Tatsumi et al. used K^+ exchange to modify TS-1, and the modified catalyst was not favorable for the oxidation of hydrophobic molecules but for the hydrophilic molecules because the polarity of the Si—O—K species changed the hydrophilicity and hydrophobicity of TS-1 [46]. Overall, the post-modification by shielding the surface hydrophilic hydroxyl groups effectively reduces the number and density of surface hydroxyl groups, which improves the surface hydrophobicity of titanosilicates.

7.3 Surface Reaction

Surface reaction is the most important part in the catalytic reaction process. Post-modification is beneficial to regulate the microenvironment of Ti active sites in titanosilicates and then to enhance the surface reaction and catalytic efficiency. We will explain the basic mechanism of acid/base treatment, metal and non-metal ion modification, and silylation and their catalytic effects from three perspectives: the content, the distribution, and the properties of Ti active sites.

7.3.1 Ti Active Sites Content

It is generally accepted that the Ti active sites in titanosilicates are highly dispersed tetrahedrally coordinated framework Ti(IV) [5]. One of the keys to improve the catalytic performance is to increase the content of framework Ti(IV). During the synthesis of MFI-type titanosilicate, the framework Ti content is thermodynamically limited up to 2.57 Ti atoms each cell, corresponding to the minimum framework Si/Ti ratio of 39, and the excess Ti species deposit on the crystal surface as Ti oxides

[47]. In order to maximize the framework Ti content, most studies tried to control the hydrolysis rate of the titanium and silicon precursors and avoid the generation of non-framework Ti during the synthesis process. The post-treatment methods to enhance the framework Ti content are relatively few, mainly focusing on acid treatment, organic alkali treatment, introducing Ti after removing heteroatoms, and ammonium salt solution treatment. However, it is difficult to break the lower limit Si/Ti ratio.

In general, the hydrothermally synthesized titanosilicates are usually treated by further acid washing to remove the non-framework Ti species, which enhances its catalytic ability. For example, the acid washing removes the non-framework Ti species in TS-1, improving the effective utilization of H_2O_2 in alkene epoxidation [48]. In addition, the acid washing also modulates the structure of Ti species and then improves the oxidation performance. In the preparation of Ti-MWW, a Ti-containing lamellar precursor with MWW structure (Ti-MWW[P] with Si/Ti ratio of ~10) was firstly synthesized with piperidine (PI) or hexamethyleneimine (HMI) as the structure-directing agent (SDA), which contained a large amount of non-framework hexa-coordinate Ti species. The acid washing could effectively remove the non-framework Ti species from the lamellar precursor Ti-MWW(P). The resulting Ti-MWW with excellent catalytic activity (with Si/Ti ratio of ~40) could be obtained by further calcination, whereas the directly calcined Ti-MWW(P) was almost inactive in catalytic oxidation [49]. Zhang et al. compared the structures of Ti species in Ti-MWW(P) before and after acid treatment and found that during the acid treatment of the Ti-MWW(P) precursor, non-framework TiO_6 species were partially removed along with part of the framework B, resulting in the formation of a certain number of Si—OH defect sites. Meanwhile, the remaining TiO_6 species were implanted to the defect sites to form framework Ti(IV) [50, 51]. Tang et al. tracked the structural changes of Ti species during the acid washing process and found that both framework Ti(IV) and non-framework Ti existed in the precursor Ti-MWW(P), and the removal of non-framework Ti during the acid washing process eliminated their detrimental effect [52].

Post-treatment of titanosilicates with organic bases is a common modification method. It can create mesoporous phase and enhance the surface hydrophobicity through a dissolution-recrystallization process. It could also be used to enhance the catalytic activity by regulating the microenvironment of Ti or to adjust the content of framework Ti. Wang et al. hydrothermally treated TS-1 with TPAOH to form a hollow structure, with the hexa-coordinate Ti species transformed into framework Ti(IV), while Si/Ti was kept unchanged [21], leading to the increase of Ti(IV) content. In contrast, Zuo et al. found that during the hydrothermal treatment of TS-1 with TPAOH, the formation of mesoporous also induced the generation of non-framework Ti species, which reduced the amount of framework Ti in TS-1 [23]. Wu et al. hydrothermally treated TS-1 with ethylamine solution or a mixture of ethylamine and TPABr, and TS-1^s with abundant defective tetradentate Ti or TS-1^+ with more hexa-coordinate Ti species was obtained, respectively, in which the Si/Ti ratio was basically unchanged. Some of the framework Ti in TS-1 were transformed into new active sites after the post-modification [41, 42]. This predicted that the alkali modification can adjust the microenvironment of Ti active sites and improve the individual catalytic ability, which broadens the idea of alkali modification of titanosilicates.

With regard to some titanosilicates that are difficult to be obtained by direct hydrothermal synthesis, they could be post-synthesized by combining the extraction of heteroatoms (B, Fe, Al, etc.) and post-implantation of Ti atoms. This is also an effective way to increase the content of framework Ti. Song et al. prepared Ti-YNU-1 with De-Fe-YNU-1 as a precursor via Ti implantation, which could increase the Si/Ti ratio from 194 to 83 (about 70% increase in Ti content) [53]. Horikawa et al. prepared Ti-Beta by adding Al source to a mixed oxide of Ti and Si (obtained by mechanical force chemistry) to promote crystallization, followed by the removal of Al atoms, which allowed the implantation of non-framework Ti into the framework to form tetrahedrally coordinated Ti(IV) species without affecting the Ti content [54]. Treating titanosilicates with ammonium salt can also enhance their catalytic performance. Vittorio et al. treated TS-1 with different types of ammonium salt solutions under refluxing condition and found that the cell volume of TS-1 increased, implying that the content of framework Ti increased, which leaded to a significant improvement of catalytic ability in the ketone ammoximation reaction [55].

7.3.2 Ti Active Sites Distribution

The distribution of Ti active sites in the titanosilicate framework is one of the key factors affecting the surface reaction [5]. At the atomic level, the Ti active sites at distinct positions show different catalytic ability. Density Functional Theory (DFT) calculations show that the structure of Ti active sites at T1 site of MWW structure synthesized with B-free or less method is $Ti(OSi)_3OH$ (with Si—OH or Si—OH nest at T6 site nearby), which are superior to the $Ti(OSi)_4$ active sites at other T sites in catalytic oxidation reactions [56]. Grosso et al. reported that the Ti active sites distributed in 12-membered ring pores of SSZ-70 had better catalytic ability, and their catalytic conversion in alkene oxidation reaction was almost five times that of the Ti active sites at the crystal surface [57]. There are three possible Ti distributions, including less Ti atoms on the surface, rich Ti atoms on the surface, and Ti atoms homogeneously distributed in the framework. Generally, titanosilicates (such as TS-1, Ti-MWW) obtained by direct hydrothermal synthesis have fewer Ti atoms and more Si atoms on the surface [58–61]. The increase of surface Ti sites enhances the accessibility of substrates and then the catalytic activity. Therefore, in order to obtain highly active titanosilicates, increasing the content of Ti active sites on the surface is one of the effective strategies. The enrichment of surface Ti species can be obtained by modification with acid or organic base.

Hadi Nur et al. showed that TS-1 possessed a higher catalytic activity for 1-octene oxidation after the modification with sulfuric acid, which was attributed to the appearance of new Ti active sites [62]. In addition, since 1-octene molecules with large size may only be converted over the surface Ti species, the improvement of catalytic activity of modified TS-1 zeolite could be attributed to the enrichment of

Ti active sites on the outer surface. Xue et al. modified TS-1 with ethanolamine and the Energy Dispersive Spectrometer (EDS) results showed that the silicon and titanium species on the TS-1 surface of the zeolite before modification were evenly distributed, while the content of Ti atoms in the middle area of the modified zeolite was significantly higher than that at the edge, indicating that the Ti species in the crystal migrated to the surface [63]. The enrichment of Ti species on zeolite surface and the generation of mesopores increased the accessibility of Ti active sites, and thus greatly improved the selectivity and lifetime in cyclohexanone ammoximation reaction. Wu et al. found that the TS-1^s obtained by hydrothermal treatment of TS-1 with ethylamine solution was rich in Ti on its surface, and the number of tetrahedrally coordinated defective Ti sites in the framework and the Lewis acidity were increased, leading to enhanced catalytic oxidation performance [42].

There are few strategies for regulating the homogeneous distribution of Ti atoms in titanosilicates. Chen et al. synthesized TS-1 with homogeneous distribution of Ti atoms by controlling the aggregation state of silicon species [61, 64]. Up to now, no post-modification techniques have been reported to achieve the homogeneous distribution.

7.3.3 Ti Active Sites Properties

With limited content of Ti active sites, it is especially important to improve their individual catalytic ability. The structure of Ti active sites and the state of their surrounding hydroxyl groups directly affect the catalytic ability of titanosilicates. The properties of Ti active sites can be modified from three aspects, including electrophilicity, coordination state, and adjacent silicon hydroxyl groups.

7.3.3.1 Electrophilicity of Ti Active Sites

The tetrahedrally coordinated Ti species (Ti(IV)) exhibit Lewis acidity due to the unsaturated coordination state. It can activate H_2O_2 to form Ti—O_α—O_β—H intermediates with electrophilic differences between O_α and O_β [65]. Therefore, the Lewis acidic strength of the Ti(IV) active sites can be adjusted by changing the charge environment, which further affects the "O" transfer process [65]. At present, the main strategy to adjust the Ti(IV) charge is to enhance the electrophilicity of Ti sites, through the post-treatments with metal (Zn, Ge, Cd, etc.), non-metal (F, N, P, etc.), organic base, and alcohol [7, 36, 38, 39, 44, 52, 66–80].

The post-treatment with metal ions is mainly based on the implantation of metal ions into the framework via different ways. The metal ions can enhance the electrophilicity of Ti by interacting with O atoms and affecting the charge distribution around the Ti active sites. Transition metals with empty *d* orbitals are generally chosen to accept lone pairs of electrons from other metal atoms. Area et al. obtained Zn-TS-1 after refluxing TS-1 in $Zn(OAC)_2 \cdot 2H_2O$ solution and calcination at different temperatures [79]. Zn^{2+} ions were implanted into the framework, forming Zn—O—Si/Ti bonds, which enhanced the electrophilicity of the Ti active sites. It has also been found by ^{29}Si MAS NMR investigation and DFT calculations that the introduction of Zn^{2+} into TS-1 resulted in stronger hydrophobicity and a more relaxed structure of the Ti active sites/Ti—O—O—H active intermediate [80]. The introduction of Ge, La, Cd, and other metal ions into the titanosilicates by ultrasonic impregnation also weakened the strength of the Ti—O bond, resulting in a more relaxed structure of Ti—O—O—H, and enhanced the electrophilicity of Ti [78, 81–83]. In general, the metal ions formed covalent bonds with the O attached to Ti/Si, which reduced the electron cloud density of Ti. Due to the coordination effect of metal ions, the coordination number of Ti species was increased, and thus a characteristic peak below 460.0 eV appeared in the X-ray Photoelectron Spectroscopy (XPS) spectrum, indicating the enhancement of the electrophilicity of Ti.

The post-treatment of titanosilicates by non-metal ions with strong electronegativity can also enhance the electrophilicity of Ti, and the most used non-metallic elements are F, N, and P. At the same time, the introduction of non-metal ions will also form hydrogen bonds with H atoms in Ti—O—O—H, thus stabilizing the active intermediate Ti—O—O—H and achieving the purpose of improving the catalytic oxidation performance of titanosilicates.

As early as 1998, Blasco et al. reported the post-treatment Ti-Beta with fluorine and proposed that the introduction of F^- enhanced the hydrophobicity of the titanosilicates. The reactant conversion and the product selectivity increased from 25.4% to 26.8% and from 90.1% to 97.9% on the epoxidation of 1-hexene [74]. In 1999, Koller et al. proposed the structure of $SiO_{4/2}F^-$ (Figure 7.1) evidenced by the ^{19}F MAS NMR and ^{29}Si MAS NMR characterization, indicating that F interacted with part of the Si atoms in the framework to form the $SiO_{4/2}F^-$ structure [75]. Fang et al. treated the Ti-MWW(P) precursor with HNO_3 solution containing NH_4F and successfully implanted F atoms into the framework of titanosilicates to form $SiO_{3/2}F$ groups for the first time [84]. The obtained F-Ti-MWW exhibited significantly improved catalytic oxidation performance. ^{19}F MAS NMR spectra showed the appearance of a resonance peak at around −153 ppm, indicating that F was implanted into the Ti-MWW framework by forming $SiO_{3/2}F$ units with Si—F bonds (Figure 7.1) [36, 44]. Due to the electron pulling effect of F, the electrophilicity of the adjacent Ti was enhanced and its Lewis acid strength was significantly enhanced. Later, this method was also successfully applied to the fluorination modification of Ti-MOR [76]. However, the fluorination modification produced the $SiO_{3/2}F$ groups along with the formation of $SiO_{4/2}F^-$ groups with a resonance peak at around −141 ppm in the ^{19}F MAS NMR spectrum, which had side effects on the catalytic performance (Figure 7.1). To address this issue, Fang et al. effectively eliminated

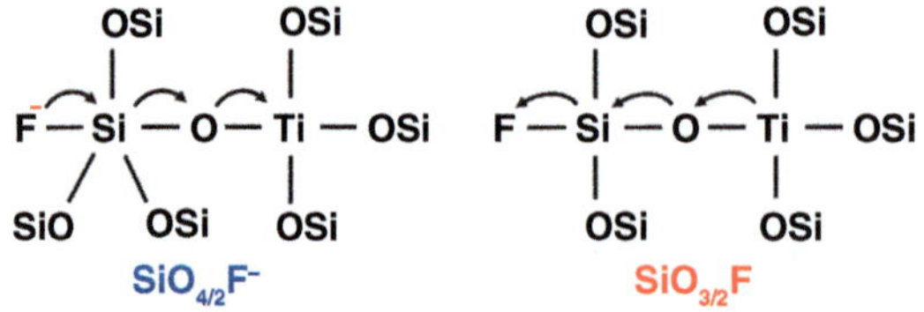

Figure 7.1 Effects of $SiO_{4/2}F^-$ units and $SiO_{3/2}F$ units on electropositivity of Ti active sites. Source: Reprinted with permission from Refs [36, 74]; © 1999 American Chemical Society and 2012 Royal Society of Chemistry.

the side effects of $SiO_{4/2}F^-$ in F-Ti-MWW by anion exchange, thus enhancing the catalytic oxidation performance while maintaining the positive effect of $SiO_{3/2}F$ [44, 85]. Lu et al. improved the fluorination modification process to inhibit the generation of $SiO_{4/2}F^-$ groups [77]. They found that the thermal treatment of Ti-MWW with the methanol solution of NH_4F could result in the F-Ti-MWW containing mainly $SiO_{3/2}F$ groups.

For the post-treatment of titanosilicates with ammonia (NH_3), N—O—Si/Ti bonds are formed in the framework mainly by refluxing in ammonia solution at high temperature. After post-treatment, the amount of acidic sites in TS-1 was reduced and PO selectivity in the propylene epoxidation reaction was improved [70, 71]. Zhang et al. showed that the ammonia treatment of titanosilicates resulted in the adsorption of ammonia onto the surface of the molecular sieves, which could inhibit the formation of coke [73]. Li et al. treated TS-1 with a high-temperature ammonia stream followed by CH_3OH washing to remove the surface unstable species and found that the introduction of N resulted in the formation of more stable Ti—O—O—H active intermediates, which improved the catalytic oxidation performance of TS-1 [72]. A similar effect was observed after modification of Ti-MWW with NH_3 [73].

Post-treatment with phosphorus is often used to modify the acid properties of silica-alumina molecular sieves such as ZSM-5. P modification can also be applied in the modification of titanosilicates, and the typical P source is acidic phosphate or phosphoric acid. Zhang et al. used acidic phosphates (NaH_2PO_4, KH_2PO_4, $NH_4H_2PO_4$) to modify TS-1, and the obtained TS-1 showed enhanced H_2O_2 conversion and PO selectivity in the propylene epoxidation reaction [69]. It is proposed that phosphate radical could interact with the Ti active sites. In contrast, the basic phosphates (Na_2HPO_4, Na_3PO_4) modification was detrimental to the activity of titanosilicates. Fan et al. added phosphoric acid in the acid washing of Ti-MWW(P) with nitric acid and found that it could significantly improve 1-hexene conversion and H_2O_2 utilization in the epoxidation reaction of 1-hexene [86]. The UV-Raman results suggested that highly active TiO_4^* species might be formed. Recently, Lin et al. used phosphoric acid solution to treat the deactivated TS-1 from the cyclohexanone ammoximation reaction [68]. ^{31}P MAS NMR and Py-IR results showed that P not only interacted with the Si—OH(Ti) groups to reduce the number and strength of Brønsted acid sites and Lewis acid sites, but also interacted with Ti—OH to transform the tetra-coordinated Ti to hexa-coordinated Ti active sites. It is predicted that the P modification can enhance its catalytic oxidation performance and also provide a feasible way to regulate the Brønsted and Lewis acids.

In the post-treatment of titanosilicates with organic alkali, defective sites are constructed in the framework by dissolving silicon through alkali etching, and the perfect tetra-coordinated Ti active sites ($Ti(OSi)_4$) are modified to defective ones ($Ti(OSi)_3OH$). In 2004, David et al. proposed the existence of two types of tetra-coordinated Ti sites in the TS-1 framework, that is, perfect $Ti(OSi)_4$ and defective $Ti(OSi)_3OH$ sites, through DFT theoretical calculations [7]. In 2008, Yang et al. characterized TS-1 by solid-sate ^{31}P MAS NMR and observed a main resonance peak at −33.7 ppm and a shoulder peak at −32.8 ppm, which demonstrated the presence of two types of Lewis acid sites in TS-1 [66]. Theoretical calculations and

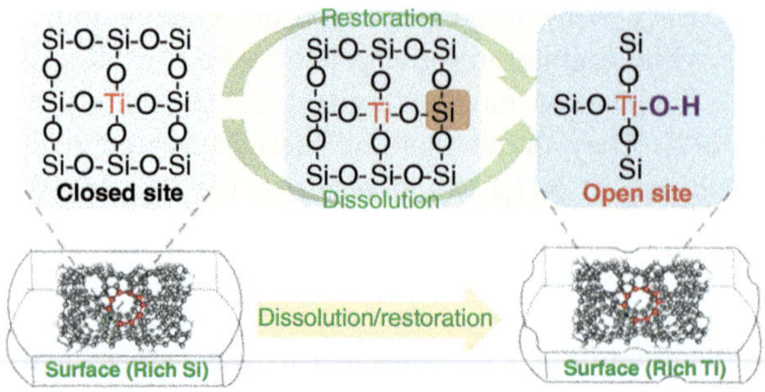

Figure 7.2 The defect $Ti(OSi)_3OH$ active site was constructed in titanosilicates. Source: Reproduced from Ref. [42]/with permission of Royal Society of Chemistry.

experimental results suggested that defective $Ti(OSi)_3OH$ had a stronger Lewis acid strength than perfect $Ti(OSi)_4$ and would show superior catalytic oxidation performance [87]. In general, the framework Ti species mainly exist as $Ti(OSi)_4$ in titanosilicates obtained by direct synthesis. Zhou et al. investigated the position and structure of the framework Ti(IV) using DFT calculations based on a simulated Ti-MWW with Si/Ti = 35 and proposed the existence of $Ti(OSi)_3OH$ active sites [67]. The modification of titanosilicates with organic base promoted the generation of defective $Ti(OSi)_3OH$. In 2018, Wu et al. modified TS-1 by hydrothermal post-treatment with ethylamine solution and successfully constructed TS-1^S with abundant defective $Ti(OSi)_3OH$ active sites by selectively dissolving Si species adjacent to Ti, while keeping the Ti content essentially unchanged (Figure 7.2) [42]. Py-IR analysis showed that the acid strength of Lewis acid sites was significantly enhanced, and DFT simulations showed that the positive charge of $Ti(OSi)_3OH$ was increased in comparison to $Ti(OSi)_4$. The Turnover Number (TON) value of TS-1^S in catalyzing 1-hexene epoxidation was substantially increased from 178–184 to 415–437. In addition, during the acid treatment of Ti-MWW(P) precursor to form Ti-MWW, Tang et al. found that a large number of framework B atoms near the Ti active sites in the form of "BO_4" species were removed to trigger the desilication as the acid treatment time proceeded [52]. This led to the change of framework Ti species from $Ti(OSi)_4$ to $Ti(OSi)_3OH$ active sties and improved the Lewis acid strength of the titanosilicates. When applied in 1-hexene epoxidation, the TON value gradually increased from 59 to 284. Recently, Xu et al. proved that Ti-MWW indeed contains $Ti(OSi)_3OH$ and $Ti(OSi)_4$ Ti active centers using combined isotope-labeled infrared spectroscopy and DFT theoretical calculations [56].

In 2008, Cheng et al. functionalized S-1 with alcohols, illustrating the enhancement of hydrophobicity by covalently linking alcohols into the internal silanol defect sites [38]. In 2019, the alcohols were again used to post-treat the TS-1 titanosilicates (HTS-1) by impregnation and refluxing at 120 °C [39]. The binding energy of Ti revealed by XPS characterication was found to decrease from 460.1 to 459.6 eV. The alcohol modification process allowed the alkoxy to covalently bind into the defect sites to enhance the hydrophobicity. The electrophilicity of the Ti active sites in HTS-1 was enhanced and obtained more electrophilic "O" as the active oxygen-donating species.

From the above post-modification methods and effects on the electrophilicity of Ti, it can be found that the chemical essence is to change the microstructure of Ti active sites to affect the charge distribution of Ti. The coordination of empty *d* orbital, the attraction of high-electronegativity atoms, the construction of defective sites, and the induced effect of organic groups affect the electrophilicity of Ti and enhance the Lewis acid strength, resulting in improved catalytic ability.

7.3.3.2 Coordinate State of Ti Active Sites

The states of Ti species in titanosilicates mainly include tetrahedrally coordinated Ti, pentahedrally coordinated Ti, six-coordinated Ti, and non-framework Ti, which are often analyzed and identified by UV–vis spectra (perfect or closed Ti(IV): ~210 nm; defective or open Ti(IV): ~220 nm; Ti(V): ~250 nm; Ti(VI): ~270 nm; anatase: ~330 nm) [5]. It is generally accepted that highly dispersed Ti(IV) plays a catalytic oxidation role [5]. However, pentahedrally coordinated or six-coordinated Ti species are also responsible for the catalytic performance [7]. Recently, Gordon et al. proposed that dinuclear Ti species played a catalytic role in TS-1 in the propylene epoxidation by solid ^{17}O NMR characterization [88]. Therefore, the study of adjusting the coordination state of Ti has attracted extensive attention.

The related modifications of Ti(IV) have been described above. In addition to the Ti(IV) species, the Ti(VI) species has also received particular attention. Balducci et al. treated TS-1 (TS-1A, which containing only Ti(IV)) with NH_4HF_2 and H_2O_2 to obtain TS-1B and found that it still showed excellent activity and selectivity in the phenol hydroxylation although a large amount of framework Ti species was lost (about 75%). A new absorption band appeared at 40 000 cm^{-1} in the UV–vis spectrum, which could be attributed to high-coordination Ti species. Guo et al. proved that this Ti species with high coordination number is a six-coordinated octahedral "TiO_6" with the structure of $Ti(OSi)_2(OH)_2(H_2O)_2$ by the combination of UV–vis, in situ UV-Raman spectroscopy and DFT theoretical calculations (Figure 7.3) [89]. These Ti species showed a characteristic band at 270–290 nm and 695 cm^{-1} in the UV–vis and UV-Raman (244 nm excitation source) spectra, respectively.

To reduce the loss of framework Ti in TS-1B, Wu et al. hydrothermally post-treated TS-1 [41]. The obtained TS-1$^+$ with high-performance "TiO_6" species had limited loss of framework Ti. The catalytic activity of TiO_6 was two to three times higher than that of $Ti(OSi)_4$, and the effective utilization of H_2O_2 was significantly increased. Xu et al. obtained Re-Ti-MWW-PI-P by the rearrangement treatment of Ti-MWW with PI and found a PI-liganded six-coordinated Ti species ($Ti(OSi)_2(OH)_2(H_2O)(PI)$) in the framework in addition to the common Ti(VI) (Figure 7.4) [90]. The insertion of PI converted part of the Ti(IV) into Ti(VI), and its TON value in the cyclopentene epoxidation was increased by 10 times.

Zuo et al. discovered an efficient Ti(V) species in TS-1 by the UV–vis and X-ray Absorption Near Edge Structure (XANES) spectral characterizations [91]. They had a high Turnover Frequency (TOF) value up to 903.8 in propylene epoxidation. Comparing the TS-1 obtained by TPAOH organic base treatment to original sample, the XANES analysis suggested that Ti(V) might increase the activity of TPAOH-treated TS-1. However, the study on Ti(V) species is still limited, and the accurate controlled synthesis of Ti(V) is also difficult. It has also been shown that the reduction of Ti(IV) in TS-1 by triethylaluminum leads to the formation of a tri-coordinated framework Ti(III) with an open shell layer [92].

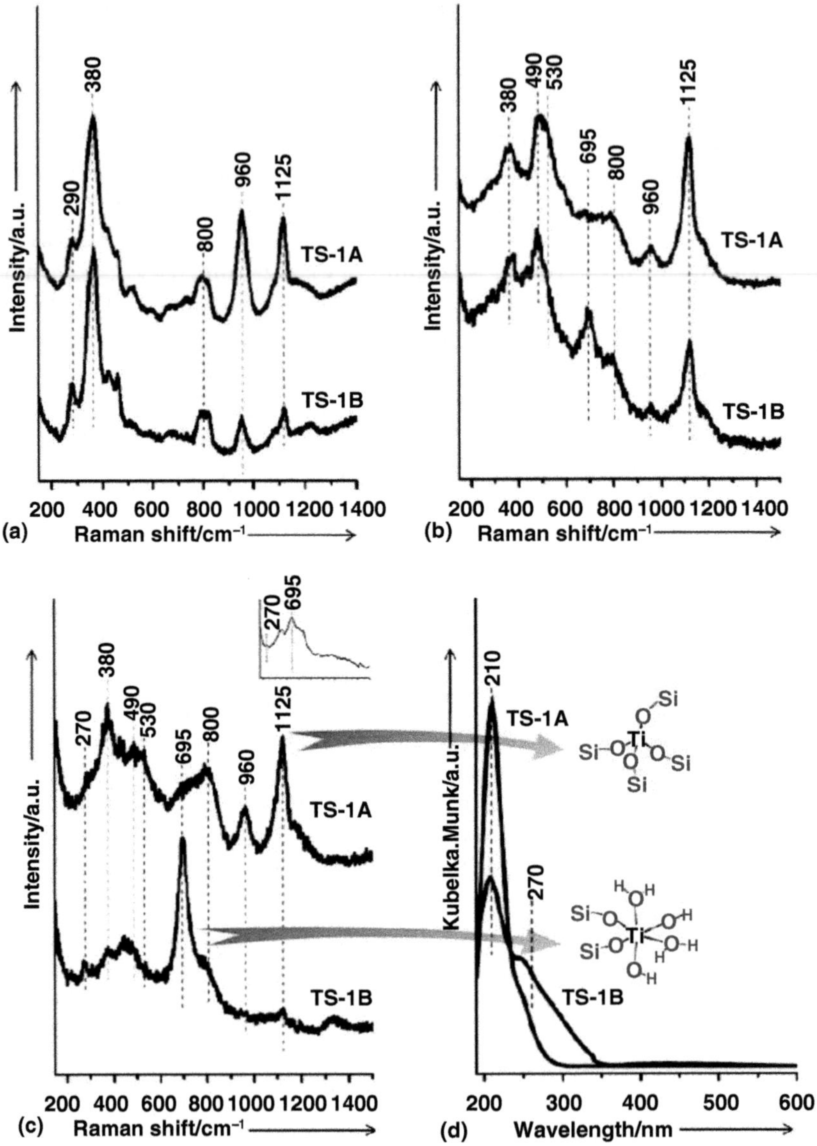

Figure 7.3 The UV-Raman spectra of TS-1A and TS-1B excited at (a) 325, (b) 244, and (c) 266 nm. (d) UV–vis spectra of TS-1A and TS-1B and the proposed structure of the extra framework titanium species. Inset: Raman spectrum of the mononuclear titanium species. Source: Reproduced from Ref. [89]/with permission of German Chemical Society.

Combining the catalytic performance of Ti active sites with different coordination states and DFT theoretical calculations, it is clear that the highly coordinated Ti tends to exhibit higher activity compared to the classical Ti(IV). And the catalytic activity of defective Ti(IV) is higher than that of the perfect one. Therefore, further study on how to construct Ti species with a specific coordination state is still desired to improve their catalytic performance.

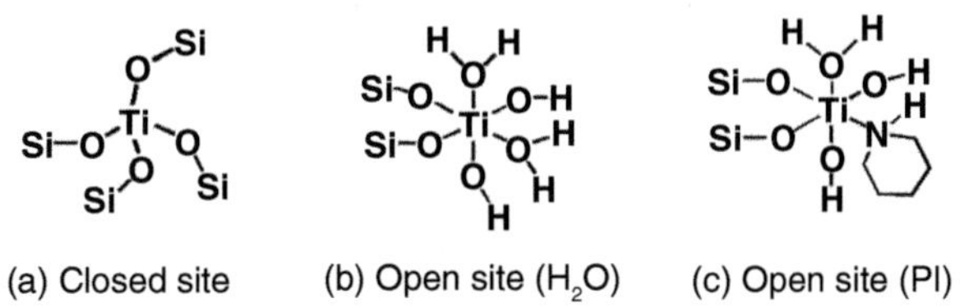

Figure 7.4 Proposed titanium active sites: (a) closed site, (b) open site (H_2O), and (c) open site (PI). Source: Reproduced from Ref. [90]/with permission of Royal Society of Chemistry.

7.3.3.3 Adjacent Silanol Groups of Ti Active Sites

The adjustment of Si—OH groups around Ti active sites could also affect the catalytic activity of titanosilicates. A major difference between mesoporous and microporous zeolites is the abundant Si—OH groups in mesoporous materials. Compared to microporous TS-1 and the homogeneous model $Ti(OSiMe_3)_4$, amorphous mesoporous titania–silica aerogels presented a lower stereoselectivity in the epoxidation of geraniol and cyclohexenol. Beck et al. proposed a "silanol-assisted" epoxidation mechanism, according to which the OH group in the reactant allylic alcohol interacted with a neighboring Si—OH group via H-bonding [93]. This interaction was expected to diminish the stereoselectivity of allylic alcohol epoxidation and interpreted the detrimental effect of water. Afterward, the adjustment of Si—OH groups around Ti active sites has received increasing interest.

Employing the theoretical calculations, Wells et al. investigated the effect of Si—OH groups around Ti active sites in propylene epoxidation [7]. The "open" $Ti(OSi)_3OH$ species with adjacent Si—OH groups was more active than the "closed" $Ti(OSi)_4$ species, and the latter was even inactive. The authors suggested that the adjacent Si—OH species could form a seven-membered ring intermediate to further stabilize Ti—O—O—H species. Furthermore, Wells et al. disclosed that the Ti active sites with various microenvironments in titanosilicates (Figure 7.5) showed different activities in propylene epoxidation with the order of $Ti—OH(HO—Si)_3 > Ti—OH > Ti—OH(HO—Si)$ [94]. Panyaburapa et al. proposed that the introduction of Si—OH groups around Ti led to the transformation from a single proton-transfer to a double proton-transfer mechanism, and the activation energy for the formation of Ti—O—O—H species through the double proton-transfer mechanism was lower [95]. Therefore, it is clear that the Si—OH microenvironment around Ti active sites plays an important role in achieving high oxidation activity.

In addition, the location of Si—OH affects the catalytic performance of titanosilicates. Yuan et al. used a full ab initio method with 40T clusters with a Si:Ti molar ratio of 39 : 1 to investigate the possible position of Si—OH groups in the MFI structure and concluded that the former four energetically favorable sites for Si vacancies are T6, T12, T4, and T8 [96]. The Si vacancy at T6 with a Ti at its neighboring T9 site (T6-def-T9-Ti pair) was the most energetically favorable one, followed by a T6-def-T5-Ti pair with a small energy gap. However, the characterization methods for the location of Si—OH and Ti in titanosilicates were rarely reported. The present studies mainly focus on the adjustment of Si—OH content around Ti sites and the effect on the catalytic activity.

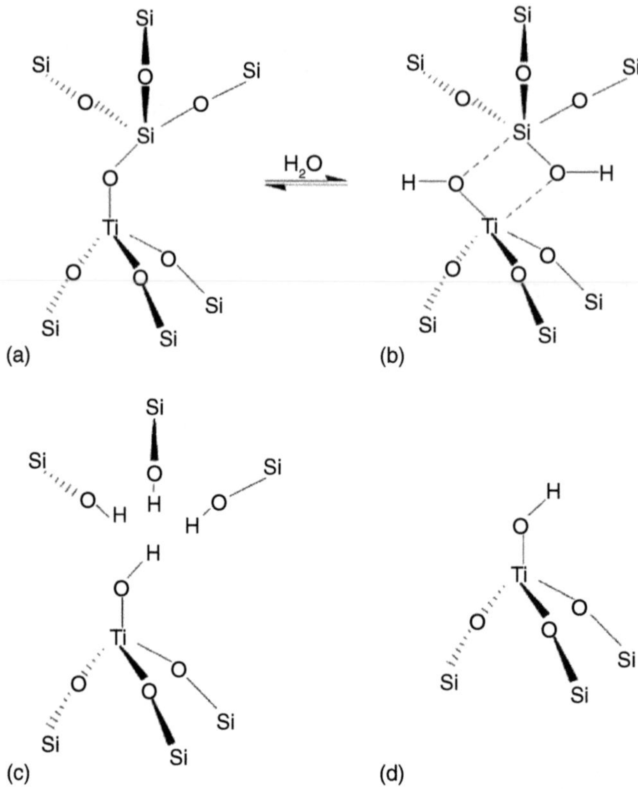

Figure 7.5 Possible defective Ti species in titanosilicates. Source: Reproduced from Ref. [94]/with permission of American Chemical Society.

The current methods to decrease Si—OH content around Ti sites mainly include fluorination and cation exchange. For the former, the F atom interacts with Si—OH groups on the surface of titanosilicates to form Si—F bonds, which can reduce the surface Si—OH content and improve the hydrophobicity of titanosilicates and the electropositive character of the Ti sites. Ryoo et al. treated nanosheet TS-1 with NH_4F to enhance the surface hydrophobicity and improved its catalytic activity for the epoxidation of bulky alkenes [13]. Fang et al. implanted fluorine species into a Ti-MWW framework through an acid treatment process in the presence of NH_4F to obtain F-Ti-MWW [36, 44, 84]. Fluorine implanting improved the surface hydrophobicity of the zeolite and altered the electropositivity nearby the tetrahedral Ti sites, which thus improved the catalytic activity of titanosilicates. Lu et al. suggested that the catalytic performance of the uncalcined F-Ti-MWW was higher than that of the calcined F-Ti-MWW-cal, due to the stronger hydrogen-bond interaction between the H in the Ti—O—O—H intermediate and the adjacent Si—F species in F-Ti-MWW [77].

Cation exchange is a convenient method to adjust the Si—OH content in titanosilicates. On one hand, cation exchange can shield the Brønsted acid sites responsible for the ring-opening reaction of the epoxides, thus improving the epoxides selectivity of alkenes epoxidation [46, 97–99]. In addition, it can improve the intrinsic catalytic ability of titanosilicates. Yamashita et al. demonstrated that by introducing alkali metal cations on ion-exchange sites of the framework, the diffusional properties of the bulky alkenes could be improved on account of the predominance of the cation–π interaction and the catalytic activity for oxidation of bulky cyclohexene and styrene could be thus enhanced [100]. In the process of H_2O_2 activation to form Ti—O—O—H, O_β in H_2O_2 (H_α—O_α—O_β—H_{end}) would form hydrogen bond with H in Si—OH(HO—Ti), and the hydrogen bond would increase the activation energy of H_2O_2 and thus inhibit the activation of H_2O_2 according to DFT calculations [94]. Based on it, Yu et al. performed the K-ion exchange of Ti-MWW via the treatment of potassium chloride (KCl) [101]. Compared to the original Ti-MWW, the K-modified Ti-MWW (Ti-MWW-K) exhibited much improved catalytic performance in the liquid-phase epoxidation of alkenes. It was revealed that the breaking of the hydrogen-bonding interactions between the O_β atom of H_2O_2 (H_α—O_α—O_β—H_{end}) and the H atom of the silanols adjacent to the Ti—OH in the Ti-MWW-K promoted the activation of H_2O_2, which was responsible for its superior catalytic oxidation activity.

To increase the Si—OH content, silylation and alkali treatment are the main methods. Dialkyl-substituted silylation reagents were often adopted. The alkyl groups were introduced into the framework and converted to hydroxyl groups after the calcination treatment to increase the hydroxyl content [27]. While these post-treatments are effective in increasing the Si—OH content, the methods are not able to ensure the location Si—OH adjacent to Ti increases.

Combining kinetic analysis and spectroscopic results, Bregante et al. suggested that increasing $(Si—OH)_4$ density near Ti active sites in Ti-Beta improved the catalytic performance in the cyclohexene epoxidation, which was attributed to the interactions among transition states, H_2O clusters, and catalyst surfaces [102]. Subsequently, further studies revealed that the increase in Si—OH content required a suitable confined space to improve the performance in olefin epoxidation reactions, through the van der Waals forces among the Si—OH walls, the solvent, and the active species [103, 104]. Giordano et al. showed that the increase in Si—OH content achieved by Calix arene-Ti(IV) on UCB-4 framework facilitated the activation of bound intermediates (e.g. for hydroperoxide binding or subsequent transfer of oxygen to epoxide), leading to excellent performance in the cyclohexene epoxidation reaction [105]. However, the amorphous Ti/SiO_2 and Ti/MCM-41 obtained by the same grafting method showed much lower catalytic activity than the crystalline Ti/UCB-4, indicating the synergistic effect of Si—OH content and confined space. Nicolás et al. designed a series of titanosilicates with different Si—OH contents and different confined spaces and systematically investigated the catalytic oxidation activity and their catalytic mechanism, which further demonstrated the synergistic effect of Si—OH and confined space [57]. Therefore, Si—OH adjacent near Ti sites have interactions with the solvent and Ti—O—O—H active intermediates, which would be described in detail in Section 7.4.

Alkali treatment method was also applied to modulate the Si—OH adjacent to Ti sites. Yu et al. adjusted the Si—OH content around the Ti sites in Ti-MWW by mild alkali treatment method, constructing Ti sites with different Si—OH, and the catalytic oxidation activities increased in the order of Ti—OH(HOSi) < Ti—OH(HOSi)$_3$ < Ti—OH (HO—Si)$_2$(KO—Si) [106]. Systematic investigation showed that the change of Si—OH content around Ti sites influenced the activation of H_2O_2 and the stability of Ti—O—O—H species, revealing that a suitable Si—OH content was required.

In summary, the fine adjustment of the Si—OH environment around Ti sites is important for the catalytic oxidation performance of titanosilicates, which affects the formation, stability, and transfer of Ti—O—O—H species. Section 7.2.2 also described the adjustment of the surface hydrophobicity of titanosilicates by changing the Si—OH amount on the surface. Although the adjustment of Si—OH content around Ti sites would change the surface hydrophobicity, it could directly affect the reaction behavior of H_2O_2 by the synergistic effect of Si—OH and confined space.

7.4 Solvent Effect

In 1890, Menschutkin found that the rates for the quaternization of triethylamine with iodoethane were closely related to the choice of solvent and proposed that protic benzyl alcohol molecules could stabilize the separation of charges within the transition state for quaternization [107]. It is generally accepted that the role of solvents in catalytic reactions includes influencing the solubility and adsorption/desorption of reactants/products, proton/electron transfer during the reaction, and the stability of reaction intermediates [108].

Zeolite catalysis is a typical heterogeneous catalytic process, and the formation of the active intermediates is critical. For the Brønsted acid-catalyzed process with aluminosilicate zeolite, the active intermediate (carbocation) is formed through the proton transfer from proton hydrogen to organic reactant molecules. Due to the strong acidity of aluminosilicate, the proton hydrogen is easily ionized and the acid-catalyzed reaction can generally occur without the introduction of a solvent [109, 110]. In the selective oxidation of hydrocarbons over TS/H_2O_2 system, the Ti—O—O—H intermediate is the real active site formed through H_2O_2 activation over Ti sites. Solvent plays a vital role in the formation, stabilization, and transfer of the Ti—O—O—H intermediate (Figure 7.6) [7, 95, 111–113]. The choice of a proper solvent in titanosilicate-catalyzed oxidation system is complex (Table 7.1). It not only depends on the properties of solvent and titanosilicates but also is closely related to reactants [111]. For example, in the oxidation reactions of alkenes, alkanes, and sulfides, the best solvent for TS-1 is protic methanol (MeOH), and the best solvent for Ti-MWW and Ti-Beta is aprotic acetonitrile (MeCN), whereas, for the aldehyde/ketone ammoximation and phenol hydroxylation, the proper solvents are protic *tert*-butanol (*t*-BuOH) and aprotic acetone (MeCOMe), respectively, irrespective of the titanosilicate topologies [24, 111, 124–131, 134, 135]. Thus, the chemical effect of solvent over TS/H_2O_2 system has been extensively

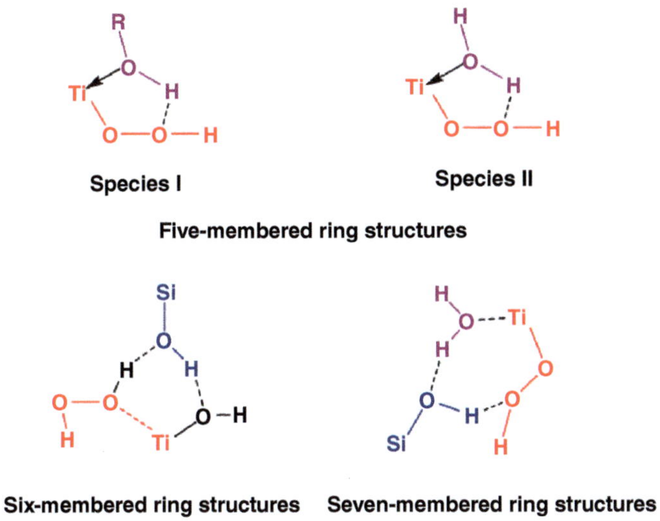

Figure 7.6 Proposed Ti—OOH active intermediates [five-membered ring (5MR, Species I and Species II)], six-membered ring (6MR), or seven-membered ring (7MR) structures. Source: Reprinted with permission from Refs [7, 24, 65, 95]; © 2004, 2007 American Chemical Society and 1993, 1996 Elsevier.

investigated, indicating that solvents can affect the catalytic oxidation performance by influencing diffusion, adsorption/desorption, and surface reactions.

7.4.1 Effect of Solvent on Diffusion

Diffusion is the basis of heterogeneous catalysis and can be evaluated by the Weisz modulus Φ [136]. When Φ is over 0.15, the reaction is affected by diffusion, while when Φ is less than 0.15, the reaction is diffusion unlimited. The effect of solvent on the diffusion can be illustrated using three typical examples including alkenes epoxidations, ammoximation, and aromatic hydroxylation.

For alkenes epoxidations, based on the analysis of the Weisz modulus, Ramachandran et al. demonstrated that in MeOH, MeCN, and MeCOMe solvents, the critical particle size with diffusion limitation (particle size of the catalyst at $\Phi = 0.15$) for the TS-1 catalyzed 1-hexene epoxidation reaction was 38 μm, 60 μm, and 55 μm, respectively [137]. Since the classical hydrothermal synthesis of TS-1 is ~300 nm (much smaller than the critical particle size for diffusion limitation), the effect of solvent on diffusion limitation is generally not considered in the catalytic epoxidation of small-molecule olefins (for straight chain olefins, carbon number < 8; for cyclic olefins, carbon number < 6).

The ketone ammoximation over titanosilicates followed the hydroxylamine mechanism, which consists of the catalytic oxidation of NH_3 to generate NH_2OH and the non-catalytic oxime reaction with ketone. It means that the diffusion of NH_3 into the channel of titanosilicates is the key process. Yip et al. suggested that the internal and external diffusion limitations were eliminated when the crystal size of TS-1 was less

Table 7.1 The favorable solvents for various oxidations reactions catalyzed by titanosilicates with H_2O_2 as the oxidant.

Catalyst	Linear alkenes	Cyclopentene	Cyclohexene	Allyl alcohol	Cyclohexanol	Diallyl ether	1,4-Dioxane	2,5-DHF	Pyridine	NH_3	Phenol	Aliphatic hydrocarbon	Sulfides	B-V oxidation
TS-1	CH_3OH	CH_3OH	CH_3CN	CH_3OH/H_2O		CH_3CN/CH_3COCH_3	Solvent-free	CH_3CN/H_2O	CH_3OH/H_2O/non-solvent	*t*-BuOH	CH_3COCH_3	*t*-BuOH		Solvent-free
Ti-MWW	CH_3CN	CH_3CN/CH_3OH	CH_3CN/CH_3OH	CH_3CN/H_2O		CH_3OH/CH_3COCH_3	Non-solvent	CH_3OH	Non-solvent	*t*-BuOH			Non-solvent	
Ti-Beta	CH_3CN	CH_3OH/C_2H_5OH	CH_3OH	CH_3OH/C_2H_5OH	CH_3CN					*t*-BuOH			CH_3CN	
Reference	[111, 114–116]	[117, 118]	[111]	[119]	[24]	[118]	[120]	[121, 122]	[123]	[124–126]	[127, 128]	[129]	[130, 131]	[132, 133]

than 500 nm and the solvent was the mixture of water and *t*-BuOH [138]. Since NH_3 is water-soluble and small, the diffusion limitation can be generally avoided even in water solvent or solvent-free system [139–142].

For aromatic hydroxylation, diffusion limitations generally existed due to the large size of aromatics [143]. For 10-MR TS-1, phenol showed low diffusion, while in 12-MR pores of Ti-MOR and Ti-Beta, phenol exhibited higher diffusion rates [143–147]. The diffusion of aromatics was also closely related to the solvent. Wilkenhoner et al. found that the diffusion rate of phenol in TS-1 was greater than that of catechol and hydroquinone when MeOH was used as the solvent, and the latter two had similar diffusion rates [146]. When H_2O was used as the solvent, the diffusion rate of phenol in TS-1 was higher than that of catechol and hydroquinone. In addition, the bulk Ti—O—O—H species formed by H_2O_2 activation also intensified the diffusion limitation of the aromatics [143]. The diffusion limitation in the aromatic hydroxylation reactions affected not only the activity but also the product selectivity. The products with less steric hindrance had a higher selectivity. For example, hydroquinone was the main product in phenol hydroxylation [112]. It is clear that the diffusion of aromatic hydroxylation reactions is not only determined by the size of substrates and the nature of titanosilicates but also closely related to the solvent property.

7.4.2 Effect of Solvent on Adsorption/Desorption

Since the chemisorption of the oxidant H_2O_2 was not dependent on solvent types, the effects of solvent on adsorption/desorption of alkenes, NH_3, and aromatic hydrocarbons are discussed [148].

In the alkene epoxidation, the solvent affects not only the adsorption of alkenes but also the desorption of products epoxides. Ramachandran et al. showed that the strong adsorption ability of 1-hexene on TS-1 contributed to the high activity and the solvent could affect the kinetics by influencing the adsorption of alkenes [148]. Jiao et al. found that the epoxidation of methallyl chloride (MAC) was much more active in *t*-BuOH solvent than in MeOH and MeCN partially because *t*-BuOH increased the adsorption amount of MAC in the channel of Ti-rich TS-1, and consequently enhanced the MAC concentration around the Ti sites [149]. Langhendries et al. investigated the adsorption ability of 1-hexene on TS-1 in different solvents using a tracer chromatographic method and the adsorption ability decreased in the order of MeOH > EtOH > MeCN > MeCOMe > 1-PrOH [150]. The corresponding olefin epoxidation reaction rates decreased in the order of MeOH ≫ MeCN, MeCOMe > EtOH > 1-PrOH, which indicated that the adsorption of alkenes affected its catalytic performance.

The desorption of the product epoxides has a very important effect on the whole catalytic cycle. The deactivation rate of 1-nonene oxidation over TS-1 was fast in MeOH because the adsorption ability of the product epoxide was strong, which inhibited the further adsorption of 1-nonene. When acetone was used as the solvent, the product desorbed easily and leading to slower deactivation rate of TS-1 [150]. By DFT theoretical calculations, Li et al. demonstrated that the desorption of the product epoxide was the rate-determining step during 1-hexene epoxidation [151]. The epoxide product showed lower desorption energy from the active site of Ti-η^2(OOH)—CH_3CN compared to Ti-η^2(OOH)—H_2O, so CH_3CN was the best solvent for Ti -MWW.

For ketone ammoximation processes, the formation of NH_2OH is critical. Bordiga et al. compared the adsorption behavior of H_2O and NH_3 on titanosilicates using UV–vis, IR, XANES, and Extended X-ray Absorption Fine Structure (EXAFS) characterizations [152, 153]. The results showed that H_2O was difficult to adsorb on Ti sites, but NH_3 was easier to adsorb, so the effect of NH_3 adsorption on the oxidation reaction was generally not considered. Fang et al. showed that free NH_2OH was easily oxidation decomposed by Ti—O—O—H species in the TS/H_2O_2 system [139]. Thus, the desorption of NH_2OH would significantly affect the utilization of NH_2OH and the economy of N atoms. The study of NH_2OH desorption in different solvents would facilitate the in-depth understanding of the oxime process and the efficient preparation of free NH_2OH.

For the hydroxylation of aromatics, Guo et al. demonstrated that TS-1 presented enhanced adsorption ability of H_2O_2, styrene, and cyclohexanone than S-1, indicating that the introduction of Ti promoted the adsorption of reaction substrates [154]. Callanan et al. investigated the effect of methanol/water ratio in the solvent mixture on the hydroxylation of phenol and proposed the highest activity in pure water, mainly due to that the methanol and phenol competed for coordination with Ti—O—O—H species [155]. Bhaumik et al. found that the oxidation activity of benzene, toluene, anisole, and benzyl chloride was 3–10 times higher in the absence of solvent than in the presence of a solvent, which was attributed to the competitive adsorption of organic solvents with the substrate [156]. Therefore, in the hydroxylation reaction of aromatics, the influence of the solvent on the adsorption/desorption process is mainly focused on the competitive adsorption between solvent and aromatics.

7.4.3 Effect of Solvent on Surface Reactions

Generally, the epoxidation consists of the chemisorption of H_2O_2 on framework Ti active sites, the further activation of H_2O_2 to form the Ti—O—O—H species, which could coordinate with solvent molecules (MeOH or H_2O) to form a five-membered ring, six-membered ring, or seven-membered ring structures (Figure 7.6), and finally the transfer of active "O" to the reactants [5, 7, 95, 111, 112]. The synergistic effect of the formation, stability, and transfer of Ti—O—O—H species was the key to achieve a high atom economy in TS/H_2O_2 systems [8, 157]. Numerous theoretical and experimental studies have been carried out to explore the effect of solvents on surface reactions including the formation, stabilization, and transfer of Ti—O—O—H species in TS/H_2O_2 systems.

7.4.3.1 Effect on the Formation on Ti—O—O—H

The solvent can directly affect H_2O_2 activation to form Ti—O—O—H species [5, 158]. Employing the in situ UV-Raman characterization for propylene epoxidation on TS-1, Nie et al. observed a consistency between the area of the characteristic peak at 837 cm^{-1} assigned to the O—O stretching mode in the Ti—O—O—H active intermediate and the formation rate of product PO [159]. Besides, solvent only affected the intensity of characteristic peak at 837 cm^{-1} without any shift. The formation of Ti—O—O—H species was faster in the solvent of CH_3OH compared to H_2O. Therefore, the solvent directly participates in the formation of Ti—O—O—H, which affected the whole catalytic oxidation process, but the chemical role of solvent remains unclear. It is generally accepted that in the activation of H_2O_2 to form Ti—O—O—H species, dissociation of H_2O_2 would occur, indicative of proton transfer [7, 65, 113, 159–162]. Both Our-own-N-layered Integrated molecular Orbital + molecular Mechanics (ONIOM) and DFT theoretical calculations indicated that the activation energy of proton transfer on the activation of H_2O_2 to form Ti—O—O—H species was high [95, 159].

Protic molecule (e.g. water, alcohols, ammonia, and nitromethane) can act as a mediator to promote proton transfer [163–167]. For the TS/H_2O_2 system, there are two kinds of proton-transfer mediators: the hydroxyl group in titanosilicates and the introduced protic solvent molecules. Theoretical calculations have shown that the hydroxyl groups in titanosilicates acted as a mediator to facilitate the proton transfer and reduced the activation energy for Ti—O—O—H formation process. Sever et al. indicated that the protic solvent could interact with the Ti—OH site and H_2O_2 by hydrogen bond according to the DFT calculations, and thus facilitate the formation of Ti—O—O—H [168]. Utilizing the 144T and 143T cluster models, Nie et al. indicated that protic CH_3OH could be co-adsorbed with H_2O_2 on Ti sites through coordination and hydrogen bonding, facilitating the proton transfer and thus reducing the activation energy required for Ti—O—O—H formation [159]. In contrast, the introduction of aprotic CH_3CN did not significantly change the reaction pathway and kinetics. They proposed a possible mechanism for the formation of Ti—O—O—H, involving the O atom in CH_3OH interacting with H_α in H_α—O_α—O_β—H_β by hydrogen bonding, and the H in CH_3OH formed hydrogen bonds with O in Ti—OH, which promoted the proton transfer and the formation of Ti—O_α—O_β—H_β. Combining isotopic labeled IR spectra and reaction kinetics investigation, Yu et al. have demonstrated that protic solvent (ROH/H_2O) acted as a mediator to promote the proton transfer on H_2O_2 activation and the solvent effect of titanosilicates was closely related to Ti active sites [113]. For $Ti(OSi)_4$ active sites, ROH was the mediator, and the solvent CH_3OH with the highest acidity was the best solvent. For $Ti(OSi)_3OH$ active sites, H_2O was the mediator, and the weak hydrogen bonding of H_2O led to the high acidity of H_2O molecules, promoting proton transfer. Solvent influenced the catalytic performance by changing the hydrogen bonding of H_2O molecules, and the strong nucleophilic ability of solvent meant a strong hydrogen bonding with H_2O, so CH_3CN with the weakest nucleophilic ability was the best solvent for $Ti(OSi)_3OH$ sites. The ability of protic solvents to promote proton transfer was related to their acidity, while aprotic solvents were related to their nucleophilic ability [113] (Figure 7.7).

When H_2O was used as a proton transfer mediator, theoretical calculations have shown that the structure of water molecules had an important effect on the activation of H_2O_2 and propylene epoxidation [103, 168, 169]. Sever et al. proposed that a single protonic molecule reduced the Gibbs free energy of the H_2O_2 activation

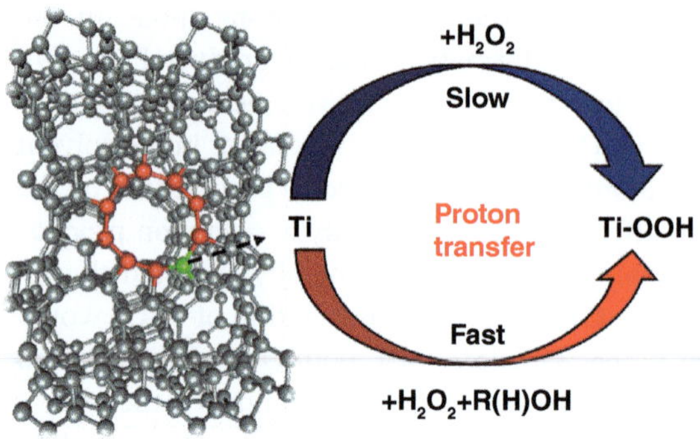

Figure 7.7 Protic solvent acts as a mediator to promote the proton transfer on H_2O_2 activation [113]. Source: Reproduced from Ref. [113]/with permission of American Chemical Society.

process by 4–8.5 kcal/mol (17–35 kJ/mol, to unify the units, kcal/mol in the original text was converted to kJ/mol), while hydrogen-bonded protonic molecules reduced 5–6 kcal/mol (21–25 kJ/mol) [168]. Through theoretical calculations, Stare et al. showed that a single water molecule could not change the reaction energy of Ti—O—O—H formation, but can effectively reduce the activation energy required for proton transfer to 25.7 kcal/mol (107 kJ/mol), thus promoting the formation of Ti—O—O—H, while the introduction of another water molecule had no further effect on the proton transfer process [169]. Bregante et al. demonstrated that in alkenes epoxidation, the reorganization of confined H_2O structures could improve alkene epoxidation rates by up to 400-fold [103]. Recently, using Raman and IR spectroscopy techniques, Yu et al. found that the catalytic performance of propylene epoxidation in H_2O solvent depended on the structure of H_2O molecules [170]. The superior H_2O structure was the one that donated two and accepted one hydrogen bond, known as the "DDA" structure. The DDA structure with a strong ability to accept an additional proton was the superior proton-transfer mediator on H_2O_2 activation. Accordingly, they proposed a highly efficient H_2O-solvent alkene epoxidation system (e.g. Ti-MWW/H_2O/H_2O_2/propylene/K^+), which possessed high activity, selectivity, and stability.

7.4.3.2 Effect on the Stability of Ti—O—O—H

Protic solvent can also promote the stability of Ti—O—O—H through coordination and hydrogen bonding, which affects the kinetics of alkenes epoxidation. Clerici et al. first proposed a five-membered ring reactive intermediate structure formed by coordination and hydrogen bonding between Ti—O—O—H and ROH (Species I, Figure 7.6), which promoted the stability of Ti—O—O—H and enhanced its electrophilic ability, thus improving the epoxidation activity [65]. Corma et al. proposed that the H_2O molecule could stabilize Ti—O—O—H species by forming a five-membered ring (Species II, Figure 7.6), which explained the highest activity of hydrophilic Ti-Beta in acetonitrile (a polar aprotic solvent) for the epoxidation of 1-hexene and cyclohexanol [24]. Vayssilov et al. systematically compared different kinds of 5MR intermediate structures formed by the interaction of Ti—O—O—H with CH_3OH, H_2O, or a silanol (Si—OH) group through DFT theoretical calculations [171]. Results showed that CH_3OH favored the formation of the 5MR structure compared to H_2O and Si—OH species; however, the formation of such a 5MR structure was difficult with the high energy barrier of 90–130 kJ/mol. Wells et al. suggested that Ti—O—O—H could form a seven-membered ring intermediate with H_2O molecule and the neighboring Si—OH (Figure 7.6), which improved the stability of the active intermediate and thus exhibited superior catalytic performance [7]. In general, protic molecules could coordinate with the active intermediate Ti—O—O—H to stabilize them, but there is no solid evidence to prove the five-membered, six-membered, or seven-membered ring structures.

Li et al. investigated the effect of solvent coordination on the stability and catalytic performance of Ti—O—O—H in Ti-MWW through DFT theoretical calculations, and the results showed that the stability decreased in the order of Ti-η^2(OOH)—H_2O > Ti-η^2(OOH)—CH_3CN > Ti-η^2(OOH) as well as Ti-η^2(OOH)—H_2O > Ti-η^1(OOH) [151]. The activation barriers of 1-hexene epoxidation over Ti6-η^2(OOH), Ti6-η^2(OOH)—H_2O, and Ti6-η^2(OOH)—CH_3CN were 33.2, 35.0, and 43.9 kJ/mol, respectively. However, these results could not explain why acetonitrile was the best solvent for Ti-MWW.

Based on the stabilization of Ti—O—O—H through hydrogen bonding, Deng et al. firstly proposed that a double hydrogen bonded reactive intermediate could be formed by introducing a hydrogen bond acceptor molecule (e.g. NH_3) to interact with H_{end} in Ti—O_α—O_β—H_{end}, generating a new hydrogen bond [172]. It further stabilized Ti—OOH species and improved the activity for alkenes epoxidation. Subsequently, researchers have also reported other ways to stabilize Ti—O—O—H by introducing hydrogen-bonding acceptors (e.g. implanting N and F atoms into the zeolite framework) [70, 77, 173]. Recently, Bregante et al. demonstrate that the solvent could affect the free energy of the reactants and transition state and thus affect the stability of epoxidation transition states, which was related to solvent properties, the zeolite frameworks, and Si—OH densities [102, 103, 108, 174].

7.4.3.3 Effect on the Transfer of Ti—O—O—H

Based on DFT theoretical calculations, Sever et al. claimed that the activation energy of the transfer of active "O" to ethylene was 25.7 kcal/mol (107.4 kJ/mol). The introduction of a single protic molecule (e.g. MeOH/H_2O) did not change the activation energy, while the introduction of multiple protic molecules was detrimental [175]. In contrast, Zou et al. demonstrated that the adsorption of solvent molecules affected the electrophilicity and catalytic activity of active oxygen species, and the adsorption of CH_3CN on Ti-η^2(OOH) reduced the activation energy of the epoxidation reaction of chloropropene [176]. Meng et al. investigated the effect of solvent on the epoxidation of propylene, showing that as the solvent polarity increased, the heterolytic cleavage of O—O bond in Ti—O—O—H intermediate for the transfer of active "O" was promoted, based on the Hughes-Ingold rules [177].

In the TS/H_2O_2 system, the solvent generally affected the catalytic activity, selectivity, and stability of titanosilicates by influencing the diffusion, adsorption/desorption of the reactants and products. More importantly, solvent can participate in the catalytic reaction through the interaction with the substrates and Ti sites. The chemical nature of solvent effect in the TS/H_2O_2 system can be summarized as follows: acting as a proton-transfer mediator to promote the proton transfer in the formation of Ti—O—O—H, stabilizing Ti—O—O—H by coordination and hydrogen bonding, and promoting the transfer of active O by affecting the electrophilicity of active "O."

7.5 Conclusions and Prospects

Post-modification of titanosilicates is an effective way to improve their catalytic performance. From the perspective of post-modification strategies, acid/base modification, metal and non-metal ion modification, and fluorination modification are commonly used, and alcohol functionalization and phosphorus modification have been recently developed. Different modification methods can be used to tailor the similar property of the titanosilicates, for example, alkali modification, metal and non-metal ion modification, fluorination modification, and phosphorus modification all can adjust the electrophilicity and Lewis acid strength of the framework Ti species. The same modification method can also be used on different structural properties of the titanosilicates, for example, the organic alkali modification can adjust the hydrophobicity, the electrophilicity of framework Ti, and the coordination state of Ti. Meanwhile, in the TS/H_2O_2 system, the solvent affects the diffusion, adsorption/desorption process, and surface reaction of reactants and products, showing obvious solvent effect.

In order to give full play to the catalytic role of titanosilicates and focus on the purpose of improving the catalytic performance, it is crucial to promote the formation, stability, and transfer of the Ti—O—O—H species. Therefore, adjusting the Ti microenvironment is the core to the post-modification of titanosilicates for achieving high performance. Although there is still no exact definition of Ti microenvironment, it can be preliminarily understood as a collection of structural factors that affect the catalytic ability of a single Ti site. The structural factors mainly include (1) the coordination status of the framework Ti, such as tetrahedrally coordinated Ti(IV), pentahedrally coordinated Ti(V), and six-coordinated Ti(VI); (2) perfect or defective Ti(IV) sites; (3) the amount of Si—OH group adjacent to framework Ti; (4) the heteroatom adjacent to framework Ti; and (5) solvent molecule adjacent to framework Ti. Although the methods and strategies for the regulation of Ti microenvironment have been rapidly developed in recent years, the systematic analysis and characterization of its structure and the understanding of structure-property relationships are not sufficient and still in the initial stage.

References

1 Taramasso M., Perego, G., Notari, B. (1983) *Preparation of porous crystalline synthetic material comprised of silicon and titanium oxides*. US, 4410501 [P].

2 Moliner, M. and Corma, A. (2014). Advances in the synthesis of titanosilicates: from the medium pore TS-1 zeolite to highly-accessible ordered materials. *Microporous Mesoporous Mater.* 189: 31–40.

3 Bai, R., Song, Y., Bai, R., and Yu, J. (2021). Creation of hierarchical titanosilicate TS-1 zeolites. *Adv. Mater. Interfaces* 8: 2001095.

4 Bellussi, G. and Millini, R. (2017). Background and recent advances in Ti-containing zeolite materials. In: *Structure and Reactivity of Metals in Zeolite Materials*, vol. 178, 1–52. Springer Verlag.

5 Ratnasamy, P., Srinivas, D., and Knozinger, H. (2004). Active sites and reactive intermediates in titanium silicate molecular sieves. *Adv. Catal.* 48: 1–169.

6 Smeets, V., Gaigneaux, E.M., and Debecker, D.P. (2022). Titanosilicate epoxidation catalysts: a review of challenges and opportunities. *ChemCatChem* 14: e202101132.

7 Wells, D.H., Delgass, W.N., and Thomson, K.T. (2004). Evidence of defect-promoted reactivity for epoxidation of propylene in titanosilicate (TS-1) catalysts: a DFT study. *J. Am. Chem. Soc.* 126: 2956–2962.

8 Yu, Y., Tang, Z., Wang, J. et al. (2020). Insights into the efficiency of hydrogen peroxide utilization over titanosilicate/H_2O_2 systems. *J. Catal.* 381: 96–107.

9 Schmidt, I., Krogh, A., Wienberg, K. et al. (2000). Catalytic epoxidation of alkenes with hydrogen peroxide over first mesoporous titanium-containing zeolite. *Chem. Commun.* 21: 2157–2158.

10 Schmidt, I., Boisen, A., Gustavsson, E. et al. (2001). Carbon nanotube templated growth of mesoporous zeolite single crystals. *Chem. Mater.* 13: 4416–4418.

11 Pei, X., Liu, X., Liu, X. et al. (2019). Synthesis of hierarchical titanium silicalite-1 using a carbon-silica-titania composite from xerogel mild carbonization. *Catalysts* 9: 672.

12 Cheneviere, Y., Chieux, F., Caps, V., and Tuel, A. (2010). Synthesis and catalytic properties of TS-1 with mesoporous/microporous hierarchical structures obtained in the presence of amphiphilic organosilanes. *J. Catal.* 269: 161–168.

13 Na, K., Jo, C., Kun, J. et al. (2011). MFI titanosilicate nanosheets with single-unit-cell thickness as an oxidation catalyst using peroxides. *ACS Catal.* 1: 901–907.

14 Du, Q., Guo, Y., Duan, H. et al. (2017). Synthesis of hierarchical TS-1 zeolite via a novel three-step crystallization method and its excellent catalytic performance in oxidative desulfurization. *Fuel* 188: 232–238.

15 Lv, G., Deng, S., Zhai, Y. et al. (2018). P123 lamellar micelle-assisted construction of hierarchical TS-1 stacked nanoplates with constrained mesopores for enhanced oxidative desulfurization. *Appl. Catal. A Gen.* 567: 28–35.

16 Xue, Y., Zuo, G., Wen, Y. et al. (2019). Seed-assisted synthesis of TS-1 crystals containing Al with high catalytic performances in cyclohexanone ammoximation. *RSC Adv.* 9: 2386–2394.

17 Song, Y., Bai, R., Zou, Y. et al. (2020). Temperature-regulated construction of hierarchical titanosilicate zeolites. *Inorg. Chem. Front.* 7: 1872–1879.

18 Chao, P., Tsai, S.T., Tsai, T.C. et al. (2009). Phenol hydroxylation over alkaline treated TS-1 catalysts. *Top. Catal.* 52: 185–192.

19 Du, S., Chen, X., Sun, Q. et al. (2016). A non-chemically selective top-down approach towards the preparation of hierarchical TS-1 zeolites with improved oxidative desulfurization catalytic performance. *Chem. Commun.* 52: 3580–3583.

20 Lin M., Shu, X., Wang, X., Zhu, B. (2002) *Titanium-silicalite molecular sieve and the method for its preparation*. JP, 2006075252 [P].

21 Wang, Y., Lin, M., and Tuel, A. (2007). Hollow TS-1 crystals formed via a dissolution–recrystallization process. *Microporous Mesoporous Mater.* 102: 80–85.

22 Wang, B., Peng, X., Zhang, W. et al. (2017). Hierarchical TS-1 synthesized via the dissolution-recrystallization process: influence of ammonium salts. *Catal. Commun.* 101: 26–30.

23 Zuo, Y., Song, W., Dai, C. et al. (2013). Modification of small-crystal titanium silicalite-1 with organic bases: recrystallization and catalytic properties in the hydroxylation of phenol. *Appl. Catal. A Gen.* 453: 272–279.

24 Corma, A., Esteve, P., and Martınez, A. (1996). Solvent effects during the oxidation of olefins and alcohols with hydrogen peroxide on Ti-beta catalyst: the influence of the hydrophilicity–hydrophobicity of the zeolite. *J. Catal.* 161: 11–19.

25 Xiao, F., Xie, B., Zhang, H. et al. (2011). Interlayer-expanded microporous titanosilicate catalysts with functionalized hydroxyl groups. *ChemCatChem* 3: 1442–1446.

26 Tanaka, Y. and Nakata, S. (1997). Stabilization of mesoporous molecular sieves by trimethylsilylation. *J. Phys. Chem. B* 101: 9436–9440.

27 Wang, L., Sun, J., Meng, X. et al. (2014). A significant enhancement of catalytic activities in oxidation with H_2O_2 over the TS-1 zeolite by adjusting the catalyst wettability. *Chem. Commun.* 50: 2012–2014.

28 Tatsumi, T., Koyano, K.A., and Igarashi, N. (1998). Remarkable activity enhancement by trimethylsilylation in oxidation of alkenes and alkanes with H_2O_2 catalyzed by titanium-containing mesoporous molecular sieves. *Chem. Commun.* 325–326.

29 D'Amore, M.B. and Schwarz, S. (1999). Trimethylsilylation of ordered and disordered titanosilicates improvements in expoxidation with aqueous H_2O_2 from micro-to meso-prous and beyond. *Chem. Commun.* 121–122.

30 Kasahara, T., Inumaru, K., and Yamanaka, S. (2004). Enhanced photocatalytic decomposition of nonylphenol polyethoxylate by alkyl-grafted TiO_2-MCM-41 organic–inorganic nanostructure. *Microporous Mesoporous Mater.* 76: 123–130.

31 Sonoda, J., Kamegawa, T., Kuwahara, Y. et al. (2010). Hydrophobic modification of Ti-containing zeolite (TS-1) and their applications in liquid-phase selective catalytic reactions. *Bull. Chem. Soc. Jpn.* 83: 592–594.

32 Lv, G., Wang, F., Zhang, X., and Binks, B.P. (2018). Surface-active hollow titanosilicate particles as a Pickering interfacial catalyst for liquid-phase alkene epoxidation reactions. *Langmuir* 34: 302–310.

33 Wang, B., Lu, L., Ge, B. et al. (2019). Hydrophobic and hierarchical modification of TS-1 and application for propylene epoxidation. *J. Porous. Mater.* 26: 227–237.

34 Aneke, L., Gerritsen, L., Van den Berg, P., and De Jong, W. (1979). The disproportionation of toluene over a HY/β-AlF3/Cu catalyst: 1. Preparation and characterization. *J. Catal.* 59: 26–36.

35 Corma, A., Domine, M., Gaona, J.A. et al. (1998). Strategies to improve the epoxidation activity and selectivity of Ti-MCM-41. *Chem. Commun.* 2211–2212.

36 Fang, X., Wang, Q., Zheng, A. et al. (2012). Fluorine-planted titanosilicate with enhanced catalytic activity in alkene epoxidation with hydrogen peroxide. *Catal. Sci. Technol.* 2: 2433–2435.

37 Rodenas, Y., Fierro, J., Mariscal, R. et al. (2019). Post-synthesis treatment of TS-1 with TPAOH: effect of hydrophobicity on the liquid-phase oxidation of furfural to maleic acid. *Top. Catal.* 62: 560–569.

38 Cheng, C.H., Bae, T.H., McCool, B.A. et al. (2008). Functionalization of the internal surface of pure-silica MFI zeolite with aliphatic alcohols. *J. Phys. Chem. C* 112: 3543–3551.

39 Wei, Y., Li, G., Su, R. et al. (2019). Ti-sites environment-mediated hierarchical TS-1 catalyzing the solvent-free epoxidation: the remarkably promoting role of alcohol modification. *Appl. Catal. A Gen.* 582: 117108.

40 Song, W., Xiong, G., Long, H. et al. (2015). Effect of treatment with different bases on the catalytic properties of TS-1/SiO_2 extrudates in propylene epoxidation. *Microporous Mesoporous Mater.* 212: 48–55.

41 Wu, L., Deng, X., Zhao, S. et al. (2016). Synthesis of a highly active oxidation catalyst with improved distribution of titanium coordination states. *Chem. Commun.* 52: 8679–8682.

42 Wu, L., Tang, Z., Yu, Y. et al. (2018). Facile synthesis of a high-performance titanosilicate catalyst with controllable defective $Ti(OSi)_3OH$ sites. *Chem. Commun.* 54: 6384–6387.

43 Wang, J., Chen, Z., Yu, Y. et al. (2019). Hollow core–shell structured TS-1@ S-1 as an efficient catalyst for alkene epoxidation. *RSC Adv.* 9: 37801–37808.

44 Fang, X., Wang, Q., Zheng, A. et al. (2013). Post-synthesis, characterization and catalytic properties of fluorine-planted MWW-type titanosilicate. *Phys. Chem. Chem. Phys.* 15: 4930–4938.

45 Goa, Y., Wu, P., and Tatsumi, T. (2004). Catalytic performance of [Ti, Al]-beta in the alkene epoxidation controlled by the postsynthetic ion exchange. *J. Phys. Chem. B* 108: 8401–8411.

46 Tatsumi, T., Koyano, K.A., and Shimizu, Y. (2000). Effect of potassium on the catalytic activity of TS-1. *Appl. Catal. A Gen.* 200: 125–134.

47 Notari, B. (1993). Titanium silicalites. *Catal. Today* 18: 163–172.

48 Fan, W., Duan, R.G., Yokoi, T. et al. (2008). Synthesis, crystallization mechanism, and catalytic properties of titanium-rich TS-1 free of extraframework titanium species. *J. Am. Chem. Soc.* 130: 10150–10164.

49 Wu, P., Tatsumi, T., Komatsu, T., and Yashima, T. (2001). A novel titanosilicate with MWW structure. I. Hydrothermal synthesis, elimination of extraframework titanium, and characterizations. *J. Phys. Chem. B* 105: 2897–2905.

50 Zhang, S., Jin, S., Tao, G. et al. (2017). The evolution of titanium species in boron-containing Ti-MWW zeolite during post-treatment revealed by UV resonance Raman spectroscopy. *Microporous Mesoporous Mater.* 253: 183–190.

51 Zhang, J., Jin, S., Deng, D. et al. (2021). Insight into the formation of framework titanium species during acid treatment of MWW-type titanosilicate and the effect of framework titanium state on olefin epoxidation. *Microporous Mesoporous Mater.* 314: 110862.

52 Tang, Z., Yu, Y., Liu, W. et al. (2020). Deboronation-assisted construction of defective $Ti(OSi)_3OH$ species in MWW-type titanosilicate and their enhanced catalytic performance. *Catal. Sci. Technol.* 10: 2905–2915.

53 Song, S., Wang, P., He, Y. et al. (2012). Preparation, characterization and catalytic properties of Ti-rich Ti-YNU-1. *Microporous Mesoporous Mater.* 159: 74–80.

54 Horikawa, H., Iida, T., Osuga, R. et al. (2018). Crystallization of Ti-rich* BEA zeolites by the combined strategy of using Ti–Si mixed oxide composites and intentional aluminum addition/post-synthesis dealumination. *Cryst. Growth Des.* 18: 2180–2188.

55 Arca V., Boscoletto, A. B., Fracasso, N., Meda, L., Ranghino, G. (2006) *Process for the activation of zeolitic catalysts containing titanium and their use in oxidation reactions.* US, 10481465 [P].

56 Xu, H., Guan, Y., Lu, X. et al. (2020). Relation of selective oxidation catalytic performance to microenvironment of Ti^{IV} active site based on isotopic labeling. *ACS Catal.* 10: 4813–4819.

57 Grosso-Giordano, N.A., Hoffman, A.S., Boubnov, A. et al. (2019). Dynamic reorganization and confinement of Ti-IV active sites controls olefin epoxidation catalysis on two-dimensional zeotypes. *J. Am. Chem. Soc.* 141: 7090–7106.

58 Tang, Z., Yu, Y., Li, L. et al. (2019). Regulation of framework titanium siting in MWW-type titanosilicates by boron and aluminum incorporation and its effect on the catalytic performance. *Microporous Mesoporous Mater.* 287: 93–100.

59 Liang, X., Peng, X., Xia, C. et al. (2021). Improving Ti incorporation into the BEA framework by employing ethoxylated chlorotitanate as Ti precursor: post-synthesis, characterization, and incorporation mechanism. *Ind. Eng. Chem. Res.* 60: 1219–1230.

60 Lin, D., Zhang, Q., Qin, Z. et al. (2021). Reversing titanium oligomer formation towards high-efficiency and green synthesis of titanium-containing molecular sieves. *Angew. Chem. Int. Ed.* 60: 3443–3448.

61 Chen, Z., Liu, X., Yu, Y. et al. (2021). Gripper-like silicon species for efficient synthesis of crystalline metallosilicates with spatially homogeneous heteroatoms in the framework. *Chem. Mater.* 33: 4988–5001.

62 Nur, H., Prasetyoko, D., Ramli, Z., and Endud, S. (2004). Sulfation: a simple method to enhance the catalytic activity of TS-1 in epoxidation of 1-octene with aqueous hydrogen peroxide. *Catal. Commun.* 5: 725–728.

63 Xue, Y., Xie, Y., Wei, H. et al. (2014). Improving the performance of TS-1 catalysts for continuous cyclohexanone ammoximation through controlment of active species distribution. *New J. Chem.* 38: 4229–4234.

64 Chen, Z., Zhang, L., Yu, Y. et al. (2021). TS-1 zeolite with homogeneous distribution of Ti atoms in the framework: synthesis, crystallization mechanism and its catalytic performance. *J. Catal.* 404: 990–998.

65 Clerici, M.G. and Ingallina, P. (1993). Epoxidation of lower olefins with hydrogen peroxide and titanium silicalite. *J. Catal.* 140: 71–83.

66 Yang, G., Zhuang, J., Ma, D. et al. (2008). A joint experimental–theoretical study on trimethylphosphine adsorption on the Lewis acidic sites present in TS-1 zeolite. *J. Mol. Struct.* 882: 24–29.

67 Zhou, D., Zhang, H., Zhang, J. et al. (2014). Density functional theory investigations into the structure and spectroscopic properties of the Ti4+ species in Ti-MWW zeolite. *Microporous Mesoporous Mater.* 195: 216–226.

68 Lin, Y., Xu, D., Chen, Z. et al. (2022). P-modified deactivated TS-1: a benign catalyst for the MTP reaction. *Catal. Today* 405: 258–266.

69 Zhang, C., Wang, G., Li, Y. et al. (2021). Effect of phosphates modification on TS-1 zeolite and its catalytic performance in epoxidation of propylene and hydrogen peroxide. *J. Mol. Catal.* 35: 1–12.

70 Li, H., Lei, Q., Zhang, X., and Suo, J. (2011). Nitrogen-incorporated TS-1 zeolite obtained by post-synthetic nitridation. *ChemCatChem* 3: 143–145.

71 Li, H., Lei, Q., Zhang, X., and Suo, J. (2012). Nitrogen-incorporated TS-1 zeolite: synthesis, characterization and application in the epoxidation of propylene. *Microporous Mesoporous Mater.* 147: 110–116.

72 Li, H., Xu, B., Deng, B. et al. (2014). Epoxidation of 1-hexene with hydrogen peroxide over nitrogen-incorporated TS-1 zeolite. *Catal. Commun.* 46: 224–227.

73 Zhang, L., Xu, L., Sun, J. et al. (2012). Enhancement of alkene epoxidation activity of titanosilicates by gas-phase ammonia modification. *Chin. J. Chem.* 30: 2205–2211.

74 Blasco, T., Camblor, M., Corma, A. et al. (1998). Direct synthesis and characterization of hydrophobic aluminum-free Ti—beta zeolite. *J. Phys. Chem. B* 102: 75–88.

75 Koller, H., Wölker, A., Villaescusa, L. et al. (1999). Five-coordinate silicon in high-silica zeolites. *J. Am. Chem. Soc.* 121: 3368–3376.

76 Yang, Y., Ding, J., Wang, B. et al. (2014). Influences of fluorine implantation on catalytic performance and porosity of MOR-type titanosilicate. *J. Catal.* 320: 160–169.

77 Lu, X., Zhou, W., Guan, Y. et al. (2017). Enhancing ethylene epoxidation of a MWW-type titanosilicate/H_2O_2 catalytic system by fluorine implanting. *Catal. Sci. Technol.* 7: 2624–2631.

78 Zicovich-Wilson, C. and Corma, A. (2000). Modifying the catalytic activity of Ti-zeolites by isomorphic substitution of Si by Ge atoms. A periodic quantum-chemical study. *J. Phys. Chem. B* 104: 4134–4140.

79 Arca, V., Boscoletto, A.B., Fracasso, N. et al. (2006). Epoxidation of propylene on Zn-treated TS-1 catalyst. *J. Mol. Catal. A Chem.* 243: 264–277.

80 Fang, X., Wu, L., Yu, Y. et al. (2018). Improving the catalytic performance of TS-1 through $Zn(AC)_2$ modification. *Catal. Commun.* 114: 1–5.

81 Wu, M., Song, H., Wang, F., and Chou, L. (2013). The positive role of cadmium in TS-1 catalyst for butadiene epoxidation. *J. Mol. Catal. A Chem.* 379: 207–212.

82 Mei, W., Lingjun, C., and Huanling, S. (2013). Effect of metals on titanium silicalite TS-1 for butadiene epoxidation. *Chin. J. Catal.* 34: 789–797.

83 Wu, M., Song, H., and Chou, L. (2013). Preparation, characterization and catalytic performance study of La-TS-1 catalysts. *RSC Adv.* 3: 23562–23570.

84 Liu Y., Fang, X., Wang, Y., Zhuo, Z., Deng, X. J., Shen, L., Lin, L., Qiu, F., He, M. (2012) *Fluorine-planted MWW-type titanosilicate, its prepraration and applications*. CN, 201210100532 [P].

85 Fang, X., Sun, L., Lin, L. et al. (2017). Enhanced catalytic oxidation performance of F-Ti-MWW through the synergistic effect of anion and cation. *Catal. Commun.* 96: 54–57.

86 Fan, X., Hu, W., Jin, S. et al. (2022). Effect of P modification on the structure and catalytic performance of Ti-MWW zeolite. *Microporous Mesoporous Mater.* 336: 111887.

87 Montejo-Valencia, B.D., Salcedo-Pérez, J.L., and Curet-Arana, M.C. (2016). DFT study of closed and open sites of BEA, FAU, MFI, and BEC zeolites substituted with tin and titanium. *J. Phys. Chem. C* 120: 2176–2186.

88 Gordon, C.P., Engler, H., Tragl, A.S. et al. (2020). Efficient epoxidation over dinuclear sites in titanium silicalite-1. *Nature* 586: 708–713.

89 Guo, Q., Sun, K., Feng, Z. et al. (2012). A thorough investigation of the active titanium species in TS-1 zeolite by in situ UV resonance Raman spectroscopy. *Chem Eur J* 18: 13854–13860.

90 Xu, L., Huang, D., Li, C. et al. (2015). Construction of unique six-coordinated titanium species with an organic amine ligand in titanosilicate and their unprecedented high efficiency for alkene epoxidation. *Chem. Commun.* 51: 9010–9013.

91 Zuo, Y., Liu, M., Zhang, T. et al. (2015). Role of pentahedrally coordinated titanium in titanium silicalite-1 in propene epoxidation. *RSC Adv.* 5: 17897–17904.

92 Morra, E., Giamello, E., and Chiesa, M. (2014). Probing the redox chemistry of titanium silicalite-1: formation of tetrahedral Ti^{3+} centers by reaction with triethylaluminum. *Chem Eur J* 20: 7381–7388.

93 Beck, C., Mallát, T., and Baiker, A. (2003). Epoxidation of allylic alcohols with TiO_2–SiO_2: hydroxy-assisted mechanism and dynamic structural changes during reaction. *Catal. Lett.* 88: 203–209.

94 Wells, D.H., Joshi, A.M., Delgass, W.N., and Thomson, K.T. (2006). A quantum chemical study of comparison of various propylene epoxidation mechanisms using H_2O_2 and TS-1 catalyst. *J. Phys. Chem. B* 110: 14627–14639.

95 Panyaburapa, W., Nanok, T., and Limtrakul, J. (2007). Epoxidation reaction of unsaturated hydrocarbons with H_2O_2 over defect TS-1 investigated by ONIOM method: formation of active sites and reaction mechanisms. *J. Phys. Chem. C* 111: 3433–3441.

96 Yuan, S., Si, H., Fu, A. et al. (2011). Location of Si vacancies and $Ti(OSi)_4$ and $Ti(OSi)_3OH$ sites in the MFI framework: a large cluster and full ab initio study. *J. Phys. Chem. A* 115: 940–947.

97 Khouw, C.B. and Davis, M.E. (1995). Catalytic activity of titanium silicates synthesized in the presence of alkali-metal and alkaline-earth ions. *J. Catal.* 151: 77–86.

98 Shetti, V.N., Srinivas, D., and Ratnasamy, P. (2004). Enhancement of chemoselectivity in epoxidation reactions over TS-1 catalysts by alkali and alkaline metal ions. *J. Mol. Catal. A Chem.* 210: 171–178.

99 Wu, L., Zhao, S., Lin, L. et al. (2016). In-depth understanding of acid catalysis of solvolysis of propene oxide over titanosilicates and titanosilicate/H_2O_2 systems. *J. Catal.* 337: 248–259.

100 Kuwahara, Y., Nishizawa, K., Nakajima, T. et al. (2011). Enhanced catalytic activity on titanosilicate molecular sieves controlled by cation–π interactions. *J. Am. Chem. Soc.* 133: 12462–12465.

101 Yu, Y., Tang, Z., Liu, W. et al. (2019). Enhanced catalytic oxidation performance of K^+-modified Ti-MWW through selective breaking of interfacial hydrogen-bonding interactions of H_2O_2. *Appl. Catal. A Gen.* 587: 117270.

102 Bregante, D.T., Johnson, A.M., Patel, A.Y. et al. (2019). Cooperative effects between hydrophilic pores and solvents: catalytic consequences of hydrogen bonding on alkene epoxidation in zeolites. *J. Am. Chem. Soc.* 141: 7302–7319.

103 Bregante, D.T., Chan, M.C., Tan, J.Z. et al. (2021). The shape of water in zeolites and its impact on epoxidation catalysis. *Nat. Catal.* 4: 797–808.

104 Bregante, D.T., Tan, J.Z., Sutrisno, A., and Flaherty, D.W. (2020). Heteroatom substituted zeolite FAU with ultralow Al contents for liquid-phase oxidation catalysis. *Catal. Sci. Technol.* 10: 635–647.

105 Grosso-Giordano, N.A., Schroeder, C., Okrut, A. et al. (2018). Outer-sphere control of catalysis on surfaces: a comparative study of Ti (IV) single-sites grafted on amorphous versus crystalline silicates for alkene epoxidation. *J. Am. Chem. Soc.* 140: 4956–4960.

106 Yu, Y., Wang, R., Liu, W. et al. (2021). Control of Ti active-site microenvironment in titanosilicate catalysts and its effect on oxidation pathways. *Appl. Catal. A Gen.* 610: 117953.

107 Menschutkin, N. (1890). Über die affinitätskoeffizienten der alkylhaloide und der amine. *Z. Phys. Chem.* 6: 41–57.

108 Potts, D.S., Bregante, D.T., Adams, J.S. et al. (2021). Influence of solvent structure and hydrogen bonding on catalysis at solid–liquid interfaces. *Chem. Soc. Rev.* 50: 12308–12337.

109 Haw, J.F., Nicholas, J.B., Xu, T. et al. (1996). Physical organic chemistry of solid acids: lessons from in situ NMR and theoretical chemistry. *Acc. Chem. Res.* 29: 259–267.

110 Van Santen, R. and Kramer, G. (1995). Reactivity theory of zeolitic broensted acidic sites. *Chem. Rev.* 95: 637–660.

111 Fan, W., Wu, P., and Tatsumi, T. (2008). Unique solvent effect of microporous crystalline titanosilicates in the oxidation of 1-hexene and cyclohexene. *J. Catal.* 256: 62–73.

112 Wilkenhöner, U., Langhendries, G., van Laar, F. et al. (2001). Influence of pore and crystal size of crystalline titanosilicates on phenol hydroxylation in different solvents. *J. Catal.* 203: 201–212.

113 Yu, Y., Fang, N., Chen, Z. et al. (2023). Evidence of solvent-mediated proton transfer during H_2O_2 activation in titanosilicate-catalyzed oxidation systems. *Phys. Chem. Chem. Phys.* 25: 12220–12230.

114 Wu, P. and Tatsumi, T. (2002). Unique trans-selectivity of Ti-MWW in epoxidation of cis/trans-alkenes with hydrogen peroxide. *J. Phys. Chem. B* 106: 748–753.

115 Wu, P. and Tatsumi, T. (2001). Extremely high trans selectivity of Ti-MWW in epoxidation of alkenes with hydrogen peroxide. *Chem. Commun.* 897–898.

116 Wang, L., Liu, Y., Xie, W. et al. (2007). Highly efficient and selective production of epichlorohydrin through epoxidation of allyl chloride with hydrogen peroxide over Ti-MWW catalysts. *J. Catal.* 246: 205–214.

117 Wu, P., Nuntasri, D., Liu, Y. et al. (2006). Selective liquid-phase oxidation of cyclopentene over MWW type titanosilicate. *Catal. Today* 117: 199–205.

118 Wu, P., Liu, Y., He, M., and Tatsumi, T. (2004). A novel titanosilicate with MWW structure: catalytic properties in selective epoxidation of diallyl ether with hydrogen peroxide. *J. Catal.* 228: 183–191.

119 Wu, P. and Tatsumi, T. (2003). A novel titanosilicate with MWW structure: III. Highly efficient and selective production of glycidol through epoxidation of allyl alcohol with H_2O_2. *J. Catal.* 214: 317–326.

120 Fan, W., Kubota, Y., and Tatsumi, T. (2008). Oxidation of 1, 4-dioxane over Ti-MWW in the presence of H_2O_2. *ChemSusChem* 1: 175–178.

121 Wu, H., Liu, Y., Wang, L. et al. (2007). Epoxidation of 2, 5-dihydrofuran to 3, 4-epoxytetrahydrofuran over Ti-MWW catalysts. *Appl. Catal. A Gen.* 320: 173–180.

122 Wu, H., Wang, L., Zhang, H. et al. (2006). Highly efficient and clean synthesis of 3, 4-epoxytetrahydrofuran over a novel titanosilicate catalyst, Ti-MWW. *Green Chem.* 8: 78–81.

123 Xie, W., Zheng, Y., Zhao, S. et al. (2010). Selective oxidation of pyridine to pyridine-N-oxide with hydrogen peroxide over Ti-MWW catalyst. *Catal. Today* 157: 114–118.

124 Roffia, P., Leofanti, G., Cesana, A. et al. (1990). Cyclohexanone ammoximation: a break through in the 6-caprolactam production process. *Stud. Surf. Sci. Catal.* 55: 43–52.

125 Song, F., Liu, Y., Wu, H. et al. (2006). A novel titanosilicate with MWW structure: highly effective liquid-phase ammoximation of cyclohexanone. *J. Catal.* 237: 359–367.

126 Wu, C., Wang, Y., Mi, Z. et al. (2002). Effects of organic solvents on the structure stability of TS-1 for the ammoximation of cyclohexanone. *React. Kinet. Catal. Lett.* 77: 73–81.

127 Thangaraj, A., Puthoor, L., and Sivasanker, S. (1994). Solvent effects in the hydroxylation of phenol with H_2O_2 over TS-1. *Indian J. Chem.* 33A: 255–258.

128 Tuel, A., Moussa-Khouzami, S., Taarit, Y.B., and Naccache, C. (1991). Hydroxylation of phenol over TS-1: surface and solvent effects. *J. Mol. Catal.* 68: 45–52.

129 Clerici, M.G. (1991). Oxidation of saturated hydrocarbons with hydrogen peroxide, catalysed by titanium silicalite. *Appl. Catal.* 68: 249–261.

130 Kon, Y., Yokoi, T., Yoshioka, M. et al. (2013). Selective oxidation of bulky sulfides to sulfoxides over titanosilicates having an MWW structure in the presence of H_2O_2 under organic solvent-free conditions. *Tetrahedron Lett.* 54: 4918–4921.

131 Hulea, V., Fajula, F., and Bousquet, J. (2001). Mild oxidation with H_2O_2 over Ti-containing molecular sieves—a very efficient method for removing aromatic sulfur compounds from fuels. *J. Catal.* 198: 179–186.

132 Xia, C., Ju, L., Zhao, Y. et al. (2015). Heterogeneous oxidation of cyclohexanone catalyzed by TS-1: combined experimental and DFT studies. *Chin. J. Catal.* 36: 845–854.

133 Bhaumik, A., Kumar, P., and Kumar, R. (1996). Baeyer-Villiger rearrangement catalysed by titanium silicate molecular sieve (TS-1)/H_2O_2 system. *Catal. Lett.* 40: 47–50.

134 Bellussi, G., Carati, A., Clerici, M.G. et al. (1992). Reactions of titanium silicalite with protic molecules and hydrogen peroxide. *J. Catal.* 133: 220–230.

135 Clerici, M.G. (2001). The role of the solvent in TS-1 chemistry: active or passive? An early study revisited. *Top. Catal.* 15: 257–263.

136 Datta, R. and Leung, S.W. (1985). Shape generalized isothermal effectiveness factor for first-order kinetics. *Chem. Eng. Commun.* 39: 155–173.

137 Ramachandran, C.E., Zhao, Q., Zikanova, A. et al. (2006). Is 1-hexene epoxidation in TS-1 diffusion limited in different solvents? *Catal. Commun.* 7: 936–940.

138 Yip, A.C.K. and Hu, X.J. (2009). Catalytic activity of clay-based titanium silicalite-1 composite in cyclohexanone ammoximation. *Ind. Eng. Chem. Res.* 48: 8441–8450.

139 Yu, H., Shang, F., Chu, Q. et al. (2020). Cleaner and atomic economy production of hydroxylamine hydrochloride under solvent-free conditions through process intensification. *J. Clean. Prod.* 269: 122187.

140 Wang, M., Song, S., Yuan, X., and Wu, J. (2017). Catalytic performance and deactivation of the shaped Ti-MWW zeolite for ammoximation of cyclohexanone in liquid-liquid-solid three phase system. *Acta Petrol. Sin.* 33: 1183–1191.

141 Guoqing, L. and Jian, W. (2012). Ammoximation of cyclohexanone to cyclohexanone oxime catalyzed by titanium silicalite-1 zeolite in three-phase system. *Chin. J. Chem. Eng.* 20: 889–894.

142 Thangaraj, A., Sivasanker, S., and Ratnasamy, P. (1991). Catalytic properties of crystalline titanium silicalites III. Ammoximation of cyclohexanone. *J. Catal.* 131: 394–400.

143 Wu, P., Komatsu, T., and Yashima, T. (1998). Hydroxylation of aromatics with hydrogen peroxide over titanosilicates with MOR and MFI structures: effect of Ti peroxo species on the diffusion and hydroxylation activity. *J. Phys. Chem. B* 102: 9297–9303.

144 Tyablikov, I.A. and Romanovsky, B.V. (2017). Mass transfer effect on the oxidation of alkenes and phenol with hydrogen peroxide using TS-1 titanosilicate as a catalyst. *Catal. Lett.* 147: 2434–2439.

145 Van der Pol, A., Verduyn, A., and Van Hooff, J. (1992). Why are some titanium silicalite-1 samples active and others not? *Appl. Catal. A Gen.* 92: 113–130.

146 Wilkenhöner, U., Duncan, W., Möller, K., and Van Steen, E. (2004). Intracrystalline diffusivity of hydroxybenzenes in TS-1 and Al-free Ti-beta. *Microporous Mesoporous Mater.* 69: 181–186.

147 Aiello, R., Testa, F., and Giordano, G. (2002). *Impact of Zeolites and Other Porous Materials on the New technologies at the Beginning of the New Millennium: Proceedings of the 2nd International FEZA Conference*, vol. 142. Elsevier 3-2027.

148 Ramachandran, C.E., Du, H., Kim, Y.J. et al. (2008). Solvent effects in the epoxidation reaction of 1-hexene with titanium silicalite-1 catalyst. *J. Catal.* 253: 148–158.

149 Jiao, W., He, Y., Li, J. et al. (2015). Ti-rich TS-1: a highly active catalyst for epoxidation of methallyl chloride to 2-methyl epichlorohydrin. *Appl. Catal. A Gen.* 491: 78–85.

150 Langhendries, G., De Vos, D.E., Baron, G.V., and Jacobs, P.A. (1999). Quantitative sorption experiments on Ti-zeolites and relation with α-olefin oxidation by H_2O_2. *J. Catal.* 187: 453–463.

151 Li, M., Wang, Y., Wu, Y. et al. (2017). Structure and catalytic activity of a newly proposed titanium species in a Ti-YNU-1 zeolite: a density functional theory study. *Catal. Sci. Technol.* 7: 4105–4114.

152 Bordiga, S., Coluccia, S., Lamberti, C. et al. (1994). XAFS study of Ti-silicalite: structure of framework Ti (IV) in the presence and absence of reactive molecules (H_2O, NH_3) and comparison with ultraviolet-visible and IR results. *J. Phys. Chem.* 98: 4125–4132.

153 Bordiga, S., Damin, A., Bonino, F. et al. (2002). Effect of interaction with H_2O and NH_3 on the vibrational, electronic, and energetic peculiarities of Ti (IV) centers TS-1 catalysts: a spectroscopic and computational study. *J. Phys. Chem. B* 106: 9892–9905.

154 Guo, X.W., Li, G., Zhang, X.F., and Wang, X.S. (1997). Synthesis of titanium silicalite-I from TPABr system. *Stud. Surf. Sci. Catal.*, Spillover and Migration of Surface Species on Catalysts 112: 499–508.

155 Callanan, L.H., Burton, R.M., Wilkenhoner, U., and Van Steen, E. (2004). Phenol hydroxylation of over aluminium-free titanium-beta using water as a solvent. *Stud. Surf. Sci. Catal.* 154: 2596–2602.

156 Bhaumik, A., Mukherjee, P., and Kumar, R. (1998). Triphase catalysis over titanium–silicate molecular sieves under solvent-free conditions: I. Direct hydroxylation of benzene. *J. Catal.* 178: 101–107.

157 Fang, N., Liu, H., Liu, D. et al. (2022). Reaction pathways in the oxidation decomposition of hydroxylamine over a titanosilicates/H_2O_2 system. *J. Phys. Chem. C* 158: 19178–19187.

158 Sinclair, P.E. and Catlow, C.R.A. (1999). Quantum chemical study of the mechanism of partial oxidation reactivity in titanosilicate catalysts: active site formation, oxygen transfer, and catalyst deactivation. *J. Phys. Chem. B* 103: 1084–1095.

159 Nie, X., Ren, X., Ji, X. et al. (2019). Mechanistic insight into propylene epoxidation with H_2O_2 over titanium silicalite-1: effects of zeolite confinement and solvent. *J. Phys. Chem. B* 123: 7410–7423.

160 Russo, V., Tesser, R., Santacesaria, E., and Di Serio, M. (2014). Kinetics of propene oxide production via hydrogen peroxide with TS-1. *Ind. Eng. Chem. Res.* 53: 6274–6287.

161 Nie, X., Ji, X., Chen, Y. et al. (2017). Mechanistic investigation of propylene epoxidation with H_2O_2 over TS-1: active site formation, intermediate identification, and oxygen transfer pathway. *Mol. Cat.* 441: 150–167.

162 Bordiga, S., Bonino, F., Damin, A., and Lamberti, C. (2007). Reactivity of Ti (IV) species hosted in TS-1 towards H_2O_2–H_2O solutions investigated by ab initio cluster and periodic approaches combined with experimental XANES and EXAFS data: a review and new highlights. *Phys. Chem. Chem. Phys.* 9: 4854–4878.

163 Haw, J.F., Xu, T., Nicholas, J.B., and Goguen, P.W. (1997). Solvent-assisted proton transfer in catalysis by zeolite solid acids. *Nature* 389: 832–835.

164 Simon, U., Flesch, U., Maunz, W. et al. (1998). The effect of NH_3 on the ionic conductivity of dehydrated zeolites Na beta and H beta. *Microporous Mesoporous Mater.* 21: 111–116.

165 Adams, J.S., Chemburkar, A., Priyadarshini, P. et al. (2021). Solvent molecules form surface redox mediators in situ and cocatalyze O_2 reduction on Pd. *Science* 371: 626–632.

166 Higazy, A.A., Kassem, M.E., and Sayed, M.B. (1992). Water-assisted and thermally-enhanced protonic conduction in HZSM-5, effect of gamma-irradiation on the electric properties. *J. Phys. Chem. Solids* 53: 549–554.

167 Jobic, H., Tuel, A., Krossner, M., and Sauer, J. (1996). Water in interaction with acid sites in H-ZSM-5 zeolite does not form hydroxonium ions. A comparison between neutron scattering results and ab initio calculations. *J. Phys. Chem.* 100: 19545–19550.

168 Sever, R.R. and Root, T.W. (2003). DFT study of solvent coordination effects on titanium-based epoxidation catalysts. Part one: formation of the titanium hydroperoxo intermediate. *J. Phys. Chem. B* 107: 4080–4089.

169 Stare, J., Henson, N.J., and Eckert, J. (2009). Mechanistic aspects of propene epoxidation by hydrogen peroxide. Catalytic role of water molecules, external electric field, and zeolite framework of TS-1. *J. Chem. Inf. Model.* 49: 833–846.

170 Yu, Y., Fang, N., Chen, Z. et al. (2022). Greening oxidation catalysis: water as a solvent for efficient alkene epoxidation over a titanosilicate/H_2O_2 system. *ACS Sustain. Chem. Eng.* 10: 11641–11654.

171 Vayssilov, G.N. and Van Santen, R.A. (1998). Catalytic activity of titanium silicalites—a DFT study. *J. Catal.* 175: 170–174.

172 Deng, X., Zhang, S., Wang, B. et al. (2013). Enhanced catalytic activity of titanosilicates controlled by hydrogen-bonding interactions. *Chem. Commun.* 49: 7504–7506.

173 Liu, Y., Zhang, X., and Suo, J. (2013). Gold supported on nitrogen-incorporated TS-1 for gas-phase epoxidation of propylene. *Chin. J. Catal.* 34: 336–340.

174 Bregante, D.T. and Flaherty, D.W. (2019). Impact of specific interactions among reactive surface intermediates and confined water on epoxidation catalysis and adsorption in Lewis acid zeolites. *ACS Catal.* 9: 10951–10962.

175 Sever, R.R. and Root, T.W. (2003). DFT study of solvent coordination effects on titanium-based epoxidation catalysts. Part two: reactivity of titanium hydroperoxo complexes in ethylene epoxidation. *J. Phys. Chem. B* 107: 4090–4099.

176 Zou, J., Fan, Z., Jiang, L., and Zhou, D. (2016). Theoretical study of the solvent adsorption on Ti-peroxo active center in Ti-MWW zeolite. *Acta Phys. -Chim. Sin.* 32: 935–942.

177 Meng, J., Gang, L.I., and Wang, X. (2003). Effect of solvent on the propylene epoxidation catalyzed by TS-1. *J. Mol. Catal. A Chem.* 106–110.

8

Spectroscopic Characterization of Heteroatom-Containing Zeolites

Guodong Qi, Jun Xu, and Feng Deng

Division of magnetic resonance for materials and chemistry, National Centre for Magnetic Resonance in Wuhan, State Key Laboratory of Magnetic Resonance and Atomic and Molecular Physics, Innovation Academy for Precision Measurement Science and Technology, Chinese Academy of Sciences, West No.30 Xiao Hong Shan, Wuhan 430071, China

8.1 X-Ray Technique

X-ray, short for X-radiation, is a penetrable high-energy electromagnetic radiation with a wavelength range of 10 pm to 10 nm. The application of X-ray to materials relies on four main interactions, namely scattering, photoelectric effect, adsorption, and emission. X-ray techniques are used in many scientific fields including chemistry, physics, engineering, geography, and biology due to their sensitivity to local structures and chemical environments [1–3]. In this section, we will introduce the most frequently used X-ray techniques for the characterization of heteroatom-containing zeolites. For example, X-ray diffraction (XRD) has been widely exploited for studying the structure of ordered materials, X-ray absorption spectroscopy (XAS) is commonly used to obtain atomic coordination information, and X-ray photoelectron spectroscopy (XPS) is powerful for the analysis of atomic valence state.

8.1.1 XRD Determination of Framework Structure and Heteroatoms in Zeolites

XRD technique takes advantage of X-ray scattering on materials having regular atom arrangements, such as zeolites, metal oxides, and organic crystals. Because the wavelength of X-ray is in the same range of interatomic distances in these materials, when the X-ray beam irradiates the materials, it can add constructively in a few directions according to Bragg's law,

$$2d\ \sin\theta = n\lambda \tag{8.1}$$

where d is the distance (Å) between parallel diffracting planes determined by the atomic arrangement in material, θ is the incident angle in degrees, n is diffraction order (any integer), and λ is the wavelength (Å) of the X-ray beam [4]. This means

Micro-Mesoporous Metallosilicates: Synthesis, Characterization, and Catalytic Applications, First Edition. Edited by Peng Wu and Hao Xu.

that the structural information of crystalline solids can be obtained from the XRD pattern. In the case of a single crystal, a Laue diagram showing a dot diffraction pattern is commonly observed, by which the atom positions can be determined. The single-crystal XRD approach has been successfully used to reveal the structures of large zeolite crystals. Aluminophosphate DAF-1 [5], boron-containing RUB-13 [6], and germanium-silicate ITQ-33 zeolite [7] are the recent examples of single-crystal XRD structure solutions of heteroatom-containing zeolites. Most zeolites are in the form of powders and difficult to form larger crystals. The structure of small zeolite crystals can be in principle determined by single-crystal XRD. The difficulty lies in collecting complete data and analyzing the multiple diffractions. The high-intensity synchrotron X-ray source is helpful, which however brings damage to zeolite structures [8, 9]. Thus, the structural determination of zeolites usually depends on the X-ray powder diffraction approach. For a powder sample with random-distributed small crystals, the dot pattern is transformed into diffraction rings related to 2θ angles. Generally, the powder XRD pattern is recorded and reported as a function of X-ray intensity against the 2θ angle [10]. Comparing with the XRD database is one of the basic ways to characterize the incorporation of heteroatoms and verify the framework structure for the known zeolitic structures [11]. When combined with lattice refinement techniques, it can be used to obtain structural information about unknown heteroatom-containing zeolites [12]. Many algorithms in powder XRD refinement have been developed in recent decades, such as the ZEFSAAII method [13–15], FOCUS [16] Fourier recycling method, simulated annealing method [13], and XLENS [17] direct method.

In the ZEFSAAII process, extra-framework organic molecules can be included. This is particularly useful in the case that the XRD intensities are significantly influenced by the structure-directing agents or other adsorbed molecules in zeolites. Changing node type parameters is allowed in the FOCUS process, which is attractive in the characterization of aluminophosphate or gallophosphate zeolites, where the framework atoms are neither octahedrally coordinated nor fourfold connected. XLENS program shows advantages for complex crystallographically ordered zeolite structures. For example, it has been employed to determine the structure of germanium-containing ITQ zeolites with repetitive clusters/cages in the unit cell (Figure 8.1) [18, 19]. Using the same approach, the structure of ITQ-22 was also determined. It is constructed by $[4^4 5^8 6^{12}]$ building blocks with 16T atoms, resulting in a rather large unit cell (a = 42.1 Å, b = 12.99 Å, c = 12.68 Å) in the space group symmetry Pbam [19]. The introduction of a set of phases into the XLENS program can reduce the peak overlap at low 2θ when simulating a complex zeolitic structure [20]. Only tetrahedra structural units (TO_4) are considered in the ZEFSAAII and FOCUS programs, which is not enough to simulate complex zeolite structures like ITQs. The DIFFaX program [21, 22] has also been used for simulating powder XRD patterns of zeolites. Zeolites often feature structural disorders, such as one-dimensional disorders arising from stacking faults of two-dimensional (2D) layers. XRD patterns of zeolites with the one-dimensional disorder are characterized by the co-existence of sharp and broadened peaks. The DIFFaX algorithm can be used to simulate the powder pattern of a proposed model for a faulted structure. By

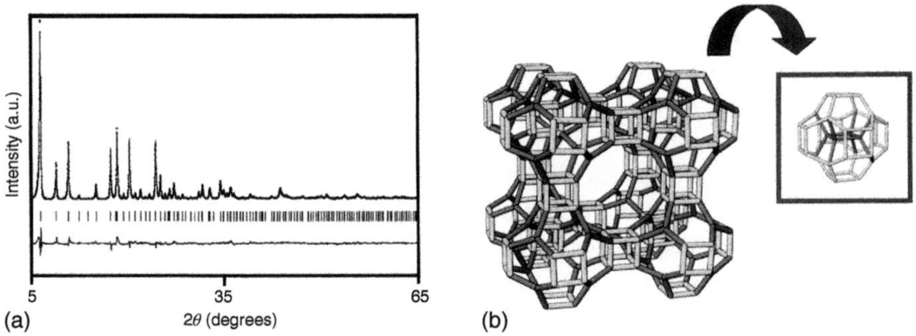

Figure 8.1 (a) Rietveld refinement of ITQ-21. The experimental (crosses) and calculated (lines) XRD patterns are shown, as well as the difference profile (bottom). The short tick marks below the pattern give the positions of the Bragg reflections. (b) Perspective view of the unit cell of ITQ-21. It shows the outer T atoms of the clusters – that is, cage $[4^6 6^{12}]$ – interconnected via double four-membered rings (D4MR). Each tetrahedral intersection represents one T site (either occupied by Si or Ge) bridged by O atoms. Inset, the detailed structure of the complete cluster shows the four-membered ring inside. This ring is assumed to be randomly oriented normally to the three cell axes. Source: Reproduced from Ref. [18]/with permission of Springer Nature.

comparing the simulated XRD patterns of different faulting probabilities with the experimental data, the faulting in a disordered zeolite can be evaluated.

In general, powder XRD is usually coupled with solid-state nuclear magnetic resonance (NMR) [23], high-resolution transmission electron microscopy (HRTEM) [12], and electron diffraction [24] to determine the structure of zeolites because it is difficult to determine the unit cell of a zeolite solely from the powder XRD pattern. Solid-state NMR is powerful in collecting information on atomic distance and connection, while HRTEM and electron diffraction can be employed to evaluate the structure of the unit cell in zeolites that define the intergrowth. These details benefit a better simulation of the powder XRD patterns for zeolite structure solution.

8.1.2 XAS Characterization of Metals in Zeolite

Electrons' emission with specific energy might be observed when materials adsorb light or X-ray irradiation. This phenomenon is called the photoelectric effect, which was theoretically explained by Einstein in 1905. The generation of the photoelectric effect depends on the particular electronic core level to be excited and the energy of the incident X-ray. If the energy of the X-ray is stronger than the binding energy of the electron, the X-ray will be absorbed and the electron will be excited out of its quantum level. If the energy of the X-ray is lower than the binding energy, the electron will not be perturbed and the X-ray will not be absorbed [25]. It is worth noting that every atom has core-level electrons (K, L, M) with known binding energies. To cover as many elements as possible, a strong X-ray beam must be generated on a synchrotron radiation (SR) source and should be tunable over a wide energy range in the X-ray absorption measurement. For a selected element, a sharp rise in absorption will be observed when the incident X-ray energy has been turned near or equal to the binding energy of a core-level electron. This energy is defined as "the absorption edge" corresponding to its core-level electrons [26]. Besides the adsorption edge, there are other two key features in XAS spectra. The energy within a range of 50–100 eV above the adsorption edge is generally referred to as the X-ray Absorption Near Edge Structure (XEANES), which depends on the local geometric structure, and the spectral shape reflects the excited electronic state. The characteristic higher energy (up to 500 eV or greater from the adsorption edge) is referred to as the Extended X-ray Absorption Fine Structure (EXAFS), which is influenced by the neighboring atoms surrounding the tested atoms. By analyzing these features and comparing with the known structures, the elemental composition, oxidation state, and electronic structure of the materials can be ascertained. The element-specific feature allows the sample with complex elemental composition to be easily measured by XAS, particularly for zeolites with the incorporation of hetero-atoms [27, 28].

In the recent 20 years, XAS has been applied to characterize a variety of heteroatoms in zeolites, such as Al [29], Ga [30], Ti [31], Ge [32], Sn [33], V [34], and Zn [35]. Taking the most common aluminum element as an example, the aluminum K-edge indicates the energy for electron transition from 1s to 3p orbit on a specific Al atom, which is sensitive to its coordination and bonding. Three-, four-, five-, and six-coordinated Al species in zeolite can be differentiated by analyzing the Al K-edge in the XEANES spectrum [29]. Typically, the K-edge energies for four-, five-, and six-coordinated Al species follow a low to high order, while the three-coordinated Al species is featured by strong pre-edge adsorption [36]. Introducing exchanged ions into the zeolite pores generally influences the local environment of Brønsted Al species, which can also be reflected by the Al K-edge energy. Fröba et al. performed XAS measurement on zeolite sodalite containing different exchanged ions (Cl^-, NO_2^-, $B(OH)_4^-$, or CrO_4^{2-}) [37]. They found that the size of the exchanged ions affects the XAS spectrum to a large extent in the energy range of 5–20 eV above the adsorption edge. This phenomenon can be explained by the influence of anions on the Al—O—Si bridging angle in the zeolite framework, while the Al—O and Si—O bond lengths are not significantly changed. Koningsberger and Miller first examined the local Al structure of the Brønsted acid sites in Y zeolite using EXAFS at the Al K-edge [38]. The local structures of the Brønsted Al atoms are different when the charge-balancing cations are H^+, Na^+, or NH_4^+. H—Y zeolite has the largest average Al—O bond length of 1.700 Å. The Al—O bond lengths in Na—Y and NH_4—Y are 1.636 and 1.620 Å, respectively. It was reported that the four Al—O bond lengths were unequal to each other when the proton acts as the charge-balancing cation for the Brønsted Al atoms in Y and MFI framework [39, 40]. The Al—O bond containing oxygen atom connected by proton has a longer bond length than the other three Al—O bonds. In H—Y zeolite, the long Al—O bond length is 1.87 Å which is slightly shorter than the length of 1.98 Å in H-MFI zeolite. The design of in situ cells facilitates XAS measurement of Al species in zeolites under working conditions [41–44]. As revealed by in situ Al K-edge spectroscopy, the local structure of H—Y zeolite was changed during oligomerization of olefins due to the formation of intermediates on the Brønsted acid sites, shortening the

Al—O(H) bond from 1.89 to 1.81 Å [39]. In situ Al K-edge XANES and EXAFS have also been employed to detect the evolution of aluminum coordination in steaming process of zeolites, showing that the exposure of an acidic zeolite to water at high temperature leads to the formation of six-coordinated aluminum, the amount of which depends on the Si/Al ratio and structure of the steamed zeolite [42].

To illustrate the structural retrieval process of hetero-atoms in zeolites, the XANES spectra of a series of Fe reference samples and Fe-silicalite-1 are shown in Figure 8.2 [45]. The Fe-silicalite-1 with template has a K-edge position at 7123.6 eV with a pre-edge peak at 7114.2 eV (Figure 8.2b). Furthermore, the sample with template shows a white line intensity similar to that for $FePO_4$ and much lower than that observed for the other Fe^{3+} atoms in the six-coordinated environment (α-Fe_2O_3 and $Fe(acac)_3$). These results indicate that Fe atoms in the sample with template are likely to be Fe^{3+} in four-coordination. Thermal treatments shifted the K-edge position from 7123.6 eV to 7122.4 and 7120.6 eV for the samples activated at 773 and 973 K, respectively. Additionally, the reduction of the pre-edge peak intensity with a gradual appearance of the shoulder peak at low energy (around 7111.8 eV) was observed, which is similar to the $FeCp_2$ reference sample. It also can be found that the treatment at increasing temperatures caused a progressive reduction in white

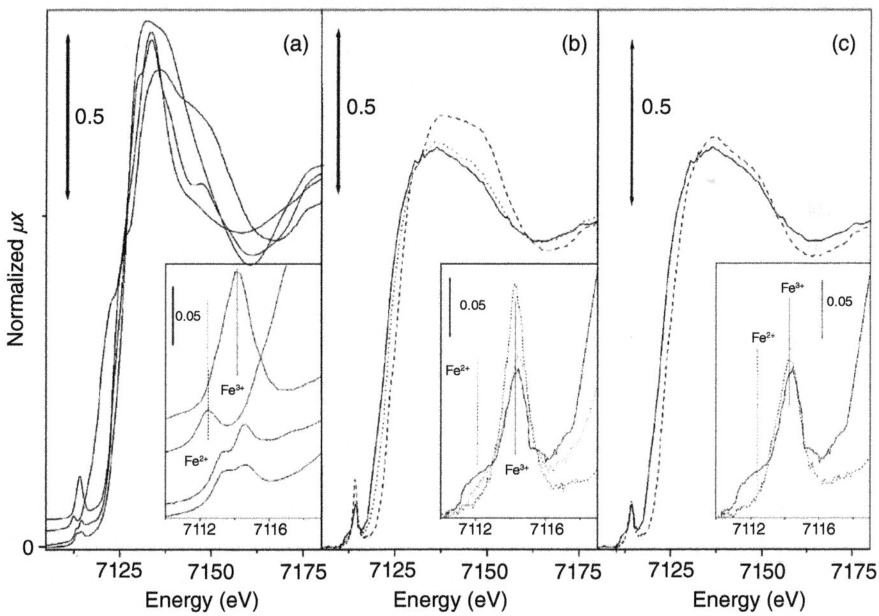

Figure 8.2 (a) XANES spectra of iron reference samples. From top to bottom: $FePO_4$, $FeCp_2$, Fe_2O_3, and $Fe(acac)_3$. (b) XANES spectra of Fe-silicalite-1 with template (dashed line) and activated at 773 (dotted line) and 973 K (solid line). (c) XANES spectra of Fe-silicalite-1 activated at 973 K (solid line, same curve as in part b, here reported again for direct comparison) and oxidized in N_2O at 523 K (dashed line). The insets of each part show the magnification of the pre-edge peak. Source: Reproduced from Ref. [45]/with permission of Elsevier.

line intensity for the Fe-silicalite-1 sample. These observations suggest that Fe^{2+} properly exists on the high-temperature treated Fe-silicalite-1. The XANES spectra of the Fe-silicalite-1 sample activated at 973 K and oxidized in N_2O at 523 K show the shift of the K-edge adsorption from 7120.6 back to 7123.1 eV with concomitant disappearance of the shoulder pre-edge peak for Fe^{2+} (Figure 8.2c). This confirms the transformation of Fe species between Fe^{3+} and Fe^{2+} during the high-temperature activation process. EXAFS can be employed to provide detailed structural information about the Fe atoms in Fe-silicalite-1 [46].

8.1.3 XPS Analysis of the Chemical State of Metal Species

XPS is an important tool for the characterization of surface properties as well as the nature, dispersion, and coordination state of metal species in zeolites and related materials [47–50]. XPS technique uses X-rays as the exciting source to measure the photoelectric effect of an atom. The binding energy of a specific electron orbit is recorded in XPS measurement, which depends on the type of element observed. For example, O(1s), $Fe(2P_{3/2})$, and $Au(4f_{7/2})$ are selected for the characteristic orbit of each element. Thus, the Mg $K\alpha$ (1253.6 eV) and Al $K\alpha$ (1486.3 eV) X-ray sources are sufficient for the XPS study of most elements. This is different from XAS in which a wide range and strong SR X-ray source are required. XPS is a surface analysis technique because the X-ray generated by Mg or Al source can pass through only several nano-meter depths of the material. The binding energy for the electron at the excited orbit is affected by the outer-sphere electrons of the atom in the examination. The existence of valency electrons on the reduction state atoms causes a lower binding energy of the electrons on their characteristic orbits than that of the oxidation state atoms. A similar effect holds for the atoms bound with an electron-donating group. However, larger binding energy can be observed for the atoms in an oxidation state or bound with an electron-withdrawing group. These features allow XPS to be widely employed in the determination of atomic composition and their chemical properties [51].

XPS analysis showed that the binding energies of Al(2p) and Si(2p) photoelectrons in zeolites decline with the increase of the Si/Al ratio, and the surface Si/Al ratio is generally equal to or slightly greater than that in the bulk for most synthetic and mineral zeolites [52]. With the help of a strong SR X-ray source and electron inelastic scattering background analysis [53], the surface Si/Al ratio analysis of zeolite can be achieved with an interval of less than 1 nm. The surface composition of Cu [54], Co [55], Ga [56], and Fe-containing [55] zeolites has been investigated by using XPS. Different metal concentration was found between the surface layer and the bulk of the zeolites.

The isomorphically substituted heteroatoms (Ti, Ge, Sn, Zr, Fe, Ga, and In) in zeolite framework T sites can be directly detected by XPS [57–59]. As shown in Figure 8.3, there are two different binding energies at 460.3 and 458.4 eV for the Ti 2p on the three Ti-silicalite-1 (TS-1) samples. The surface Si/Ti ratio for r23 sample was determined to be 10, much lower than the targeted atomic ratio of 23 used in sample synthesis. The Si/Ti surface atomic ratio for the r50 sample is estimated to

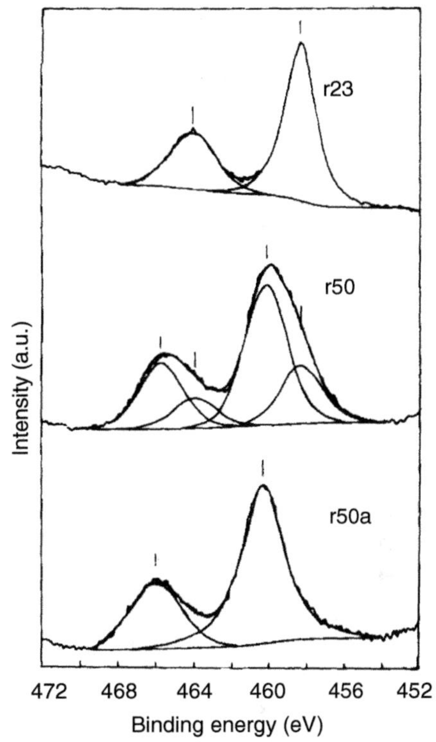

Figure 8.3 Ti 2p region for Ti-silicalites with two different Si/Ti ratios of 23 (r23) and 50 (r50). The r50a is obtained by acid washing the r50 sample. Source: Reproduced from Ref. [57]/with permission of John Wiley & Sons.

be 90, higher than the value used for sample preparation. The Ti species at 460.3 eV is attributed to framework Ti, which accounts for 70% of the total Ti loading on the r50 surface, while the dominant peak at low Ti $2p_{3/2}$ binding energy (458.4 eV) on the r23 sample may arise from TiO_2 species on the external surface. The spectrum of r50a after acid washing shows that the non-framework Ti species was removed, whereas the framework Ti species was kept unchanged.

The nature and electronic state of metal atoms are directly related to the catalytic performance of metal-containing zeolites. For Cu-modified zeolites, XPS measurement reveals that four types of Cu species can be formed including Cu^{2+}, Cu^{+}, small metal clusters, and bulk metallic Cu species, depending on the sample preparation and treatment. Cu^{2+} is predominantly relevant to the selective catalytic reduction of nitrogen monoxide (NO) in engine exhaust gas purification [60–62]. Rare-earth La-exchanged Y zeolite (La—Y) shows superb stability in fluid catalytic cracking (FCC). A significant difference in La 3d line shape and binding energy is found on the XPS of La—Y and La_2O_3 reference sample, revealing the hybridization of La valence states in zeolites [63]. The metallic noble metals on zeolite, such as Pt, Pd, Rh, and Au, are often considered to be the active sites in hydrogenation and selective oxidation reactions. Probing the metal valence state during the heating treatment is important for understanding the catalytic behavior. Rh 3d line in XPS analysis shows that Rh^{3+} in Y zeolite can be easily reduced to rhodium metal under remarkably mild conditions (around 573 K), while the reduction of Rh^{3+} ions is not found in A zeolite under the same conditions [64]. The lack of catalytic activity of Rh-modified A zeolite for many hydrogenation and selective oxidation reactions is attributed to the absence of metallic Rh species.

8.2 Ultraviolet–Visible-Near Infrared (UV–VIS–NIR) Spectroscopy

Ultraviolet–visible–near infrared spectroscopy (UV–VIS–NIR) is a technique used to probe the optical and electronic properties (reflectance, absorbance, and transmittance) of various liquids and solids. For UV–VIS spectroscopy, the light source covers the wavelengths from 200 to 800 nm with energies from 1.5 to 6.2 eV, which leads to electron transitions in molecules from a low level to an excited level [65]. Measurements in the NIR spectroscopy use wavelengths from 800 to 2500 nm, in which the absorbance is sensitive to molecular overtone and combination vibrations [66]. The transition metal, rare-earth metal ions, and adsorbed species are the subjects of UV–VIS study on heteroatom-containing zeolites [67], which provide rich information including the coordination states of metals in zeolite frameworks, the specific active metal species and their evolution upon activation or in reaction, and the location of metals in zeolites channels.

8.2.1 UV–VIS–NIR Characterization of Framework and Non-Framework Metal Species

Tetravalent metals (Ti, Sn, Zr, and Ta) can generate Lewis acid sites in zeolites, resulting in distinct activity in many reactions, such as selective oxidization [68–70], reduction [71, 72], aldol condensation [73, 74], and biomass conversion [75]. Both framework and non-framework metal species can be formed during zeolite preparation and post-treatment. UV–VIS spectroscopy is the most commonly used method to determine the coordination state of tetravalent metal atoms in zeolites [76–79]. Figure 8.4a shows the UV–VIS spectra of TS-1 zeolites obtained by different preparation and post-treatment processes [80]. The bands at 200–220 nm and 220–280 nm are attributed to the electron transfer between O^{2-} and Ti^{4+} in the framework "TiO_4" and "TiO_6" species, respectively [79]. However, the bands at 280–375 nm belong to the extra-framework TiO_2 (anatase) [76, 77]. For the conventionally hydrothermal-synthesized samples (TS-1-C_{20}, TS-1-C_{60}, and TS-1-C_{120}), the framework Ti species are almost unchanged when the amount of incorporated Ti is increased; however, the amount of extra-framework TiO_2 species is significantly increased. Adding NH_4F mineralizer during the preparation results in the disappearance of extra-framework TiO_2 species and increase of framework Ti species on TS-1-C_F sample.

The TS-1-C_{INJ} sample was prepared by a slow injection rate of titanium source, where a slow hydrolysis rate of Ti source benefits the incorporation of framework Ti species [82, 83]. Coating a silicalite-1 layer onto the TS-1-C_{INJ} surface

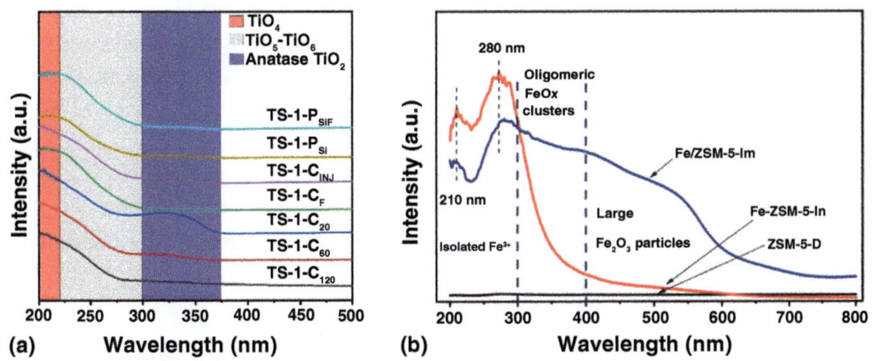

Figure 8.4 UV–VIS spectra of (a) TS-1 samples: TS-1-C_{120}, TS-1-C_{60}, and TS-1-C_{20} are synthesized by the conventionally hydrothermal-method with the Si/Ti ratio of 120, 60, and 20. TS-1-C_F and TS-1-C_{INJ} with a Si/Ti ratio of 60 are obtained by the addition of NH_4F mineralizer and injection of Ti source at a slow rate during hydrothermal synthesis. TS-1-C_{Si} and TS-1-C_{SiF} are post-treated TS-1-C_{INJ} by SiO_2 coating and $(NH_4)_2SiF_6$ washing. Source: Reproduced from Ref. [80]/with permission of Elsevier. (b) Fe-ZSM-5 samples prepared by dry gel conversion (DGC) (Fe-ZSM-5-In) and impregnation method (Fe-ZSM-5-Im) and H-type ZSM-5 zeolite (ZSM-5-D). Source: Reproduced from Ref. [81]/with permission of Springer Nature.

caused a reduction in the UV–VIS band intensities of both framework and extra-framework Ti species in the TS-1-P_{Si} sample. The $(NH_4)_2SiF_6$ treatment removed the extra-framework TiO_2 species on TS-1-P_{SiF}. The location and distribution of Ti species have a significant influence on the activity of TS-1. For example, UV–VIS in combination with Raman spectroscopy indicated that the amount of active intermediate Ti–OOH for propylene epoxidation correlates well with the concentration of framework Ti species [84]. It was reported that the presence of extra-framework Ti species is detrimental to propylene epoxidation. The catalytic performance of boron-containing Ti-MWW (B-Ti-MWW) zeolite can be greatly improved by acid treatment before calcination [85]. The role of acid treatment and the evolution of titanium species during post-treatment have attracted much attention [86]. As indicated by UV–VIS analysis, the acid treatment of as-synthesized B-Ti-MWW zeolite not only removed about half of the dominant non-framework TiO_6 species but also made the residues converted into framework TiO_4 species. However, the same acid treatment performed on the calcined B-Ti-MWW sample resulted in a massive transformation of non-framework TiO_6 species into anatase TiO_2. Importantly, only the sample obtained from acid treatment of as-synthesized B-Ti-MWW and subsequent calcination exhibited catalytic activity in 1-hexene epoxidation, demonstrating that the remarkable enhancement of catalytic performance is ascribed to the transformation of inactive non-framework TiO_6 species into active framework TiO_4 species.

The development of a sustainable route for the introduction of trivalent metals such as Fe and Ga into ZSM-5 zeolite is important due to their outstanding activity and propylene selectivity in the FCC process [87, 88]. An efficient and eco-friendly dry gel conversion method was reported for the synthesis of Fe-ZSM-5 zeolite in one pot without the addition of any template and seed crystal [81]. As revealed by UV–VIS spectra (Figure 8.4b), the Fe species in the one-pot synthesized Fe-ZSM-5 zeolite (Fe-ZSM-5-In) are mainly in the form of framework Fe^{3+} (280 nm), while the Fe species in the Fe/ZSM-5 zeolite prepared by impregnation (Fe/ZSM-5-Im) mostly exist on the external surface of zeolite in the form of non-framework Fe such as oligomeric FeO_x clusters and Fe_2O_3 particles (c. 300 and 400 nm, respectively). Due to the selective incorporation of Fe into the framework sites, Fe-ZSM-5-In has more acid sites and better Fe distribution than Fe/ZSM-5-Im, resulting in enhanced propylene yield and lower coke production in the FCC process.

8.2.2 UV–VIS–NIR Characterization of Metal Species on Ion Exchange Sites of Zeolites

Highly dispersed metal ions, metal oxides, and metal hydroxyl species can be found at the ion-exchanged sites or on the external surface of zeolite, depending on the conditions used for the introduction of metals including the pH value, preparation, and activation procedure. For example, in the study of the coordination of Cu^{2+} to lattice oxygen of CuCsY zeolite, the Cu loading was carefully controlled to avoid the formation of CuO_x clusters [89]. The CuCsY zeolites exhibit three d-d bands of Cu—O coordination peak positions at around 10 800, 12 500, and 14 500 cm^{-1} in the UV–VIS–NIR spectrum. A linear relationship is present for the band intensities at 10 800 and 14 500 cm^{-1} against the amount of Cu^{2+}, which provides evidence for the existence of two coordination environments of Cu^{2+} (or two kinds of Cu^{2+} species) in Y zeolite. However, these two Cu^{2+} environments are not observed in high silica zeolites such as mordenite and ZSM-5, indicating the influence of zeolite structure on Cu^{2+} coordination to the lattice oxygen in the zeolite framework [90]. Moreover, no such linear relationships are found between d–d band intensities and the amount of metal loading in Ni^{2+} or Co^{2+} modified zeolites [90, 91]. This indicates that the coordination sites in the zeolite framework are quite different for different metals. Transition metals can be present in different oxidation states in the zeolite framework, particularly during sample activation and catalytic reaction. UV–VIS–NIR alone is usually insufficient to differentiate oxidation states of metal species in zeolites because of the low peak-resolution and serious peak-overlapping. Thus complementary techniques such as electron spin spectroscopy, XPS, and XAS are often applied jointly with UV–VIS–NIR spectroscopy. For instance, a broad unresolved band of d–d transitions is observed in the UV–VIS–NIR spectra of calcined Cu-ZSM5 zeolite [92]. However, after O_2 activation, three well-resolved d–d bands of Cu^{2+} at 14 400, 22 700, and 40 000–45 000 cm^{-1} are observed in UV–VIS–NIR spectra of Cu-ZSM5 (Figure 8.5) [93]. Interestingly, the band at 22 700 cm^{-1} is not found on any Cu zeolite without O_2 activation, and its band intensity increases with the addition of Cu. This band also can be observed on Cu-ZSM-5 treated with N_2O at a lower temperature. It disappears upon a high-temperature treatment in a helium environment and reappears upon subsequent O_2 treatment. These results imply that the extra-lattice oxygen atoms coming from O_2 or N_2O are involved in the formation of Cu species occurring at 22 700 cm^{-1}. With the help of XAS, Raman,

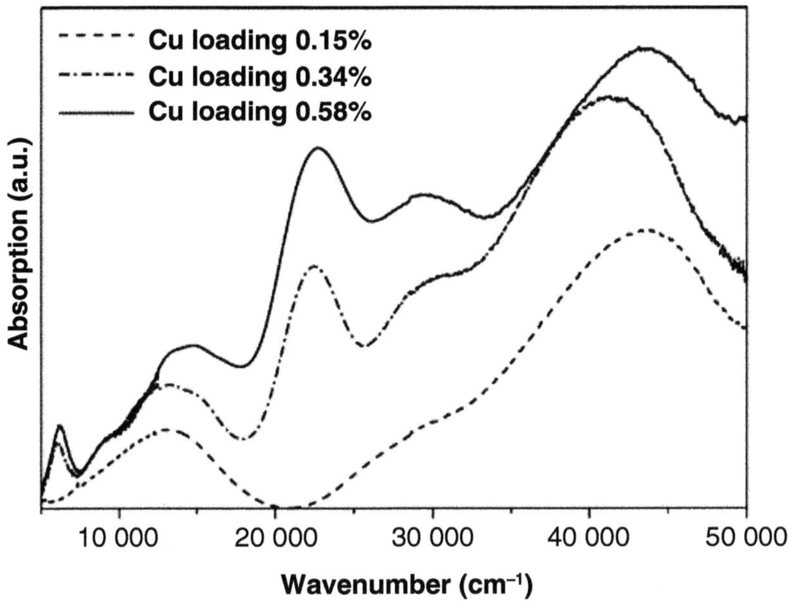

Figure 8.5 UV–VIS–NIR spectra of O_2-treated Cu-ZSM-5 samples with different Cu loadings. Source: Reproduced from Ref. [93]/with permission of American Chemical Society.

and theoretical analysis, this specific band is attributed to a dinuclear mono-oxo Cu species [Cu—O—Cu]. By using in situ UV–VIS measurements on Cu-ZSM-5 during the reaction, it was found that the dinuclear mono-oxo Cu species are the catalytically active sites involved in the catalytic decomposition of NO_x and NO as well as in the stoichiometric oxidation of methane into methanol [93, 94].

UV–VIS-NIR spectroscopy is a useful method for determining the location of metal ions in zeolites [69, 95–98]. For example, the siting of Co^{2+} in dehydrated Beta zeolite was identified by the spectroscopic features of Co^{2+}. Three different exchanged sites with Co^{2+} ions are revealed in Beta zeolite, denoted as α-, β-, and γ-sites. The α-site is formed in polymorph C of Beta structure by an elongated six-membered ring composed of twofold connected five-membered rings. The α-Co^{2+} site exhibits a characteristic band at 14 600 cm^{-1} in the UV–VIS–NIR spectrum. The β-site is formed by a planar deformed six-membered ring of the hexagonal cage both in polymorphs A and B. The Co^{2+} ions siting on the β-sites are characterized by four UV–VIS-NIR bands at 15 500, 16 300, 17 570, and 21 700 cm^{-1}. The γ-site occurring in polymorphs A and B of Beta zeolite is located inside the hexagonal cage. The Co^{2+} ions exchanged at the γ-site can be reflected by the doublet UV–VIS bands at 18 500 and 20 100 cm^{-1}.

8.3 Raman Spectroscopy

Raman spectroscopy has evolved into one of the most powerful techniques for catalyst characterization. Particularly, the switch of the light source from visible to ultraviolet allows UV Raman to significantly enhance the Raman scattering intensity and solve the problem of fluorescence interference that commonly occurs in visible excitation Raman. UV Raman spectroscopy becomes an efficient spectroscopic method for the investigation of the synthesis mechanism and assembly of metal-zeolites by identification of primary structural-units and key assembly-intermediates as well as determination of the incorporation of transition metal atoms and the role of the organic template in framework formation [35, 91, 99–102].

8.3.1 Raman Study of Synthesis Mechanism and Assembly of Metal-Zeolites

Understanding the isomorphous substitution of the zeolite framework atoms by transition metals is one of the most challenging issues in zeolite synthesis [103]. The crystallization of Fe-ZSM-5 (Si/Fe = 152) was monitored by Raman spectroscopy with a 244 nm light source. A band at 1165 cm^{-1} was observed after six hours of crystallization and became narrower and stronger with increasing crystallization time. Periodical DFT calculations showed that this band corresponds to asymmetric stretching of framework Fe—O—Si units, confirming the incorporation of Fe atoms into the zeolite framework. Based on this observation, a crystallization mechanism of Fe-ZSM-5 was proposed [100]. The framework Ti—O—Si species have a strong UV adsorption centered at 220 nm due to the pπ-dπ charge-transfer transition between the titanium and oxygen atoms. The assembly of titanium atoms into titanium-substituted silicalite (TS-1) can be also identified by UV Raman spectroscopy. Li and coworkers observed three featured Raman bands at 1125, 530, and 490 cm^{-1} when a 244 nm wavelength laser source was used. These Raman bonds disappeared when the wavelength of the excitation source was changed to 325 or 488 nm, clearly showing their UV resonance-enhanced nature. [104] The laser source of 244 nm is very close to the center of the UV adsorption of TS-1, while the sources at 325 and 488 nm are far away, implying that these UV resonance-enhanced Raman bands come from the framework Ti species.

8.3.2 Raman Characterization of Active Metal-Oxygen Species in Zeolites

The catalytic activities of metal-containing zeolites often come from only a few metal species which have distinct structures and properties compared with bulk metals [105]. A verity of metal species including cationic ions, nanoparticles, and metal-oxide clusters can be formed at the metal–zeolite support interface. The molecular structures of the surface metal oxide species are reflected in the terminal M=O and bridging M—O—M vibrations in Raman spectroscopy [106–108]. Resonance Raman [109], surface-enhanced Raman [110, 111], and in situ Raman [112, 113] are powerful methods for providing insights into the nature of metal species in zeolites such as the structural determination of Mo=O [114, 115], Fe-OOH [116], and Cu—O—Cu species [117].

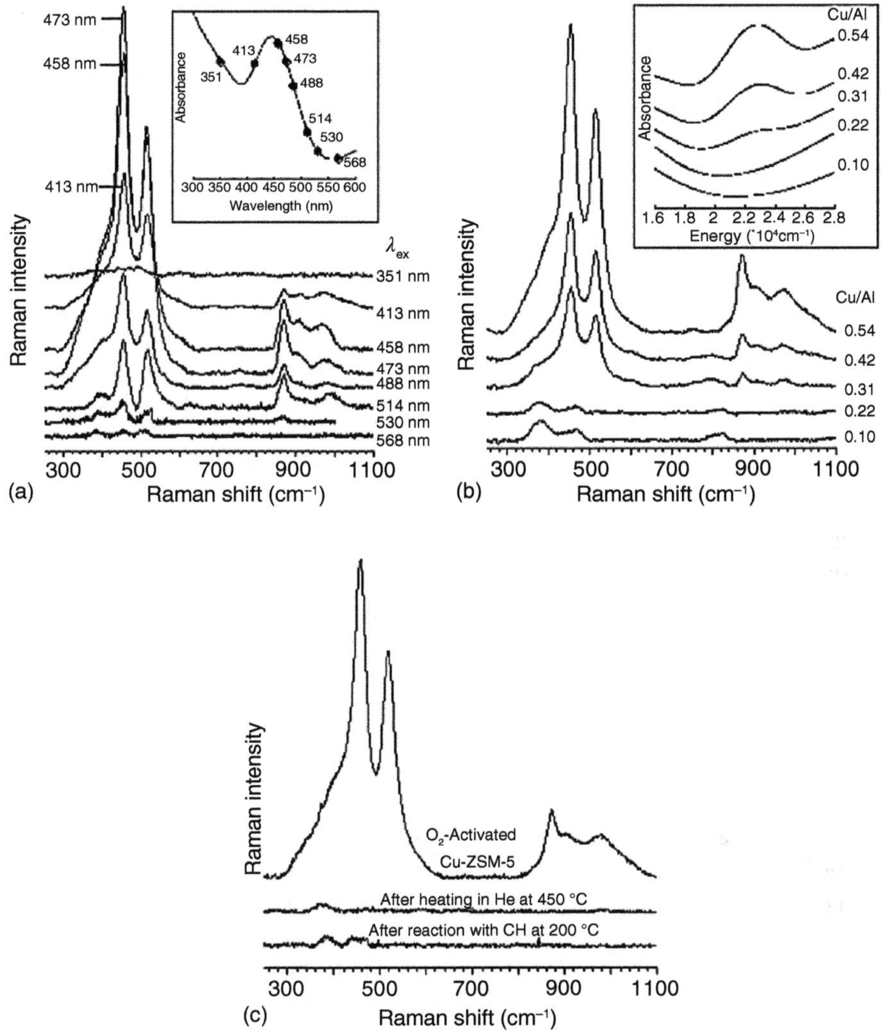

Figure 8.6 Resonance Raman spectra of O_2-activated Cu-ZSM-5. (a) Recorded on O_2-activated Cu-ZSM-5 (Cu/Al = 0.54) using λ_{ex} from 351 to 568 nm with corresponding absorption spectrum (inset). (b) Recorded on O_2-activated Cu-ZSM-5 with Cu/Al ratios from 0.10 to 0.54 using λ_{ex} = 457.9 nm with corresponding absorption spectra (inset). (c) Recorded on Cu-ZSM-5 (Cu/Al = 0.54, pretreated in O_2 at 450 °C) before and after heating in He at 723 K and after reaction with CH_4 at 473 K using λ_{ex} = 457.9 nm. Source: Reproduced from Ref. [117]/with permission of National Academy of Science.

Cu-zeolites have been demonstrated to be promising catalysts for methane partial oxidation to oxygenates under mild conditions. A bis (μ-oxo) dicopper structure was tentatively proposed to be the active Cu site in Cu-ZSM-5 zeolites by UV–VIS–NIR spectroscopy [118] and EXAFS analysis [93]. Resonance Raman provides an unambiguous identification of the geometric of the active Cu-oxygen species, showing that the active site has a mono (μ-oxo) dicopper rather than the bis (μ-oxo) dicopper structure [117, 119–121]. As shown in Figure 8.6a, using a Raman light source with wavelengths from 351 to 568 nm on Cu-ZSM-5 resulted in resonance enhancement of Raman vibrations associated only with the active site for methanol oxidization [117]. The maximum resonance is observed using λ_{ex} = 457.9 nm, with vibrations at 237, 456, 870, and 974 cm^{-1}. These vibration intensities grew with the increasing Cu/Al ratio. The same tend was found for the absorption band at 22 700 cm^{-1} in the UV–VIS spectra (inset in Figure 8.6b), which was not observed on the sample after being reacted with CH_4 (at 200 °C) or heated in He at 350 °C (Figure 8.6c). These results confirmed that these resonances Raman vibrations came from the active Cu sites. The bis (μ-oxo) dicopper complexes on Cu-ZSM-5 were reported to have an isotope-sensitive vibration at ~600 cm^{-1} [118]. However, isotope $^{16/18}O_2$ resonance Raman experiments showed the absence of isotope-sensitive features at ~600 cm^{-1}. This excludes the existence of bis (μ-oxo) dicopper species on O_2-activated Cu-ZSM-5 zeolite. Such Raman profile has been observed on mono (μ-oxo) diferric complex [122]. Therefore, a mono (μ-oxo) dicopper active site for one oxygen bridging two Cu structures was determined, showing high-intensity symmetric stretch (ν_{sym}) vibration at 456 cm^{-1} and a weak antisymmetric stretch (ν_{asym}) vibration at 870 cm^{-1}. While the second quantum antisymmetric ($2\nu_{asym}$) stretch was observed at 1725 cm^{-1}. Analogously, the mono (μ-oxo) dicopper active site was also identified on both Cu-MOR [120] and Cu-CHA [121] zeolites after O_2 activation at high temperatures.

8.4 Solid-State NMR Spectroscopy

NMR technique is based on the magnetic properties of certain nuclei, whose spin (I) is not equal to zero. In the absence of a strong external magnetic field, all the nuclei have the same energy. Zeeman effect will be observed when these nuclei are exposed to a strong external magnetic field, resulting in the nuclear spin energy split to $(2I + 1)$ levels. Bearing electromagnetic radiation, transitions between these energy levels are observed by NMR spectroscopy that are closely related to nuclei's chemical environment [123]. Solid-state NMR has been developed to be one of the most powerful analytic tools for the atomic-level analysis of solid-state materials [124–126]. Solid-state NMR is featured by its high sensitivity to structural geometries and orderings in the short or medium range, which complements the XRD technique focusing on the determination of long-range ordering structures. Due to the variety of elements in heteroatom-containing zeolites, many NMR-visible nuclei such as ^1H, ^{11}B, ^{27}Al, ^{29}Si, ^{31}P, $^{47/49}Ti$, ^{51}V, ^{91}Zr, ^{95}Mo, and ^{119}Sn can be detected for their structural characterization [127–132]. Information on the chemical composition, local environment, and coordination network of heteroatoms can be extracted from various NMR parameters. Besides, solid-state NMR spectroscopy possesses an advantage in the detection of internuclear proximity and intermolecular interactions in zeolites. In situ NMR technique is becoming increasingly important in the study of heterogeneous catalysis thanks to its ability to observe the active reaction intermediates for elucidating the reaction mechanisms.

8.4.1 Solid-State NMR Characterization of Metal Elements in Zeolites

Many metal nuclei in zeolites bring challenges for NMR analysis. For example, ^{119}Sn and ^{195}Pt ($I = 1/2$) feature large spectral widths (up to thousands of ppm). The quadrupolar nuclei, such as $^{47/49}$Ti (I = 5/2 and 7/2), ^{51}V (I = 7/2), ^{67}Zn (I = 5/2), ^{91}Zr (I = 5/2), and ^{95}/^{97}Mo (I = 5/2 and 5/2), are characterized by low Larmor frequency (low-γ) and large quadrupolar moments. Besides, the difficulty in NMR observations is exaggerated by the low loading amount of metal species on zeolites and the low NMR-active isotope abundance. The development of NMR instrumentation, methodologies, and sample preparation protocols has contributed to the increasingly successful application of solid-state NMR to heteroatom-containing zeolites [129, 133, 134].

Zn, Ga, and Mo-modified zeolites show superior catalytic performance in the conversion of light alkanes [135, 136]. NMR observation of these metal species in zeolites is not easy because their NMR-active isotopes (e.g., ^{67}Zn and ^{95}Mo) are low-γ and quadrupole nuclei. To address this problem, the sensitivity-enhanced method Hyperbolic Secant and Quadrupolar Carr–Purcell–Meiboom–Gill (HS-QCPMG) was used to acquire ^{67}Zn NMR spectrum on Zn-modified ZSM-5 zeolites [137]. The combination of HS-QCPMG with ^{67}Zn isotope enrichment allows for direct observation of different Zn species in ZSM-5 with 2 wt% Zn loading, namely Zn^{2+} ions and ZnO clusters. Regarding the Ga species in zeolites, Wideband Uniform-Rate Smooth Truncation (WURST) and QCPMG NMR (WURST-QCPMG) can be used to significantly enhance the ^{71}Ga NMR signal. As an example, highly dispersed Ga_2O_3 clusters and cationic GaO^+ species were asserted on Ga/ZSM-5 zeolite [138]. Most recently, the combination of WURST-QCPMG with an ultra-high magnetic field of 35.2T (1.5 GHz) was applied to follow the formation and transformation of active Mo-carbide species on ZSM-5 zeolite during the methane dehydroaromatization (MDA) reaction [139].

Framework Lewis acid (Ti, Sn, Zr, etc.) containing zeolites exhibit advantages in biomass conversions in terms of distinct activity and selectivity [140–142]. Among these zeolites, tin-substituted β (Sn-β) is identified as the most promising Lewis acid catalyst [75, 143–146]. Open ($(SiO)_3$Sn-OH) and closed ($(SiO)_4$Sn) Sn sites can be generated by the substitution of framework Si by Sn atoms on β zeolite [147, 148]. The open Sn sites were evidenced to be more active, for example, in the Baeyer–Villiger oxidation of cyclic ketones [148] and the Meerwein–Ponndorf–Verley–Oppenauer (MPVO) reaction of acetone with cyclohexanol [149]. Spectroscopic discrimination of the two Sn sites is difficult due to their very similar chemical environments in the zeolite framework. ^{119}Sn isotope enrichment is often required for NMR observation because of the low Sn loading (usually lower than 2 wt%) and low natural abundance of NMR-observable ^{119}Sn isotope (8.6%) [69, 145]. The sensitivity-enhanced techniques, such as Carr–Purcell–Meiboom–Gill [150, 151] and dynamic nuclear polarization [152], allow it possible for NMR observation of Sn sites in natural abundance. However, the broad distribution of Sn sites on various framework T sites and the existence of extra-framework SnO_2 species make the speciation determination complicated [153–155]. The cross-polarization

(CP) NMR experiment from ^{1}H to ^{119}Sn can be used to detect the Sn sites with spatially close H atoms, which however does not provide solid evidence for the open Sn sites with Sn-OH structure. This is because all the proton-containing groups such as Si-OH and water can act as proton sources to enhance the ^{119}Sn signal in the CP experiment [147, 156–160]. To make a direct identification of active open Sn site, proton-detected 2D ^{1}H–^{119}Sn Dipolar-mediated Heteronuclear Multiple Quantum Correlation (D-HMQC) NMR technique has been employed on ^{119}Sn-enriched Sn-β zeolite [161]. As shown in Figure 8.7a,b, both the hydrated and 298 K dehydrated ^{119}Sn-β exhibit a ^{1}H–^{119}Sn correlation peak at (5.4, −686) ppm in the 2D ^{1}H–^{119}Sn D-HMQC NMR spectra, revealing the presence of water molecule-bound six-coordinated Sn sites. At higher dehydration temperature (393 K), 2 four-coordinated Sn sites at −443 and −429 are observed with correlation to H species at 0.1–0.4 ppm (Figure 8.7c). The application of ^{119}Sn decoupling during the 2D NMR experiment resulted in the merger of four correlation peaks into two ones centered at (0.28, −443) and (0.26, −429) ppm, confirming the presence of J-coupling between ^{1}H and ^{119}Sn nuclei (Figure 8.7d). The J-coupling constant was

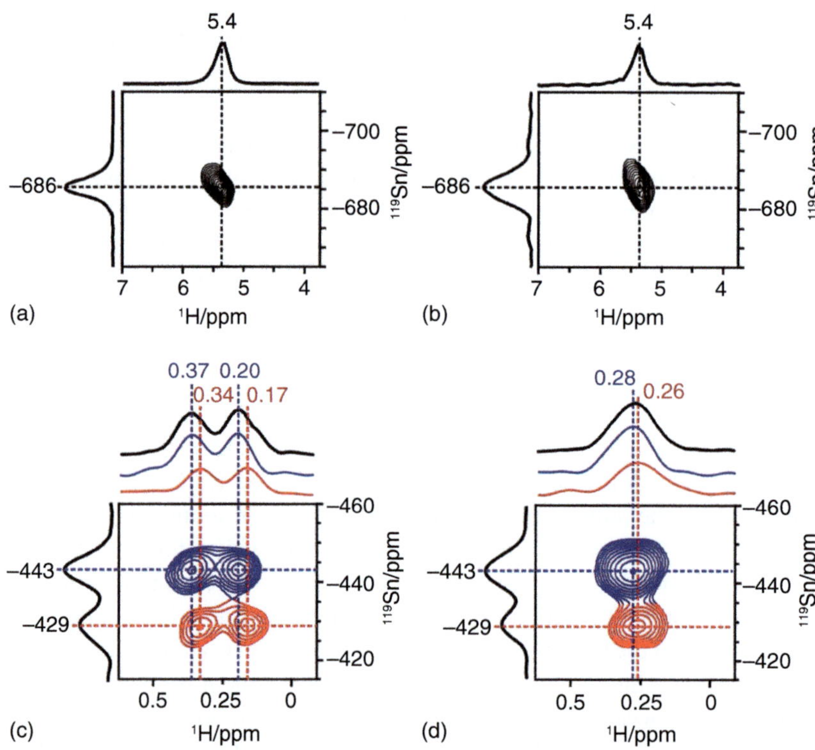

Figure 8.7 Two-dimensional ^{1}H−^{119}Sn HMQC MAS NMR spectra of ^{119}Sn-β without dehydration (a), dehydrated at 298 K (b), dehydrated at 393 K without ^{119}Sn decoupling (c), and dehydrated at 393 K with ^{119}Sn decoupling (d). Projections of ^{1}H and ^{119}Sn dimensions are shown in black. Representative slices along −429 ppm (red) and −443 ppm (blue) in the F1 dimension are also displayed in c and d. Source: Reproduced from Ref. [161]/with permission of Springer Nature.

calculated to be ~136 Hz, which is very close to the ^2J (^{119}Sn–^1H) constant of 130 Hz for Sn–OH species. These results provide direct and ambiguous evidence for the presence of two types of open Sn sites on Sn-β zeolite. In principle, this technique also can be applied to elucidate Ti and Zr metal active sites in zeolites, both of which have been intensively investigated as solid Lewis acids relevant to biomass transformations.

Besides direct observation, NMR probe molecule techniques are frequently used to detect the metal active sites in zeolites. Pyridine, acetone, trimethyl phosphate (TMP), and trimethylphosphoxy (TMPO) are common choices among various probe molecules [162–164]. Among them, TMP and TMPO are very suitable for the detection of metal species as acid sites because ^{31}P nuclei have a wide range of chemical shifts, high sensitivity, and natural abundance. Ti atoms contain two NMR-active isotopes, namely ^{47}Ti and ^{49}Ti, which have close Larmor frequencies, for example, 45.102 and 45.112 MHz for ^{47}Ti and ^{49}Ti, respectively, at 18.8 T and natural abundances (7.44% and 5.41% for ^{47}Ti and ^{49}Ti, respectively). These features bring significant difficulties in direct NMR observation and differentiation of Ti species on zeolites. This was demonstrated on titanium-substituted Y zeolite [165], which shows large NMR spectral width and seriously overlapped resonances of the two Ti isotopes even at a strong magnet field of 18.8 T [124, 129, 133]. Solid-state NMR of the TMP probe molecule was used to identify Ti sites on TS-1 [166], showing that most of the Brønsted and Lewis acid sites in TS-1 zeolites originate from the incorporation of Ti species. In the NMR investigation of Brønsted acid dominated zeolites by using phosphorus-based probe molecules, TMP is sensitive to Lewis acid sites [167], while TMPO is favored for Brønsted acid sites [164, 168, 169]. For Lewis acid sites containing zeolites, ^{31}P NMR of TMPO adsorption can also be employed in discriminating and quantifying the relative concentration of various Lewis acid sites in heteroatom-containing zeolites, including Sn- Ti-, Hf-, and Zr-β [170]. For example, three Sn Lewis acid sites with ^{31}P chemical shifts at 54.9, 55.8, and 58.6 ppm were determined after the adsorption of TMPO on Sn-β zeolite. Quantitative analysis showed that they are responsible for different catalytic reactions. The reaction rate of glucose isomerization was directly correlated with the two Sn Lewis acid sites with ^{31}P chemical shifts at 55.8 and 54.9 ppm These two signals were attributed to TMPO adsorption on the open Sn Lewis sites, which are commonly predicted to be more active toward glucose isomerization [171]. It is worth noting that their content determined by TMPO adsorption is 12–33% of total Sn sites, very close to the previously reported value for the open Sn sites [148, 161]. On the other hand, the Sn Lewis acid sites with ^{31}P chemical shift at 58.6 ppm were assigned to the closed Sn sites, the amount of which showed a liner correlation with the benzaldehyde-acetone aldol condensation reaction.

8.4.2 Solid-State Correlation NMR Measurement of Active Site Proximity and Host–Guest Interactions

Active site synergy is usually assumed to be responsible for enhanced catalytic performance on bi-functional catalysts. Direct detection of the spatial proximity and interaction between active sites is critical to a comprehensive understanding of their synergistic effect. Solid-state correlation NMR techniques including double-resonance, homonuclear, and heteronuclear correlation that are based on the coupling of nuclear spins via J-coupling or dipolar interaction can provide useful information on the through-bond and through-space connectivity of the observed nuclei [172]. For example, 2D ^{1}H–^{1}H and ^{27}Al–^{27}Al double-quantum single-quantum correlation magic angle spinning (MAS) NMR method allows the identification of spatial proximity between Brønsted (SiOHAl) and Lewis acids (extra-framework AlOH group) in dealuminated Y zeolites by measuring the ^{1}H–^{1}H and ^{27}Al–^{27}Al internuclear distances, which are closely related to the geometric structure of the synergistic Brønsted/Lewis acid sites [173, 174]. In metal-containing zeolites, ^{1}H-M (M: metal) double-resonance NMR technique provides a versatile approach to analyze the acid-metal synergy in zeolites. The direct observation of the active sites synergy between acidic protons and zinc species in Zn/ZSM-5 zeolite was achieved by using ^{1}H-{^{67}Zn} S-RESPDOR NMR [137]. This type of synergist effect was also observed on Ga-modified zeolites in the methanol-to-aromatics reaction [138, 175]. Detailed information on the formation and development of the synergetic active sites in catalytic reactions is important for understanding the catalytic process. Spatial interactions between active Mo sites and various protons including Brønsted acid sites (4.0 ppm), olefins (6.4 ppm), and aromatics (8.1 ppm) in ZSM-5 channels were measured by the ^{1}H-{^{95}Mo} S-RESPDOR NMR method (Figure 8.8a–c) [139]. In principle, the ΔS can be used to measure the degree of

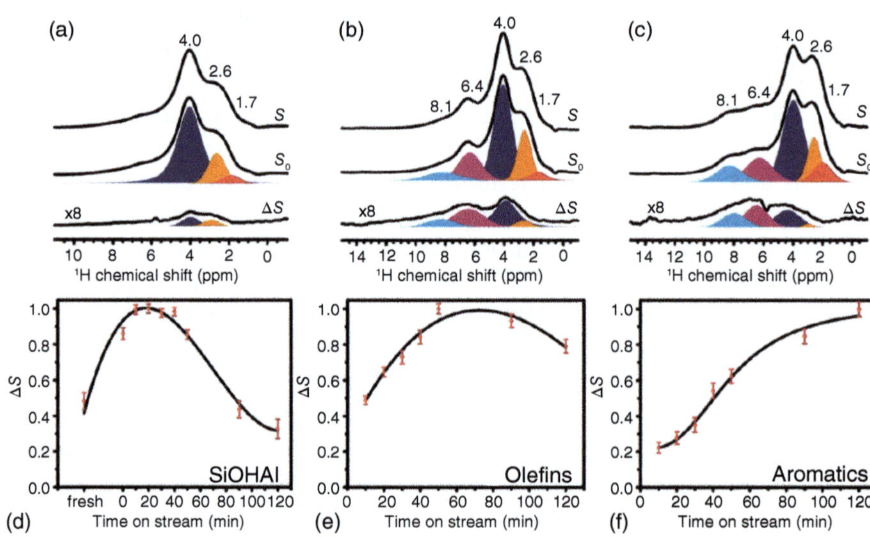

Figure 8.8 ^{1}H-{^{95}Mo} S-RESPDOR NMR spectra of (a) fresh 5Mo/ZSM-5, (b) 5Mo/ZSM-5 reacted for 30 minutes and (c) for 120 minutes of MDA reaction at 973 K. S and S_0 represent the signal intensity with and without ^{95}Mo irradiation, respectively. $\Delta S = S_0 - S$. Normalized ΔS of (d) Brønsted acid sites at 4.3 ppm, (e) olefins at 6.4 ppm, and (f) aromatics at 8.1 ppm versus MDA reaction time. Source: Reproduced from Ref. [139]/with permission of John Wiley & Sons.

proton-Mo proximity during MDA reaction. A significant increase of the ΔS for the acidic $^{1}H-^{95}Mo$ pair was observed in the initial reaction period (before 30 minutes) and a rapid reduction occurred at the long-term reaction, indicating that the acidic $^{1}H-^{95}Mo$ pair was gradually formed at the beginning of the MDA reaction and consumed in the later stages (Figure 8.8d). Note that the ΔS for the $^{1}H-^{95}Mo$ pair associated with olefins showed a similar trend (Figure 8.8e), implying the intermediate role of olefins in the MDA reaction. However, a different behavior was observed for aromatics (Figure 8.8f), which kept increasing with reaction time, reflecting the formation and accumulation of aromatics and their derivatives in the zeolite channels. The results provide new insights into the understanding of the active sites in the Mo/ZSM-5 catalyst as well as the MDA reaction mechanism.

8.4.3 In Situ Solid-State NMR for the Study of Reaction Mechanisms

In situ solid-state NMR is one of the most important methods for elucidating reaction mechanisms in heterogenous catalysis [131, 176–179]. In situ NMR measurement can be conducted under two types of conditions, namely, batch reaction and continuous-flow (CF) reaction. The batch reaction is typically performed in a small glass ampule containing catalyst and reactants, which can be fitted into the NMR rotor. The ampule can be heated inside or outside the probe at different temperatures for a fixed period (or for a different period at a fixed temperature) and the NMR spectra are recorded under MAS. The batch mode in situ NMR is widely employed to identify the reaction intermediates in many catalytic processes [180, 181]. For example, in situ ^{13}C MAS NMR experiment was used to monitor the MPVO reaction of acetone with cyclohexanol on Sn-β zeolite under batch condition [149]. A stable surface gem-diols-type species bound to the open Sn site was identified, which provided an energy-feasible pathway for the MPVO reaction. Besides the gas–solid phase reaction, in situ NMR was also applied in the liquid-phase catalytic conversion of carbohydrates over NaX and NaY zeolites under batch condition [182]. In situ ^{13}C MAS NMR analysis of the adsorption and reaction of carbohydrates showed that the solvent composition (water/γ-valerolactone ratio) has a strong influence on the mobility and adsorption behavior of the carbohydrates on zeolites.

TS-1 is a key catalyst that can catalytically epoxidize olefins with hydrogen peroxide, which produces almost one million tons of propylene oxide per year [183]. The active sites on TS-1 zeolite are generally assigned to the isolated framework Ti(IV). However, the detailed structure of these active Ti(IV) sites under the reaction condition and the relevant reaction mechanism remain unclear [184]. In the work by Gordon et al., a batch condition in situ ^{17}O MAS NMR joint with other spectroscopic methods and theoretical calculations were used to characterize the highly active and selective Ti species in a series of TS-1 catalysts [185]. As shown in Figure 8.9a,b, after the two TS-1 samples with different Ti loadings contacted with ^{17}O-labeled H_2O_2, both samples showed identical ^{17}O NMR line-shape to the bridging peroxo-species on the dinuclear titanium reference structure (Figure 8.9c), which is different from $H_2{}^{17}O_2$ (Figure 8.9e). This observation confirmed that the active oxygen species originated from the activation of $H_2{}^{17}O_2$ on Ti sites. To support the assignment of

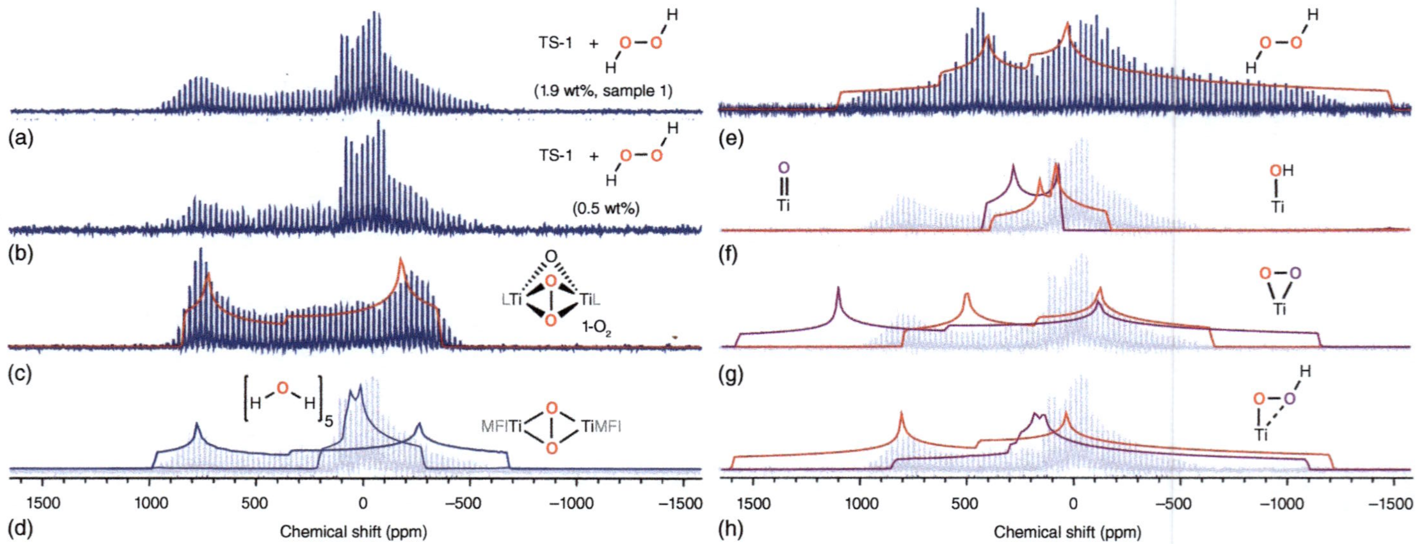

Figure 8.9 (a) ^{17}O NMR spectrum of sample 1 (1.9 wt% Ti) contacted with $H_2{}^{17}O_2$ (1 equiv. with respect to Ti). (b) The corresponding spectrum of the sample with 0.5 wt% Ti. (c) Experimental ^{17}O NMR (blue) and calculated (red) spectra of the dinuclear molecular $\mu^2\eta^2$-peroxy compound 1-O_2. (d) DFT-calculated ^{17}O NMR signature of a dinuclear Ti-peroxo species and $H_2{}^{17}O$ (solid blue lines on top of the shaded experimental spectrum). (e) Experimental (blue) and calculated (red) spectra of $H_2{}^{17}O_2$. (f) Calculated spectra for Ti-oxo and Ti-hydroxo species in the MFI framework. (g) Calculated spectra for Ti-peroxo (the two oxygen atoms are inequivalent in the MFI structure) and (h) Ti-hydroperoxo species in the MFI framework. The experimental spectrum of the TS-1/H_2O_2 system is shaded in the background. Source: Reproduced from Ref. [185]/with permission of Springer Nature.

the observed ^{17}O NMR signals, theoretical calculations were employed to simulate the ^{17}O NMR signatures of various proposed intermediates in the TS-1/H_2O_2 system. The calculated line-shapes of 1-O_2 and H_2O_2 were in good agreement with their experimental spectra. In comparison, the DFT-calculated line-shape of Ti sites associated with oxo- (Figure 8.9d) and OH groups (Figure 8.9f) exhibited much smaller anisotropies than the experimental spectra of TS-1 contacted with $H_2{}^{17}O_2$ (Figure 8.9a,b). On the other hand, the side-on peroxo (Figure 8.9g) or side-on hydroperoxy ligands (Figure 8.9h) have much broader spectral line shapes. These results strongly suggest the formation of bridging peroxo-species on the dinuclear titanium sites. Further theoretical calculations indicated that the two cooperative titanium atoms enable effective propylene epoxidation via a low-energy reaction pathway involving a key oxygen-transfer transition state, analogous to olefin epoxidation by peracids. It was proposed that the dinuclear titanium sites rather than isolated titanium atoms in the framework are responsible for the high efficiency of TS-1 in propylene epoxidation with H_2O_2.

It is unable to continuously introduce reactants and remove products in the batch reactors. To address this problem, pulse-quench and operando approaches were developed to monitor the CF reactions. The pulse-quench method allows the NMR study of heterogeneous reactions under actual reaction conditions at a short time scale [186, 187]. Instrumentally, this approach uses a standard fix-bed micro-reactor on which continuous and/or pulse introduction of reactants is conducted and the volatile products out of the reactor can be online analyzed by gas chromatography (GC) or GC–MS. The most significant feature of the pulse-quench approach is that it enables a rapid switch of the gas stream that flows over the catalyst bed to cryogenically cooled nitrogen, by which the reaction is quenched by rapid reduction of the catalyst temperature within 200 ms. The frozen catalyst can be moved out of the micro-reactor and transformed into solid-state NMR rotors under an inert atmosphere for detailed NMR measurements. This approach is preferable for the investigation of catalytic reactions involving highly active and unstable intermediates. ^{13}C MAS NMR experiment using this pulse quench approach was applied in the investigation of methanol-to-olefins (MTO) reaction over ZSM-5 zeolite, revealing cyclopentenyl cations as the active carbon species for the formation of olefins [186]. By using this approach in combination with solid-state and solution ^{13}C NMR, a series of methylcyclopentenyl cations and ethylcyclopentenyl cations intermediates have been observed in the MTO reaction over zeolites, allowing the establishment of an exact pathway for the formation of ethene and propene via the carbenium ions mediated hydrocarbon pool mechanism [188, 189].

Note that the NMR measurement using the pulse-quench method is performed off-line rather than under the catalyst working condition. The Operando CF NMR approach allows for NMR experiments at reaction conditions identical to those utilized for flow reactors [190]. Although it is expensive and difficult to conduct such experiment due to the use of a special NMR probe and a large number of isotopic labeling reactants, numerous attempts have been made in the investigation of various catalytic reactions [191]. In the work by Hunger et al., methylation of aniline by methanol on zeolite H–Y has been investigated by using an in situ stopped-flow MAS NMR probe [192]. The in situ ^{13}C CF MAS NMR experiments were performed at reaction temperatures between 473 and 523 K. By following the formation and decomposition of *N,N,N*-trimethylanilinium cations on zeolite, the chemical equilibrium between the anilinium cations carrying different numbers of methyl groups was found to dominate the distribution of products in the gas phase. In the MTO reaction over zeolites, the catalytically active role of the hydrocarbon pool was revealed by using this in situ NMR approach [193]. Recently, triethyloxonium ions were observed in an ethanol dehydration reaction over ZSM-5 zeolite by using a CF ^{13}C MAS NMR experiment [194]. Ethylation of the zeolite by triethyloxonium ions can occur at a lower temperature, leading to the formation of surface ethoxy species and then ethene. As supported by theoretical calculations, the triethyloxonium ions-ethoxide pathway is found to be energetically preferable for the dehydration of ethanol to ethene in the initial stage.

8.5 Conclusions

In this chapter, we introduced various spectroscopic techniques including X-ray, UV–VIS, Raman, and solid-state NMR that are frequently used for the characterization of heteroatom-containing zeolites (Table 8.1). XRD has been routinely used for the analysis bulk zeolite crystal, while XAS (EXANES and EXAFS) techniques are sensitive to the coordination and bond number of metal atoms in zeolite. The valence states of heteroatoms in zeolite can be detected by XPS method. UV–VIS is suitable for the detection of metal speciation and location in zeolites. Different metal species and structure units in zeolite can be reflected by distinct Raman features, which is high valuable for the investigation of zeolite synthesis and assembly mechanism. Solid-state NMR can provide atomic-level information about the structure of heteroatoms, atomic distances, and host–guest interactions of adsorbed molecules in zeolites. The application and impact of these spectroscopic techniques will be broadened and strengthened in the field of zeolites and related catalysis with

Table 8.1 Spectroscopic techniques for the characterization of heteroatom-containing zeolites.

Information	Technique					
	XRD	XAS	XPS	UV–VIS	Raman	NMR
Bulk framework structure	+	–	–	–	–	–
Elemental analysis	–	–	+	–	–	–
Metal speciation, location	–	–	–	+	+	–
Metal valence state	–	–	+	–	–	–
Local structure, dynamics	–	–	–	–	–	+
Bond distance, coordination number	–	+	–	–	–	+

+ and – denote the most favorable and unfavorable method, respectively.

advances in the instrumentation and in situ approaches. Rich information obtained by the combination of various spectroscopic techniques allows for a deep understanding of the structure–property relationships of heteroatom-containing zeolites, which is highly relevant to their synthesis and applications.

Acknowledgments

This work was supported by the National Natural Science Foundation of China (grants 22272185, 22225205, 22320102002 and 22127801), the Magnetic Resonance Technology Alliance of the Chinese Academy of Sciences (2021gzl004), the Strategic Priority Research Program of the Chinese Academy of Sciences (XDB0540000) and the Natural Science Foundation of Hubei Province of China (S22H120101).

References

1 Tangcharoen, T., Klysubun, W., Kongmark, C., and Pecharapa, W. (2014). Synchrotron X-ray absorption spectroscopy and magnetic characteristics studies of metal ferrites (metal = Ni, Mn, Cu) synthesized by sol–gel auto-combustion method. *Phys. Status Solidi A Appl. Mater. Sci.* 211: 1903–1911.

2 Tangcharoen, T., Klysubun, W., and Kongmark, C. (2019). Synchrotron X-ray absorption spectroscopy and cation distribution studies of $NiAl_2O_4$, $CuAl_2O_4$, and $ZnAl_2O_4$ nanoparticles synthesized by sol-gel auto combustion method. *J. Mol. Struct.* 1182: 219–229.

3 Evans, J. (2018). *X-Ray Absorption Spectroscopy for the Chemical and Materials Sciences*, 1–8. John Wiley & Sons Ltd.

4 Tatlier, M. (2011). Artificial neural network methods for the prediction of framework crystal structures of zeolites from XRD data. *Neural Comput. Appl.* 20: 365–371.

5 Wright, P.A., Jones, R.H., Natarajan, S. et al. (1993). Synthesis and structure of a novel large-pore microporous magnesium-containing aluminophosphate (DAF-1). *J. Chem. Soc, Chem. Commun.* 633–635.

6 Vortmann, S., Marler, B., Daniels, P., Dierdorf, I. and Gies, H. (1995). In *Studies in Surface Science and Catalysis*; H. G. Karge, J. Weitkamp, Eds.; Elsevier. 98: 262–263.

7 Liu, L.F., Yu, Z.B., Chen, H. et al. (2013). Disorder in extra-large pore zeolite ITQ-33 revealed by single crystal XRD. *Cryst. Growth Des.* 13: 4168–4171.

8 Park, S.-H., Parise, J.B., Gies, H. et al. (2000). A new porous lithosilicate with a high ionic conductivity and ion-exchange capacity. *J. Am. Chem. Soc.* 122: 11023–11024.

9 Corma, A., Puche, M., Rey, F. et al. (2003). A zeolite structure (ITQ-13) with three sets of medium-pore crossing channels formed by 9- and 10-rings. *Angew. Chem. Int. Ed.* 42: 1156–1159.

10 Burton, A.W. (2004). Structure solution of zeolites from powder diffraction data. *Z. Kristallogr.* 219: 866–880.

11 Blasco, T., Corma, A., Diaz-Cabanas, M.J. et al. (2004). Synthesis, characterization, and framework heteroatom localization in ITQ-21. *J. Am. Chem. Soc.* 126: 13414–13423.

12 Gramm, F., Baerlocher, C., McCusker, L.B. et al. (2006). Complex zeolite structure solved by combining powder diffraction and electron microscopy. *Nature* 444: 79–81.

13 Deem, M.W. and Newsam, J.M. (1989). Determination of 4-connected framework crystal structures by simulated annealing. *Nature* 342: 260–262.

14 Deem, M.W. and Newsam, J.M. (1992). Framework crystal structure solution by simulated annealing: test application to known zeolite structures. *J. Am. Chem. Soc.* 114: 7189–7198.

15 Falcioni, M. and Deem, M.W. (1999). A biased Monte Carlo scheme for zeolite structure solution. *J. Chem. Phys.* 110: 1754–1766.

16 Grosse-Kunstleve, R.W., McCusker, L.B., and Baerlocher, C. (1997). Powder diffraction data and crystal chemical information combined in an automated structure determination procedure for zeolites. *J. Appl. Crystallogr.* 30: 985–995.

17 Rius, J. (1999). XLENS, a direct methods program based on the modulus sum function: its application to powder data. *Powder Diffract.* 14: 267–273.

18 Corma, A., Díaz-Cabañas, M.J., Martínez-Triguero, J. et al. (2002). A large-cavity zeolite with wide pore windows and potential as an oil refining catalyst. *Nature* 418: 514–517.

19 Corma, A., Rey, F., Valencia, S. et al. (2003). A zeolite with interconnected 8-, 10- and 12-ring pores and its unique catalytic selectivity. *Nat. Mater.* 2: 493–497.

20 Rius, J. (1993). Derivation of a new tangent formula from Patterson-function arguments. *Acta Crystallogr. A* 49: 406–409.

21 Treacy, M.M.J., Newsam, J.M., and Deem, M.W. (1988). Diffraction from zeolites containing planar faults. *MRS Online Proc. Library* 138: 497–502.

22 Treacy, M.M.J., Vaughan, D.E.W., Strohmaier, K.G., and Newsam, J.M. (1996). Intergrowth segregation in FAU-EMT zeolite materials. *Proc. R. Soc. A – Math. Phys. Eng. Sci.* 452: 813–840.

23 Jordá, J.L., McCusker, L.B., Baerlocher, C. et al. (2003). Structure analysis of the novel microporous aluminophosphate IST-1 using synchrotron powder diffraction data and HETCOR MAS NMR. *Microporous Mesoporous Mater.* 65: 43–57.

24 Wagner, P., Terasaki, O., Ritsch, S. et al. (1999). Electron diffraction structure solution of a nanocrystalline zeolite at atomic resolution. *J. Phys. Chem. B* 103: 8245–8250.

25 de Groot, F. (2001). High-resolution X-ray emission and X-ray absorption spectroscopy. *Chem. Rev.* 101: 1779–1808.

26 Wende, H. (2004). Recent advances in x-ray absorption spectroscopy. *Rep. Prog. Phys.* 67: 2105–2181.

27 van Bokhoven, J.A. and Lamberti, C. (2014). Structure of aluminum, iron, and other heteroatoms in zeolites by X-ray absorption spectroscopy. *Coord. Chem. Rev.* 277–278: 275–290.

28 Bordiga, S., Groppo, E., Agostini, G. et al. (2013). Reactivity of surface species in heterogeneous catalysts probed by in situ X-ray absorption techniques. *Chem. Rev.* 113: 1736–1850.

29 Bokhoven, J.A.v., Nabi, T., Sambe, H. et al. (2001). Interpretation of the Al K- and $L_{II/III}$-edges of aluminium oxides: differences between tetrahedral and octahedral Al explained by different local symmetries. *J. Phys. Condens. Matter* 13: 10247–10260.

30 Al-majnouni, K.A., Hould, N.D., Lonergan, W.W. et al. (2010). High-temperature decomposition of Brønsted acid sites in gallium-substituted zeolites. *J. Phys. Chem. C* 114: 19395–19405.

31 Bordiga, S., Coluccia, S., Lamberti, C. et al. (1994). XAFS study of Ti-silicalite: structure of framework Ti(IV) in the presence and absence of reactive molecules (H2O, NH3) and comparison with ultraviolet-visible and IR results. *J. Phys. Chem.* 98: 4125–4132.

32 Tuilier, M.H., Lopez, A., Guth, J.L., and Kessler, H. (1991). EXAFS study of germanium-rich MFI-type zeolites. *Zeolites* 11: 662–665.

33 Bare, S.R., Kelly, S.D., Sinkler, W. et al. (2005). Uniform catalytic site in Sn-β-zeolite determined using X-ray absorption fine structure. *J. Am. Chem. Soc.* 127: 12924–12932.

34 Anpo, M., Zhang, S.G., Higashimoto, S. et al. (1999). Characterization of the local structure of the vanadium silicalite (VS-2) catalyst and its photocatalytic reactivity for the decomposition of NO into N_2 and O_2. *J. Phys. Chem. B* 103: 9295–9301.

35 Fan, F., Feng, Z., and Li, C. (2010). UV Raman spectroscopic study on the synthesis mechanism and assembly of molecular sieves. *Chem. Soc. Rev.* 39: 4794–4801.

36 van Bokhoven, J.A., van der Eerden, A.M.J., and Koningsberger, D.C. (2003). Three-coordinate aluminum in zeolites observed with in situ X-ray absorption near-edge spectroscopy at the Al K-edge: flexibility of aluminum coordinations in zeolites. *J. Am. Chem. Soc.* 125: 7435–7442.

37 Fröba, M., Wong, J., Behrens, P. et al. (1995). Correlation of multiple scattering features in XANES spectra of Al and Si K edges to the Al—O—Si bond angle in aluminosilicate sodalites: an empirical study. *Physica B* 208–209: 65–67.

38 Koningsberger, D.C. and Miller, J.T. (1994). Local structure determination of aluminum in Y zeolite: application of low energy X-ray absorption fine structure spectroscopy. *Catal. Lett.* 29: 77–90.

39 van Bokhoven, J.A., van der Eerden, A.M.J., and Prins, R. (2004). Local structure of the zeolitic catalytically active site during reaction. *J. Am. Chem. Soc.* 126: 4506–4507.

40 Joyner, R.W., Smith, A.D., Stockenhuber, M., and van den Berg, M.W.E. (2004). The local structure of aluminium sites in zeolites. *Phys. Chem. Chem. Phys.* 6: 5435–5439.

41 van der Eerden, A.M.J., van Bokhoven, J.A., Smith, A.D., and Koningsberger, D.C. (2000). Apparatus for in situ x-ray absorption fine structure studies on catalytic systems in the energy range $1000 \text{ eV} < E < 3500 \text{ eV}$. *Rev. Sci. Instrum.* 71: 3260–3266.

42 Agostini, G., Lamberti, C., Palin, L. et al. (2010). In situ XAS and XRPD parametric Rietveld refinement to understand dealumination of Y zeolite catalyst. *J. Am. Chem. Soc.* 132: 667–678.

43 van Bokhoven, J.A., van der Eerden, A.M.J., Smith, A.D., and Koningsberger, D.C. (1999). Design and construction on an in situ cell for catalytic studies on elements with $12 < Z < 20$. *J. Synchrotron Radiat.* 6: 201–203.

44 Drake, I.J., Zhang, Y., Gilles, M.K. et al. (2006). An in situ Al K-edge XAS investigation of the local environment of H^+- and Cu^+-exchanged USY and ZSM-5 zeolites. *J. Phys. Chem. B* 110: 11665–11676.

45 Berlier, G., Spoto, G., Bordiga, S. et al. (2002). Evolution of extraframework iron species in Fe silicalite: 1. Effect of Fe content, activation temperature, and interaction with redox agents. *J. Catal.* 208: 64–82.

46 Berlier, G., Spoto, G., Fisicaro, P. et al. (2002). Co-ordination and oxidation changes undergone by iron species in Fe-silicalite upon template removal, activation and interaction with N_2O: an in situ X-ray absorption study. *Microchem. J.* 71: 101–116.

47 Grünert, W. and Schlögl, R. (2004). *Characterization I*, Molecular Sieves—Science and Technology, vol. 4 (ed. H.G. Karge and J. Weitkamp), 467–515. Berlin, Heidelberg: Springer Berlin Heidelberg.

48 Gruenert, W., Muhler, M., Schroeder, K.-P. et al. (1994). Investigations of zeolites by photoelectron and ion scattering spectroscopy. 2. A new interpretation of XPS binding energy shifts in zeolites. *J. Phys. Chem.* 98: 10920–10929.

49 Barr, T.L., Chen, L.M., Mohsenian, M., and Lishka, M.A. (1988). XPS valence band study of zeolites and related systems. 1. General chemistry and structure. *J. Am. Chem. Soc.* 110: 7962–7975.

50 Stöcker, M. (1996). X-ray photoelectron spectroscopy on zeolites and related materials. *Microporous Mater.* 6: 235–257.

51 Czanderna, A.W. (2012). *Methods of Surface Analysis*. Elsevier.

52 Kaushik, V.K., Bhat, S.G.T., and Corbin, D.R. (1993). Surface composition and electronic structure of zeolites using X-ray photoelectron spectroscopy. *Zeolites* 13: 671–677.

53 Sato, K., Nishimura, Y., Imamura, M. et al. (2002). Application of synchrotron radiation XPS to the in-depth profiling of HY zeolite particles. *Anal. Sci.* 17: i1061–i1064.

54 Knecht, J. and Stork, G. (1977). Quantitative Analyse von Zeolithen mit Hilfe der Röntgenphotoelektronenspektroskopie. *Fresenius Z. Anal. Chem.* 283: 105–108.

55 Knecht, J. and Stork, G. (1977). Quantitative Analyse von Zeolithen mit Hilfe der Röntgenphotoelektronenspektroskopie. *Fresenius Z. Anal. Chem.* 286: 47–49.

56 Kanazirev, V., Price, G.L., and Tyuliev, G. (1992). XPS study on the effect of hydrogen treatment on the state of gallium in Ga_2O_3/HZSM-5 mechanically mixed catalysts. *Zeolites* 12: 846–850.

57 Grohmann, I., Pilz, W., Walther, G. et al. (1994). XPS-investigation of titanium modified MFI-type zeolites. *Surf. Interface Anal.* 22: 403–406.

58 On, D.T., Bonneviot, L., Bittar, A. et al. (1992). Titanium sites in titanium silicalites: an XPS, XANES and EXAFS study. *J. Mol. Catal.* 74: 233–246.

59 Moretti, G., Salvi, A.M., Guascito, M.R., and Langerame, F. (2004). An XPS study of microporous and mesoporous titanosilicates. *Surf. Interface Anal.* 36: 1402–1412.

60 Narayana, M., Contarini, S., and Kevan, L. (1985). X-ray photoelectron and electron spin resonance spectroscopic studies of Cu-NaY zeolites. *J. Catal.* 94: 370–375.

61 Contarini, S. and Kevan, L. (1986). X-ray photoelectron spectroscopic study of copper-exchanged X- and Y-type sodium zeolites: resolution of two cupric ion components and dependence on dehydration and X-irradiation. *J. Phys. Chem.* 90: 1630–1632.

62 Shpiro, E.S., Grünert, W., Joyner, R.W., and Baeva, G.N. (1994). Nature, distribution and reactivity of copper species in over-exchanged Cu-ZSM-5 catalysts: an XPS/XAES study. *Catal. Lett.* 24: 159–169.

63 Gruenert, W., Sauerlandt, U., Schloegl, R., and Karge, H.G. (1993). XPS investigations of lanthanum in faujasite-type zeolites. *J. Phys. Chem.* 97: 1413–1419.

64 Kuznicki, S.M. and Eyring, E.M. (1980). An ESCA study of rhodium(III)-exchanged zeolite catalysts. *J. Catal.* 65: 227–230.

65 Schoonheydt, R.A. (2010). UV-VIS-NIR spectroscopy and microscopy of heterogeneous catalysts. *Chem. Soc. Rev.* 39: 5051–5066.

66 Giordanino, F., Vennestrøm, P.N., Lundegaard, L.F. et al. (2013). Characterization of Cu-exchanged SSZ-13: a comparative FTIR, UV-Vis, and EPR study with Cu-ZSM-5 and Cu-β with similar Si/Al and Cu/Al ratios. *Dalton Trans.* 42: 12741–12761.

67 Howe, R.F. (1996). *Studies in Surface Science and Catalysis*, vol. 102 (ed. H. Chon, S.I. Woo, and S.E. Park), 97–139. Elsevier.

68 Huybrechts, D.R.C., Debruycker, L., and Jacobs, P.A. (1990). Oxyfunctionalization of alkanes with hydrogen peroxide on titanium silicalite. *Nature* 345: 240–242.

69 Corma, A., Nemeth, L.T., Renz, M., and Valencia, S. (2001). Sn-zeolite beta as a heterogeneous chemoselective catalyst for Baeyer-Villiger oxidations. *Nature* 412: 423–425.

70 Bordiga, S., Lamberti, C., Bonino, F. et al. (2015). Probing zeolites by vibrational spectroscopies. *Chem. Soc. Rev.* 44: 7262–7341.

71 Corma, A., Domine, M.E., Nemeth, L., and Valencia, S. (2002). Al-free Sn-Beta zeolite as a catalyst for the selective reduction of carbonyl compounds (Meerwein–Ponndorf–Verley reaction). *J. Am. Chem. Soc.* 124: 3194–3195.

72 Sushkevich, V.L., Ivanova, I.I., Tolborg, S., and Taarning, E. (2014). Meerwein-Ponndorf-Verley-Oppenauer reaction of crotonaldehyde with ethanol over Zr-containing catalysts. *J. Catal.* 316: 121–129.

73 Van de Vyver, S. and Roman-Leshkov, Y. (2015). Metalloenzyme-like zeolites as Lewis acid catalysts for C-C bond formation. *Angew. Chem. Int. Ed.* 54: 12554–12561.

74 Palagin, D., Sushkevich, V.L., and Ivanova, I.I. (2016). C–C coupling catalyzed by zeolites: is enolization the only possible pathway for aldol condensation? *J. Phys. Chem. C* 120: 23566–23575.

75 Holm, M.S., Saravanamurugan, S., and Taarning, E. (2010). Conversion of sugars to lactic acid derivatives using heterogeneous zeotype catalysts. *Science* 328: 602–605.

76 Bai, R., Qiming, S., Song, Y. et al. (2018). Intermediate-crystallization promoted catalytic activity of titanosilicate zeolites. *J. Mater. Chem. A* 6: 8757–8762.

77 Sanz, R., Serrano, D.P., Pizarro, P., and Moreno, I. (2011). Hierarchical TS-1 zeolite synthesized from SiO_2 TiO_2 xerogels imprinted with silanized protozeolitic units. *Chem. Eng. J.* 171: 1428–1438.

78 Dugkhuntod, P., Maineawklang, N., Rodaum, C. et al. (2022). Synthesis and characterization of Sn, Ge, and Zr isomorphous substituted MFI nanosheets for glucose isomerization to fructose. *ChemPlusChem* 87: e202100289.

79 Lizhi, W., Xiujuan, D., Shufang, Z. et al. (2016). Synthesis of a highly active oxidation catalyst with improved distribution of titanium coordination states. *Chem. Commun.* 52: 8679–8682.

80 Zhao, P., Li, Z., Zhang, Y. et al. (2022). Tuning Lewis acid sites in TS-1 zeolites for hydroxylation of anisole with hydrogen peroxide. *Microporous Mesoporous Mater.* 335: 111840.

81 Wang, Y., Li, T., Li, C. et al. (2021). One-pot green synthesis of Fe-ZSM-5 zeolite containing framework heteroatoms via dry gel conversion for enhanced propylene selectivity of catalytic cracking catalyst. *J. Mater. Sci.* 56: 18050–18060.

82 Fan, W.B., Fan, B.B., Shen, X.H. et al. (2009). Effect of ammonium salts on the synthesis and catalytic properties of TS-1. *Microporous Mesoporous Mater.* 122: 301–308.

83 Huang, M., Wen, Y., Wei, H. et al. (2021). The clean synthesis of small-particle TS-1 with high-content framework Ti by using NH_4HCO_3 and suspended seeds as an assistant. *ACS Omega* 6: 13015–13023.

84 Xiong, G., Cao, Y., Guo, Z. et al. (2016). The roles of different titanium species in TS-1 zeolite in propylene epoxidation studied by in situ UV Raman spectroscopy. *Phys. Chem. Chem. Phys.* 18: 190–196.

85 Wu, P., Tatsumi, T., Komatsu, T., and Yashima, T. (2001). A novel titanosilicate with MWW structure. I. Hydrothermal synthesis, elimination of extraframework titanium, and characterizations. *J. Phys. Chem. B* 105: 2897–2905.

86 Zhang, S., Jin, S., Tao, G. et al. (2017). The evolution of titanium species in boron-containing Ti-MWW zeolite during post-treatment revealed by UV resonance Raman spectroscopy. *Microporous Mesoporous Mater.* 253: 183–190.

87 Jin, Y., Asaoka, S., Zhang, S. et al. (2013). Reexamination on transition-metal substituted MFI zeolites for catalytic conversion of methanol into light olefins. *Fuel Process. Technol.* 115: 34–41.

88 Zhou, X., Liu, Y., Meng, X. et al. (2013). Synthesis and catalytic cracking performance of Fe/Ti-ZSM-5 zeolite from attapulgite mineral. *Chinese J. Catal.* 34: 1504–1512.

89 Schoonheydt, R.A. (1989). Combined ESR-DRS spectroscopies of transition metal ions and metal ion clusters in zeolites. *J. Phys. Chem. Solids* 50: 523–539.

90 Scarano, D., Zecchina, A., Bordiga, S. et al. (1993). Fourier-transform infrared and Raman spectra of pure and Al-, B-, Ti- and Fe-substituted silicalites: stretching-mode region. *J. Chem. Soc. Faraday Trans.* 89: 4123–4130.

91 Fan, F., Feng, Z., and Li, C. (2010). UV Raman spectroscopic studies on active sites and synthesis mechanisms of transition metal-containing microporous and mesoporous materials. *Acc. Chem. Res.* 43: 378–387.

92 Tavernier, S.D. and Schoonheydt, R.A. (1991). Coordination of Cu^{2+} in synthetic mordenites. *Zeolites* 11: 155–163.

93 Groothaert, M.H., van Bokhoven, J.A., Battiston, A.A. et al. (2003). Bis(μ-oxo) dicopper in Cu-ZSM-5 and its role in the decomposition of NO: a combined in situ XAFS, UV–Vis–near-IR, and kinetic study. *J. Am. Chem. Soc.* 125: 7629–7640.

94 Smeets, P.J., Groothaert, M.H., and Schoonheydt, R.A. (2005). Cu based zeolites: a UV–vis study of the active site in the selective methane oxidation at low temperatures. *Catal. Today* 110: 303–309.

95 Zholobenko, V., Freitas, C., Jendrlin, M. et al. (2020). Probing the acid sites of zeolites with pyridine: quantitative AGIR measurements of the molar absorption coefficients. *J. Catal.* 385: 52–60.

96 Najmi, S., So, J., Stavitski, E. et al. (2021). In-situ IR spectroscopy study of reactions of C3 oxygenates on heteroatom (Sn, Mo, and W) doped BEA zeolites and the effect of Co-adsorbed water. *ChemCatChem* 13: 445–458.

97 Hadjiivanov, K. (2014). *Advances in Catalysis*, vol. 57 (ed. F.C. Jentoft), 99–318. Academic Press.

98 Bellmann, A., Rautenberg, C., Bentrup, U., and Brückner, A. (2020). Determining the location of Co^{2+} in zeolites by UV-vis diffuse reflection spectroscopy: a critical view. *Catalysts* 10: 123.

99 Ricchiardi, G., Damin, A., Bordiga, S. et al. (2001). Vibrational structure of titanium silicate catalysts. A spectroscopic and theoretical study. *J. Am. Chem. Soc.* 123: 11409–11419.

100 Fan, F., Sun, K., Feng, Z. et al. (2009). From molecular fragments to crystals: a UV Raman spectroscopic study on the mechanism of Fe-ZSM-5 synthesis. *Chem. Eur. J.* 15: 3268–3276.

101 Li, Y., Feng, Z., Lian, Y. et al. (2005). Direct synthesis of highly ordered Fe-SBA-15 mesoporous materials under weak acidic conditions. *Microporous Mesoporous Mater.* 84: 41–49.

102 Zhang, W.-H., Lu, J., Han, B. et al. (2002). Direct synthesis and characterization of titanium-substituted mesoporous molecular sieve SBA-15. *Chem. Mater.* 14: 3413–3421.

103 Hartmann, M. and Kevan, L. (1999). Transition-metal ions in aluminophosphate and silicoaluminophosphate molecular sieves: location, interaction with adsorbates and catalytic properties. *Chem. Rev.* 99: 635–664.

104 Li, C., Xiong, G., Xin, Q. et al. (1999). UV resonance Raman spectroscopic identification of titanium atoms in the framework of TS-1 zeolite. *Angew. Chem. Int. Ed.* 38: 2220–2222.

105 Guzman, J. and Gates, B.C. (2003). Supported molecular catalysts: metal complexes and clusters on oxides and zeolites. *Dalton Trans.* 3303–3318.

106 Fierro-Gonzalez, J.C., Kuba, S., Hao, Y., and Gates, B.C. (2006). Oxide- and zeolite-supported molecular metal complexes and clusters: physical characterization and determination of structure, bonding, and metal oxidation state. *J. Phys. Chem. B* 110: 13326–13351.

107 Wachs, I.E. (1996). Raman and IR studies of surface metal oxide species on oxide supports: supported metal oxide catalysts. *Catal. Today* 27: 437–455.

108 Stair, P. C. (2007). In *Advances in Catalysis*; B. C. Gates, H. Knözinger, eds.; Academic Press. 51: 75–98.

109 Kim, H., Kosuda, K.M., Van Duyne, R.P., and Stair, P.C. (2010). Resonance Raman and surface- and tip-enhanced Raman spectroscopy methods to study solid catalysts and heterogeneous catalytic reactions. *Chem. Soc. Rev.* 39: 4820–4844.

110 Langer, J., Jimenez de Aberasturi, D., Aizpurua, J. et al. (2020). Present and future of surface-enhanced Raman scattering. *ACS Nano* 14: 28–117.

111 Baker, G.A. and Moore, D.S. (2005). Progress in plasmonic engineering of surface-enhanced Raman-scattering substrates toward ultra-trace analysis. *Anal. Bioanal. Chem.* 382: 1751–1770.

112 Li, W., Gibbs, G.V., and Oyama, S.T. (1998). Mechanism of ozone decomposition on a manganese oxide catalyst. 1. In situ Raman spectroscopy and ab initio molecular orbital calculations. *J. Am. Chem. Soc.* 120: 9041–9046.

113 Lee, E.L. and Wachs, I.E. (2008). In situ Raman spectroscopy of SiO_2-supported transition metal oxide catalysts: an isotopic $^{18}O-^{16}O$ exchange study. *J. Phys. Chem. C* 112: 6487–6498.

114 Li, W., Meitzner, G.D., Borry, R.W., and Iglesia, E. (2000). Raman and X-ray absorption studies of Mo species in Mo/H-ZSM5 catalysts for non-oxidative CH_4 reactions. *J. Catal.* 191: 373–383.

115 Gao, J., Zheng, Y., Jehng, J.-M. et al. (2015). Identification of molybdenum oxide nanostructures on zeolites for natural gas conversion. *Science* 348: 686–690.

116 Hammond, C., Hermans, I., and Dimitratos, N. (2015). Biomimetic oxidation with Fe-ZSM-5 and H_2O_2? Identification of an active, extra-framework binuclear core and an Fe^{III}-OOH intermediate with resonance-enhanced Raman spectroscopy. *ChemCatChem* 7: 434–440.

117 Woertink, J.S., Smeets, P.J., Groothaert, M.H. et al. (2009). A $[Cu_2O]^{2+}$ core in Cu-ZSM-5, the active site in the oxidation of methane to methanol. *Proc. Natl. Acad. Sci. U. S. A.* 106: 18908–18913.

118 Henson, M.J., Mukherjee, P., Root, D.E. et al. (1999). Spectroscopic and electronic structural studies of the $Cu(III)_2$ bis-μ-oxo core and its relation to the side-on peroxo-bridged dimer. *J. Am. Chem. Soc.* 121: 10332–10345.

119 Himes, R.A. and Karlin, K.D. (2009). A new copper-oxo player in methane oxidation. *Proc. Natl. Acad. Sci. U. S. A.* 106: 18877–18878.

120 Vanelderen, P., Snyder, B.E.R., Tsai, M.-L. et al. (2015). Spectroscopic definition of the copper active sites in mordenite: selective methane oxidation. *J. Am. Chem. Soc.* 137: 6383–6392.

121 Rhoda, H.M., Plessers, D., Heyer, A.J. et al. (2021). Spectroscopic definition of a highly reactive site in Cu-CHA for selective methane oxidation: tuning a mono-μ-oxo dicopper(II) active site for reactivity. *J. Am. Chem. Soc.* 143: 7531–7540.

122 Czernuszewicz, R.S., Sheats, J.E., and Spiro, T.G. (1987). Resonance Raman spectra and excitation profile for bis(acetato)bis(hydrotripyrazolylborato) oxodiiron, a hemerythrin analog. *Inorg. Chem.* 26: 2063–2067.

123 Duer, M.J. (2008). *Solid State NMR Spectroscopy: Principles and Applications*. John Wiley & Sons.

124 Jakobsen, H.J. (2007). High speed MAS of half-integer quadrupolar nuclei in solids. *eMagRes* https://doi.org/10.1002/9780470034590.emrstm0212.

125 Ashbrook, S.E. and Sneddon, S. (2014). New methods and applications in solid-state NMR spectroscopy of quadrupolar nuclei. *J. Am. Chem. Soc.* 136: 15440–15456.

126 Klinowski, J. (2004). *New Techniques in Solid-State NMR*. Springer.

127 Brouwer, D.H., Darton, R.J., Morris, R.E., and Levitt, M.H. (2005). A solid-state NMR method for solution of zeolite crystal structures. *J. Am. Chem. Soc.* 127: 10365–10370.

128 Bakhmutov, V.I. (2011). Strategies for solid-state NMR studies of materials: from diamagnetic to paramagnetic porous solids. *Chem. Rev.* 111: 530–562.

129 Li, S., Lafon, O., Wang, W. et al. (2020). Recent advances of solid-state NMR spectroscopy for microporous materials. *Adv. Mater.* 32: 2002879.

130 Xu, J., Wang, Q., and Deng, F. (2019). Metal active sites and their catalytic functions in zeolites: insights from solid-state NMR spectroscopy. *Acc. Chem. Res.* 52: 2179–2189.

131 Xu, J., Wang, Q., Li, S., and Deng, F. (2019). *Solid-State NMR in Zeolite Catalysis*, vol. 103. Springer.

132 Qi, G., Wang, Q., Xu, J., and Deng, F. (2021). Solid-state NMR studies of internuclear correlations for characterizing catalytic materials. *Chem. Soc. Rev.* 50: 8382–8399.

133 Fernandez, C. and Pruski, M. (2012). Probing quadrupolar nuclei by solid-state NMR spectroscopy: recent advances. *Top. Curr. Chem.* 306: 119–188.

134 Siegel, R., Nakashima, T.T., and Wasylishen, R.E. (2004). Signal enhancement of NMR spectra of half-integer quadrupolar nuclei in solids using hyperbolic secant pulses. *Chem. Phys. Lett.* 388: 441–445.

135 Ono, Y. and Kanae, K. (1991). Transformation of butanes over ZSM-5 zeolites. Part 2.—Formation of aromatic hydrocarbons over Zn-ZSM-5 and Ga-ZSM-5. *J. Chem. Soc. Faraday Trans.* 87: 669–675.

136 Wang, L., Tao, L., Xie, M. et al. (1993). Dehydrogenation and aromatization of methane under non-oxidizing conditions. *Catal. Lett.* 21: 35–41.

137 Qi, G., Wang, Q., Xu, J. et al. (2016). Synergic effect of active sites in zinc-modified ZSM-5 zeolites as revealed by high-field solid-state NMR spectroscopy. *Angew. Chem. Int. Ed.* 55: 15826–15830.

138 Gao, P., Wang, Q., Xu, J. et al. (2018). Brønsted/Lewis acid synergy in methanol-to-aromatics conversion on Ga-modified ZSM-5 zeolites, as studied by solid-state NMR spectroscopy. *ACS Catal.* 8: 69–74.

139 Gao, W., Qi, G., Wang, Q. et al. (2021). Dual active sites on molybdenum/ZSM-5 catalyst for methane dehydroaromatization: insights from solid-state NMR spectroscopy. *Angew. Chem. Int. Ed.* 60: 10709–10715.

140 Ennaert, T., Van Aelst, J., Dijkmans, J. et al. (2016). Potential and challenges of zeolite chemistry in the catalytic conversion of biomass. *Chem. Soc. Rev.* 45: 584–611.

141 Dapsens, P.Y. and Mondelli, C. (2015). Design of Lewis-acid centres in zeolitic matrices for the conversion of renewables. *Chem. Soc. Rev.* 44: 7025–7043.

142 Kubička, D., Kubičková, I., and Čejka, J. (2013). Application of molecular sieves in transformations of biomass and biomass-derived feedstocks. *Catal. Rev.* 55: 1–78.

143 Moliner, M., Román-Leshkov, Y., and Davis, M.E. (2010). Tin-containing zeolites are highly active catalysts for the isomerization of glucose in water. *Proc. Natl. Acad. Sci. U. S. A.* 107: 6164–6168.

144 Román-Leshkov, Y., Moliner, M., Labinger, J.A., and Davis, M.E. (2010). Mechanism of glucose isomerization using a solid Lewis acid catalyst in water. *Angew. Chem. Int. Ed.* 49: 8954–8957.

145 Bermejo-Deval, R., Assary, R.S., Nikolla, E. et al. (2012). Metalloenzyme-like catalyzed isomerizations of sugars by Lewis acid zeolites. *Proc. Natl. Acad. Sci. U. S. A.* 109: 9727–9732.

146 Nikolla, E., Roman-Leshkov, Y., Moliner, M., and Davis, M.E. (2011). "One-pot" synthesis of 5-(hydroxymethyl)furfural from carbohydrates using tin-beta zeolite. *ACS Catal.* 1: 408–410.

147 Bare, S.R., Kelly, S.D., Sinkler, W. et al. (2005). Uniform catalytic site in Sn-beta-zeolite determined using X-ray absorption fine structure. *J. Am. Chem. Soc.* 127: 12924–12932.

148 Boronat, M., Concepción, P., Corma, A. et al. (2005). Determination of the catalytically active oxidation Lewis acid sites in Sn-beta zeolites, and their optimisation by the combination of theoretical and experimental studies. *J. Catal.* 234: 111–118.

149 Qi, G., Chu, Y., Wang, Q. et al. (2020). Gem-diol-type intermediate in the activation of a ketone on Sn-β zeolite as studied by solid-state NMR spectroscopy. *Angew. Chem. Int. Ed.* 59: 19532–19538.

150 Kolyagin, Y.G., Yakimov, A.V., Tolborg, S. et al. (2016). Application of ^{119}Sn CPMG MAS NMR for fast characterization of Sn sites in zeolites with natural ^{119}Sn isotope abundance. *J. Phys. Chem. Lett.* 7: 1249–1253.

151 Padovan, D., Botti, L., and Hammond, C. (2018). Active site hydration governs the stability of Sn-Beta during continuous glucose conversion. *ACS Catal.* 8: 7131–7140.

152 Gunther, W.R., Michaelis, V.K., Caporini, M.A. et al. (2014). Dynamic nuclear polarization NMR enables the analysis of Sn-Beta zeolite prepared with natural abundance Sn-119 precursors. *J. Am. Chem. Soc.* 136: 6219–6222.

153 Wolf, P., Valla, M., Rossini, A.J. et al. (2014). NMR signatures of the active sites in Sn-beta zeolite. *Angew. Chem. Int. Ed.* 53: 10179–10183.

154 Wolf, P., Liao, W.-C., Ong, T.-C. et al. (2016). Identifying Sn site heterogeneities prevalent among Sn-Beta zeolites. *Helv. Chim. Acta* 99: 916–927.

155 Wolf, P., Valla, M., Núñez-Zarur, F. et al. (2016). Correlating synthetic methods, morphology, atomic-level structure, and catalytic activity of Sn-β catalysts. *ACS Catal.* 6: 4047–4063.

156 Bermejo-Deval, R., Gounder, R., and Davis, M.E. (2012). Framework and extraframework tin sites in zeolite beta react glucose differently. *ACS Catal.* 2: 2705–2713.

157 Roy, S., Bakhmutsky, K., Mahmoud, E. et al. (2013). Probing Lewis acid sites in Sn-Beta zeolite. *ACS Catal.* 3: 573–580.

158 Tang, B., Dai, W., Wu, G. et al. (2014). Improved postsynthesis strategy to Sn-Beta zeolites as Lewis acid catalysts for the ring-opening hydration of epoxides. *ACS Catal.* 4: 2801–2810.

159 Wang, L., Zhang, J., Wang, X. et al. (2014). Creation of Brønsted acid sites on Sn-based solid catalysts for the conversion of biomass. *J. Mater. Chem. A* 2: 3725–3729.

160 Dijkmans, J., Dusselier, M., Gabriëls, D. et al. (2015). Cooperative catalysis for multistep biomass conversion with Sn/Al Beta zeolite. *ACS Catal.* 5: 928–940.

161 Qi, G., Wang, Q., Xu, J. et al. (2018). Direct observation of tin sites and their reversible interconversion in zeolites by solid-state NMR spectroscopy. *Commun. Chem.* 1: 22.

162 Rothwell, W.P., Shen, W.X., and Lunsford, J.H. (1984). Solid-state phosphorus-31 NMR of a chemisorbed phosphonium ion in HY zeolite: observation of proton-phosphorus-31 coupling in the solid-state. *J. Am. Chem. Soc.* 106: 2452–2453.

163 Lunsford, J.H., Rothwell, W.P., and Shen, W. (1985). Acid sites in zeolite Y: a solid-state NMR and infrared study using trimethylphosphine as a probe molecule. *J. Am. Chem. Soc.* 107: 1540–1547.

164 Rakiewicz, E.F., Peters, A.W., Wormsbecher, R.F. et al. (1998). Characterization of acid sites in zeolitic and other inorganic systems using solid-state ^{31}P NMR of the probe molecule trimethylphosphine oxide. *J. Phys. Chem. B* 102: 2890–2896.

165 Ganapathy, S., Gore, K.U., Kumar, R., and Amoureux, J.P. (2003). Multinuclear (Al-27, Si-29, Ti-47,Ti-49) solid-state NMR of titanium substituted zeolite USY. *Solid State Nucl. Magn. Reson.* 24: 184–195.

166 Zhuang, J., Yan, Z., Liu, X. et al. (2002). NMR study on the acidity of TS-1 zeolite. *Catal. Lett.* 83: 87–91.

167 Chen, L., Wang, Q., Hu, B. et al. (2010). Measurement of hetero-nuclear distances using a symmetry-based pulse sequence in solid-state NMR. *Phys. Chem. Chem. Phys.* 12: 9395–9405.

168 Seo, Y., Cho, K., Jung, Y., and Ryoo, R. (2013). Characterization of the surface acidity of MFI zeolite nanosheets by 31P NMR of adsorbed phosphine oxides and catalytic cracking of decalin. *ACS Catal.* 3: 713–720.

169 Zheng, A., Huang, S.-J., Liu, S.-B., and Deng, F. (2011). Acid properties of solid acid catalysts characterized by solid-state ^{31}P NMR of adsorbed phosphorous probe molecules. *Phys. Chem. Chem. Phys.* 13: 14889–14901.

170 Lewis, J.D., Ha, M., Luo, H. et al. (2018). Distinguishing active site identity in Sn-Beta zeolites using P-31 MAS NMR of adsorbed trimethylphosphine oxide. *ACS Catal.* 8: 3076–3086.

171 Yang, G., Pidko, E.A., and Hensen, E.J. (2013). The mechanism of glucose isomerization to fructose over Sn-BEA zeolite: a periodic density functional theory study. *ChemSusChem* 6: 1688–1696.

172 Levitt, M.H. (2013). *Spin Dynamics: Basics of Nuclear Magnetic Resonance.* John Wiley & Sons.

173 Li, S., Zheng, A., Su, Y. et al. (2007). Brønsted/Lewis acid synergy in dealuminated HY zeolite: a combined solid-state NMR and theoretical calculation study. *J. Am. Chem. Soc.* 129: 11161–11171.

174 Yu, Z., Zheng, A., Wang, Q. et al. (2010). Insights into the dealumination of zeolite HY revealed by sensitivity-enhanced ^{27}Al DQ-MAS NMR spectroscopy at high field. *Angew. Chem. Int. Ed.* 49: 8657–8661.

175 Gao, P., Xu, J., Qi, G. et al. (2018). A mechanistic study of methanol-to-aromatics reaction over Ga-modified ZSM-5 zeolites: understanding the dehydrogenation process. *ACS Catal.* 8: 9809–9820.

176 Haw, J.F., Goguen, P.W., Xu, T. et al. (1998). In situ NMR investigations of heterogeneous catalysis with samples prepared under standard reaction conditions. *Angew. Chem. Int. Ed.* 37: 948–949.

177 Hunger, M. and Weitkamp, J. (2001). In situ IR, NMR, EPR, and UV/Vis spectroscopy: tools for new insight into the mechanisms of heterogeneous catalysis. *Angew. Chem. Int. Ed.* 40: 2954–2971.

178 Ivanova, I.I. and Kolyagin, Y.G. (2010). Impact of in situ MAS NMR techniques to the understanding of the mechanisms of zeolite catalyzed reactions. *Chem. Soc. Rev.* 39: 5018–5050.

179 Zhang, W., Xu, S., Han, X., and Bao, X. (2012). In situ solid-state NMR for heterogeneous catalysis: a joint experimental and theoretical approach. *Chem. Soc. Rev.* 41: 192–210.

180 Anderson, M.W. and Klinowski, J. (1989). Direct observation of shape selectivity in zeolite ZSM-5 by magic-angle-spinning NMR. *Nature* 339: 200–203.

181 Carpenter, T.A., Klinowski, J., Tennakoon, D.T.B. et al. (1986). Sealed capsules for convenient acquisition of variable-temperature controlled-atmosphere magic-angle-spinning NMR-spectra of solids. *J. Magn. Reson.* 68: 561–563.

182 Qi, L., Alamillo, R., Elliott, W.A. et al. (2017). Operando solid-state NMR observation of solvent-mediated adsorption-reaction of carbohydrates in zeolites. *ACS Catal.* 7: 3489–3500.

183 Lin, M., Xia, C., Zhu, B. et al. (2016). Green and efficient epoxidation of propylene with hydrogen peroxide (HPPO process) catalyzed by hollow TS-1 zeolite: a 1.0 kt/a pilot-scale study. *Chem. Eng. J.* 295: 370–375.

184 Gamba, A., Tabacchi, G., and Fois, E. (2009). TS-1 from first principles. *J. Phys. Chem. A* 113: 15006–15015.

185 Gordon, C.P., Engler, H., Tragl, A.S. et al. (2020). Efficient epoxidation over dinuclear sites in titanium silicalite-1. *Nature* 586: 708–713.

186 Goguen, P.W., Xu, T., Barich, D.H. et al. (1998). Pulse-quench catalytic reactor studies reveal a carbon-pool mechanism in methanol-to-gasoline chemistry on zeolite HZSM-5. *J. Am. Chem. Soc.* 120: 2650–2651.

187 Haw, J.F. (1999). In situ NMR of heterogeneous catalysis: new methods and opportunities. *Top. Catal.* 8: 81–86.

188 Wang, C., Chu, Y., Zheng, A. et al. (2014). Frontispiece: new insight into the hydrocarbon-pool chemistry of the methanol-to-olefins conversion over zeolite H-ZSM-5 from GC-MS, solid-state NMR spectroscopy, and DFT calculations. *Chem. Eur. J.* 20: 12432–12443.

189 Wang, C., Yi, X., Xu, J. et al. (2015). Experimental evidence on the formation of ethene through carbocations in methanol conversion over H-ZSM-5 zeolite. *Chem. Eur. J.* 21: 12061–12068.

190 Hunger, M., Seiler, M., and Horvath, T. (1999). A technique for simultaneous in situ MAS NMR and on-line gas chromatographic studies of hydrocarbon conversions on solid catalysts under flow conditions. *Catal. Lett.* 57: 199–204.

191 Hunger, M. (2008). In situ flow MAS NMR spectroscopy: state of the art and applications in heterogeneous catalysis. *Prog. Nucl. Magn. Reson. Spectrosc.* 53: 105–127.

192 Wang, W., Seiler, M., Ivanova, I.I. et al. (2002). Formation and decomposition of N, N, N-trimethylanilinium cations on zeolite H–Y investigated by in situ stopped-flow MAS NMR spectroscopy. *J. Am. Chem. Soc.* 124: 7548–7554.

193 Seiler, M., Wang, W., Buchholz, A., and Hunger, M. (2003). Direct evidence for a catalytically active role of the hydrocarbon pool formed on zeolite H-ZSM-5 during the methanol-to-olefin conversion. *Catal. Lett.* 88: 187–191.

194 Zhou, X., Wang, C., Chu, Y. et al. (2019). Observation of an oxonium ion intermediate in ethanol dehydration to ethene on zeolite. *Nat. Commun.* 10: 1961.

9

Theoretical Calculations of Heteroatom Substituted Zeolites

Xin Yu[1,3], *Wenjun Dong*[1,2], *Wei Chen*[1], *and Anmin Zheng*[1]

[1] *State Key Laboratory of Magnetic Resonance and Atomic and Molecular Physics, National Center for Magnetic Resonance in Wuhan, Wuhan Institute of Physics and Mathematics, Innovation Academy for Precision Measurement Science and Technology, Chinese Academy of Sciences, West No.30 Xiao Hong Shan, Wuhan 430071, P.R. China*
[2] *School of Materials Science and Engineering, Zhengzhou University, No.100 Science Avenue, Zhengzhou, Henan 450001, P.R. China*
[3] *University of Chinese Academy of Sciences, No.1 Yanqihu East Rd, Beijing 100049, P.R. China*

9.1 Introduction

Since Taramasso et al. first discovered titanium silicalite-1 (TS-1) in 1983 [1], various zeolites with different substituted heteroatoms (Sn, Fe, Ga, Pb, etc.) have been reported. The vacant orbitals of these highly dispersed metal cations and alterable Brønsted/Lewis acidic strength in heteroatom substituted zeolites lead to the unique catalytic performance different from the traditional aluminosilicates. Thus, they gradually become a promising class of catalytic materials with widely applications under mild reaction conditions, especially the biomass conversion and alkene epoxidation.

There are two types of isomorphous substitution, including M^{3+}/Si^{4+} and M^{4+}/Si^{4+} (Scheme 9.1). In the first case, incorporating trivalent ions (B^{3+}, Al^{3+}, Ga^{3+}, Fe^{3+}, etc.) into zeolites will create negative charges that compensated by other extra cations (alkali metal ion or proton); therein, proton is frequently used as the Brønsted acidic site. Therefore, materials are equipped with Brønsted acidity by bridging hydroxyl group, and the acidic strength of Brønsted acidic site is also directly influenced by the dopant metal ions in M^{3+}-doped zeolites. Wang et al. have concluded that the Brønsted acidity increases in the order of B(OH)Si < In(OH)Si < Fe(OH)Si < Ga(OH)Si < Al(OH)Si based on the calculated O—H bond stretching frequency, ionicity of hydrogen, and geometrical structures [2, 3]. Herein, the acidic strength of Brønsted acidic site in M^{3+}-doped zeolites is strongly related to the strength of O(H)-M bond, and the stronger O(H)-M bond indicates more electron of O inclining to M but less electron to attract hydrogen (i.e. the smaller proton affinity), therefore, leading to the stronger acidic strength.

The second scenario involves replacing tetravalent silicon with metal ions that have the same covalent state (Ti^{4+}, Sn^{4+}, Zr^{4+}, Ge^{4+}, Pb^{4+}, etc.). With respect to this

Micro-Mesoporous Metallosilicates: Synthesis, Characterization, and Catalytic Applications, First Edition. Edited by Peng Wu and Hao Xu.

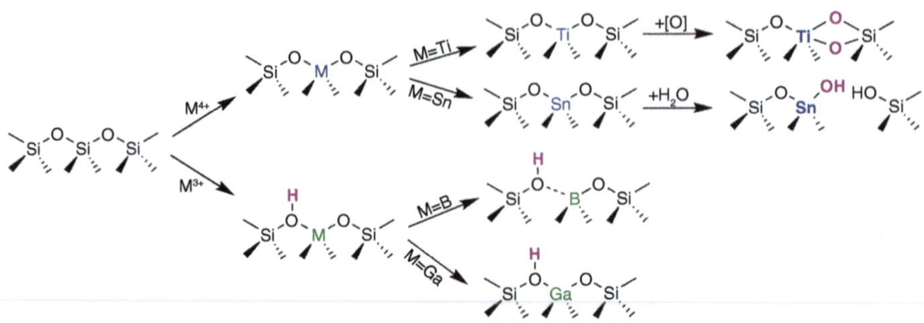

Scheme 9.1 Common heteroatom substituted zeolites.

case, due to the different electronegativity and coordination form between M^{4+} and Si^{4+}, the isomorphously substituted zeolites will possess Lewis acid properties, and the interactions of M^{4+}-substituted zeolites with probe molecules (NH_3 and H_2O) were used to evaluate Lewis acidity of the framework heteroatoms, and Lewis acidic strength of M^{4+}-substituted MFI zeolite is followed as Ge < Ti < Pb < Sn < Zr by Yang et al. [4, 5]. Companied by their water-resistant catalytic behavior, they are widely used to catalyze oxidations [6–8], isomerizations [9, 10], epimerizations [11, 12], and some other reactions [13–15]. Among them, Ti and Sn-containing zeolites are widely investigated.

In this chapter, we will mainly review the theoretical reports about M^{4+}-incorporated zeolites (such as TS-1, Sn-*BEA*) as well as M^{3+}-incorporated zeolites (such as BS-1). Specific contents will include the structure of active sites, the Brønsted/Lewis acidity of acid sites, the reaction mechanism about some industrial reactions catalyzed by these zeolites, and so on.

9.2 Ti-Doped Zeolites

When it comes to the Ti-incorporated zeolites, titanium silicalite-1 (TS-1) with the MFI topology is the most popular material for its excellent catalytic property in aqueous environment with hydrogen peroxide, which is even considered as one milestone in heterogeneous catalysis of zeolites [16]. Due to the relatively small pore diameter, this material is suitable for catalytic applications involving small- and medium-sized substrates. Moreover, the Ti-containing crystalline silicates with 12-membered ring (MR) channels (Ti-BEA (BEA), Ti-MCM-22 (MWW), and Ti-MCM-68 (MSE)) have been successfully synthesized and exhibited outstanding catalytic oxidation performance in processing the bulky molecules [17–20]. Among them, Ti-MCM-22 (Ti-MWW) zeolite, as a good catalyst for oxidative desulfurization, alkene epoxidation, and ketone ammoximation, is equipped with two independent interlayer and intralayer 10MR channels, one of which includes super cages of 12MR, plus side cups of 12MR on the external surface [21]. Ti-MCM-68 (Ti-MSE) containing a 12 × 10 × 10-ring pore system could catalyze the phenol oxidation and olefin epoxidation with high activity and selectivity [22]. In short, the development of Ti-containing zeolites extends the application of zeolites to selectively catalytic oxidations.

9.2.1 Preferred Tetrahedral (T) Sites for Substitution

The content of Ti in Ti-containing zeolites is relatively low, for example, less than 3.0 wt% in TiO_2 incorporated into MFI framework [23–25]. It is of great interest to explore the distribution of Ti atoms that has an impact on the stability and catalytic performance of materials because the different Ti distribution will lead to the distinct chemical environment and accessibility of active sites. However, the experimental characterization of Ti distribution was limited by the similarity between Ti and Si on electron density, as well as the long-range disorder of Ti distribution in the framework of Ti-containing zeolites. Therefore, the preferential T sites obtained by experimental studies are greatly different [23, 26–29]. To further correlate the structural-activity relationship of zeolites, several theoretical studies have been employed, but the theoretical determination of Ti distribution can only be limited to thermodynamics rather than kinetics because of the indistinct synthesis mechanism of Ti-containing zeolites.

The preferential T sites can be directly decided by the relative energy of all non-equivalent zeolites, but the substitution energy (ΔE_{sub}) has the higher physical significance to explain the Si/Ti substitution in zeolite based on the reaction energy of $Si(MFI) + Ti(OH)_4 \rightarrow Ti(zeolite) + Si(OH)_4$.

$$\Delta E_{sub} = E^{opt}(Ti - zeolite) - E^{opt}(Si - zeolite) + E[Si(OH)_4] - E[Ti(OH)_4]$$

$E^{opt}(Ti - zeolite)$ and $E^{opt}(Si - zeolite)$ are the energies of optimized Ti-zeolite and Si-zeolite structures, and $E[Si(OH)_4]$ and $E[Ti(OH)_4]$ are the energies of $Si(OH)_4$ and $Ti(OH)_4$.

Deka et al. [30] explored the Ti distribution of TS-1 zeolite by using a hybrid quantum mechanics/molecular mechanics (QM/MM) approach. The calculated results show that the substitution of Ti in MFI framework is thermodynamically favorable, and the substitution energy of the preferred T12 position is higher than the least stable position T2. Additionally, the energy gap among the four sites, including T12, T5, T7, and T8, are only ca. 10 kJ/mol, which indicated the thermodynamically favorable substitution of all these four T sites. However, the inconsistence between experiment and calculation on Ti location in TS-1 indicated that the distribution of Ti centers in TS-1 is not determined by the thermodynamic stability of the pure material but the kinetic preferability of Ti substituted process. Gamba et al. [31] performed periodic density function theory on two models with different Ti loading (1.35% and 2.70% in weight of TiO_2). At low Ti content (1.35%), either T3 or T4 site is preferred to be substituted, while at high Ti content (2.70%), the presence of a Ti–O–Ti bridge with the Ti—Ti distance of 3.5 Å is relatively stable, and one such structure is only 11.3 kJ/mol higher in energy with respect to the most stable T3-T4 with a distance of 7.3 Å. Moreover, the solid-state ^{17}O nuclear magnetic resonance experiments confirmed the existence of dinuclear Ti–O–Ti species in TS-1 zeolite by Gordon et al. [32]. Dong et al. [33] further explored the location of isolated Ti atoms

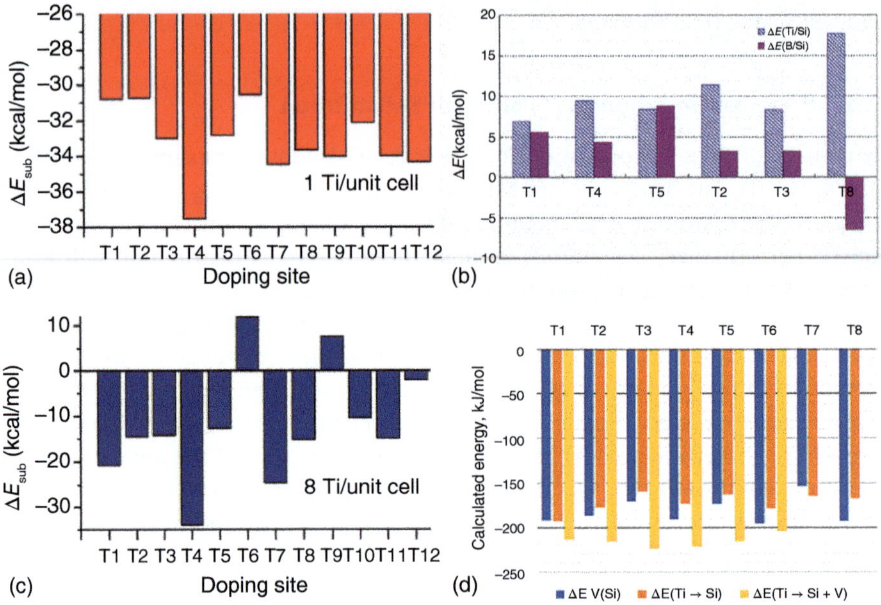

Figure 9.1 ΔE_{sub} for the substitution reaction by Ti replacing (a) one Si atom and (c) eight Si atoms in MFI. Source: Reproduced from Dong et al. [33]/with permission of American Chemical Society. (b) One Si atom in MWW. Source: Adopted from reference Zhou et al. [21] with the permission of Elsevier, copyright [2014]; (d) one Si atom in MSE unit cell. Source: Petkov et al. [22]/MDPI/CC BY 4.0.

in the high-Ti-loaded TS-1 zeolite (2.7 wt% in TiO_2). Surprisingly, they achieved a consistent picture between experiments and theoretical calculations that T4 is the most preferred site as displayed in Figure 9.1a,c. Recently, Xu et al. [34] used periodic Perdew–Burke–Ernzerhof (PBE) functional to calculate the substitution energy of TS-1, in which T1 site was the most energetically favorable site to be substituted by Ti. Subsequently, they partly compared the distribution of dinuclear titanium in TS-1, and T6/T9 sites are the favorite substitution sites for two adjacent Ti atoms.

Notably, when the Ti loading is <2 atoms/cell, an activated TS-1 sample (2.6 wt% in TiO_2) undergoes a *Pnma* to $P2_1/n$ at 170 K, and the number of independent Ti sites will be doubled in the silicalite framework [27]. T9, T10, and T15 are the preferential sites based on the substitution energies at the level of all electron periodic B3LYP-D2 by Signorile et al. [35]. Herein, the substitution energy was calculated according to the reaction of Silicalite-1 + TiO_2 (rutile) → TS-1 + SiO_2 (α-quartz).

Besides the common Ti^{IV} in Ti-containing zeolite, cyclic voltammetry experiments have confirmed the reduction of Ti^{IV} to Ti^{III} in zeolite lattices. Herein, the Ti^{III} sites exist in the Ti^{3+}-O^{1-} form, and the calculation of Ti^{III}-zeolite can set the spin multiplicities to triplet. Yang et al. [36] compared the distribution of Ti^{IV} and Ti^{III} ions in MCM-22 zeolite. Different from the relatively small difference of substitution energy in TS-1, the Ti^{IV} substitution on different T sites in MCM-22 zeolite varies greatly (about 100 kJ/mol). MCM-22 framework contains eight crystallographically distinct T sites, and the related stability of different Ti^{IV} sites follows the order: T8 > T2 > T1 > T5 > T4 > T6 > T7 > T3. The Ti^{III} substitution in MCM-22 zeolite shows the similar order to Ti^{IV} substitution, and T3 is the most probable site for both Ti^{IV} and Ti^{III} incorporation. Unlike other Ti-containing zeolites, the synthesis of Ti-MCM-22 requires the boric acid. Wang et al. [2] explored the influence of boron on Ti substitution in MCM-22 zeolite. Since the substitution of Si by B would result in a trigonal BO_3 unit and a terminal silanol group, the substitutions of Si/Ti in $(H_3SiO)_3$-T-OH and $(H_3SiO)_4$-T are studied. Results showed that the Ti substitution energy of the former is 17 kcal/mol lower than that of the latter, namely, the presence of boron is conducive to Ti insertion. Furthermore, Zhou et al. [21] compared the substitution energy of the Ti in perfect sites and defect sites in Ti-MWW zeolite. For the former, the T1 and T3 sites are energetically favorable for Ti substitution, which locates on the edge of 12MR cavity and the intersection of the intralayer 10MR channels, respectively. When the silanol nests exist in the lattice, incorporating Ti into zeolite framework is more endothermic, but the preferred T sites are the same. Additionally, they also explored the substitution of B for Si and further explained the role of boron on Ti substitution using some large models. The related stability of different B sites follows the order: T8 > T2 ≈ T3 > T4 > T1 > T5. Since T2 is adjacent to T3 and T4 is adjacent to T1, they confirm that Ti atoms are indeed preferred to replace Si atoms nearby B atoms (Figure 9.1b).

Petkov et al. [22] reported the stability of different Ti positions in the MCM-68 (MSE) zeolite. Among the eight different T sites, T1 is the most stable position for Ti substitution, followed by T6, T2, and T4 (Figure 9.1d). The energy gap between the preferred site and the least is about 33 kJ/mol and all substitution energies are exothermic. Furthermore, they also found that the formation of silanol nests in the vicinity of the Ti is able to stabilize the zeolite structure due to the reduced local framework strain.

In conclusion, incorporating Ti atom into different zeolite frameworks is usually energetically favorable, even though the framework structure would distort due to the longer Ti—O bond than Si—O bond. Besides, the distribution of Ti in titanium-containing silicates are not random, and they preferentially occupy some positions. Although theoretical studies with different calculation models and methods could not get a unified result on the preferred sites for Ti substitution, they can discriminate the substitution cases on different T sites qualitatively.

9.2.2 Lewis Acid

Lewis acidity of Ti-doped zeolite is an important factor to evaluate catalytic activity, and their Lewis acidic strength can be determined by the adsorption energy of probe molecules and calculated descriptors of Ti. In Ti-substituted zeolites, it is now widely accepted that the Ti^{4+} in the lattice is responsible for catalytic performance because the d^0 character of Ti^{4+} to accept electron results in the Lewis acid nature of Ti-substituted zeolites. Through NMR experiment and adsorption energy calculation of trimethylphosphine, Yang et al. [37] proposed two Lewis acidic sites in T12 of TS-1 zeolite, i.e. $Ti(OSiO_3)_4$ (closed site) and $(OSiO_3)_3TiOH$ (open site)

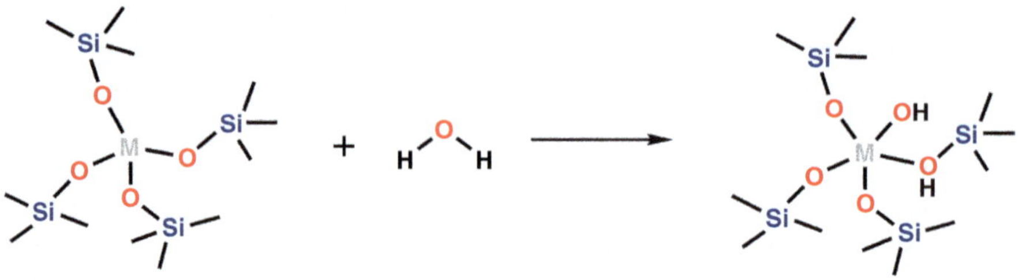

Scheme 9.2 Hydrolysis reaction of the M^{IV}-substituted zeolite.

(Scheme 9.2), but the further hydrolysis of Si–O–Ti links to form $(SiO)_2Ti(OH)_2$ or $(SiO)Ti(OH)_3$ species was not observed in TS-1 zeolite. Furthermore, Yang and co-workers [5] compared the acidity of TS-1 zeolite and other M^{4+}-doped MFI zeolites. Using different descriptors such as LUMO energy, fukui functions (f_M), absolute electronegativity (χ), and the adsorption structure of zeolites with NH_3, some slightly different sequences of Lewis acidity are obtained [38]. Due to the Lewis acidity of the M^{4+}-doped zeolite should be more accurately defined as the local sites around the M^{4+} ions rather than the M^{4+} ions merely, the acidic strength is more strongly related to the strength of M-N bond and the adsorption energies of NH_3 on M^{4+}-doped zeolites. Thus, the corresponding acidic strength order follows as silicalite-1 << Ge-MFI < Ti-MFI < Pb-MFI < Sn-MFI < Zr-MFI. Montejo-Valencia and co-workers [39] theoretically explored the acidity of a series of Ti-doped zeolites with different topologies (Ti-MFI, Ti-BEC, Ti-FAU, and Ti-BEA) by NH_3 binding energy, and Ti-MFI has the strongest Lewis acidity that evidenced by the low binding energy of NH_3, the low hardness, and the highest charge transfer from NH_3 to zeolite (Figure 9.2). Additionally, the formation of the open sites in the zeolites will give different strength of Lewis acid sites, and the lower LUMO energies of open sites in Ti-MFI, Ti-BEC, and Ti-BEA indicate the stronger ability to accept electrons, i.e. the higher Lewis acidity.

Yang et al. [36] analyzed the Lewis acidity of Ti^{IV} and Ti^{III} cations in MCM-22 zeolite by calculating the NH_3 adsorption of two cations. Usually, the Ti^{IV} could transfer to Ti^{III} ions, but the Ti^{III} site shows no Lewis acidity toward NH_3. It is evidenced by the shorter length of formed Ti^{IV}-N bond than Ti^{III}-N with adsorption energies of −23.9 and − 96.5 kJ/mol, respectively. Additionally, Ti^{III} site can interact with NH_3 through radical transfer from lattice O to N atoms, and its adsorption energy is −76.3 kJ/mol.

Clearly, the type of heteroatoms alone cannot account for all the observed differences in Lewis acidity, and there are further effects including the topology of zeolites, type of framework species, and choice of T sites.

9.2.3 Active Site with H_2O_2

It is well known that titanium-containing silicates are active in catalyzing some oxidative reactions using hydrogen peroxide as an oxidant. Actually, the oxidative activity of titanium-containing silicates is originated from the peroxide of Ti by

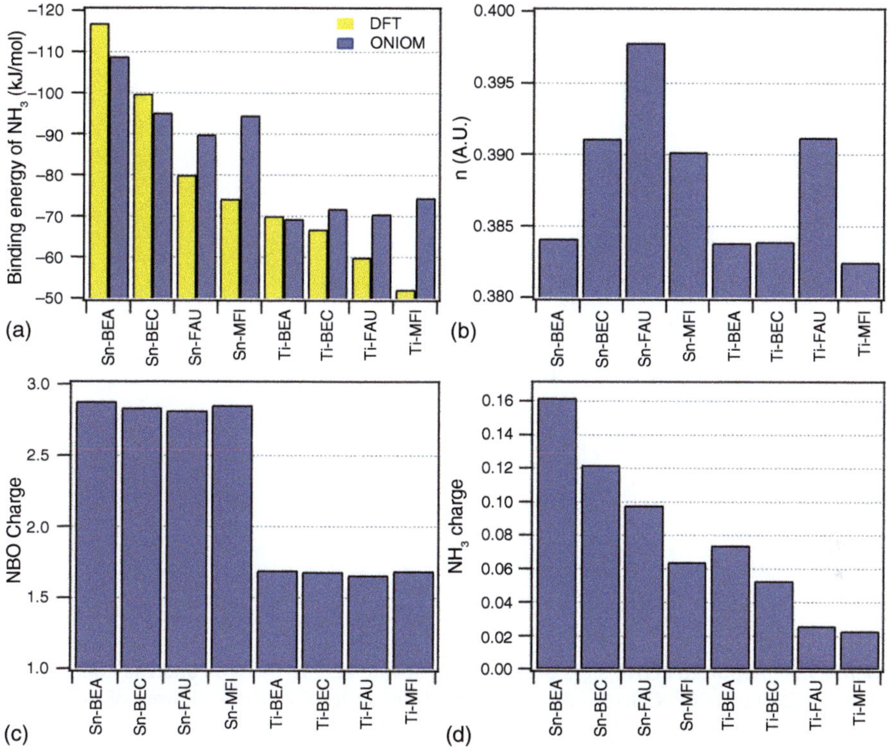

Figure 9.2 Electronic properties of metal-substituted zeolites: (a) binding energies of NH_3, (b) hardness, (c) NBO charge of the metal atom, and (d) charge of NH_3 upon adsorption. Source: Reproduced from Montejo-Valencia et al. [39]/with permission of American Chemical Society.

the reaction between defect/perfect Ti site and H_2O_2. However, discussing the structure of Ti active species in oxidations is relatively difficult because the H_2O_2 adsorption and reaction is strongly related to the local microenvironment of Ti. During last several decades, some theoretical studies about Ti active species in oxidative reactions were presented [40–42], while controversies still exist concerning the structure of these species. The common active sites reported by previous studies have been summarized in Scheme 9.3, Ti^{IV} in zeolite has three forms, i.e. [≡Ti–O–Si≡], [≡Ti–OH… (HO–Si≡)$_3$], and [≡Ti–O–Ti≡], respectively. These forms can be further oxidized to different active sites by H_2O_2. [≡Ti–O–Si≡] will be firstly hydrolyzed to [≡Ti–OH…HO–Si≡] and one O of H_2O_2 transfer to ≡Ti–OH with the formation of H_2O and two kinds of Ti-OOH sites, i.e. monodentate Ti-(η^1-OOH) or bidentate Ti-(η^2-OOH). The two active sites can be flexibly inter-transformed, and the different chemical environments around Ti atom and reaction environments will decide their stability and distribution. [≡Ti–O–Si≡] can also be oxidized to [≡Ti-(OO)-Si≡] by one H_2O_2 and [≡Ti-(OOH)$_2$-Si≡] by two H_2O_2 [44], as well as the direct oxidization to [≡Ti-(OOH)(OH)-Si≡]. [≡Ti–OH… (HO–Si≡)$_3$] is analogous to the silanol nest, which may originate from the desilication of [≡Ti–O–Si≡] and the

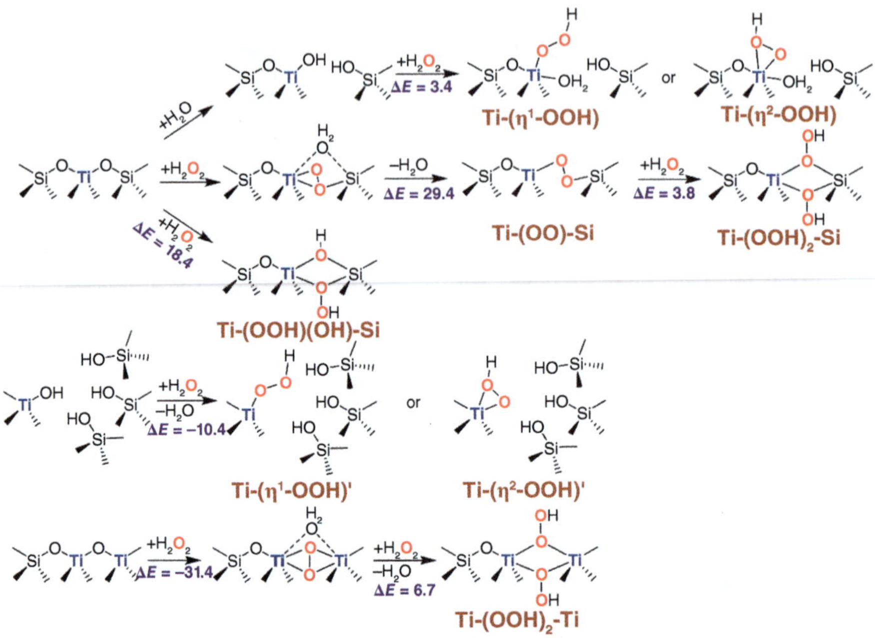

Scheme 9.3 Reported active sites of titanium silicalite zeolite by H_2O_2 and the energy difference (ΔE) of Ti-(η^1-OOH)/Ti-(η^2-OOH) and Ti-(η^1-OOH)'/Ti-(η^2-OOH)' were obtained by a 144T cluster model at ωB97X-D/6-31G(d, p) level of theory, Source: Reproduced from reference Nie et al. [43] with the permission of American Chemical Society, copyright [2019], and the ΔE of Ti-$(OOH)_2$-Si, Ti-(OOH)(OH)-Si, and Ti-$(OOH)_2$-Ti were obtained by periodic model at RPBE-D3/PAW level of theory. Source: Adapted from [44].

defects formed in titanium-silicalite synthesis, and the oxidization of [≡Ti–OH… (HO–Si≡)$_3$] will also lead to the formation of Ti-(η^1-OOH) or Ti-(η^2-OOH), but two additional silanols around Ti-(OOH) may affect the reaction process, named as Ti-(η^1-OOH)' or Ti-(η^2-OOH)'. Herein, Ti–OOH is the generally accepted active site responsible for the catalytic activity, but it is hard to experimentally determine its origin. From the perspective of theoretical calculations, the constructions of Ti-(η^1-OOH) and Ti-(η^2-OOH) in zeolite are too congested to allow the reaction occurring, and Ti-(η^1-OOH)' and Ti-(η^2-OOH)' are considered as the active sites in many computational studies of Ti-containing zeolites. Moreover, the formation of Ti-(η^1-OOH) and Ti-(η^2-OOH) by [≡Ti–O–Si≡] was required to overcome a barrier of 23.8 kcal/mol with the increasing energy of 3.4 kcal/mol, but there is only a much smaller barrier of 7.2 kcal/mol to overcome in the formation of Ti-(η^1-OOH)' and Ti-(η^2-OOH)' [43].

Therefore, the formation of Ti-(η^1-OOH)' and Ti-(η^2-OOH)' is thermodynamically and kinetically more favorable than Ti-(η^1-OOH) and Ti-(η^2-OOH), originating from [≡Ti–OH… (HO–Si≡)$_3$] and [≡Ti–O–Si≡], respectively. The formation of [≡Ti-(OO)-Si≡] and [≡Ti-$(OOH)_2$-Si≡] was calculated to be thermodynamically unfavorable based on ΔE [32], and the formation of [≡Ti-(OO)-Si≡] needs to overcome a barrier of 25.4 kcal/mol, which is much larger than the formation of Ti-(η^1-OOH)' (8.1 kcal/mol) [45].

9.2.4 Reaction Mechanism

9.2.4.1 Epoxidation of Olefins

As one of the most important applications of titanium silicalite zeolites, ethene and propene can be effectively oxidized to oxirane and epoxypropane as the useful commercial products in the petrochemical and fine chemicals industry. The general route of this epoxidation on different active sites of titanium silicalite zeolites is the reaction of ethene/propene and peroxide as shown in Scheme 9.4. In all Ti–OOH active sites, ethene and propene will be epoxidized by the peroxy oxygen atom closer to Ti, and the formation of oxirane and epoxypropane will be accompanied by the recovering of Ti–OH site [43]. However, the epoxidation on Ti-$(OOH)_2$-Si and Ti-$(OOH)_2$-Ti will optimally consume the peroxy oxygen atom closer to H, and the formed Ti-(OOH)(O)-Si and Ti-(OOH)(O)-Ti can further epoxidize ethene and propene [32]. In Ti–OO–Si active site, ethene and propene will be epoxidized by one of two peroxy oxygen atoms with the recovering of perfect [≡Ti–O–Si≡]

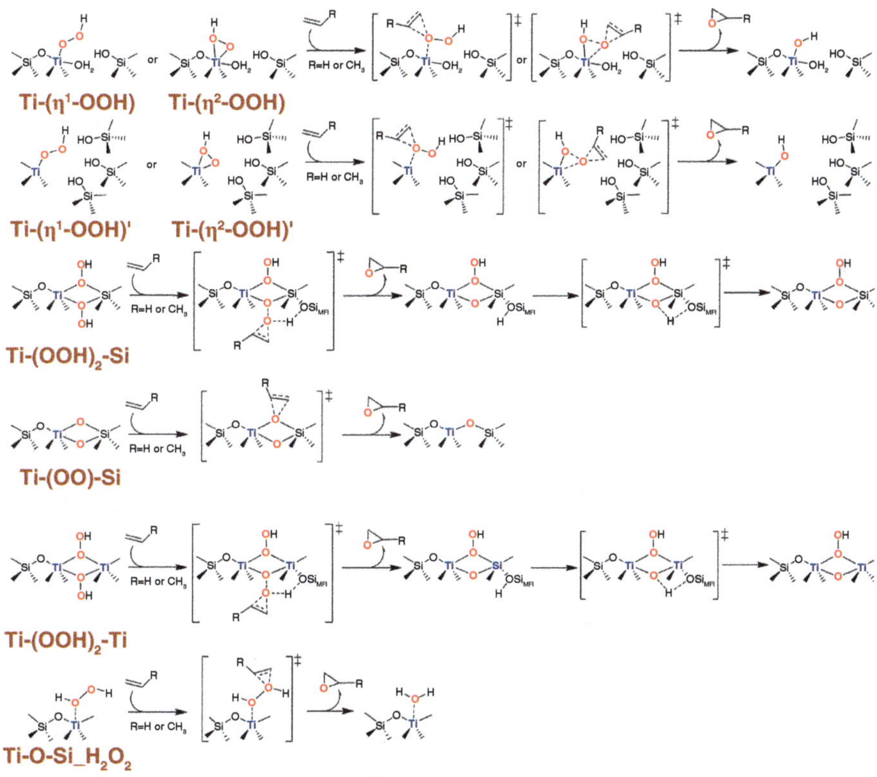

Scheme 9.4 Mechanism for the epoxidation of ethene and propene on different active sites of titanium silicalite zeolites.

structure [44]. Besides the oxidized sites in titanium silicalite zeolites, the reaction between adsorbed H_2O_2 and ethene can also occur by overcoming a barrier of 20.1 kcal/mol [46], and ethene will be epoxidized by the free peroxy oxygen atom with the formation of H_2O and oxirane.

The reported energy barriers of ethene and propene epoxidations in titanium silicalite zeolites are summarized in Table 9.1, and the different calculated methods and models lead to great difference on barriers. But the ethene epoxidation is basically hard to occur than propene epoxidation because of the weakening effect of methyl group to double bond. Notably, the difference of ethene epoxidation barriers on the different active sites is not larger than that of propene epoxidation. Ti-(η^1-OOH) shows the highest activity to ethene epoxidation with a barrier of 17.4 kcal/mol, and Ti-(η^1-OOH)', Ti-$(OOH)_2$-Ti, and Ti-(η^2-OOH)' have the higher catalytic activity than other active sites by overcoming a barrier of 8.1–11.1 kcal/mol in propene epoxidation. But the low activity of both Ti-$(OOH)_2$-Si and Ti-(OOH)(OH)-Si to propene epoxidation explains their low possibility to be the active sites. More interestingly, the barriers were slightly influenced by the pore confinement effect of TS-1 based on the calculated results of 15T and 145T cluster models because the non-covalent interaction between framework atom and the small substrate of both H_2O_2 and ethene/propene is relative weak. Based on all obtained barriers

Table 9.1 Energy barriers (kcal/mol) of alkenes epoxidation over TS-1.

Method	T site	Model	Active sites	Alkenes	Barriers[a)]	Reference
BPW91/LANL2DZ/D95	T6	8T	Ti-(η^1-OOH)	Ethene	17.4	[47]
BP	None	5T	Ti-HOOH		20.1	[46]
B3LYP/6-31G(d)	T12	Embedded 5T	Ti-(η^2-OOH)		26.1	[48]
	T12	Bare 5T	Ti-(η^2-OOH)		23.1	[48]
GGA/BP/DNP	T1	5T	Ti-OO-Si		21.0	[44]
GGA/B3LYP/DNP	None	4T	Ti-(η^1-OOH)		23.0	[49]
B3LYP/6-31G(d, p)	T10	144T	Ti-(η^1-OOH)	Propene	20.1	[43]
	T10	16T	Ti-(η^1-OOH)		18.4	[43]
	T10	143T	Ti-(η^1-OOH)'		8.1	[43]
	T10	15T	Ti-(η^1-OOH)'		13.5	[43]
ωB97XD/6-31G(d, p)	T10	144T	Ti-(η^1-OOH)		12.6	[43]
	T10	16T	Ti-(η^1-OOH)		17.4	[43]
	T10	143T	Ti-(η^1-OOH)'		13.5	[43]
	T10	15T	Ti-(η^1-OOH)'		11.1	[43]
RPBE-D3/PAW	T7/T11	Periodicity	Ti-$(OOH)_2$-Ti		8.1	[32]
	T7/T11	Periodicity	Ti-$(OOH)_2$-Si		33.2	[32]
	T7/T11	Periodicity	Ti-(OOH)(OH)-Si		33.6	[32]
	T7/T11	Periodicity	Ti-(η^2-OOH)'		8.3	[32]

a) Barrier was defined to the maximum energy span on potential energy surfaces.

(ca. 20 kcal/mol) of ethene/propene epoxidation by all possible active sites on TS-1, the epoxidation process is feasible at relative mild temperature as consistent to experimental temperature of <100 °C.

9.2.4.2 Ammoximation and Oxidation of Cyclohexanone

The liquid-phase ammoximation of cyclohexanone over TS-1 in the presence of NH_3 and H_2O_2 to synthesize oxime is one of important processes attracting the industrial interests as cyclohexanone oxime is the key intermediate in the manufacture of caprolactam which can be used to produce Nylon-6. Different from the alkenes epoxidation occurring in the neighboring of Ti in TS-1 zeolites, the key step of cyclohexanone ammoximation is the formation of hydroxylamine (NH_2OH), and the reaction of cyclohexanone and NH_2OH to cyclohexanone oxime is spontaneous and non-catalytic. Moreover, the molecule size of cyclohexanone (about 4.3×5.0 Å) is only slightly smaller than the sizes of TS-1 pores of $5.4 \sim 5.6$ Å (straight channel) and $5.1 \sim 5.5$ Å (zigzag channel). So it is expected that NH_2OH can diffuse more easily out of the pores into the solution than cyclohexanone diffusing into the pores since the molecular size of NH_2OH (2.5 Å) is much smaller than the pore sizes of TS-1 [50]. Therefore, the mechanistic study to cyclohexanone ammoximation in TS-1 was focused on the reaction of $H_2O_2 + NH_3 \rightarrow NH_2OH + H_2O$ catalyzed by TS-1 [44, 45, 50]. The DFT calculation reported by Jakkapan and Jumras indicates that the defective Ti species (Ti-OH) in the TS-1 zeolite are the dominant active sites for H_2O_2 decomposition rather than perfect Ti (Ti-OO-Si) species, leading to the formation of Ti–OOH species as oxygen-donating intermediates for NH_3 oxidation reaction [45]. The potential energy surfaces for the ammonia oxidation over the TiOOH species and the catalytic effect of water were fully investigated, consisting of three proposed mechanisms, i.e. Route A, Route B, and Route A + H_2O as displayed in Figure 9.3. The most favored pathway was found to be the Route A + H_2O in Figure 9.3b: the adsorption of ammonia (NH_3/Ti-η^1-OOH) → ammonia oxide complex (NH_3O/Ti-OH) → hydrated→titanium-oxyamine species (H_2O/Ti-ONH_2) → hydroxylamine product (NH_2OH/Ti-OH), in which the highest energy barrier is 16.3 kcal/mol. Besides the hydrolysis of titanium-oxyamine species, NH_2OH was also generated through the second H_2O_2 decomposition over the titanium-oxyamine species, whereas the activation energy for this step was slightly decreased to be 15.7 kcal/mol. Notably, the involved H_2O in reaction mechanism will significantly reduce the activation energy of NH_2OH formation over both Ti-OO-Si and Ti-OOH active sites. In this catalytic reaction, the water and the Ti-OH play two roles in the formation of NH_2OH: (i) the proton transfer of ammonia oxide in the presence of the water molecule as the proton-shuttling center proceeds through a much lower activation barrier than that for the self-transformation process; (ii) the Ti-OH moiety could participate as the proton acceptor in the proton transfers, leading to the formation of oxyamine-Ti species and the water molecule. The analogous promoting effect of water was also proposed by Chu et al. [50], and the calculated potential energy surface of the non-catalytic oximation of NH_2OH and cyclohexanone indicated that two barriers of 12.8 kcal/mol and 23.1 kcal/mol are required to overcome with the assistance of five waters.

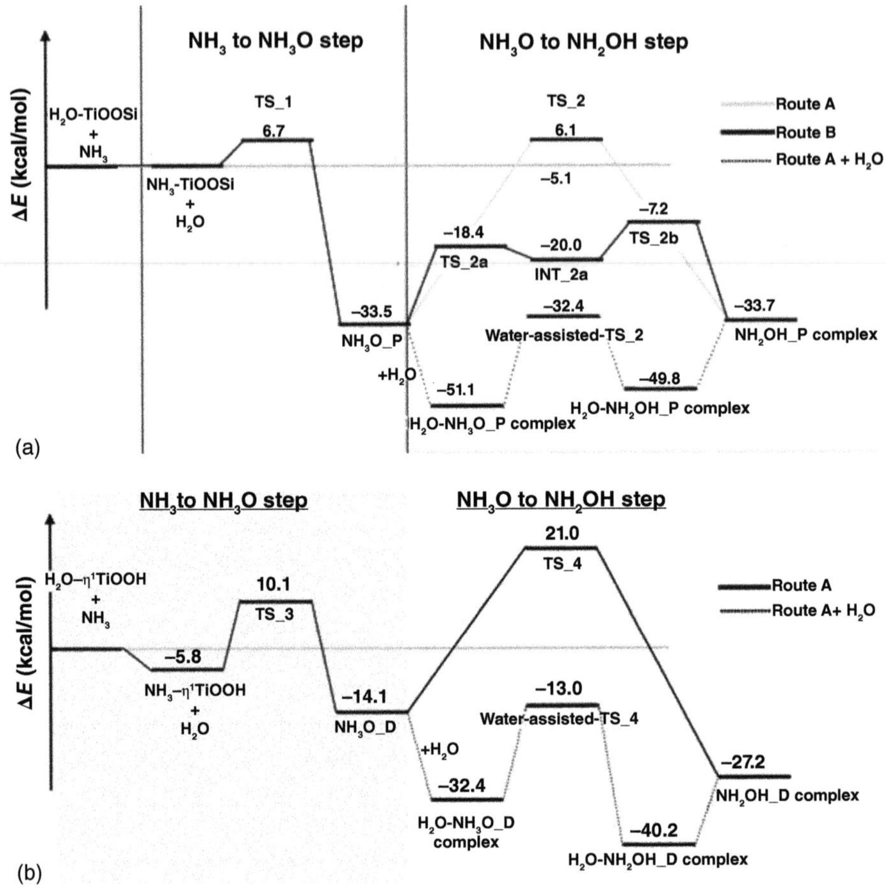

Figure 9.3 Potential energy surface of NH_3 oxidation to NH_2OH over the (a) perfect (Ti-OO-Si) and (b) defective Ti species (Ti-OH) of TS-1 zeolite. Source: Adopted from reference Sirijaraensre and Limtrakul [45] with permission from the Royal Society of Chemistry, copyright [2013].

Besides cyclohexanone ammoximation, cyclohexanone can also be directly oxidized to caprolactone, 6-hydroxyhexanoic acid, and 6-hydroxyhexanoic acid by TS-1 zeolite via Baeyer–Villiger (BV) oxidation. This oxidation is an important organic reaction by which esters or lactones are produced from ketones, and Sn-BEA zeolite was reported as the active catalyst by Corma et al. [51]. Xia et al. [52] found TS-1 is also a good catalyst for BV oxidation of cyclohexanone and proposed a different mechanism for the heterogeneous oxidation of cyclohexanone catalyzed by TS-1 zeolite (Figure 9.4). Given that TS-1 zeolite has a relatively weaker Lewis acidity than Sn-BEA, they firstly compared the adsorption of cyclohexanone and hydrogen peroxide on Ti center. The adsorption energy of the unconstrained $Ti(OSiH_3)_4$ cluster is −32.3 kcal/mol for cyclohexanone and − 91.9 kcal/mol for hydrogen peroxide, which indicated that cyclohexanone is less likely to be absorbed on the Ti active sites than

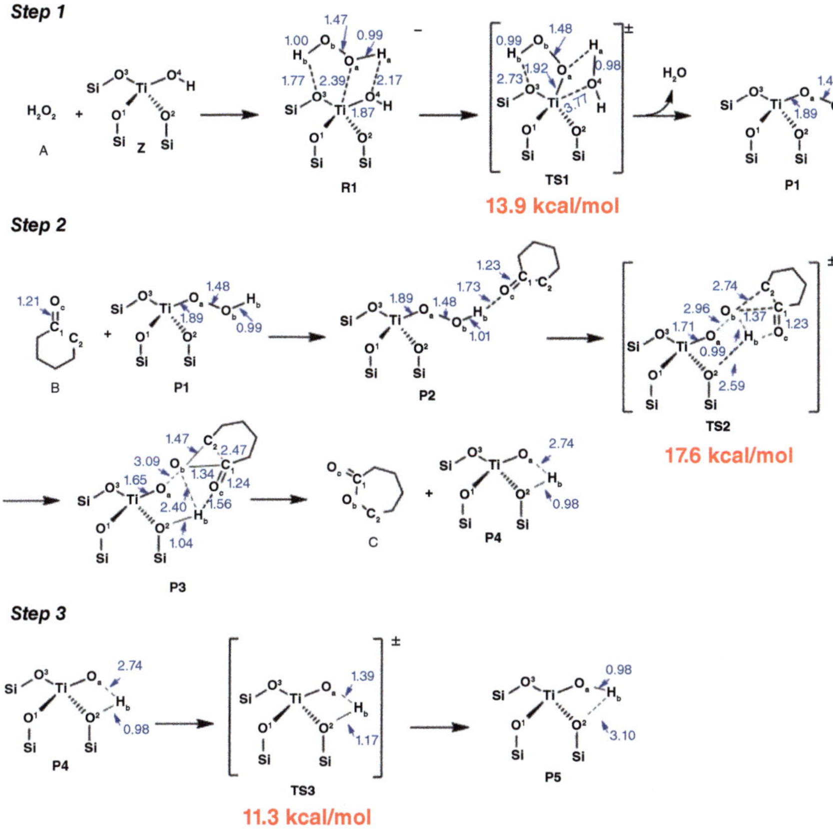

Figure 9.4 The mechanism of the oxidation of cyclohexanone catalyzed by TS-1 zeolite using hydrogen peroxide as the oxidant. Source: Adopted from reference Xia et al. [52] with the permission of Elsevier, copyright 2015.

H_2O_2. Therefore, the BV oxidation of cyclohexanone will start from the formation of Ti-OOH by H_2O_2 and defective Ti-OH site as the Step 1. Next, cyclohexanone was adsorbed on Ti-OOH active site by a hydrogen bond interaction between H of the Ti-OOH group and the carbonyl group of cyclohexanone. Subsequently, the terminal O atom of Ti-OOH with a strong negative charge can impose a nucleophilic attack of the C—C bond close to carbonyl group of cyclohexanone by overcoming a barrier of 17.6 kcal/mol, and the caprolactone molecule will be generated by this oxygen insertion into cyclohexanone. Finally, the Ti-OH will be recovered by a hydrogen transfer reaction.

9.2.4.3 Oxidation Desulfurization Reactions

Oxidation desulfurization (ODS) technology is considered as a promising process to gain fuels with ultra-low sulfur, especially by using the titanosilicate/H_2O_2 catalytic system. Organic sulfur compounds (like thiophene (Tp), benzothiophene (BT), and dibenzothiophene (DBT)) in gasoline and diesel can be selectively oxidized under mild conditions to the corresponding sulfoxide and sulfone,

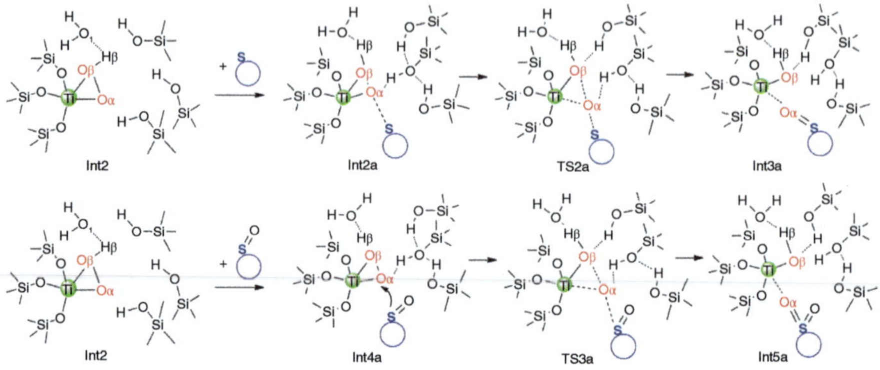

Figure 9.5 Reaction pathway of the oxidative desulfurization with H_2O_2 over the defect model, Source: Reproduced from Wang et al. [55]/with permission of American Chemical Society.

which can be removed readily by further extraction or adsorption methods to achieve deep removal. Due to the large size of organic sulfur compounds, TS-1 zeolite with the small pore size cannot accommodate these bulk compounds. To overcome this shortcoming, Ti-containing zeolites with 12-ring pore channels, such as Ti-BEA [53] and Ti-MWW [54], have been developed successively. Like the mechanism of above oxidations in Ti-containing zeolites, the critical step of oxidation desulfurization is the reaction between Ti peroxide site and organic sulfur compounds. As summarized in Scheme 9.3, the perfect [≡Ti-O-Si≡] site and defective [≡Ti-OH… (HO-Si≡)$_3$] site of Ti-MWW were considered as the active sites for oxidation desulfurization reactions by Wang et al. [55]. In the perfect model, [≡Ti-O-Si≡] is not oxidized to [≡Ti-OO-Si≡] by H_2O_2. Ti-O-Si provides the active site to adsorb H_2O_2, and the subsequent oxidations will occur around Ti-O-Si. The barriers of Tp, BT, and DBT oxidation desulfurization over this active site are 47.1, 50.0, and 50.8 kcal/mol, which are even higher than the non-catalytic process, i.e. 48.6, 47.2, and 46.9 kcal/mol. However, in the defective model, the reaction is initiated by forming a hydroperoxy intermediate Ti-(η^2OOH), and the barriers of Tp, BT, and DBT oxidation desulfurization were greatly decreased to 16.2, 13.8, and 14.3 kcal/mol, respectively. The reaction pathway of aromatic sulfides oxidation desulfurization was displayed in Figure 9.5, one aromatic sulfide compound will directly interact with the O_α of Ti-(η^2OOH) (Int2) with the help of additional H_2O, and the formed sulfoxide complex Int3a will be further oxidized to the final product sulfones by reacting with Ti-(η^2OOH) (Int2). Herein, the first step of oxidation desulfurization of Tp, BT, and DBT is slightly more difficult than the second step based on the differences on barriers of 1.4, 3.4, and 4.1 kcal/mol. Subsequently, they compared the reactivity for sulfides oxidation with or without any catalyst by the apparent activation energies. The apparent activation energies of different sulfides (Tp, BT, and DBT) are very close, and the catalysis of Ti-(η^2OOH) over Ti-BEA will reduce the activation energy from 46.9 to 48.6 kcal/mol (without any catalyst) to 27.4 ~ 28.5 kcal/mol [56].

Due to the unique MWW framework affording enough reaction spaces to process bulky aromatic sulfides, Ti-MWW zeolite exhibits outstanding catalytic oxidative performance on the refractory sulfur compounds. Bian et al. [57, 58] insightfully revealed the catalytic activity of different active centers and the reaction mechanism of oxidative desulfurization over Ti-MWW zeolite. Different solvents were employed to explore their influences on the catalytic activity of Ti-OOH in Tp oxidative desulfurization, and the energy barriers are in the order of Ti-(η^2-OOH)-CH_3CN (13.7 kcal/mol) < Ti-(η^2-OOH)-CH_3COCH_3 (13.8 kcal/mol) < Ti-(η^2-OOH)-CH_3OH (13.9 kcal/mol) < Ti-(η^2-OOH)-H_2O (14.0 kcal/mol) < Ti-(η^2-OOH) (16.0 kcal/mol) < Ti-(η^1-OOH) (23.6 kcal/mol). Obviously, Ti-(η^2-OOH) has much higher catalytic activity than Ti-(η^1-OOH), and the solvent can slightly enhance the catalytic activity of Ti-(η^2-OOH). Notably, the linear correlation between energy barriers and the Hirshfeld charge of O_α indicates that the promotion of solvent to the catalytic activity is induced by electron-withdrawing effect of solvent molecules. Moreover, the first step energy barriers of the four aromatic sulfur compounds are consistent with their sizes, i.e. Tp (16.0 kcal/mol) > BT (15.0 kcal/mol) > DBT (11.5 kcal/mol) > 4,6-DMDBT (8.0 kcal/mol). Consistently with the reaction over Ti-BEA, the second step of oxidative desulfurization is easier than the first step. Subsequently, the formation energy and energy barrier of different active sites are calculated to determine the real active center of oxidative desulfurization over Ti-MWW zeolite [58]. The thermodynamical stability of Ti-(η^2-OOH)-H_2O at both T1 and T3 sites is higher than Ti-(η^1-OOH) and Ti-(η^2-OOH) based on the formation energy, and Ti-(η^2-OOH)-H_2O also has the highest catalytic activity due to its lowest energy barrier of oxidative desulfurization.

9.3 Sn-Doped Zeolites

Sn-containing zeolites have been widely applied to Baeyer–Villiger oxidation [51], Meerwein–Ponndorf–Verley (MPV) reductions, Oppenauer oxidations [59–61], and some biomass conversions [62–65], including the isomerization of glucose to fructose, isomerization of glyceraldehyde to dihydroxyacetone (DHA), and C—C coupling between DHA and formaldehyde. Especially, Sn-BEA zeolite provides the large pores to process the adsorption, reaction, and diffusion of bulky reactant species. Theoretical calculations play a significant role in understanding the reaction mechanism of the above catalytic reactions, catalytic activity, active sites of Sn-containing zeolites, and structure–performance relationship between acidity/zeolite topology and catalytic performance [66].

9.3.1 Preferred Substitution T Sites and Acidity

The determination of the location and distribution of Sn in zeolites is the first step to ensure the reasonability and accuracy of subsequent mechanistic study of different catalytic reactions. In Sn-zeolites, two separate framework species, a fully framework connected (closed) $(SiO)_4Sn$ and a partially hydrated (open) $(SiO)_3SnOH$

species, are assumed as active sites (Scheme 9.3). Boronat et al. reported two corresponding bands at 2316 and 2308 cm^{-1} in the deuterated acetonitrile adsorbed IR spectra for Sn-BEA, confirming their existences [67]. Due to the coexistence of close $(SiO)_4Sn$ site and hydrated $(SiO)_3SnOH$ site in Sn-zeolite, $(SiO)_4Sn$ site is a Lewis acid but $(SiO)_3SnOH$ can act as both Lewis acid and Brønsted acid [39]. With hydrated $(SiO)_3SnOH$ site, Sn-BEA shows the much higher Brønsted acidic strength than Sn-FAU, Sn-BEC, and Sn-MFI based on the calculated deprotonation energy. Because the charge transfer from small NH_3/H_2O to the zeolite is proportional to the Lewis acidity of the system, the Lewis acidity of different close T sites in Sn-BEA follow the order of T7 > T8 > T9 > T5 > T6 > T3 > T1 > T2 > T4 based on their adsorption energy to water [4], but the difference of water adsorption energy on different T sites is smaller than 14 kJ/mol. Analogously, for different Sn-containing zeolites, the difference of Lewis acidity is smaller than 20 kJ/mol, according to the NH_3 adsorption energy on T2 site [39], and the Lewis acidic strength of the closed site in Sn-BEA is also stronger than Sn-MFI, Sn-FAU, and Sn-BEC. Due to the disturbance of -OH group in open $(SiO)_3SnOH$ to the adsorption of NH_3, the binding energy of NH_3 on a hydrolyzed metal-substituted zeolite cannot correlate to the Lewis acidity of the system, and the LUMO energy is the better descriptor to reflect the Lewis acidity of $(SiO)_3SnOH$ in different Sn-zeolites based on their ability to accept electron pair, i.e. the lower LUMO energy and the higher Lewis acidity. In this context, it is interesting to compare the Lewis acidic strength of open and closed sites of Sn-zeolites. The open sites of Sn-BEA have higher acidic strength than its closed sites, but in contrast to Sn-BEC, Sn-FAU, and Sn-MFI. Despite closed and open Sn sites of Sn-zeolites, Sn-BEA shows the unique acidic performance compared to other Sn-zeolites, exhibiting high activity on broad application of different catalytic reactions.

Except the mono Sn substitution, double Sn substitution is also possible in Sn-zeolites, and a double Sn substitution in the BEA framework was found to have the higher Lewis acidity than the mono Sn substitution based on the lower NH_3 bind energy and the lower LUMO energy [4]. Moreover, the hydrolysis of double Sn substitution in Sn-BEA is thermodynamically more favorable than the mono one, but the hydrated $(SiO)_3SnOH$ of double substitution in Sn-BEA has the similar Brønsted acidic strength to the mono one [39].

9.3.2 Reaction Mechanism

9.3.2.1 Glucose Isomerization to Fructose and Epimerization to Mannose

With the closed Sn sites, the isomerization of glucose to fructose over Sn-zeolites was consisted by eight steps as displayed in Figure 9.6a [68]: (i) O_1H of glucose was deprotonated by a lattice oxygen with the Sn-(H)OSi bond rupture; (ii) H of Sn...(H)OSi was transferred to O_5 of glucose (hydride shift) with the ring-opening of glucose and the recovery of Sn-OSi bond; (iii) the second O_2H group of glucose will approach Sn site with the bidentate coordination to the Lewis acidic site, and ready to the second proton transfer; (iv) the proton transfer from O_2 to the other Sn-OSi bond, (v) intramolecular hydrogen transfer of two carbon in glucose; (vi)

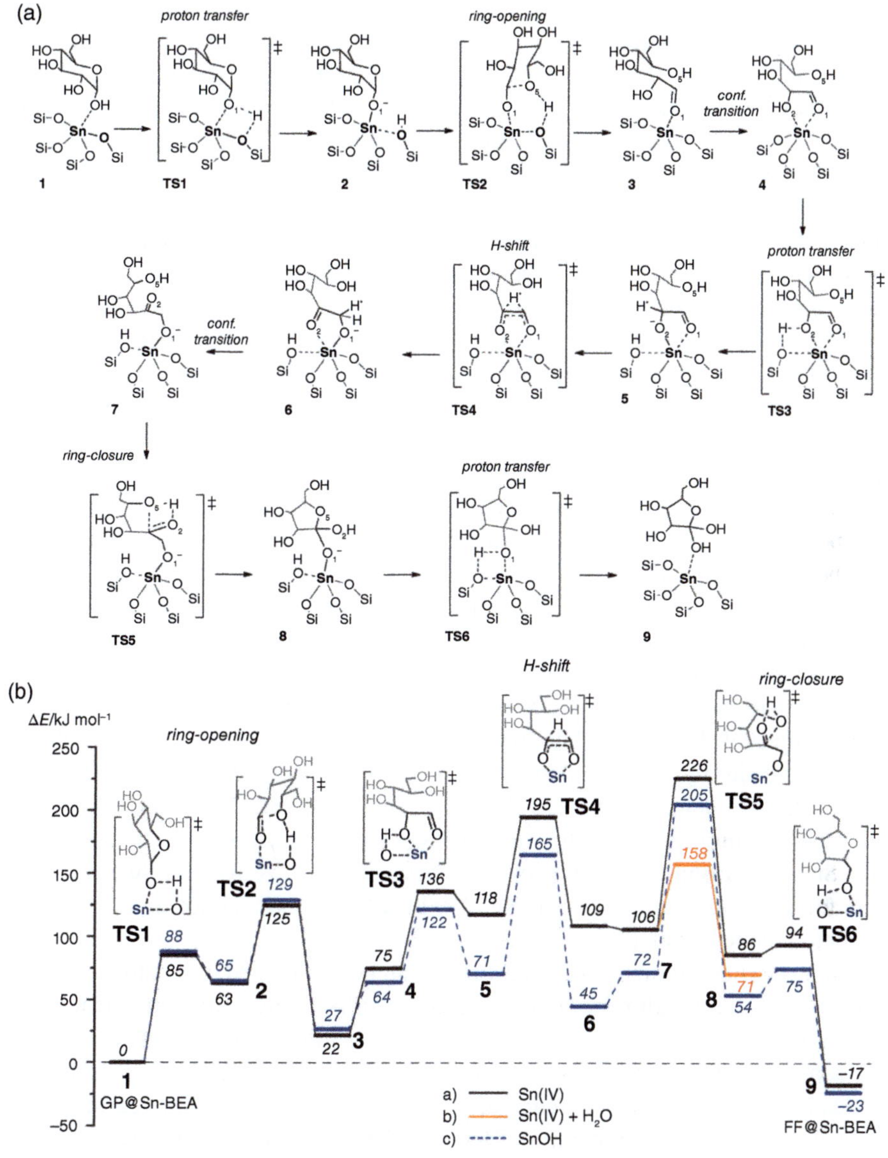

Figure 9.6 (a) Reaction route and (b) potential energy surface of glucose to fructose on Sn site of Sn-BEA zeolite. Source: Reproduced from Yang et al. [68]/with permission of American Chemical Society.

configuration transition from bidentate coordination to monodentate coordination; (vii) intramolecular hydrogen transfer (hydride transfer) from O_5 to O_2 with the ring-closure by carbonyl C and O_5; and (viii) H of Sn…(H)OSi is transferred to O_1 of glucose with the formation of fructose. Over the close Sn sites, all elementary reaction steps proceed with moderate barriers lower than 90 kJ/mol, but the ring-closure step (**7** → **[TS5]** → **8**) has a barrier of 120 kJ/mol because of the non-catalytic process. However, the water as the solvent can promote the proton transfer and ring-closure with a decreased barrier of 52 kJ/mol as displayed in Figure 9.6b, and the rate-determining step will transfer to the hydride transfer (**5** → [**TS4**] → **6**). All elementary reactions over closed and open Sn sites are basically similar, but the deprotonation of the acyclic glucose (**4** → [**TS3**] → **5**) over the closed Sn site is promoted by the extraframework SnOH hydroxyl of open Sn site. Although the defect site SnOH allows a slightly better stabilization of the key anionic sugar intermediates **5**, **6**, **7**, and **8**, the computed barriers are similar for the elementary steps for the closed and open Sn sites with the maximum difference of 17 kJ/mol. Ricardo et al. [69] calculated the enthalpy profile of glucose–fructose isomerization catalyzed by open site of Sn-BEA with a 5T cluster model. The rate-determining step is **5** → [**TS4**] → **6** with a enthalpy barrier of 82 kJ/mol. Contrarily, the ring-opening of glucose (**1** → [**TS1**] → **2**) is the rate-determining step over close Sn site of Sn-zeolites with a enthalpy barrier of 123 kJ/mol, and the same conclusion was also drawn based on the QM/MM model by Li et al. [70].

As the key step of glucose–fructose isomerization over small cluster, the hydride shift between C_2 and C_1 atoms of the sugar was further studied to investigate the synergy between Lewis acid sites and hydroxyl groups over different Sn-zeolites by Li et al. [71]. The proximate hydroxyl groups derived from co-adsorbed water molecules or internal silanols were found to strongly promote the hydride shift. Although the different zeolite topology can greatly influence the carbohydrate adsorption dominated by van der Waals interaction, the intrinsic reactivity of open Sn sites was slightly affected. Moreover, Bell and coworkers analyzed the effect of substituting Sn in active site by Ti, Zr, V, Nb, Si, and Ge [70], and the apparent reaction barriers of the rate-determining step (hydride transfer) are 32.1, 23.5, 39.0, 25.7, 37.4, and 32.2 kcal/mol, respectively. All these tetravalent metal atoms are less activity than Sn substitution (23.3 kcal/mol). Analogously, Sn-BEA displays the lowest barriers for the ring opening of glucose (**2** → [**TS2**] → **3**) than Ti, Zr, and Hf-BEA by Montejo-Valencia et al. [72].

Besides isomerization to fructose, glucose can be epimerized into mannose via the direct 1,2→carbon shift over Na-exchanged Sn-BEA as the competitive reaction of isomerization to fructose [73], and Sn-BEA zeolite in combination with the borate salt can also realize the epimerization of glucose to mannose [11, 74]. To mechanistically understand the influence of Na and borate to the conversion of glucose in Sn-BEA zeolite, several computational studies were carried out. Chethana et al. [12] investigated the detailed mechanism of glucose epimerization to mannose catalyzed by Sn-BEA-borate catalyst. First, they elucidated the role of borate salts in promoting the epimerization. In an aqueous solution, at neutral pH, borate salts exist as trigonal borate [$B(OH)_3$] and tetrahedral borate ions [$B(OH)_4^-$]. Therein, the tetrahedral borate [$B(OH)_4^-$] prefers to complex with sugar through proton transfer, while [$B(OH)_3$] prefers to form a complex with the stannanol group of the Sn-BEA active site and with the silanol group adjacent to Sn through sp^3d hybridization. Thus, tetrahedral [$B(OH)_4^-$] blocks the C_1 and C_2 positions of the sugar molecular to prevent Sugar-Sn-BEA complexation, and [$B(OH)_3$] blocks the active sites on Sn-BEA, which leads to the inaccessibility of active sites in Sn-BEA by the pure

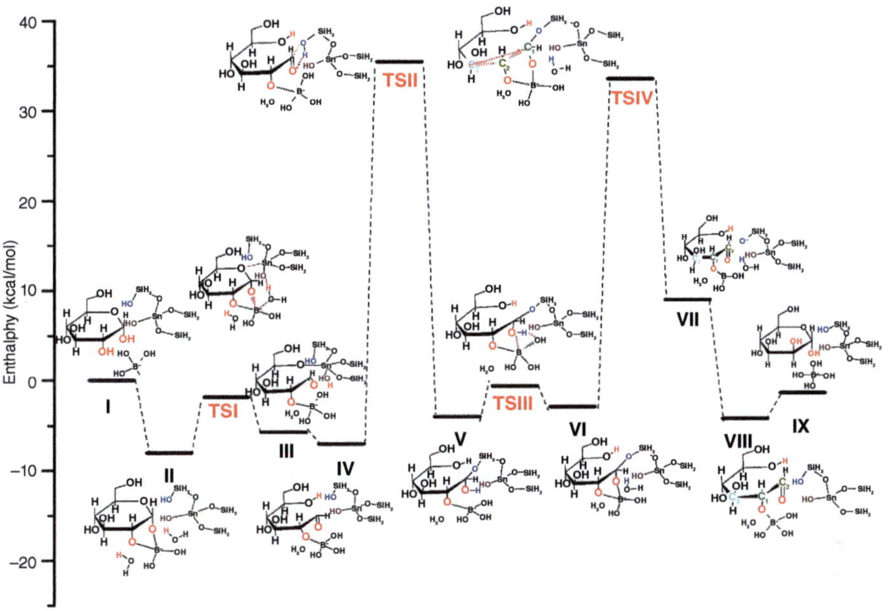

Figure 9.7 Enthalpy profile for the epimerization of glucose catalyzed by Sn-BEA zeolite and borate salts. Source: Adopted from reference Chethana and Mushrif [12] with the permission of Elsevier, copyright [2015].

glucose, further indirectly promoting epimerization. In this context, the epimerization of glucose to mannose will go through five steps as displayed in Figure 9.7: (i) formation of the sugar-borate ester complex, (ii) glucose ring opening, (iii) keto-enol transformation, (iv) Bilik type intramolecular carbon–carbon rearrangement, and (v) ring closure of the epimer product with the formation of mannose. In this mechanism, the Lewis acid site of Sn-OH catalyzes the ring opening of glucose with the assistance of water and [$B(OH)_4^-$], and the silanol group in the zeolite donates a proton for the keto-enol transformation, thus acting as a Brønsted acid site. Therefore, Sn-BEA zeolite could act as a Lewis acid catalyst for the ring-opening reaction and a Brønsted acid catalyst for the epimerization reaction as well. Notably, Sn substitution shows the higher catalytic activity than Zr and Ti substitution of BEA zeolite based on their activation enthalpies of the rate-determining step, 26.3, 27.5, and 29.8 kcal/mol for Sn, Zr, and Ti-BEA, respectively.

The catalytic activity of glucose epimerization to mannose in Na exchanged Sn-BEA zeolites can be greatly influenced by the solvent and proximal silanol groups [75], and Na^+ exchange will reverse the relative stabilities of critical intermediates for the epimerization vs. isomerization paths and drive the reaction toward the epimerization path with production of mannose. The proximal silanol groups will participate in the synchronous proton transfer of sugar fragment, which help to promote the reaction thermodynamics and reduce the activation barriers. The activation barriers are generally lowered with the increase of solvent (water/methanol) contents, and 4 ~ 6 water molecules achieve comparable catalytic results as 3 methanol molecules. Li et al. compared the reaction pathways in both the absence and presence of water molecules for Sn-BEA and Na-exchanged Sn-BEA [76]. Water not only influences the energetics of an individual pathway but also effectively determines the selectivity, especially in the case of Na-Sn-BEA. Jeffrey et al. concluded that Na exchange influences the mechanism by electrostatic stabilization of carbon shift relative to hydride transfer and that this effect is relatively independent of geometric and flexibility constraints [77]. The higher activity of methanol solvent than water was explained by their different in solvation of the hydrophobic pores.

9.3.3 Other Catalytic Reactions

Due to the Lewis acidity of Sn site and Brønsted acidity of Sn-OH, Sn-BEA can also be used to catalyze the alkane dehydrogenation. Yue et al. [78] explored two mechanisms of propane dehydrogenation catalyzed by Sn-BEA as displayed in Figure 9.8. In both two pathways, one of the H atoms in propane was firstly abstracted by O atom adjacent to Lewis Sn site, and the remaining C_3H_7 motif bonds with Sn site. Then, in the direct way, the H atom in the C_3H_7 motif attacks the H atom in the framework to directly produce propene and H_2 molecule. In the indirect pathway, the second H atom in the C_3H_7 motif attacks the Sn atom to form

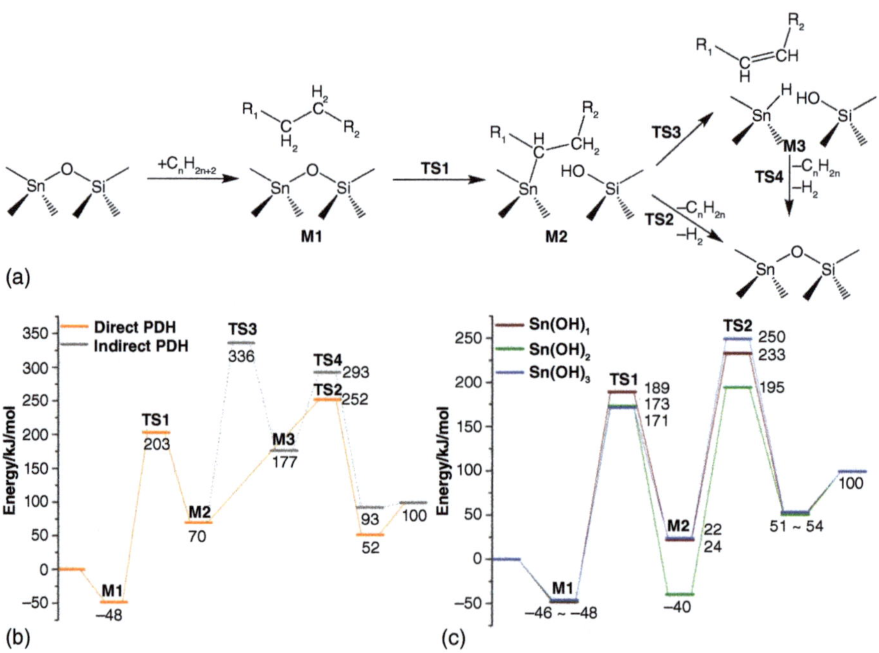

Figure 9.8 The energy profile of propane dehydrogenation catalyzed by Sn-BEA. (a) Reaction pathway including direct and indirect mechanisms, (b) potential energy surface of direct and indirect mechanisms, and (c) the direct mechanisms over different Sn locations. Source: Adopted from reference Yue et al. [78] with the permission of Elsevier, copyright [2021].

the Sn—H moiety and propene, and finally H_2 was produced by the coupling of two H atoms at Sn and O sites (Figure 9.8a). Based on the potential energy surface of two pathways, the direct pathway is energetically more favorable than the indirect one (Figure 9.8b), and the catalytic activity of Sn-BEA was greatly influenced by the different Sn locations (Figure 9.8c). Analogously, isolated Co-motifs into purely siliceous MFI zeolite framework (Co-MFI) displayed the similar mechanism on propane dehydrogenation [79]. But in H-Fe-ZSM-5 zeolite, the propane dehydrogenation proceeds through radical cation-like intermediates, and propane will directly bind to the acidic proton with the formation of H_2 and $C_3H_7^+$, and $C_3H_7^+$ will be induced to radical cation by Fe atom [80, 81].

Treu et al. [82] found that Sn-BEA zeolite catalysts can be unprecedently used for the Diels-Alder cycloaddition reaction between bio-available methyl acrylate and isoprene affording intermediates to bio-terephthalates, and Sn-BEA more selectively leads to the desired products via specific isomers than Al-BEA and non-catalysis system by comparing the energy barriers of different reaction pathways.

9.4 Other Metal-Substituted Zeolites

In addition to Ti^{4+} and Sn^{4+}, other tetravalent mental ions including Zr^{4+} and Hf^{4+} have also been incorporated into different silicate matrixes. As mentioned above, Ti-containing zeolites are more active for the epoxidation of olefins and Sn-doped zeolites are preferred to catalyze the BV oxidation and some biomass conversions. For Lewis acidic Zr- and Hf-containing zeolites, they are equally or more active for intermolecular hydride shifts, etherifications, and aldol condensations [83]. Therein, Wang et al. [84] demonstrated that Hf-BEA zeolite is active for the self-aldol condensation of ethyl pyruvate in toluene to produce itaconic acid eater analogues. Besides, Rojas-Buzo et al. [85] found that both Hf- and Zr-BEA were the most active catalysts for the tandem N-alkylation reaction of aniline with benzyl alcohol, in both cases the selectivity toward the corresponding N-benzylaniline higher than 97%. Combined with experimental and theoretical analyses, a reaction mechanism through a hydrogen borrowing pathway was proposed as displayed in Figure 9.9, in which the open sites are the active sites. First, the benzyl alcohol adsorbed on the Hf Lewis center in two forms, the H atom of hydroxyl group (A) and the methylene group of the reactant (D). Thus, for the next step, there is a competitive reaction between these forms. The former dehydrogenates to benzaldehyde (**A** → **C**), while the latter transforms to an alkoxide with the formation of H_2O (**D** → **E**). Then, aniline adsorbs close to benzaldehyde companied by a nucleophilic addition reaction (**F** → **G**) that requires an activation energy of 23.1 kcal/mol. However, the dehydration of this intermediate to yield the corresponding imine cannot occur directly on the Hf-hydride site but requires its desorption and migration to a different Hf open site. Again, the basic hydroxyl group attached to Hf atom in a concerted way through a cyclic transition state with an activation energy of 12.6 kcal/mol (**H** → **I**). Finally, the resultant imine is hydrogenated on the Hf-hydride site with a low barrier of 4.8 kcal/mol, yielding the highly stable N-benzylaniline product (**J** → **K**).

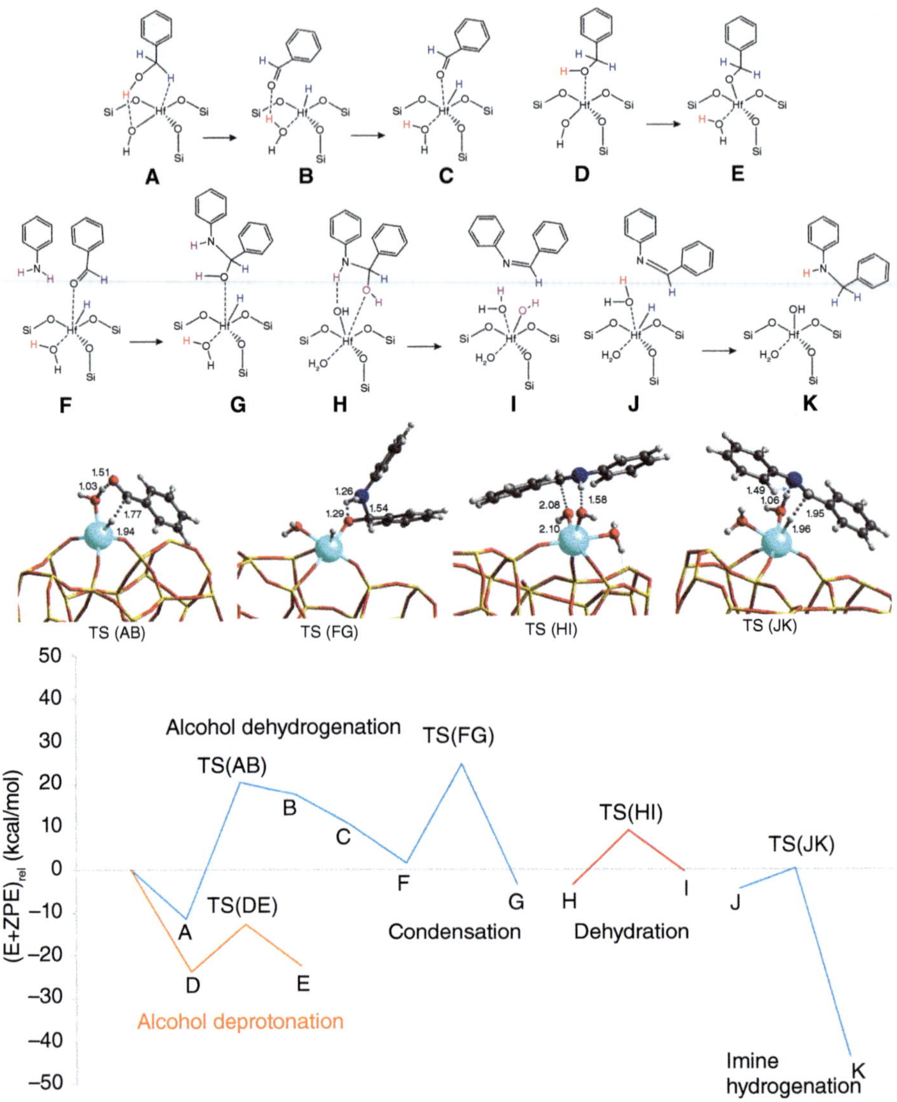

Figure 9.9 (Top) Structures involved in the mechanism of N-alkylation of aniline with benzoyl alcohol on an open site model of Hf-BEA zeolite, (middle) transition state structures, and (bottom) potential energy surface with zero-point energy correction of N-alkylation of aniline with benzoyl alcohol. Source: Reproduced from Rojas-Buzo et al. [85]/with permission of American Chemical Society / CC BY 4.0.

Additionally, Zhang et al. [86] explored the reaction of ethanol to 1,3-butadiene (ETB) over Zr-BEA zeolite by DFT calculations (Figure 9.10). The closed site and open site are considered, and the main steps are similar. But due to the participation of Zr-OH, the energy barrier of each step is generally lower than that over closed site, which indicated the better catalytic performance of open site in Zr-BEA. Over

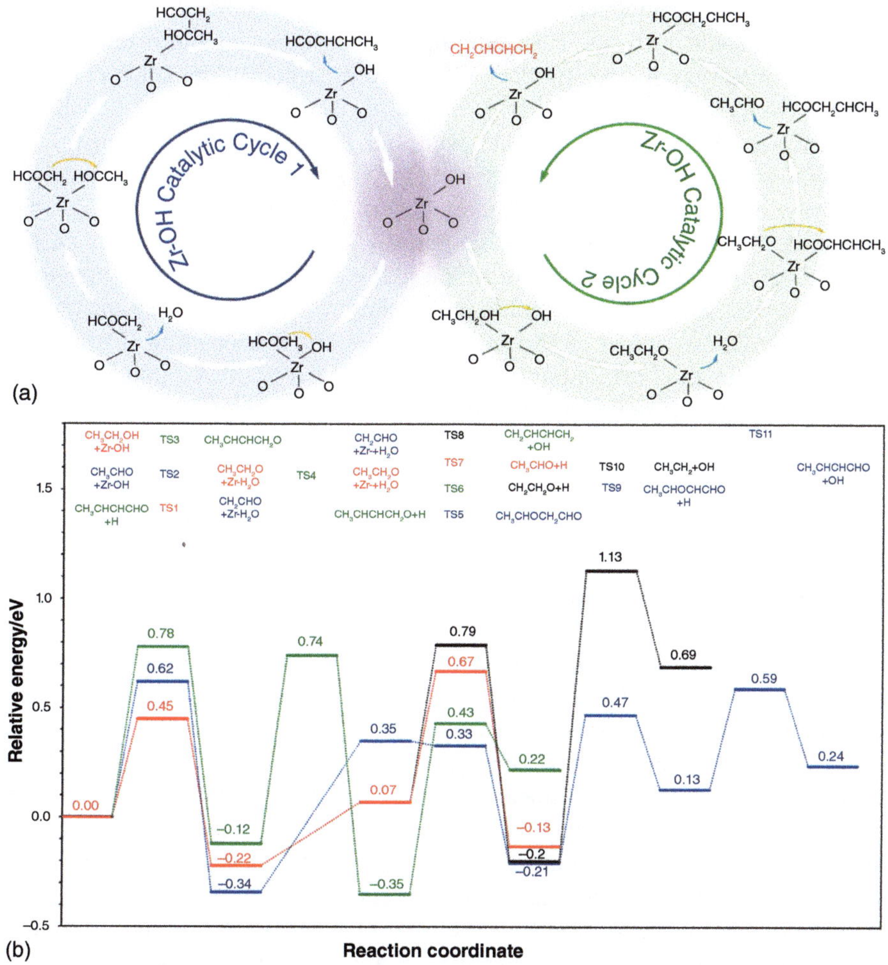

Figure 9.10 (a) Schematic diagram of Zr-OH participating in ETB process at Zr-BEA open site, (b) energy profiles diagram for the Aldol condensation and crotonaldehyde formation (blue lines), MPV reaction and butadiene formation (green lines), acetaldehyde formation (red lines), and ethylene formation (black lines) at Zr-BEA open site in the ETB process. Source: Adopted from reference Zhang et al. [86] with the permission of Elsevier, copyright [2022].

open site, the dehydration and dehydrogenation of ethanol were firstly studied, and they are multi-step instead of one-step mechanism (Figure 9.10a). There are four steps: (i) dehydrogenation of ethanol to ethoxy, (ii) further dehydrogenation of ethoxy to acetaldehyde, (iii) removing α-H of acetaldehyde, and (iv) breaking of C—O bond in acetaldehyde to form ethylene. Next process involves the aldol condensation of two acetaldehyde molecules co-adsorbed onto the Zr centers and the Meerwein–Ponndorf–Verley (MPV) reaction between ethoxy and crotonaldehyde species. Finally, 1,3-butadiene is produced successfully. Herein, MPV reaction at open site is the rate-determining step with the E_a of 0.78 eV (Figure 9.10b). Mesoporous zeolites with the large pore have better longevity and reactivity than microporous zeolites and are harder to be coked by deactivation. Dong et al. [87] explored two possible mechanisms of ethanol to ETB over Zr-SBA-15, including Toussaint's aldol condensation mechanism and Fripiat's Prins mechanism. The direct reaction of ethylene and acetaldehyde in Prins condensation proceeds with a free energy barrier of 2.90 eV, but the rate-determining step of two acetaldehydes reaction in Toussaint's mechanism is kinetically more favorable with a barrier of 2.15 eV. Montejo-Valencia and Curet-Arana [72] compared the catalytic activity of M-BEA (M = Hf, Zr, Ti, and Sn) zeolites on the ring-opening of fructose and glucose by DFT calculations. Based on the energy barriers, the ring opening of fructose is integrally harder than the one of glucose over all four M-BEA zeolites. Hf-BEA and Zr-BEA have the higher activity on ring opening of fructose than Sn-BEA and Ti-BEA, but Sn-BEA shows the highest activity than Zr-BEA, Hf-BEA, and Ti-BEA in ring opening of glucose.

9.5 Summary and Outlook

In this chapter, we have highlighted the specific reaction mechanisms of the heteroatom-containing zeolites by calculations. As a useful tool, theoretical calculation is playing a more and more important role in catalytic research of heteroatom substituted zeolites. Through theoretical calculations, we can obtain an in-depth and systematic understanding of catalysts, including the structure of active sites, acid–base properties, reaction mechanisms, and catalytic activity, thus providing reasonable and accurate guidance for designing the catalysts. However, there are many limitations of theoretical calculations that impeded the development of their application in the field of heteroatom-containing zeolites. (i) The kinetics of heteroatom substitution are still unclear, and the location of heteroatom substitution only relies on the thermodynamics. (ii) The structures of active sites are long-debate, and the interconversion of different active sites increase the difficulty to illustrate catalytic reaction mechanism. (iii) The influence of solvent molecules and additional reactants on the catalytic activity and catalytic reaction pathway is significant but not be well considered. (iv) The relationship between acidity and catalytic activity is hard to be established because of the complicated and changeable environment of heteroatoms in zeolite.

Acknowledgments

This research was supported by the National Key Research and Development Program of China (2021YFA15026009) and the National Natural Science Foundation of China (22125304, 22032005, 22002174, and 21991092).

References

1 Taramasso, M., Perego, G., and Notari, B. (1983). Preparation of porous crystalline synthetic material comprised of silicon and titanium oxides. In: U.S. Patent No. 4,410,501. Washington, DC: U.S. Patent and Trademark Office.
2 Wang, Y., Zhou, D.H., Yang, G. et al. (2004). A DFT study on isomorphously substituted MCM-22 zeolite. *J. Phys. Chem. A* 108: 6730–6734.
3 Li, H.C., Zhou, D.H., Tian, D.X. et al. (2014). Framework stability and bronsted acidity of isomorphously substituted interlayer-expanded zeolite COE-4: a density functional theory study. *ChemPhysChem* 15: 1700–1707.
4 Yang, G., Zhou, L.J., and Han, X.W. (2013). Structure, stability, and Lewis acidity of mono and double Ti, Zr, and Sn framework substitutions in BEA zeolites: a periodic density functional theory study. *J. Phys. Chem. C* 117: 3976–3986.
5 Yang, G., Zhou, L.J., and Han, X.W. (2012). Lewis and Brönsted acidic sites in M^{4+}-doped zeolites (M=Ti, Zr, Ge, Sn, Pb) as well as interactions with probe molecules: a DFT study. *J. Mol. Catal. A Chem.* 363–364: 371–379.
6 Kang, Z.Z., Fang, G.Y., Ke, Q.P. et al. (2013). Superior catalytic performance of mesoporous zeolite TS-1 for the oxidation of bulky organic sulfides. *ChemCatChem* 5: 2191–2194.
7 Lin, J.Z., Xin, F., Yang, L.B., and Zhuang, Z. (2014). Synthesis, characterization of hierarchical TS-1 and its catalytic performance for cyclohexanone ammoximation. *Cat. Com.* 45: 104–108.
8 Clerici, M.G. (2009). Titanium silicalite-1. *Metal Oxide Catal.* 2: 705–754.
9 Moliner, M., Román-Leshkov, Y., and Davis, M.E. (2010). Tin-containing zeolites are highly active catalysts for the isomerization of glucose in water. *Proc. Natl. Acad. Sci. U. S. A.* 107: 6164–6168.
10 Román-Leshkov, Y., Moliner, M., Labinger, J.A., and Davis, M.E. (2010). Mechanism of glucose isomerization using a solid Lewis acid catalyst in water. *Angew. Chem.-Int. Ed.* 49: 8954–8957.
11 Gunther, W.R., Wang, Y., Ji, Y.W. et al. (2012). Sn-beta zeolites with borate salts catalyse the epimerization of carbohydrates via an intramolecular carbon shift. *Nat. Commun.* 3: 1109.
12 Chethana, B.K. and Mushrif, S.H. (2015). Brønsted and Lewis acid sites of Sn-beta zeolite, in combination with the borate salt, catalyze the epimerization of glucose: a density functional theory study. *J. Catal.* 323: 158–164.
13 Vyver, S.V.D., Odermatt, C., Romero, K. et al. (2015). Solid Lewis acids catalyze the carbon–carbon coupling between carbohydrates and formaldehyde. *ACS Catal.* 5: 972–977.
14 Deshpande, N., Parulkar, A., Joshi, R. et al. (2019). Epoxide ring opening with alcohols using heterogeneous Lewis acid catalysts: Regioselectivity and mechanism. *J. Catal.* 370: 46–54.
15 Martausová, I., Spustováa, D., Cvejnc, D. et al. (2019). Catalytic activity of advanced titanosilicate zeolites in hydrogen peroxide S-oxidation of methyl(phenyl)sulfide. *Catal. Today* 324: 144–153.

16 Ratnasamy, P., Srinivas, D., and Knozinger, H. (2004). Active sites and reactive intermediates in titanium silicate molecular sieves. *Adv. Catal.* 1–169.

17 Inagaki, S., Tsuboi, Y., Sasaki, M. et al. (2016). Enhancement of Para-selectivity in the phenol oxidation with H_2O_2 over Ti-MCM-68 zeolite catalyst. *Green Chem.* 18: 735–741.

18 Kubota, Y., Koyama, Y., Yamada, T. et al. (2008). Synthesis and catalytic performance of Ti-MCM-68 for effective oxidation reactions. *Chem. Commun.* 6224–6226.

19 Pitinova-Stekrova, M., Eliasova, P., Weissenberger, T. et al. (2018). Highly selective synthesis of campholenic aldehyde over Ti-MWW catalysts by α- pinene oxide isomerization. *Cat. Sci. Technol.* 8: 4690–4701.

20 Sasidharan, M., Wu, P., and Tatsumi, T. (2002). Epoxidation of α,β-unsaturated carbonyl compounds over various titanosilicates. *J. Catal.* 205: 332–338.

21 Zhou, D.H., Zhang, H.J., Zhang, J.J. et al. (2014). Density functional theory investigations into the structure and spectroscopic properties of the Ti4+ species in Ti-MWW zeolite. *Micropor. Mesopor. Mater.* 195: 216–226.

22 Petkov, P.S., Simeonova, K., Koleva, I.Z. et al. (2021). Defect formation, T-atom substitution and adsorption of guest molecules in MSE-type zeolite framework-DFT modeling. *Molecules* 26: 7296.

23 Lamberti, C., Bordiga, S., Zecchina, A. et al. (2001). Ti location in the MFI framework of Ti-silicalite-1: a neutron powder diffraction study. *J. Am. Chem. Soc.* 123: 2204–2212.

24 Ricchiardi, G., Damin, A., Bordiga, S. et al. (2001). Vibrational structure of titanium silicate catalysts. A spectroscopic and theoretical study. *J. Am. Chem. Soc.* 123: 11409–11419.

25 Fan, W.B., Duan, R.-G., Yokoi, T. et al. (2008). Synthesis, crystallization mechanism, and catalytic properties of titanium-rich TS-1 free of extraframework titanium species. *J. Am. Chem. Soc.* 130: 10120–10164.

26 Lamberti, C., Bordiga, S., Zecchina, A. et al. (1999). Structural characterization of Ti-silicalite-1: a synchrotron radiation x-ray powder diffraction study. *J. Catal.* 183: 222–231.

27 Marra, G.L., Artioli, G., Fitch, A.N. et al. (2000). Orthorhombic to monoclinic phase transition in high-Ti-loaded TS-1: An attempt to locate Ti in the MFI framework by low temperature XRD. *Micropor. Mesopor. Mater.* 40: 85–94.

28 Hijar, C.A., Jacubinas, R.M., Eckert, J. et al. (2000). The siting of Ti in TS-1 is non-random. Powder neutron diffraction studies and theoretical calculations of TS-1 and FeS-1. *J. Phys. Chem. B* 104: 12157–12164.

29 Henry, P.F., Weller, M.T., and Wilson, C.C. (2001). Structural investigation of TS-1: determination of the true nonrandom titanium framework substitution and silicon vacancy distribution from powder neutron diffraction studies using isotopes. *J. Phys. Chem. B* 105: 7452–7458.

30 Deka, R.C., Nasluzov, V.A., Shor, E.A.I. et al. (2005). Comparison of all sites for Ti substitution in zeoliteTS-1 by an accurate embedded-cluster method. *J. Phys. Chem. B* 109: 24304–24310.

31 Gamba, A., Tabacchi, G., and Fois, E. (2009). TS-1 from first principles. *J. Phys. Chem. A* 113: 15006–15015.

32 Gordon, C.P., Engler, H., Tragl, A.S. et al. (2020). Efficient epoxidation over dinuclear sites in titanium silicalite-1. *Nature* 586: 708–713.

33 Dong, J.C., Zhu, H.L., Xiang, Y.J. et al. (2016). Toward a unified identification of Ti location in the MFI framework of high-Ti-loaded TS-1: combined EXAFS, XANES, and DFT study. *J. Phys. Chem. C* 120: 20114–20124.

34 Xu, H., Xu, H.X., and Cheng, D.J. (2022). Resolving the reaction mechanism for oxidative hydration of ethylene toward ethylene glycol by titanosilicate catalysts. *ACS Catal.* 12: 9446–9457.

35 Signorile, M., Damin, A., Bonino, F. et al. (2018). Computational assessment of relative sites stabilities and site-specific adsorptive properties of titanium silicalite-1. *J. Phys. Chem. C* 122: 1612–1621.

36 Yang, G., Zhou, L.J., Liu, X.C. et al. (2011). Density functional calculations on the distribution, acidity, and catalysis of Ti^{IV} and Ti^{III} ions in MCM-22 zeolite. *Chem. A Eur. J.* 17: 1614–1621.

37 Yang, G., Zhuang, J.Q., Ma, D. et al. (2008). A joint experimental–theoretical study on trimethylphosphine adsorption on the Lewis acidic sites present in TS-1 zeolite. *J. Mol. Struct.* 882: 24–29.

38 Sastre, G. and Corma, A. (1999). Relation between structure and Lewis acidity of Ti-beta and TS-1 zeolites – a quantum-chemical study. *Chem. Phys. Lett.* 302: 447–453.

39 Montejo-Valencia, B.D., Salcedo-Pérez, J.L., and Curet-Arana, M.C. (2016). DFT study of closed and open sites of BEA, FAU, MFI, and BEC zeolites substituted with tin and titanium. *J. Phys. Chem. C* 120: 2176–2186.

40 Barker, C.M., Gleeson, D., Kaltsoyannis, N. et al. (2002). On the structure and coordination of the oxygen-donating species in Ti-MCM-41↑TBHP oxidation catalysts: a density functional theory and EXAFS study. *Phys. Chem. Chem. Phys.* 4: 1228–1240.

41 Chaudhari, K., Srinivas, D., and Ratnasamy, P. (2001). Reactive oxygen species in titanosilicates TS-1 and TiMCM-41: An in situ EPR spectroscopic study. *J. Catal.* 203: 25–32.

42 Sever, R.R. and Root, T.W. (2003). DFT study of solvent coordination effects on titanium-based epoxidation catalysts. Part one: formation of the titanium hydroperoxo intermediate. *J. Phys. Chem. B* 107: 4080–4089.

43 Nie, X.W., Ren, X.X., Ji, X.J. et al. (2019). Mechanistic insight into propylene epoxidation with H_2O_2 over titanium silicalite-1: effects of zeolite confinement and solvent. *J. Phys. Chem. B* 123: 7410–7423.

44 Munakata, H., Oumi, Y., and Miyamoto, A. (2001). A DFT study on peroxo-complex in titanosilicate catalyst: hydrogen peroxide activation on titanosilicalite-1 catalyst and reaction mechanisms for catalytic olefin epoxidation and for hydroxylamine formation from ammonia. *J. Phys. Chem. B* 105: 3493–3501.

45 Sirijaraensre, J. and Limtrakul, J. (2013). Mechanisms of the ammonia oxidation by hydrogen peroxide over the perfect and defective Ti species of TS-1 zeolite. *Phys. Chem. Chem. Phys.* 15: 18093–18100.

46 Vayssilov, G.N. and Santen, R.A.V. (1998). Catalytic activity of titanium silicalites – a DFT study. *J. Catal.* 175: 170–174.

47 Wells, D.H. Jr., Delgass, W.N. Jr., and Thomson, K.T. (2004). Evidence of defect-promoted reactivity for epoxidation of propylene in titanosilicate (TS-1) catalysts: a DFT study. *J. Am. Chem. Soc.* 126: 2956–2962.

48 Limtrakul, J., Inntam, C., and Truong, T.N. (2004). Density functional theory study of the ethylene epoxidation over Ti-substituted silicalite (TS-1). *J. Mol. Catal. A Chem.* 207: 139–148.

49 Karlsen, E. and Schoffel, K. (1996). Titanium-silicalite catalyzed epoxidation of ethylene with hydrogen peroxide – a theoretical study. *Catal. Today* 32: 107–114.

50 Chu, C.Q., Zhao, H.T., Qi, Y.Y., and Xin, F. (2013). Density functional theory studies on hydroxylamine mechanism of cyclohexanone ammoximation on titanium silicalite-1 catalyst. *J. Mol. Model.* 19: 2217–2224.

51 Corma, A., Nemeth, L.T., Renz, M., and Valencia, S. (2001). Sn-zeolite beta as a heterogeneous chemoselective catalyst for Baeyer-Villiger oxidations. *Nature* 412: 423–425.

52 Xia, C.J., Ju, L., Zhao, Y. et al. (2015). Heterogeneous oxidation of cyclohexanone catalyzed by TS-1: combined experimental and DFT studies. *Chinese J. Catal.* 36: 845–854.

53 Sengupta, A., Kamble, P.D., Basu, J.K., and Sengupta, S. (2012). Kinetic study and optimization of oxidative desulfurization of benzothiophene using mesoporous titanium silicate-1 catalyst. *Ind. Eng. Chem. Res.* 51: 147–157.

54 Cheng, S.F., Liu, Y.M., Gao, J.B. et al. (2006). Catalytic oxidation of benzothiophene and dibenzothiophene in model light oil over Ti-MWW. *Chinese J. Catal.* 27: 547–549.

55 Wang, H.L., Yu, M., and Lin, Z.G. (2019). A DFT study on catalytic oxidative desulfurization with H_2O_2 over Ti-MWW zeolite. *J. Mol. Model.* 25: 106.

56 Wang, H.L., Deng, Y.Q., and Zhou, R.J. (2018). Aromatic sulfur compounds oxidation with H_2O_2 over fully coordinated and defect sites in Ti-beta zeolites: evaluation by density functional theory. *Theoret. Chem. Acc.* 137: 66.

57 Bian, H., Zhang, H.H., Li, D.Z. et al. (2020). Insight into the oxidative desulfurization mechanism of aromatic sulfur compounds over Ti-MWW zeolite: a computational study. *Micropor. Mesopor. Mater.* 294: 109837.

58 Bian, H., Wang, F., Wu, S. et al. (2021). A DFT study on the mechanism of catalytic oxidation desulfurization over Ti-MWW zeolite. *J. Cluster Sci.* 3: 2103–2112.

59 Corma, A., Domine, M.E., and Valencia, S. (2003). Water-resistant solid Lewis acid catalysts: Meerwein–Ponndorf–Verley and oppenauer reactions catalyzed by tin-beta zeolite. *J. Catal.* 215: 294–304.

60 Corma, A., Domine, M.E., Nemeth, L., and Valencia, S. (2002). Al-free Sn-beta zeolite as a catalyst for the selective reduction of carbonyl compounds (Meerwein–Ponndorf–Verley reaction). *J. Am. Chem. Soc.* 124: 3194–3195.

61 Boronat, M., Corma, A., and Renz, M. (2006). Mechanism of the Meerwein–Ponndorf–Verley–Oppenauer (MPVO) redox equilibrium on Sn- and Zr-beta zeolite catalysts. *J. Phys. Chem. B* 110: 21168–21174.

62 Holm, M.S., Pagán-Torres, Y.J., Saravanamurugan, S. et al. (2012). Sn-beta catalysed conversion of hemicellulosic sugars. *Green Chem.* 14: 702–706.

63 Mahmoud, E., Yu, J., Gorte, R.J., and Lobo, R.F. (2015). Diels–Alder and dehydration reactions of biomass-derived furan and acrylic acid for the synthesis of benzoic acid. *ACS Catal.* 5: 6946–6955.

64 Vyver, S.V.D. and Román-Leshkov, Y. (2015). Metalloenzyme-like zeolites as Lewis acid catalysts for c-c bond formation. *Angew. Chem.-Int. Ed.* 54: 12554–12561.

65 Li, H., Yang, S., Saravanamurugan, S., and Riisager, A. (2017). Glucose isomerization by enzymes and chemo-catalysts: status and current advances. *ACS Catal.* 7: 3010–3029.

66 Ennaert, T., Van Aelst, J., Dijkmans, J. et al. (2016). Potential and challenges of zeolite chemistry in the catalytic conversion of biomass. *Chem. Soc. Rev.* 45: 584–611.

67 Boronat, M., Concepción, P., Corma, A. et al. (2005). Determination of the catalytically active oxidation Lewis acid sites in Sn-beta zeolites, and their optimisation by the combination of theoretical and experimental studies. *J. Catal.* 234: 111–118.

68 Yang, G., Pidko, E.A., and Hensen, E.J. (2013). The mechanism of glucose isomerization to fructose over Sn-BEA zeolite: a periodic density functional theory study. *ChemSusChem* 6: 1688–1696.

69 Bermejo-Devala, R., Assary, R.S., Nikolla, E. et al. (2012). Metalloenzyme-like catalyzed isomerizations of sugars by Lewis acid zeolites. *Proc. Natl. Acad. Sci. U. S. A.* 109: 9727–9732.

70 Li, Y.-P., Head-Gordon, M., and Bell, A.T. (2014). Analysis of the reaction mechanism and catalytic activity of metal-substituted beta zeolite for the isomerization of glucose to fructose. *ACS Catal.* 4: 1537–1545.

71 Li, G.N., Pidko, E.A., and Hensen, E.J. (2014). Synergy between Lewis acid sites and hydroxyl groups for the isomerization of glucose to fructose over Sn-containing zeolites: a theoretical perspective. *Cat. Sci. Technol.* 4: 2241–2250.

72 Montejo-Valencia, B.D. and Curet-Arana, M.C. (2019). Periodic DFT study of the opening of fructose and glucose rings and the further conversion of fructose to trioses catalyzed by M-BEA (M= Sn, Ti, Zr, or Hf). *J. Phys. Chem. C* 123: 3532–3540.

73 Bermejo-Deval, R., Orazov, M., Gounder, R. et al. (2014). Active sites in Sn-beta for glucose isomerization to fructose and epimerization to mannose. *ACS Catal.* 4: 2288–2297.

74 Rai, N., Caratzoulas, S., and Vlachos, D.G. (2013). Role of silanol group in Sn-beta zeolite for glucose isomerization and epimerization reactions. *ACS Catal.* 3: 2294–2298.

75 Yang, G. and Zhou, L.J. (2018). Glucose conversions catalyzed by zeolite Sn-bea: synergy among Na^+ exchange, solvent, and proximal silanol nest as well as critical specifics for catalytic mechanisms. *ACS Catal.* 8: 6691–6698.

76 Li, S., Josephson, T., Vlachos, D.G., and Caratzoulas, S. (2017). The origin of selectivity in the conversion of glucose to fructose and mannose in Sn-BEA and Na-exchanged Sn-BEA zeolites. *J. Catal.* 355: 11–16.

77 Christianson, J.R., Caratzoulas, S., and Vlachos, D.G. (2015). Computational insight into the effect of Sn-beta Na exchange and solvent on glucose isomerization and epimerization. *ACS Catal.* 5: 5256–5263.

78 Yue, Y.Y., Fu, J., Wang, C.M. et al. (2021). Propane dehydrogenation catalyzed by single Lewis acid site in Sn-beta zeolite. *J. Catal.* 395: 155–167.

79 Hu, Z.-P., Qin, G., Han, J. et al. (2022). Atomic insight into the local structure and microenvironment of isolated co-motifs in MFI zeolite frameworks for propane dehydrogenation. *J. Am. Chem. Soc.* 144: 12127–12137.

80 Yun, J.H. and Lobo, R.F. (2014). Radical cation intermediates in propane dehydrogenation and propene hydrogenation over H-[Fe] zeolites. *J. Phys. Chem. C* 118: 27292–27300.

81 Yun, J.H. and Lobo, R.F. (2014). Catalytic dehydrogenation of propane over iron-silicate zeolites. *J. Catal.* 312: 263–270.

82 Treu, P., Huber, P., Plessow, P.N. et al. (2022). Lewis acid Sn-bea catalysts for the cycloaddition of isoprene and methyl acrylate: a greener route to bio-derived monomers. *Cat. Sci. Technol.* 12: 7439–7447.

83 Luo, H.Y., Lewis, J.D., and Roman-Leshkov, Y. (2016). Lewis acid zeolites for biomass conversion: perspectives and challenges on reactivity, synthesis, and stability. *Ann. Rev. Chem. Biomol. Eng.* 7: 27.21–27.30.

84 Wang, Y.R., Lewis, J.D., and Román-Leshkov, Y. (2016). Synthesis of itaconic acid ester analogues via self-aldol condensation of ethyl pyruvate catalyzed by hafnium BEA zeolites. *ACS Catal.* 6: 2739–2744.

85 Rojas-Buzo, S., Concepción, P., Corma, A. et al. (2021). In-situ-generated active Hf-hydride in zeolites for the tandem N-alkylation of amines with benzyl alcohol. *ACS Catal.* 11: 8049–8061.

86 Zhang, M.H., Guan, X.Y., Zhuang, J.Y., and Yu, Y.Z. (2022). Insights into the mechanism of ethanol conversion into 1,3-butadiene on Zr-β zeolite. *Appl. Surf. Sci.* 579: 152212.

87 Dong, X.Q., Lu, J., Yu, Y.Z., and Zhang, M.H. (2018). A DFT study on Zr-SBA-15 catalyzed conversion of ethanol to 1,3-butadiene. *Phys. Chem. Chem. Phys.* 20: 12970–12978.

10

Catalytic Ammoximation of Ketones or Aldehydes Using Titanosilicates

Rusi Peng, Hao Xu, Chengwei Zhai, Mingyuan He, and Peng Wu

East China Normal University, Shanghai Key Laboratory of Green Chemistry and Chemical Processes, School of Chemistry and Molecular Engineering, North Zhongshan Road 3663, Shanghai 200062, China

10.1 Introduction

The discovery of the first titanosilicate TS-1 with the MFI (structure code certified by International Zeolite Association) topology by ENI (Enichem company) in the 1980s has opened new opportunities for using zeolites as selective oxidation catalysts [1]. The tetrahedral Ti sites isomorphously substituted in the framework can activate aqueous hydrogen peroxide under mild conditions, catalyzing selective oxidation of various substrates including alkenes, alkanes, ketones/aldehydes, aromatics, organic amines, and sulfides [2–10]. In particular, TS-1/H_2O_2 catalytic system-based epoxidation of propylene to propylene oxide, hydroxylation of phenol to diphenol, and ammoximation of cyclohexanone to cyclohexanone oxime have been commercialized on a very large industrial scale, establishing a new milestone in zeolite catalysis after Y zeolite-derived fluid catalytic cracking (FCC) and ZSM-5-based shape-selective reactions.

Among these selective oxidations, the titanosilicate-catalyzed ammoximation of cyclohexanone in the presence of NH_3 and H_2O_2 is an important reaction to synthesize cyclohexanone oxime [11], the key intermediate for the manufacture of downstream product of ε-caprolactam (CPL) through Beckmann rearrangement. As a bulk chemical, the annual production of CPL is expected to reach 7 500 000 and 4 200 000 tons in 2024 for global and China markets, respectively. Compared with the conventional commercial processes mainly by non-catalytic oximation of ketone with hydroxylamine derivatives, this novel catalytic technology offers outstanding advantages for oxime production, requiring only a single step and avoiding the use of environmentally undesirable chemicals like oleum, halides, and oxide of nitrogen. Moreover, only a limited amount of byproducts formed in this novel catalytic process. Further combining the TS-1/H_2O_2/NH_3-based ammoximation with the heterogeneous vapor-phase Beckmann rearrangement of cyclohexanone oxime over highly siliceous ZSM-5 (Silicalite-1), Sumitomo Chemical commercialized in 2003 a revolutionary zero-emission CLP process with a 60 000-ton capacity, which is

Micro-Mesoporous Metallosilicates: Synthesis, Characterization, and Catalytic Applications, First Edition. Edited by Peng Wu and Hao Xu.

recognized as a model of green chemical processes [11]. The success of TS-1 has stimulated extensive researchers to find more efficient titanosilicate catalysts with high performance and even new abilities that are not shared by TS-1. The need to commercialize the ammoximation of ketones other than cyclohexanone as well as aldehydes were also waiting for new selective catalysts. In this sense, appropriate post-chemical modification and improved synthesis methods have been proposed to effectively enhance the ammoximation performance of TS-1. In addition, a series of titanosilicates with topologies different from MFI-type TS-1 have been developed either by direct hydrothermal synthesis or by post-synthesis, including Ti-Beta, Ti-MOR, Ti-MWW, Ti-MSE, and Ti-FER [12–16]. In comparison to 10-MR TS-1, these later-emerged materials usually possess larger pores, which are expected to exhibit an inbuilt advantage to fasten intrapore or intracrystal adsorption, desorption, and diffusion limitations for reactant and product molecules. Indeed, several titanosilicates like Ti-MWW [11] and Ti-MOR [17] with larger porosity are proved to be superior to TS-1 in liquid-phase ammoximation not only in catalytic activity but also in oxime selectivity.

The ammoximation involves three reactants of ketone or aldehyde, hydrogen peroxide, and ammonia, which turns out to be a more complicated reaction in comparison to other oxidations such as epoxidation of alkenes and hydroxylation of aromatics. Understanding the solvent effect and reaction mechanism of titanosilicate-catalyzed ammoximation and establishing new strategies that control oxime selectivity and suppress side reactions are critical to developing more innovative catalytic technologies. With respect to the ammoximation for practical applications, it is needed to enhance the activity of titanosilicates and to prolong their catalyst life as possible by precisely modifying the microenvironment and physicochemical properties of Ti active sites [18, 19].

This chapter focuses on the introduction of representative titanosilicates, typical synthesis and post-modification methods, ammoximation characteristics, and state of the art commercial technologies. As well-known TS-1 has achieved great success in the manufacture of cyclohexanone oxime in many industrial processes, special attention is paid to summarize the progress of developing promising ammoximation titanosilicates beyond TS-1 and the ammoximation process of other linear ketones and aldehydes. A brief summary will also be given to titanosilicate-based bifunctional catalysts with the potential for establishing process-intensified ammoximation and novel tandem reaction process.

10.2 The Development of Titanosilicates in Ammoximation of Ketones and Aldehydes

Oximes are a kind of important fine chemicals widely applied in petrochemical industry and medical and agriculture fields. And cyclohexanone oxime, especially, is the key intermediate in the production of CPL, the important monomer of nylon-6, which is widely applied in transportation, machinery, and textile fields. Taking cyclohexanone oxime as an example, the conventional and representative production process once unexceptionally relied on the non-catalytic oximation of cyclohexanone and hydroxylamine derivates in strong acidic media, which has the drawbacks such as complex process, highly corrosive system, and huge amount of low value-added byproducts (ammonium salts).

The discovery of the first titanosilicate TS-1 has greatly stimulated the researchers to broaden the application scope of the unique titanosilicate/H_2O_2 catalytic system. In 1988, TS-1 has been firstly applied in cyclohexanone ammoximation by direct synthesis of hydroxylamine, which is of great industrial importance as intermediate in the production of cyclohexanone oxime, by means of direct oxidation of ammonia and hydrogen peroxide over TS-1 [20]. The discovery of titanosilicates with different topologies and structures provides possibility of developing more efficient ammoximation system. The ammoximation using various titanosilicates with different topology structures is summarized in Table 10.1. Up to now, the catalytic ammoximation activities over eight kinds of microporous titanosilicates and three kinds of mesoporous titanosilicates have been investigated. Compared with 10-membered ring (MR) TS-1, these later-emerged titanosilicates usually possess larger pores, which favors the adsorption and diffusion of reactants and products. It is found that **MOR** and **MWW**-type titanosilicates exhibit comparable and even better catalytic ammoximation performance than the conventional TS-1 catalysts, showing the potential for industrial application. For the sake of achieving the clean production of various oximes, the catalytic activity over titanosilicates in the ammoximation of various aldehydes and ketones, such as acetone, methyl ethyl ketone (MEK), and furfural, has been investigated. In comparison with the traditional TS-1, Ti-MWW and Ti-MOR show higher activity, oxime selectivity, and catalytic stability in various ammoximation reactions. The manufacture of MEK oxime through ammoximation based on Ti-MWW/H_2O_2 and Ti-MOR/H_2O_2 systems has been industrialized in China.

In the liquid-phase catalytic ammoximation system, solvents have effect on the catalytic properties and titanosilicates with different structures exhibiting distinct solvent effects. *t*-Butanol has been verified to be the optimal solvent for ammoximation process in the titanosilicate/H_2O_2 system, and nearly all the current oximes industrial production process catalyzed by titanosilicates employs *t*-butanol as solvent. During the alkaline ammoximation process, the presence of organic solvents remarkably limits the damaging effect of ammonia, which tends to destroy the active sites and —Ti—O—Si— structure in titanosilicates [27]. In addition, there is discrepancy of the solubility of the reactants in the various organic solvents, and with *t*-butanol as solvent could change the reaction system from organic liquid-aqueous liquid-solid to one liquid phase-solid, which favors the contact with reactants for improved ammoximation performances [28]. However, different from TS-1, which shows excellent ammoximation activity only with *t*-butanol as solvent, Ti-MOR and Ti-MWW exhibit desired catalytic performance with H_2O solvent (form aqueous H_2O_2), showing potential to develop the green manufacture of oximes in the organic solvent-free system. It will cut the energy cost required to evaporate and recycle the used *t*-butanol solvent in the practical production process, which will be elaborated in detail in Section 10.5.

Table 10.1 Summary of titanosilicates with different structures in the ammoximation of ketones or aldehydes in various solvents.

Year[a)]	Ti-zeolite	Topology	Channels (MR)	**Ammoximation**		Ref.
				Substrate	**Solvent**	
Microporous titanosilicates						
1988	TS-1	**MFI**	10 × 10	Cyclohexanone	*t*-BuOH	[2, 18, 20–73]
					H_2O	[17, 59, 74–82]
					i-PrOH	[83]
					MeOH	[84]
					MeCN	[85]
					Cyclohexane	[86]
					Dimethyl carbonate	[49]
					Acetic acid	[76]
				Methyl ethyl ketone	*t*-BuOH	[44, 53, 87–95]
					i-PrOH	[94]
					H_2O	[94, 95]
					MeOH	[96]
				Acetaldehyde	*t*-BuOH	[89, 97, 98]
					H_2O	[99]
				Acetone	*t*-BuOH	[44, 53, 89, 100–106]
					H_2O	[100, 101, 103, 106–108]
					MeOH	[96, 101]
					i-PrOH	[103]
				Citronellal	*t*-BuOH	[109]
				Benzaldehyde	H_2O	[110, 111]
				Furfural	H_2O	[112]
				5-Isooctyl salicylaldehyde	*t*-BuOH	[113]
1996	Ti-Beta	Intergrowth family	12 × 12 × 12	Cyclohexanone	*t*-BuOH	[2, 87]
		Beta		Benzaldehyde	H_2O	[110]
				Hydroxyacetophenones	*t*-BuOH	[2]
1996	Ti-ZSM-48	Intergrowth family	10	Cyclohexanone	*t*-BuOH	[2]
		ZSM-48		Hydroxyacetophenones	*t*-BuOH	[2]

1997	Ti-MOR	**MOR**	12–8	Cyclohexanone	*t*-BuOH	[17, 37, 41, 48, 87, 114, 115]
					H_2O	[18, 115–117]
				Methyl ethyl ketone	H_2O	[88, 115]
				Acetaldehyde	*t*-BuOH	[97]
				Benzaldehyde	H_2O	[110]
				Acetone	*t*-BuOH	[100, 101, 115]
					H_2O	[100, 101, 115]
					MeOH	[100]
				Furfural	H_2O	[112]
2000	Ti-ZSM-5	**MFI**	10 × 10	Cyclohexanone	H_2O	[118]
2005	Ti-MWW	**MWW**	10–10	Cyclohexanone	*t*-BuOH	[37, 41, 44, 48, 67, 119–124]
					H_2O	[11, 125–127]
					Dimethyl carbonate	[49]
				Methyl ethyl ketone	*t*-BuOH	[44, 89, 93, 95, 128, 129]
					H_2O	[87, 88, 95, 127]
				Citronellal	*t*-BuOH	[109]
				Acetaldehyde	*t*-BuOH	[89, 97]
				Benzaldehyde	H_2O	[110]
				Acetone	*t*-BuOH	[44, 89, 100]
					H_2O	[100]
					MeOH	[101]
				Furfural	H_2O	[112]
2015	TS-2	**MEL**	10	Cyclohexanone	*t*-BuOH	[44, 130]
				Methyl ethyl ketone	*t*-BuOH	[44]
				Acetone	*t*-BuOH	[44]
2018	TAPO-5	**AFI**	12	Cyclohexanone	*t*-BuOH	[131]
2020	Ti-MCM-68	**MSE**	12 × 10 × 10	Cyclohexanone	*t*-BuOH	[48]
2022	Ti-ERB-1	**MWW**	10–10	Cyclohexanone	*t*-BuOH	[67]
Mesoporous titanosilicates						
2004	Ti-DMS1			Cyclic ketones	*t*-BuOH	[132]
2004	Ti-MCM-41		1D hexagonal	Cyclic ketones	*t*-BuOH	[132]
2017	Ti-PMO		1D hexagonal	Cyclohexanone	*t*-BuOH	[133]

a) The year firstly reported for the catalytic ammoximation with a specific titanosilicate catalyst.

10.3 Ammoximation Mechanism and Product Distributions of Representative Ketones and Aldehydes

The ammoximation of ketones or aldehydes with hydrogen peroxide and ammonia on titanosilicates produces corresponding oximes from simple one-pot systems, in which the reaction occurs in a relatively complicated way as three reactants are involved. Understanding its mechanism is helpful for developing more effective catalysts and more importantly controllable engineering technology useful to practical applications. With respect to the ammoximation of specific carbonyl compounds using different titanosilicate catalysts, the reaction pathways and catalytic performances greatly differ.

10.3.1 Titanosilicate-Catalyzed Ammoximation Mechanism

The titanosilicate/hydrogen peroxide systems usually exhibit an obvious shape-selective catalysis feature in the oxidation of various alkenes, amines, and organic sulfides, that is, more active for small and linear substrates but showing greatly retarded activity for aromatics, branched and cyclic molecules with bulky dimensions. The shape-selectivity and molecular size effect, however, are not prominent in ammoximation, as cyclic and aromatic ketones and even with branched substituent groups are also readily converted to oximes [17], implying a different mechanism from that of alkene epoxidation where alkene molecules need to enter the micropores and reach the Ti sites or the Ti—OOH species (after interaction with hydrogen peroxide). Up to date, two possible ammoximation mechanisms have been proposed (Figure 10.1), that is, so-called imine route and hydroxylamine route. The ammoximation via imine intermediate has been hypothesized because the carbonyl groups of ketones or aldehydes react with ammonia to give imines readily even in the absence of any catalyst. Once imines are formed, catalytic oxidation on the Ti/H_2O_2 or TiOOH active species possibly converts them into the oxime products (Figure 10.1a). In this case, the imine intermediates have to enter into the pores, and then obvious shape-selectivity phenomena should exist, which is in conflict with the experimental results actually observed in above-mentioned ammoximation. The cyclohexylimine formed through the reaction between cyclohexanone and ammonia has been confirmed over TS-1 with IR spectroscopy [21, 134, 135], which is thus taken as the evidence to support the imine route mechanism. In addition, the imines are more easily formed when reactive aldehydes are employed in the ammoximation. However, an IR study using Ti-MOR indicates that the formation of cyclohexylimine from the interaction of adsorbed cyclohexanone molecules with gas-phase ammonia is exactly possible, while once water vapor is introduced cyclohexylimine is completely transformed back to cyclohexanone through reversible hydrolyzation [17]. This study excludes the possible imine route as water from aqueous hydrogen peroxide is always present in the ammoximation reaction system, suppressing the formation of imines.

The hydroxylamine route [136], firstly proposed by ENI group on the ammoximation of cyclohexanone on TS-1, is becoming a widely acceptable and well-clarified

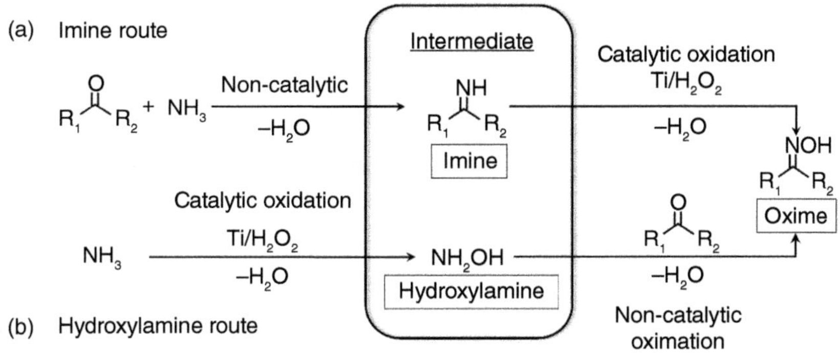

Figure 10.1 Two possible reaction mechanisms proposed for the titanosilicate-catalyzed ammoximation with (a) imine and (b) hydroxylamine as intermediate, respectively.

mechanism. As shown in Figure 10.1b, ammonia is first catalytically oxidized by the Ti/H_2O_2 or TiOOH active species to yield free hydroxylamine, which subsequently reacts with the ketones or aldehydes through non-catalytic oximation to produce corresponding oximes. As hydroxylamine is a relatively small molecule, it is formed readily inside zeolite pores and then can react with those carbonyl compounds diffusing into the pores, or hydroxylamine itself diffuses out of the zeolite pores to react with the carbonyl molecules in external reaction media. The oximation in both ways contributes to the oxime production. The carbonyl molecules are not necessary to enter the pores, which reasonably explains why the ammoximation does not obey a strict reactant shape-selective catalysis, and the catalytic activity of oxime production is still high for cyclohexanone or other bulkier ketones that are hardly accommodated by medium-pores of TS-1. The hydroxylamine mechanism is proposed and well-studied with TS-1, and thereafter, it was verified on the ammoximation with the titanosilicate catalysts of Ti-MWW and Ti-MOR with opener or larger porosity.

The formation of hydroxylamine has been experimentally confirmed in the oxidation of ammonia with hydrogen peroxide over TS-1, Ti-MWW, and Ti-MOR either in the presence or absence of ketones or aldehydes [23, 70]. Nevertheless, free hydroxylamine is much less stable than its chemically available salts of chloride and sulfate that are employed as oximation agents in the conventional oxime manufacture processes. With the presence of ketones, free hydroxylamine rapidly oximize them into oximes, enhancing the whole ammoximation efficiency (Scheme 10.1). In the absence of ketones, that is, the single oxidation of ammonia with hydrogen peroxide on titanosilicate, free hydroxylamine formed is easily decomposed or over-oxidized into N_2, NO_x, nitrates, or nitrites either in a catalytic or non-catalytic way. This would lower the utilization efficiency of both ammonia and hydrogen peroxide and the free hydroxylamine yield. Ti-MOR and TS-1 prove to be extremely effective for the oxidation of ammonia with hydrogen peroxide to form hydroxylamine; meanwhile, the hydroxylamine decomposition takes place relatively difficult on these two titanosilicates in comparison to Ti-MWW [37]. Thus, Ti-MOR and TS-1 can achieve high hydroxylamine yields such as 82.7–95.1%. In contrast, although serving as an

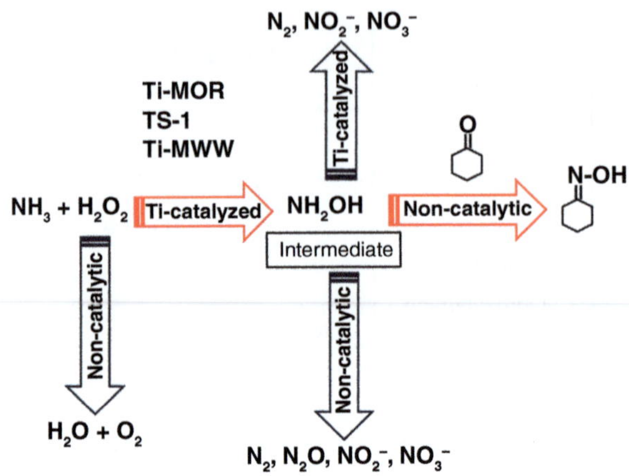

Scheme 10.1 Overall reaction network of titanosilicate-catalyzed ketone ammoximation. Source: Reprint from Ref. [37] with the copyright permission from Elsevier.

effective ammoximation catalyst, Ti-MWW is less efficient in terms of the hydroxylamine yield because of its high catalytic activity for the decomposition of hydroxylamine. Ion-exchange with alkali metal cations removes the hydroxyl-related acid sites on titanosilicates, significantly improving its catalytic activity in both reactions of hydroxylamine formation via ammonia oxidation with hydrogen peroxide and oxime production via ammoximation.

10.3.2 Product Distributions for Ammoximation of Representative Carbonyl Compounds

The ammoximation of ketones or aldehydes principally follows the mechanism with hydroxylamine as intermediate irrespective of the topologies of titanosilicates, possibly achieving desirable conversion and oxime selectivity by optimizing the operating conditions. The catalytic performances and product distributions for specific ketones and aldehydes, however, are different among the three representative titanosilicates of TS-1, Ti-MWW, and Ti-MOR. They are capable of providing comparably high cyclohexanone oxime selectivity (>99%) in the ammoximation of cyclohexanone, generally taking a catalytic activity order of Ti-MOR > Ti-MWW > TS-1 in terms of catalyst loading. The residual byproducts, although extremely low in absolute amount, are diverse in category. Taking the Ti-MWW deactivated in cyclohexanone ammoximation as an example, the GC-MS analysis detects several representative organic byproducts after dissolving the used catalyst in HF solution as listed in Table 10.2 [119]. They originate from either the reactant cyclohexanone, the product oxime, or both of them by tandem catalytic reactions of rearrangement, condensation, etc.

In the case of ammoximation of linear ketones such as MEK and dimethyl ketone (acetone, DMK), more complicated product distributions are observed as these ketones and corresponding oximes are more active, making the absolute

Table 10.2 Major by-products detected in cyclohexanone ammoximation by GC-MS.

No.	Molecular weight	Compound	Formula
1	100	Cyclohexanol	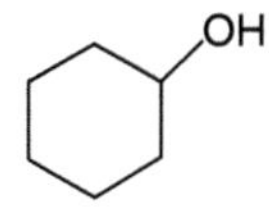
2	127	Cyclohexanecarboxamide	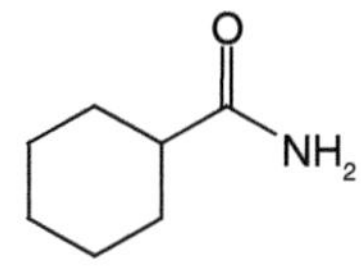
3	180	2-Cyclohexylcyclohexanone	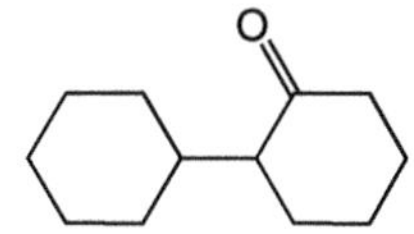
4	179	*N*-Cyclohexylidenecyclohexanamine	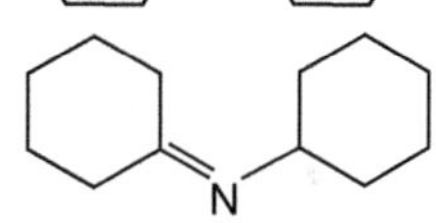
5	113	ε-Caprolactam	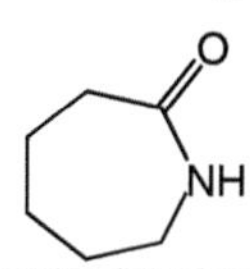

Source: Reproduced from [119]/with permission of Elsevier.

amount of the formed byproducts greatly relay on the choice of titanosilicates. The reaction pathways and product distributions of MEK and DMK ammoximation are summarized in Scheme 10.2 and Scheme 10.3, respectively. MEK oxime is produced through first formation of hydroxylamine and then non-catalytic oximation of MEK. Both Ti-MWW and Ti-MOR possibly give much a higher methyl ethyl ketoxime (MEKO) selectivity (>99%) at low catalyst loading, when compared with TS-1 [87, 88]. As shown in Scheme 10.2, the hydroxylamine intermediate may be decomposed into N_2, NO_x, nitrate, etc., while the MEKO product could return to MEK by acid-catalyzed hydrolysis. These two reactions mainly lower the efficiency of reactant utilization and hinder the reaction process. A dominant byproduct 2-nitrobutane as a result of Ti-catalyzed over-oxidation of MEKO by H_2O_2 is usually observed on TS-1, and then the MEKO selectivity is lowered to 75.3–80%. This may induce various problems such as consumption of materials, MEKO product quality, and even safety as 2-nitrobutane is a chemically explosive compound.

Scheme 10.3 summarizes the reaction pathways in the ammoximation of DMK with a more active nature than MEK. Non-catalytic condensation of DMK with NH_3 and H_2O_2 takes place relatively easily, possibly forming 2,2,6,6-tetramethyl-piperidin-4-one (TMPDO), a mixture of diacetone amine, triacetone diamine, triacetone amine, and other products. TMPDO can further undergo ammoximation like conventional ketones to yield 2,2,6,6-tetramethylpiperidin-4-one oxime

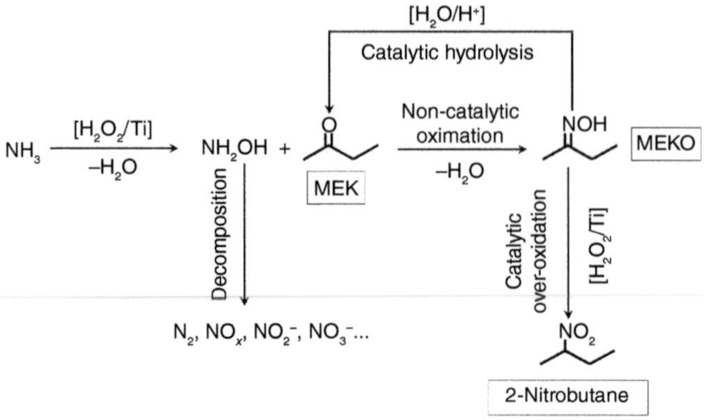

Scheme 10.2 Reaction pathway and product distribution of MEK ammoximation.

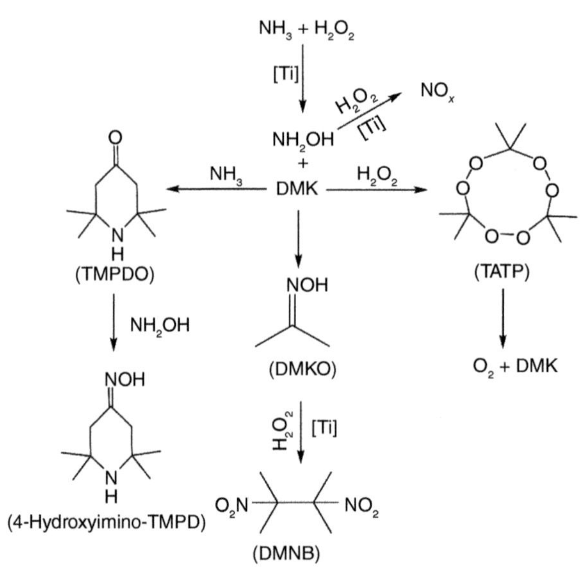

Scheme 10.3 A summary of reaction pathways involved in dimethyl ketone (DMK) ammoximation. Source: Reprinted from Ref. [100] with the copyright permission from Elsevier.

(4-hydroxyimino-TMPD). The reaction between H_2O_2 and DMK may produce a primary dangerous explosive peroxide of 3,3,6,6,9,9-hexamethyl-1,2,4,5,7,8-hexaoxonane (i.e. triacetone triperoxide, TATP), which could further decompose into O_2 and DMK. In addition, once dimethyl ketone oxime (DMKO) is formed, it can also go through deoximation or deep oxidative over H_2O_2/titanosilicate to give a major byproduct 2,3-dimethyl-2,3-dinitrobutane (DMNB). The DMK ammoximation, tested in a batchwise reactor with *t*-BuOH as solvent, indicates that the performance in terms of the DMK conversion and DMKO selectivity follows the order of Ti-MOR (99.2%, 99.5%) > Ti-MWW (98.9%, 97.9%) ≫ TS-1 (97.8%, 78.1%) [100].

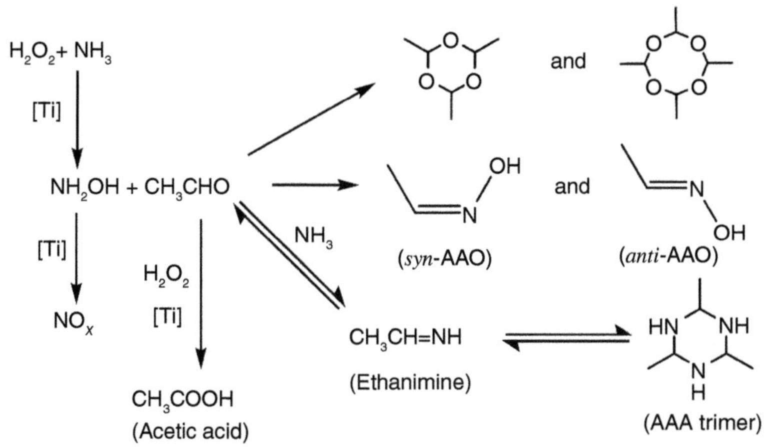

Scheme 10.4 A summary of reaction pathways involved in acetaldehyde (AA) ammoximation. Source: Reprinted from Ref. [97] with the copyright permission from The Royal Society of Chemistry.

Aldehydes are more chemically reactive than ketones, making the selective production of aldehyde oximes difficult with the ammoximation over conventional titanosilicates. Scheme 10.4 summaries the reaction pathways in the ammoximation of acetaldehyde (AA). Non-catalytic polymerization of AA is known to occur at room temperature when AA is stocked in aqueous solution, forming trimer or tetramer. In addition, even in the absence of any catalyst, non-catalytic reaction of AA with ammonia takes place easily to yield ethanimine which possibly further undergoes trimerization to form acetaldehyde-ammonia (AAA) trimer. Moreover, the oxidization of AA by H_2O_2 on the active Ti sites may yield acetic acid since AA with a small molecular dimension could enter readily into the channels of titanosilicates. All these side reactions may decrease the efficiency of AA ammoximation and the acetaldehyde oxime (AAO) selectivity. After optimizing the conditions of AA ammoximation, 12-MR large-pore Ti-MOR shows more attractive catalytic properties than TS-1 and Ti-MWW, providing AA conversion and AAO selectivity both high as 99% [89, 97–99], while the maximum AA conversion of TS-1 and Ti-MWW is in a lower range of 75–78% as these two titanosilicates are more active to the oxidation of AA to acetic acid. The AAO products have two configurational isomers, and their distributions achieved with TS-1, Ti-MWW, and Ti-MOR are very comparable, showing similar $anti/(anti + syn)$ ratio of 0.41–0.43. Ti-MOR possesses another advantage for the AA ammoximation process, in that its AA conversion and AAO selectivity are less influenced by the addition methods of AA, ammonia, and hydrogen peroxide. When one-pot mixing three reactants directly or adding each of reactants dropwise, the AA conversion and the AAO selectivity of are always maintained high as the formation of acetic acid is extensively suppressed with Ti-MOR as the catalyst. This is extremely important as easy handling catalytic technology and processes are always welcomed by industrial applications.

Obtaining value-added chemical products from renewable biomass sources is an important approach for developing environmentally friendly and sustainable processes, and the same is true for oxygen and nitrogen-containing products. Furfural, produced by acid-catalyzed dehydration of cellulose biomass, has attracted much attention in past years and employed as a platform chemical to produce biofuels and downstream chemicals. Furfural's another derivative, i.e. furfural oxime (FO), is an important intermediate chemical useful to produce biologically active compounds with a wide range of medicinal properties. Moreover, FO has gained attention as a versatile platform chemical to produce 2-furylamide, 2-furonitrile, 2-furfurylamine, and tetrahydro-2-furfurylamine through selective hydrogenation and rearrangement reactions (Scheme 10.5).

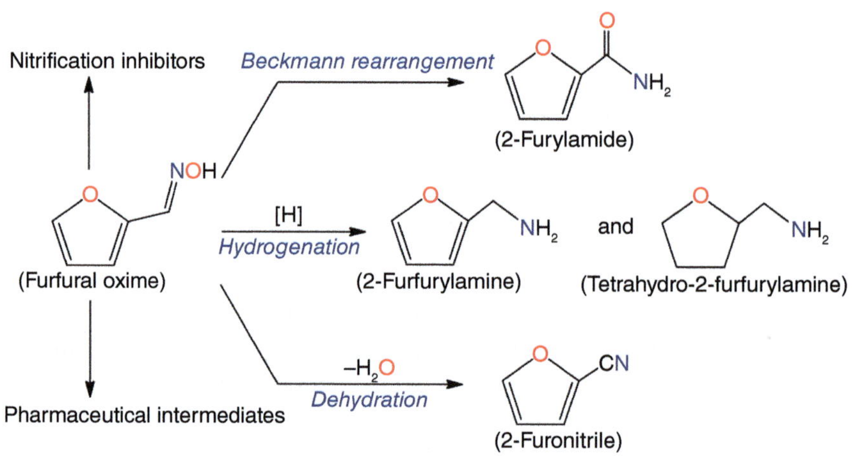

Scheme 10.5 Downstream chemicals and potential applications of biomass-derived furfural oxime. Source: Reprinted from Ref. [112] with the copyright permission from The Royal Society of Chemistry.

The ammoximation of biomass furfural shows possible reaction pathways as shown in Scheme 10.6, in which the imine route and the hydroxylamine route may parallelly occur to give a complicated product distribution [113]. Furfural reacts with ammonia to yield corresponding imine, which is further oxidized by hydrogen peroxide to produce 2-furylamide and 2-furoic acid. Following the hydroxylamine route, FO is produced from the main catalytic reaction, and it is possibly converted to the main by-product of 2-furonitrile due to the acid-catalyzed dehydration of FO. Investigating the solvent effects on the ammoximation of furfural indicates that Ti-MOR is superior to TS-1 and Ti-MWW in both furfural conversion and FO selectivity. Under optimized conditions, Ti-MOR can provide furfural conversion and oxime selectivity both above 97% and serves as a robust and regenerable catalyst showing unchanged catalytic properties in the green solvent of water. The performance of furfural ammoximation is closely related to the activity of titanosilicates in catalyzing hydroxylamine decomposition. The decomposition of hydroxylamine and the non-catalytic oxidation of furfural can be effectively suppressed in Ti-MOR

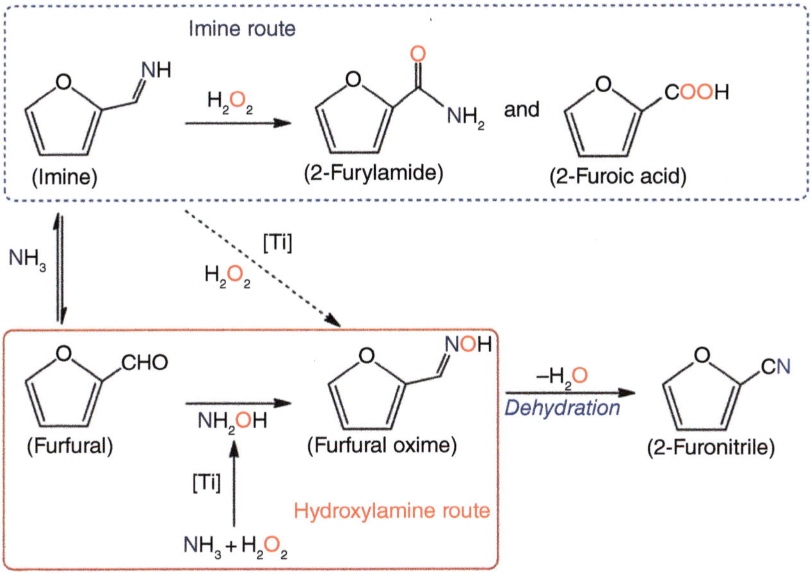

Scheme 10.6 Possible reaction pathways in furfural ammoximation. Source: Reprinted from Ref. [112] with the copyright permission from The Royal Society of Chemistry.

catalyzed system when employing water as solvent and adding H_2O_2 dropwise into the reaction system.

10.4 Enhancing Ammoximation Performances in Titanosilicate/H_2O_2 System

Scheme 10.7 summarizes the factors that influence the performance of titanosilicate/H_2O_2 catalytic ammoximation system. First, no matter whether the ammoximation reaction is based on hydroxylamine mechanism or imine mechanism, it is no doubt that the activation of hydrogen peroxide at Ti sites and transfer of active oxygen is the core of titanosilicate/H_2O_2 catalytic system. Although the acid sites in titanosilicates do not directly participate the ammoximation process, they will affect the apparent activity by influencing the catalytic behavior of hydrogen peroxide and reaction intermediates. In addition, as a typical heterogeneous reaction, the ammoximation performance over titanosilicates is not only affected by the intrinsic catalytic ability of Ti active sites but also influenced by the diffusion properties of guest molecules, which depends on the pore structure and morphology. And the hydrophilicity of catalysts will also affect the adsorption and desorption of guest molecules and thus influence the catalytic performance.

Ammonia serves as one of the important reactants in the ammoximation reaction, resulting in a strongly alkaline reaction medium. This will dissolve the skeleton of titanosilicates and leach the active sites and then accelerate the decline of

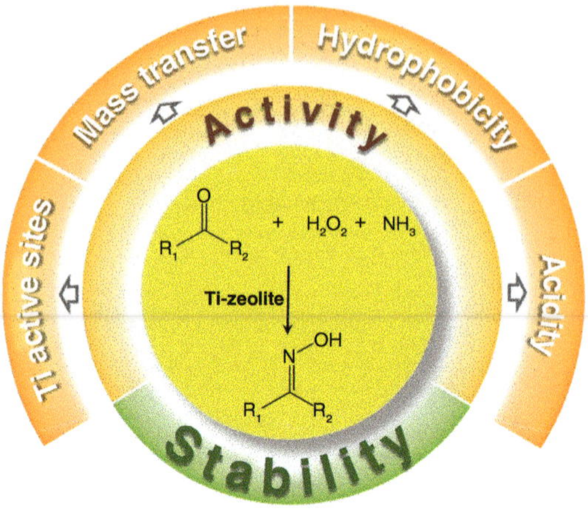

Scheme 10.7 Influencing factors involved in ammoximation catalyzed by Ti-zeolite/H_2O_2 system.

catalytic stability. Therefore, enhancing the catalytic stability of titanosilicate/H_2O_2 catalytic ammoximation system is also a key to improve the catalytic ammoximation performance.

10.4.1 Improvement of Catalytic Ammoximation Activity

10.4.1.1 Regulation of Ti Active Sites

Ti active sites are known to play the essential role in titanosilicates. The performance of catalytic ammoximation can be improved by means of adjusting the chemical state and distribution of Ti sites (Figure 10.2). The framework Ti contributes to the weak acid sites, while the extra-framework Ti contributes to the strong acid sites, which generally catalyze undesirable side reactions such as ineffective decomposition of hydrogen peroxide and peroxidation of products. It has been found by Huang et al. that the employment of ethanol instead of *iso*-propanol in the crystallization gel promoted the incorporation of Ti into the framework of TS-1 and retarded the formation of anatase. The prepared TS-1 sample with a large content of framework Ti but without extra-framework Ti and anatase exhibited improved catalytic performance in cyclohexanone ammoximation [31]. Besides, modification with ethanolamine can alter the distribution of Ti species in TS-1 (Figure 10.2a). The uniform distribution of Ti species resulted in an improved catalytic performance, and thus-synthesized M-TS-1 stably maintained the performance at a high level when the cyclohexanone ammoximation was carried out for 410 hours [40]. In addition, Tang et al. reported that the direct synthesis of Ti-MWW containing Al and B atoms could regulate the even distribution of Ti species on the surface of Ti-MWW, derived from the competitive occupancy of the Al, B, and Ti heteroatoms over T sites. It could enhance the accessibility of Ti sites and thus improve the catalytic ammoximation activity [122].

Based on the catalytic nature of titanosilicates and the ammoximation mechanism, the Lewis acidic strength of Ti active sites is confirmed to be responsible for the

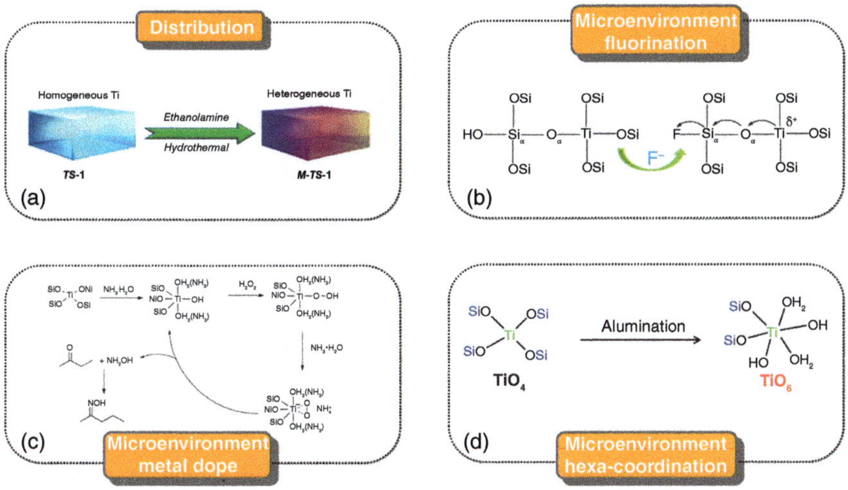

Figure 10.2 The presentative methods about regulating the distribution (a) and microenvironment of Ti active sites including fluorination (b), Ni dope (c), and construction of hexa-coordinated Ti sites (d) to enhance the catalytic ammoximation activities. Source: (a) Reproduced from [40]/with permission of Royal Society of Chemistry. (b) Reproduced from [115]/with permission of Elsevier. (c) [92]/MDPI/CC BY 4.0.

high selectivity of oxime. It has been found by Zhuo et al. that stronger Lewis acid promoted the formation of NH_2OH, while weaker acid strength is less able to suppress the free H_2O_2-induced side reactions [89]. Therefore, regulating the Lewis acidity of titanosilicates could become a crucial guideline to design or modify catalysts for highly efficient synthesis of oximes. Yang et al. have introduced F^- into the framework of Ti-MOR by a hydrothermal treatment with ammonium fluoride/methanol solution (Figure 10.2b). The implantation of fluorine increases the electropositivity of Ti active sites through forming $SiO_{3/2}F$ units in the neighborhood, which remarkably enhanced the catalytic performance in the ammoximation of cyclohexanone [115]. Analogously, the fluorine modification over TAPO-5 increases the surface hydrophobicity and the Lewis acidity of Ti active sites through forming Al_xF_y units in the neighborhood, making these fluorine modified TAPO-5 sieves exhibit good catalytic performance in the ammoximation of cyclohexanone [131].

In addition, metal doping can also regulate the microenvironment of Ti active sites in titanosilicates. It was demonstrated that the skeleton structure of TS-1 was basically maintained and the electron environment of the Ti active site is changed after the nickel modification, which can optimize the adsorption capacity and the activation for H_2O_2 (Figure 10.2c). Ni-TS-1 exhibited a satisfactory conversion of MEK (99%) and a high selectivity of MEKO (99.3%), which was 6% higher than that of TS-1 under the same condition [92]. During the hydrothermal post-treatment with a aqueous solution containing aluminum salts over Ti-MOR, it has been found that part of tetrahedrally coordinated Ti species rearranged to hexa-coordinated Ti species $Ti(OSi)_2(OH)_2(H_2O)_2$ (TiO_6) (Figure 10.2d). The newly formed open TiO_6 species favored the formation of hydroxylamine compared to tetra-coordinated Ti species, resulting in obviously enhanced catalytic performance in the cyclohexanone ammoximation reaction [18].

10.4.1.2 Enhancement of Diffusion Properties

Compared with the gas-phase reaction under high temperature, the liquid-phase catalytic ammoximation under mild conditions faces problems of more obvious mass transfer resistance, which results in the block of organic compounds and coke deposition in micropores. Except for the employment of titanosilicates with larger channels, shortening the diffusion path is also an effective strategy to alleviate the diffusion resistance.

First, the diffusion path can be shortened by accurately controlling the morphology of titanosilicates. By means of altering silicon source and alkali source to control the crystallization process of **MFI**-type titanosilicate, and lamellar-stacked TS-1 can be thus obtained, which effectively avoid the diffusion resistance of large reactant molecules into zeolite channels, showing better catalytic performance for butanone ammoximation with H_2O_2 than conventional TS-1 [91]. It is generally believed that the 12-MR channels along *c*-axis of **MOR** zeolite are mainly in charge of the diffusion of guest molecules. A series of Ti-MOR catalysts with different morphologies and lengths along the *c*-axis were prepared by Yang et al. It is found that the formation of tetrahedrally coordinated Ti in $TiCl_4$ vapor treatment, the physicochemical properties, and catalytic activity of Ti-MOR exhibit regular changes with increasing crystal length along the *c*-axis, and a shorter length along the *c*-axis of Ti-MOR catalysts leaded to higher catalytic activity in the cyclohexanone ammoximation reaction [116].

In addition, it is a feasible strategy to construct hierarchical structure in titanosilicates to improve mass-transfer properties and the ammoximation activity. The introduction of hard templates and inorganic nanocrystals in the crystallization system of titanosilicates could construct hierarchical materials. The introduction of $Mg(OH)_2$ nanocrystals in the synthetic gel endows the as-synthesized TS-1 with a much more rugged surface and some voids and slits (Figure 10.3a). Benefiting from the unique morphology, TS-1 exhibited improved cyclohexanone conversion and oxime selectivity in batchwise cyclohexanone ammoximation [72]. Besides, the hydrothermal treatment of TS-1 with quaternary ammonium base such as tetrapropylammonium hydroxide (TPAOH) is one of the most classical strategies to construct hierarchical pore systems in titanosilicates. During the treatment process, hydroxide ions diffuse into the interior of crystals to dissolve the skeleton, while quaternary ammonium ions cannot diffuse into the 10-MR channels of **MFI** structure. The quaternary ammonium ions not only inhibit the excessive dissolution of Si on the outer surface but also direct the dissolved Ti and Si species to reconstruct **MFI** framework. In this dissolution-recrystallization process, the morphology of TS-1 generally changed and hierarchical or even hollow structure was constructed, leading to the improved catalytic ammoximation activity [38, 55, 56, 69, 78]. Analogically, the hydrothermally alkali treatment with the mixture of piperidine (PI), tetrabutyl ammonium hydroxide (TBAOH), and NH_4Cl could increase the catalytic performance of TS-2. The conversion and selectivity could reach >99% for post-treated MS-TS-PI-Na catalyst

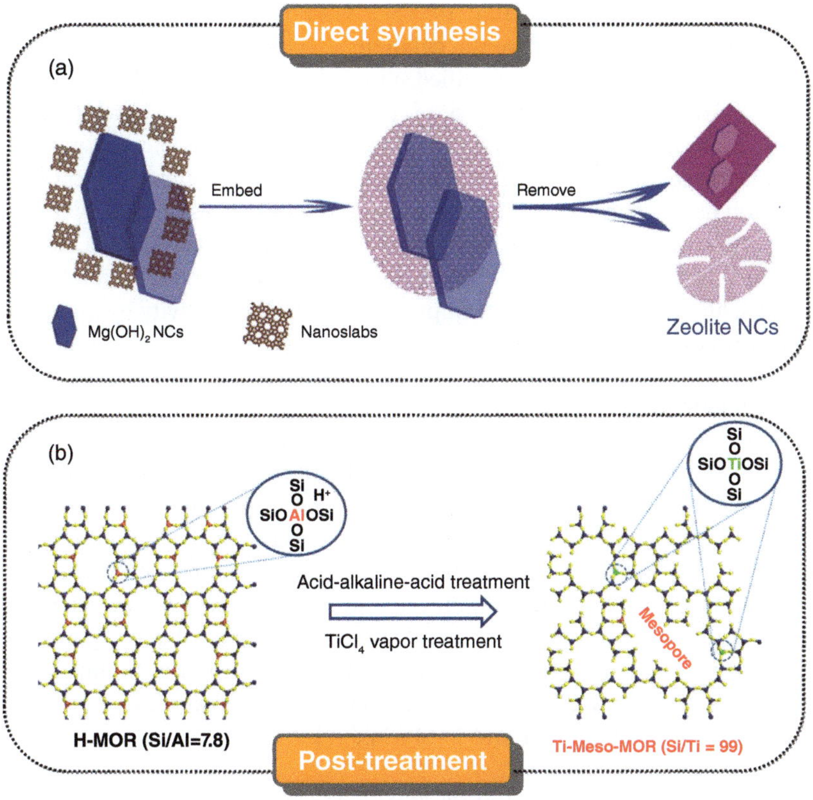

Figure 10.3 The representative methods for enhancing the diffusion properties and the catalytic ammoximation activity through direct synthesis (a) and post-treatment strategy (b). Source: (a) Reproduced from [72]/with permission of Royal Society of Chemistry. (b) Reproduced from [114]/with permission of Elsevier.

and the lifetime is comparable to the industrial TS-1 in the cyclohexanone ammoximation reaction [130]. The employment of alkali to dissolve Si in titanosilicates can achieve the goal of constructing hierarchical structures as well. After hydrothermally treated TS-1 with aqueous solutions of triethylamine, some irregular hollow cavities are created, leading to increased catalytic ammoximation performance [42]. Xu et al. have treated Al-MOR with sequential acid, alkaline, acid, and $TiCl_4$ vapor treatments (Figure 10.3b). Alkaline treatment-induced desilication introduced many intracrystal mesopores. Ti-Meso-MOR thus prepared exhibited improved catalytic activity in the batchwise and continuous ammoximation of cyclohexanone reactions [114].

10.4.1.3 Improvement of Hydrophobicity

In the liquid-phase reaction, the hydrophilicity of catalyst, mainly derived from the hydroxyl groups represented by Si—OH, influences the catalytic activity through affecting the enrichment of reactants near Ti sites. The fluorine modification was used to quench the hydroxyl species in titanosilicates to form hydrophobic species such as Si—F, and thus improved the catalytic ammoximation activity [115, 131]. The modification of TS-1 with organosilanes such as triethoxyfluorosilane can also improve the catalytic activity in the ammoximation of cyclohexanone compared to the unmodified sample [30, 80]. During the hydrothermal dissolution-recrystallization process of TS-1, the hierarchical pore structure formed to favor mass transfer, while the defects in the skeleton were partially repaired to achieve higher hydrophobicity, resulting in an improved catalytic ammoximation performance over TS-1 [78]. Wang et al. have treated Ti-MWW with PI or hexamethyleneimine (HMI) to remove the internal silanols by ca. 40%, leading to a defect-less Ti-MWW catalyst with a more rigid and hydrophobic framework. This kind of structural rearrangement improved the catalytic activity by up to 20% in the ammoximation of ketones [127].

10.4.1.4 Regulation of Acid Sites

In addition to Si—OH and Ti—OH groups, the acid sites in Ti-containing zeolites related to framework heteroatoms such as Al and B may lead to side reactions, such as the deep oxidation of hydroxylamine and oxime. Taking Ti-MOR as an example (Scheme 10.8), ion exchange with neutral alkali salts selectively removed the acid sites related to Si(OH)Al and Si—OH groups, which enhanced the catalytic activity for hydroxylamine formation and cyclohexanone ammoximation. Ion exchange with basic alkali salts, on the other hand, may cause non-selective poisoning between acid and Ti sites, leading to an irreversible deactivation of titanosilicates [18, 37]. The method of selectively quenching adverse acid sites with appropriate alkali salts is also suitable for TS-1 and Ti-MWW [37].

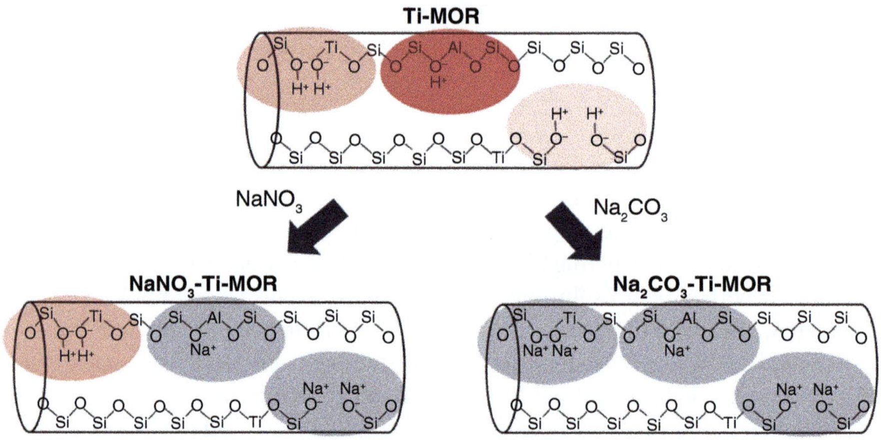

Scheme 10.8 Post-treatment of Ti-MOR with different sodium salts. Source: Reprinted from Ref. [37] with the copyright permission from Elsevier.

10.4.2 Improvement of Catalytic Ammoximation Stability

Except for the reversible deactivation of coke deposition, the irreversible deactivation, that is, the skeleton desilication accompanied by the leaching of Ti active sites arouses more attention during the long-time industrial production of cyclohexanone oxime under the alkali reaction system. As mentioned in Section 10.4.1, the diffusion efficiency can be improved by adjusting the morphology of titanosilicates and constructing hierarchical pore structure to prolong the catalytic lifetime by depressing the coke deposition. From the viewpoint of retarding the irreversible deactivation, enhancing the anti-corrosion properties of catalysts and constructing an alkaline corrosion-resistant ammoximation system are of great effective.

Zhao et al. suppressed Si leaching in Ti-MWW/H_2O_2/*t*-BuOH system by continuously feeding 5 ppm colloidal silica solution into the reaction mixture, which prolonged the ammoximation lifetime from 93 to 105 hours [119]. Apart from adding the additives into the reactor, Xu et al. fabricated a core-shell-structured Ti-MWW with meso-SiO_2 as a shell (Figure 10.4a), which served as a sacrificial lamb to protect the

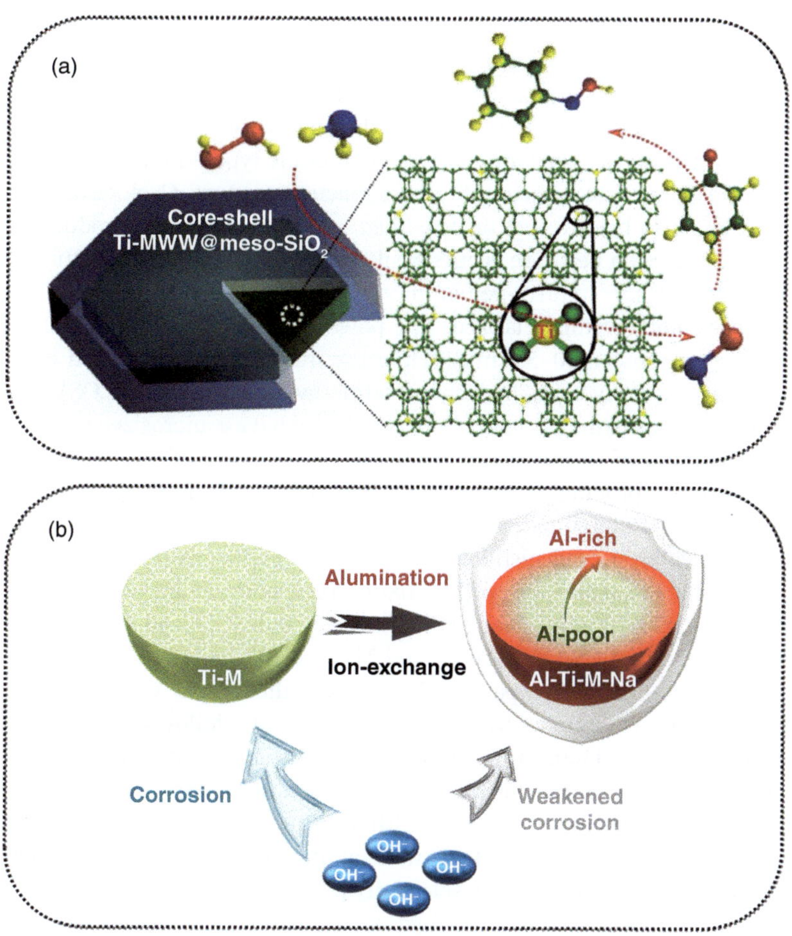

Figure 10.4 The strategies of fabricating meso-SiO_2 shell on Ti-MWW (a) and Al-rich shell on Ti-MOR (b) to enhance the anti-corrosion properties of titanosilicates. Source: (a) Reproduced from [121]/with permission of American Chemical Society. (b) Reproduced from [18]/with permission of Royal Society of Chemistry.

active component of zeolite core against rapid desilication, leading to a prolonged lifetime of 198 hours compared to 132 hours for pristine Ti-MWW [121]. Recently, our group proposed a strategy to implant Al into Ti-MOR framework to improve the anti-desilication performance through a hydrothermal treatment with Al salts (Figure 10.4b). The introduced Al species enriched on the crystal shell, serving as a protective shield against alkaline corrosion. And the Al-modified Ti-MOR showed superior stability than the pristine Ti-MOR and TS-1, contributing to a tremendously prolonged lifetime in the continuous ammoximation of cyclohexanone [18].

10.5 Ketone Ammoximation Technology for Industrial Processes

Titanosilicate-catalyzed ammoximation is highly efficient in terms of ketone conversion, oxime selectivity, and hydrogen peroxide utilization efficiency, meeting the demand of commercial applications. The reaction is conducted using continuous slurry-bed reactors in industrial plants, using either nanosized powder or microsphere titanosilicate catalysts depending on the membrane types employed for the separation of solid catalyst from the liquid-phase reaction mixture. Cyclohexanone ammoximation has also been tested on fixed-bed reactors, but the unproductive decomposition of hydrogen peroxide occurs easily due to its less effective diffusion to the Ti active sites [41]. For slurry-bed ammoximation, TS-1 and Ti-MOR power catalysts have been both commercialized. The powder catalysts are composed of nanosized primary crystals (200–500 nm), while the microspheres, prepared by a spray-drying method usually using hydrolyzed tetraethyl orthosilicate (TEOS) as a binder, are the aggregates of the primary nanocrystals and have a microsized particle distribution in the range of 10–50 μm (Figure 10.5).

The process of oxime production comprised three main workshop sections, that is, titanosilicate-based ammoximation (slurry bed reactor), membrane separation for recycling solid catalyst, and distillation for recycling unconverted ammonia, reaction solvent and extractant, as well as oxime separation (Figure 10.6). The ammoximation of cyclohexanone and MEK has been industrialized using commercial catalysts including TS-1, Ti-MWW, and Ti-MOR titanosilicates. As shown in Figure 10.7, a reasonable comparison by the so-called rapid detection test indicates that the ammoximation performance (duration life) follows the order of Ti-MOR > Ti-MWW > TS-1 [100, 119]. The variation of titanosilicates will provide users with more catalyst choices to explore the processes with improved cost performance, tackling the ever-increasing competition in the economy. In particular, Ti-MOR and Ti-MWW are superior to TS-1 in the ammoximation of linear ketones and various aldehydes in terms of oxime selectivity [87, 88, 97, 100, 119, 128].

In the industrial ammoximation process, hydrogen peroxide (fed with an oxidant/ketone molar ratio of ~1.1) is generally converted as completely as possible in the catalytic reaction section, with the purpose to avoid a possible explosion in the separation section as residual hydrogen peroxide may decompose to oxygen in distillation at elevated temperature. To make the process run continuously, catalyst recycling is

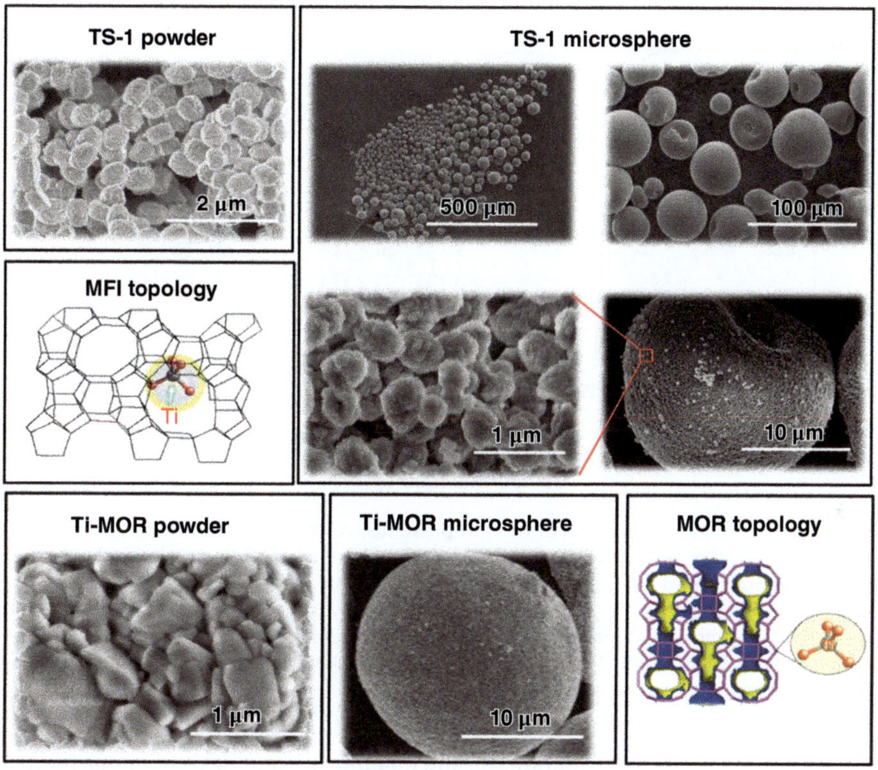

Figure 10.5 SEM images of typical ammoximation catalysts of TS-1 and Ti-MOR titanosilicates in the forms of powder and microsphere.

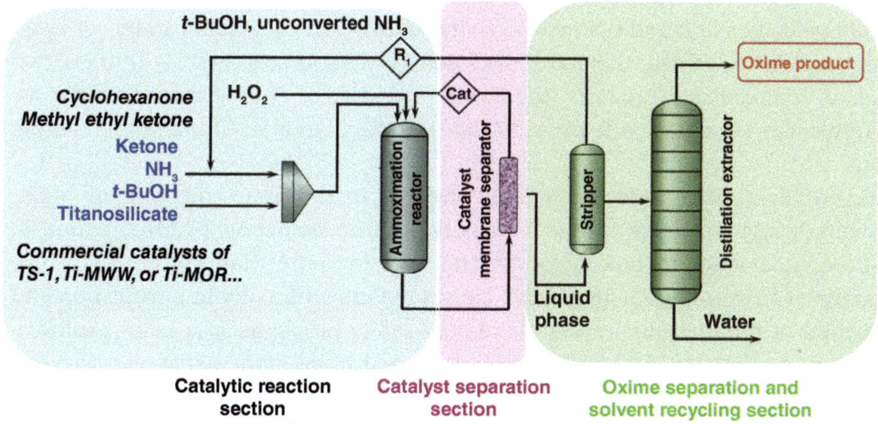

Figure 10.6 Simplified commercial process for oxime production by ammoximation.

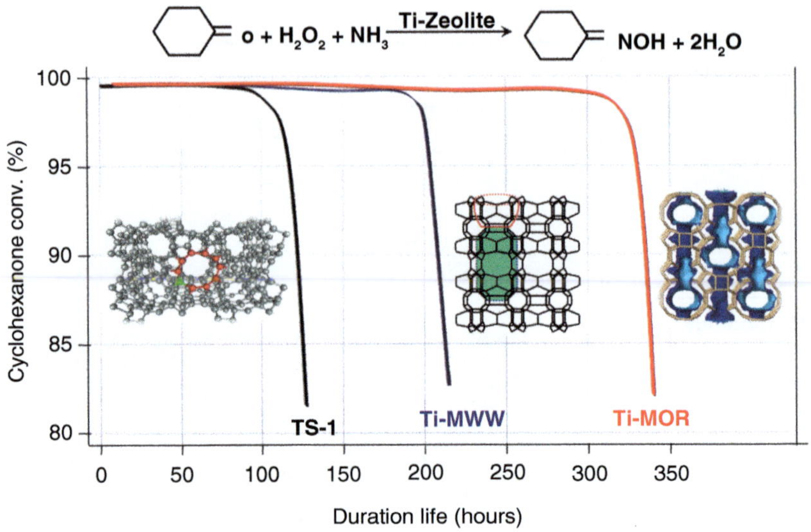

Figure 10.7 The catalyst duration of three representative titanosilicates in cyclohexanone ammoximation was tested with rapid reaction method under constant reaction conditions. The reactions are carried out using a fixed amount of titanosilicate without a national supply, while the reactants of cyclohexanone, ammonia, hydrogen peroxide, and *t*-BuOH solvent are continuously fed into the reactor. The oxime selectivity is comparably high (>99%) for these three titanosilicates, but the duration of ketone conversion is greatly dependent on their crystalline topologies.

indispensable. Ceramic oxide membrane and metallic membrane are suitable for the separation of powder-type catalysts and shaped microsphere catalysts of titanosilicates, respectively. Ketone is also completely converted in ammoximation as possible, particularly in the case of cyclohexanone, to reduce burdens on its recycling and possible byproduct formed by the side reaction of intermolecular ketone condensation. Most of the ammoximation processes are currently running at high cyclohexanone conversion and oxime selectivity (both >99.9%), which can be achieved with the classical TS-1 or thereafter disclosed Ti-MOR as a catalyst. With these two kinds of titanosilicate catalysts, the annual capacity of cyclohexanone oxime (or ε-caprolactam) is expected to exceed 4 000 000 metric tons in China in the coming years.

Nevertheless, TS-1 is facing a fatal problem in the ammoximation of MEK, showing a relatively low oxime selectivity as deep oxidation of MEK oxime to 2-nitrobutane tends to take place on this medium-pore titanosilicate [87]. The formation of 2-nitrobutane byproduct causes not only difficulty in purification and reduction of product quality but also latent safety problems as it is an explosive chemical. As both Ti-MWW and Ti-MOR proved to be more active and selective than TS-1 in MEK ammoximation, the first process for clean production of MEK oxime (5000 tons scale) has been commercialized with Ti-MWW catalyst in 2009 in China. Thereafter, several MEK ammoximation processes have been developed in China using more efficient Ti-MOR catalysts with high-cost performance, and some of them with larger scales of 300 000–400 000 t/a. The total capacity of MEK oxime produced by innovative ammoximation technology exceeds 100 000 metric tons in 2022. These processes produce the MEK oxime product with 99.9% purity, the quality of which is comparable to or even better than that produced by conventional non-catalytic but environmentally unfriendly oximation method with hydroxylamine sulfate.

Different from MEK oxime, an end product with various uses, cyclohexanone oxime is further converted to ε-caprolactam used as the raw material for nylon-6 synthesis. Thus, the current cyclohexanone ammoximation is without expectation connected to the Beckmann rearrangement process, where pure oxime solid produced by ammoximation and further distillation separation is subjected to gas-phase rearrangement with Silicalite-1 as a heterogeneous catalyst or conventional homogeneous liquid-phase rearrangement with oleum catalyst, as shown in Figure 10.8a. The gas-phase rearrangement has been commercialized only by Sumitomo, while tens of ε-caprolactam processes with several million tons scales have employed new ammoximation technology but still use existing oleum rearrangement apparatuses in China.

To make more profits from the ε-caprolactam process with the combination of titanosilicate-ammoximation and oleum-Beckmann rearrangement, a new idea is proposed to perform the oleum-catalyzed rearrangement without using pure oxime but the oxime dissolved in a suitable organic solvent like inert alkane (Figure 10.8b). This idea is based on the performance of last-step ammoximation in the non-organic solvent of water, which is always present in aqueous hydrogen

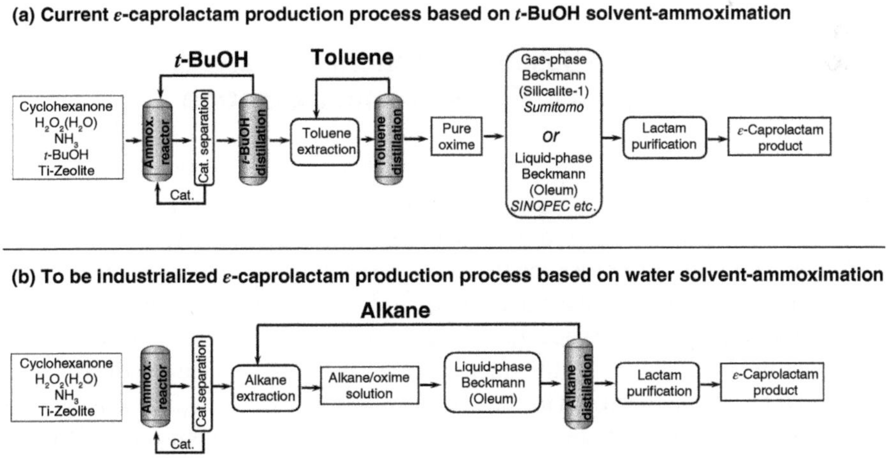

Figure 10.8 A comparison of ε-caprolactam production based on the cyclohexanone ammoximation with the solvent of t-BuOH (a) and water (b). Replacing t-BuOH with water (from aqueous hydrogen peroxide itself) in ammoximation and combining the oleum-catalyzed Beckmann rearrangement of oxime/alkane solution, an innovation ε-caprolactam process featured by less equipment investment and energy-saving is expectable.

peroxide and does not require any external addition. Replacing the typical solvent of *t*-BuOH, when the ammoximation is operated at complete cyclohexanone conversion (>99.9%), the reaction mixture contains only oxime and water. Oxime extracted into an alkane, forming a so-called oxime/alkane solution, is then subjected to oleum-catalyzed Beckmann rearrangement to produce ε-caprolactam. Thereafter, the alkane extractant is recovered for cycling. In comparison to the above-mentioned ε-caprolactam processes needing *t*-BuOH as the reaction solvent and toluene as an extractant for separation, only one organic solvent extractant of an alkane is used in the novel improved process. The alkane extractant tightly links the cyclohexanone ammoximation and the oxime rearrangement together, forming an integrated ε-caprolactam production process, contrary to the current process where ammoximation and rearrangement are operated relatively independently.

With this new ammoximation and rearrangement technique, the apparatus investment and running cost of solvent separation and energy-consuming distillation are estimated to be reduced by a third, which is expected to greatly increase the economy and competition of the whole process. Undoubtedly, the final success depends on whether the titanosilicate catalysts possess a preferable water solvent effect in ammoximation and simultaneously whether they are robust enough against desilication in water. For practical applications, TS-1 seems to be inferior to Ti-MOR in the two important characters. A pilot test has demonstrated that Ti-MOR-catalyzed ammoximation in water is also capable of providing desirable cyclohexanone conversion and oxime selectivity (both close to 99.9%), very comparable to the ammoximation in *t*-BuOH. More importantly, Ti-MOR runs steadily and continuously without obvious deactivation. After further suitable catalyst modification, the commercialization of this Ti-MOR-based new process is close at hand in the coming year.

10.6 Titanosilicate-Based Bifunctional Catalysts for Process Intensified or Tandem Ammoximation Reactions

The extensive basic and R&D (research and development) on the titanosilicate/H_2O_2 catalytic systems have achieved great success in developing various commercialized ammoximation processes for the clean production of ketone oximes. Taking the advantage of this unique catalytic feature of titanosilicates, the industrial sides are always waiting for more advanced and eco-efficient processes to produce bulk chemicals in cleaner ways. As the titanosilicates inevitably need a relatively expensive oxidant of H_2O_2, the in situ formation of H_2O_2 from hydrogen and oxygen is thus considered to combine with the titanosilicate-catalyzed oxidation reactions. Designing a bifunctional catalyst is essential to achieve the two goals. This idea has been first proposed to realize the epoxidation of propylene with O_2 as oxidant and H_2 to produce propylene oxide [137–139]. Supporting Au, Pd, and their alloy nanoparticles with precisely controlled size onto Ti-MCM-41 [140, 141], chemically modified or micro-mesostructured TS-1 [141] and Ti-MWW catalysts [142, 143] have been reported to be useful bifunctional catalysts for PO production from propylene, H_2, and O_2. In addition, physically mixed bifunctional catalysts have been patented to direct produce PO from propylene, H_2, and O_2 in liquid-phase by Simitomo Chemical Co., Ltd., combining hydrogen peroxide synthesis catalyst (Pd/carbon) with epoxidation catalyst (Ti-MWW) [143]. This system seems to be superior to the gas-phase bifunctional Au/titanosilicate system in terms of activity and PO selectivity. The detailed and new progress of this subject is summarized in Chapter 14.

Based on titanosilicates, bifunctional catalytic systems have also been constructed to develop ammoximation-related new oxidation technologies with process intensification features, including direct ammoximation of ketone, ammonia, H_2, and O_2 instead of H_2O_2, and one-pot synthesis of amide from aldehyde, ammonia, and H_2O_2 (Figure 10.9).

Hutchings et al. have firstly reported an AuPd alloy-supported TS-1 as a catalyst to generate cyclohexanone oxime via in-situ generated H_2O_2, which is comparable to the current industrial route with >95% selectivity, and the reaction route can greatly save the cost of industrial production (Figure 10.10) [64]. Among various kinds of alloys, the alloys formed by immobilizing Pd and other metals such as Au, Pt, Ni, and Sn on the TS-1 carrier can effectively catalyze cyclohexanone ammoximation process. The considerable conditional gap existing between two key different reaction pathways of the in situ method, that is, the direct synthesis of H_2O_2 and the ketone ammoximation, can be bridged. It is found that PdAu nanoalloys exhibit the

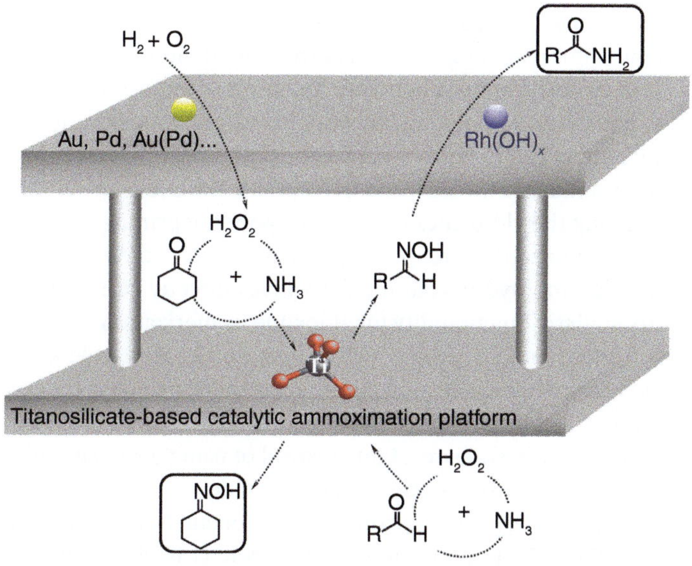

Figure 10.9 Taking the advantage of titanosilicate-catalyzed ammoximation, bifunctional catalytic concepts have been proposed to one-pot synthesize oxime from ketone, ammonia, hydrogen, and oxygen (left), as well as one-pot synthesize amide from primary aldehyde, ammonia, and hydrogen peroxide in liquid-phase (right). Both catalytic systems are featured by process intensification with precious metal nanoparticles or species highly dispersed on specific titanosilicate crystals. More detailed progress on these two reactions is summarized as follows.

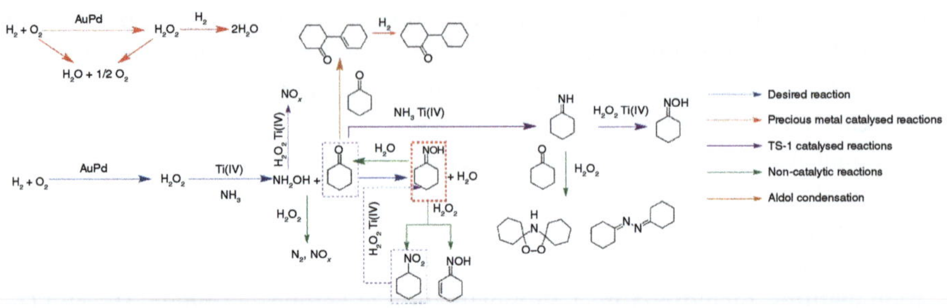

Figure 10.10 The key reaction pathways in the ammoximation of cyclohexanone to cyclohexanone oxime through in situ H_2O_2 synthesis. Source: Reprinted from Ref. [64] with the copyright permission from American Association for the Advancement of Science.

best catalytic activity and stability beyond Pd-based materials and single-metal Pd analogs [144]. In addition, the participation of Pt in alloys can reduce the size of the alloy particles and guarantee a better activity site division, leading to the increased the Pd^0 content and thus enhancing catalytic activity [66].

Cyclohexanone oxime produced by titanosilicate/H_2O_2 system is needed to rearrange to ε-caprolactam, the raw material for nylon-6 synthesis, by Beckmann rearrangement, which is conventionally carried out in liquid-phase using a sulfuric acid catalyst, or by innovative gas-phase reaction using a solid-acid catalyst of Silicalite-1 [145]. No matter what kind of Beckmann rearrangement reaction is, the ammoximation and isomerization reactions are separately operated. Moreover, acid-catalyzed Beckmann rearrangement is not useful for the transformation of aldoximes to primary amides [146, 147]. In terms of simple chemistry and process intensification, it is desirable to produce oximes from ketones or aldehydes in one pot. Precious metals like Rh, Ru, Pd, and Ir are known to efficiently catalyze the liquid-phase transformation of aldoximes to primary amides [148–155], and the reusable $Rh(OH)_x/Al_2O_3$ catalyst was capable of converting the aldoximes to the corresponding primary amides in water solvent without using hazardous reactants [155] and even one-pot synthesizing amides from aldehydes and hydroxylamine derivatives directly [156]. Thus, the $Rh(OH)_x$ species with catalytic functionality for aldoxime isomerization are supported on typical titanosilicates like TS-1, Ti-MWW, Ti-MOR, and Ti-Beta to develop environmentally benign chemical processes for the production of primary amides [110]. With these bifunctional catalysts, aldehydes, ammonia, and hydrogen peroxide are employed in a one-pot synthesis as reactants instead of using expensive and explosive aldoximes and sulfate or hydrochloride hydroxylamine.

To achieve highly dispersed $Rh(OH)_x$ species on Al_2O_3, aqueous NaOH solution was usually used to adjust the pH value in impregnation. NaOH is obviously not suitable for supporting $Rh(OH)_x$ on titanosilicates, as alkali-metal cations (Na^+ and K^+) are well known to easily poison the Ti active sites by forming closely linked ≡Si—O^- M^+ (M^+ =Na^+, K^+, etc.) moieties. In fact, $Rh(OH)_x$/Ti-MWW prepared by impregnation in aqueous NaOH solution showed only 10.0% benzaldehyde conversion and 76.3% oxime selectivity, much lower than those of pristine Ti-MWW (93.2% conversion and 98.7% oxime selectivity).

Table 10.3 One-pot synthesis of benzamide from benzaldehyde over different $Rh(OH)_x$/titanosilicate catalysts.[a)]

Cat. [b)]	Aldehyde conv. (%)	Nitrile sel. (%)	Oxime sel. (%)	Amide sel. (%)
$Rh(OH)_x$/Ti-MWW(38)	84.9	6.9	0.9	86.9
$Rh(OH)_x$/TS-1(40)	74.8	16.6	4.1	70.4
$Rh(OH)_x$/Ti-MOR(92)	83.1	17.1	44.4	31.2
$Rh(OH)_x$/Ti-Beta(60)	40.5	17.3	40.5	21.6

a) Reaction conditions: cat., 0.2 g; benzaldehyde, 2 mmol; solvent (water), 1.5 g; ammonia aqueous solution (25 wt.%), 2.5 mmol; hydrogen peroxide aqueous solution (5 wt.%), 2.5 mmol. The reaction was carried out at 343 K by feeding H_2O_2 continuously within 30 minutes and then at 413 K for another 7 hours. The Rh loading was in the range of 1.5–1.8 wt.%.

b) The number in parentheses indicates the Si/Ti ratio of titanosilicates. Source: Reprinted from Ref. [110] with the copyright permission from Wiley.

Replacing NaOH solution, a new impregnation strategy is proposed to support $Rh(OH)_x$ on titanosilicates with the $RhCl_3$ source in organic quaternary ammonium hydroxide solution, including tetramethylammonium hydroxide (TMAOH), tetraethylammonium hydroxide (TEAOH), or TPAOH. In this way, the purposes of $Rh(OH)_x$ dispersion and Ti site preservation are simultaneously realized. The investigations with UV/visible and X-ray photoelectron spectroscopies verify that the Rh species exist in the oxidation state of +3. The $Rh(OH)_x$/titanosilicate catalysts thus prepared are evaluated in one-pot reaction of benzaldehyde, ammonia, and H_2O_2 in water. $Rh(OH)_x$/Ti-MWW provides 84.9% aldehyde conversion and 86.9% amide selectivity, more efficient than other titanosilicate supports of TS-1, Ti-MOR, and Ti-Beta (Table 10.3). As shown in Scheme 10.9, this one-pot reaction actually takes place in a tandem way, that is, firstly the oxidation of ammonia by H_2O_2 to form hydroxylamine on the tetrahedral Ti sites in titanosilicate, then non-catalytic oximation of aldehyde to aldoxime, and subsequent rearrangement of aldoxime to amide on the supported $Rh(OH)_x$ species. Aldoxime from ammoximation is the key intermediate to produce amide; meanwhile, parallel dehydration of benzaldehyde oxime may also occur to a certain extent, forming benzonitrile as the main by-product. In addition to benzaldehyde, $Rh(OH)_x$/Ti-MWW works well in the one-pot conversion of various linear or other aromatic primary aldehydes to corresponding amides.

The $Rh(OH)_x$/titanosilicate catalysts operated in a basic reaction medium with the presence of ammonia inevitably suffer from the leaching problems of $Rh(OH)_x$ active species. Supporting $Rh(OH)_x$ species on core-shell structured TS-1@KCC-1 prepared in a microemulsion system gives a material with high hydrothermal stability [157]. The resulting bifunctional $Rh(OH)_x$/TS-1@KCC-1 catalyst is active to synthesize primary amides directly from corresponding aldehydes, ammonia, and hydrogen peroxide through one-pot reaction combining ammoximation and oxime rearrangement together. In the case of benzaldehyde, $Rh(OH)_x$/TS-1@KCC-1 provides 94.9% conversion and 94.7% amide selectivity. The KCC-1 mesosilica shell

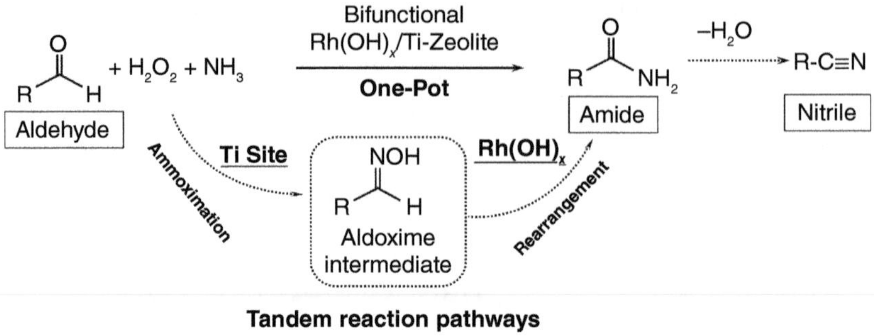

Scheme 10.9 One-pot synthesis of primary amide from aldehyde, hydrogen peroxide, and ammonia using bifunctional $Rh(OH)_x$/titanosilicate catalysts. The reaction takes place in a tandem way by combining the ammoximation of an aldehyde with the rearrangement of oxime, in which aldehyde oxime proves to be intermediate.

assembled over TS-1 crystal shows confining and stabilizing effects on the $Rh(OH)_x$ species, giving a more stable bifunctional catalyst than the above-mentioned $Rh(OH)_x$/Ti-MWW.

10.7 Conclusions and Perspectives

The liquid-phase catalytic ammoximation of ketones or aldehydes with ammonia and hydrogen peroxide is a unique one-pot reaction that produces corresponding oximes in a green way. The ammoximation is actively and selectively catalyzed with microporous titanosilicates as catalysts, while their performance and practical applicability closely depend on the crystalline structures of specific zeolites. Beyond the versatile medium-pore TS-1 titanosilicate, both post-synthesized Ti-MOR and hydrothermally synthesized Ti-MWW with larger porosity prove to be even more active and selective, particularly in the ammoximation of linear ketones and chemically reactive aldehydes. Up to date, only three titanosilicates with the **MFI**, **MWW**, and **MOR** topologies have been applied to commercial processes. In addition to zeolite topologies, suitable chemical modifications, like controlled fluorination, alkali metal ion-exchange, and treatment with organic amine or quaternary ammonium, significantly improve the ammoximation performance. The ammoximation takes place in a tandem way, including Ti site-catalyzed oxidation of ammonia with hydrogen peroxide to form hydroxylamine as intermediate and subsequent non-catalytic oximation of ketones or aldehydes to oximes, and this reaction mechanism has nothing to do with the structure types of titanosilicates. Titanosilicates suffer irreversible deactivation due to desilication in basic reaction media, which can be weakened by precisely controlled alumination, self-assembly of protective mesosilica shell covering on the titanosilicate crystals, or direct addition of silica source into the reaction system. Taking advantage of the catalytic platforms of titanosilicate-based ammoximation, it is possible to design the bifunctional catalysts of titanosilicate supported precious metal, which are useful to develop promising process-intensification technologies for oxime synthesis using hydrogen and oxygen as an oxidant instead of hydrogen peroxide, or one-pot production of amides from aldehydes, ammonia, and hydrogen peroxide.

R&D on titanosilicate-based ammoximation has made a great breakthrough in the clean production of ε-caprolactam and MEK oxime, leading to many world-class commercial processes. The industrial world always seeks for more economic, eco-friendly, and energy-saving catalytic technology to reduce waste dispose and increase economic benefits. The current commercial liquid-phase ammoximation processes operated in tert-butyl alcohol solvent is needed to be replaced by an innovative one with water only from aqueous hydrogen peroxide, which would increase production efficiency and reduce equipment investment and energy consumption. In this sense, it is needed to explore more robust titanosilicates that possess the favorable solvent effects of water and enhanced durability against desilication. In addition, the clean production of aldehyde oximes from the ammoximation aldehydes is still waiting for a more efficient ammoximation technique as chemically active aldehydes are faced with the problem of low oxime selectivity. The concepts of bifunctional ammoximation catalytic systems have been proved to work well for the synthesis of cyclohexanone oxime with hydrogen and oxygen as oxidants, as well as for the one-pot synthesis of amides from aldehydes. However, there is still a large room to increase the activity and more importantly the product selectivity by choosing appropriate bifunctional titanosilicate systems that have synergistic interaction with precious metal species and are capable of suppressing their leaching in basic reaction systems.

Acknowledgments

The authors would like to acknowledge the financial supports from the China Ministry of Science and Technology (2023YFB3810602, 2021YFA1501401), the National Natural Science Foundation of China (22273023, 22222201, and 21972044), and the Research Funds of Happiness Flower ECNU (2020ST2203).

References

1 Taramasso, M., Perego, G., and Notari, B. (1983). *Preparation of Porous Crystalline Synthetic Material Comprised of Silicon and Titanium Oxides*. US Patent, 4410501.

2 Le Bars, J., Dakka, J., and Sheldon, R.A. (1996). Ammoximation of cyclohexanone and hydroxyaromatic ketones over titanium molecular sieves. *Appl. Catal. A Gen.* 136 (1): 69–80.

3 Thangaraj, A. (1991). Ammoximation of cyclohexanone and hydroxyaromatic ketones over titanium molecular sieves. *J. Catal.* 131 (1): 294–297.

4 Spinace, E.V., Pastore, H.O., and Schuchardt, U. (1995). Cyclohexane oxidation catalyzed by titanium silicalite (TS-1): overoxidation and comparison with other oxidation systems. *J. Catal.* 157 (2): 631–635.

5 Kong, L., Li, G., and Wang, X. (2004). Mild oxidation of thiophene over TS-1/H_2O_2. *Catal. Today* 93–95: 341–345.

6 Maspero, F. and Romano, U. (1994). Oxidation of alcohols with H_2O_2 catalyzed by titanium silicalite-1. *J. Catal.* 146 (2): 476–482.

7 Huybrechts, D.R.C., Buskens, P.L., and Jacobs, P.A. (1992). Physicochemical and catalytic properties of titanium silicalites. *J. Mol. Catal.* 71 (1): 129–147.

8 Reddy, J.S. and Jacobs, P.A. (1996). Selective oxidation of secondary amines over titanium silicalite molecular sieves, TS-1 and TS-2. *Catal. Lett.* 37 (3–4): 213–216.

9 Jiao, W., He, Y., Li, J. et al. (2015). Ti-rich TS-1: a highly active catalyst for epoxidation of methallyl chloride to 2-methyl epichlorohydrin. *Appl. Catal. A Gen.* 491: 78–85.

10 Bhaumik, A., Mukherjee, P., and Kumar, R. (1998). Triphase catalysis over titanium–silicate molecular sieves under solvent-free conditions: I. Direct hydroxylation of benzene. *J. Catal.* 178 (1): 101–107.

11 Song, F., Liu, Y., Wu, H. et al. (2006). A novel titanosilicate with MWW structure: highly effective liquid-phase ammoximation of cyclohexanone. *J. Catal.* 237 (2): 359–367.

12 Camblor, M.A., Costantini, M., Corma, A. et al. (1996). Synthesis and catalytic activity of aluminium-free zeolite Ti-β oxidation catalysts. *Chem. Commun.* 11: 1339–1340.

13 Wu, P., Komatsu, T., and Yashima, T. (1996). Hydrothermal synthesis of a novel titanosilicate with MWW topology. *J. Phys. Chem.* 100 (24): 10316–10322.

14 Yamamoto, K. and Tatsumi, T. (2000). Remarkable improvement in hydrothermal stability of MCM-41 by surface modification with grignard reagents. *Chem. Lett.* 29 (6): 624–625.

15 Kubota, Y., Koyama, Y., Yamada, T. et al. (2008). Synthesis and catalytic performance of Ti-MCM-68 for effective oxidation reactions. *Chem. Commun.* (46): 6224.

16 Ahedi, R.K. and Kotasthane, A.N. (1998). Synthesis of FER titanosilicates from a non-aqueous alkali-free seeded system. *J. Mater. Chem.* 8 (8): 1685–1686.

17 Wu, P., Komatsu, T., and Yashima, T. (1997). Ammoximation of ketones over titanium mordenite. *J. Catal.* 168 (2): 400–411.

18 Peng, R., Wan, Z., Lv, H. et al. (2021). Al-modified Ti-MOR as a robust catalyst for cyclohexanone ammoximation with enhanced anti-corrosion performance. *Catal. Sci. Technol.* 11 (22): 7287–7299.

19 Xu, L., Huang, D.D., Li, C.G. et al. (2015). Construction of unique six-coordinated titanium species with an organic amine ligand in titanosilicate and their unprecedented high efficiency for alkene epoxidation. *Chem. Commun.* 51 (43): 9010–9013.

20 Roffia, P., Padovan, M., Leofanti, G., Mantegazza, M., de Alberti, G. and Tauszik, G.R. (1988). *Catalytic Process for the Manufacture of Oximes*. US Patent, 4794198.

21 Thangaraj, A. (1991). Catalytic properties of crystalline titanium silicalites III. Ammoximation of cyclohexanone. *J. Catal.* 131 (2): 394–400.

22 Mantegazza, M.A., Cesana, A., and Pastori, M. (1996). Ammoximation of ketones on titanium silicalite. A study of the reaction byproducts. *Top. Catal.* 3 (3–4): 327–335.

23 Mantegazza, M.A., Leofanti, G., Petrini, G. et al. (1994). Selective oxidation of ammonia to hydroxylamine with hydrogen peroxide on titanium based catalysts. *Stud. Surf. Sci. Catal.* 82: 541–550.

24 Anastas, P.T. and Williamson, T.C. (1996). *Green Chemistry: Designing Chemistry for the Environment*, ACS Symposium, vol. 626, 81–91. American Chemical Society.

25 Cesana, A., Mantegazza, M.A., and Pastori, M. (1997). A study of the organic by-products in the cyclohexanone ammoximation. *J. Mol. Catal. A Chem.* 117 (1–3): 367–373.

26 Mantegazza, M.A., Petrini, G., Spanò, G. et al. (1999). Selective oxidations with hydrogen peroxide and titanium silicalite catalyst. *J. Mol. Catal. A Chem.* 146 (1–2): 223–228.

27 Wu, C., Wang, Y., Mi, Z. et al. (2002). Effects of organic solvents on the structure stability of TS-1 for the ammoximation of cyclohexanone. *React. Kinet. Catal. Lett.* 77 (1): 73–81.

28 Wang, Y., Pan, M., Wu, C. et al. (2002). Ammoximation of cyclohexanone catalyzed by TS-1. *J. Sichuan Univ. Eng. Sci. Ed.* 34: 20–23.

29 Yip, A.C.K. and Hu, X. (2009). Catalytic activity of clay-based titanium silicalite-1 composite in cyclohexanone ammoximation. *Ind. Eng. Chem. Res.* 48 (18): 8441–8450.

30 Sonoda, J., Kamegawa, T., Kuwahara, Y. et al. (2010). Hydrophobic modification of Ti-containing zeolite (TS-1) and their applications in liquid-phase selective catalytic reactions. *Bull. Chem. Soc. Jpn.* 83 (5): 592–594.

31 Huang, D.G., Zhang, X., Chen, B.H., and Chao, Z.S. (2010). Ethanol-assistant synthesis of TS-1 containing no extra-framework Ti species. *Catal. Today* 158 (3–4): 510–514.

32 Liu, Y., Zong, L., Zhang, Y. et al. (2011). Synthesis of dendritic TS-1 film on honeycomb cordierite. *Mater. Lett.* 65 (3): 497–499.

33 Qi, Y., Ye, C., Zhuang, Z., and Xin, F. (2011). Preparation and evaluation of titanium silicalite-1 utilizing pretreated titanium dioxide as a titanium source. *Microporous Mesoporous Mater.* 142 (2–3): 661–665.

34 Huang, D.G., Zhang, X., Liu, T.W. et al. (2013). Synthesis of high-performanced titanium silicalite-1 zeolite at very low usage of tetrapropyl ammonium hydroxide. *Ind. Eng. Chem. Res.* 52 (10): 3762–3772.

35 Shen, L., Deng, X., and Liu, Y. (2013). Synthesis and catalytic oxidation performance of Al–TS-1. *Chin. J. Catal.* 34 (6): 1232–1241.

36 Shen, C., Wang, Y.J., Dong, C., and Luo, G.S. (2014). In situ growth of TS-1 on porous glass beads for ammoximation of cyclohexanone. *Chem. Eng. J.* 235: 75–82.

37 Xu, L., Ding, J., Yang, Y., and Wu, P. (2014). Distinctions of hydroxylamine formation and decomposition in cyclohexanone ammoximation over microporous titanosilicates. *J. Catal.* 309: 1–10.

38 Wu, X., Wang, Y., Zhang, T. et al. (2014). Effect of TS-1 treatment by tetrapropyl ammonium hydroxide on cyclohexanone ammoximation. *Catal. Commun.* 50: 59–62.

39 Yao, P., Wang, Y., Zhang, T. et al. (2014). Effect of sodium ions in synthesis of titanium silicalite-1 on its catalytic performance for cyclohexanone ammoximation. *Front. Chem. Sci. Eng.* 8 (2): 149–155.

40 Xue, Y., Xie, Y., Wei, H. et al. (2014). Improving the performance of TS-1 catalysts for continuous cyclohexanone ammoximation through controlment of active species distribution. *New J. Chem.* 38 (9): 4229.

41 Wu, P., Ding, J., Wu, J. et al. (2015). Green production of cyclohexanone oxime with a fixed-bed reactor using shaped titanosilicate catalysts. *Chin. Sci. Bull.* 60 (16): 1538–1545.

42 Zhang, T., Wang, Y., Wang, S. et al. (2015). Effect of triethylamine treatment of titanium silicalite-1 on cyclohexanone ammoximation in a continuous system. *React. Kinet. Mech. Catal.* 114 (2): 735–752.

43 Dong, C., Wang, K., Zhang, J.S., and Luo, G.S. (2015). Reaction kinetics of cyclohexanone ammoximation over TS-1 catalyst in a microreactor. *Chem. Eng. Sci.* 126: 633–640.

44 Zhuo, Z., Wang, L., Zhang, X. et al. (2015). Insights into the key to highly selective synthesis of oxime via ammoximation over titanosilicates. *J. Catal.* 329: 107–118.

45 Xue, T., Liu, H., Wang, Y. et al. (2015). Seed-induced synthesis of small-crystal TS-1 using ammonia as alkali source. *Chin. J. Catal.* 36 (11): 1928–1935.

46 Hu, Y., Du, C., Wang, T., and Luo, G. (2019). Highly efficient and greener synthesis of TS-1 in a flow system by recycling the mother liquid. *Microporous Mesoporous Mater.* 288: 109585.

47 Zhang, M., Wen, Y., Zong, L. et al. (2020). Improved Ti species distribution and hierarchical pores in TS-1: towards regeneration of TS-1 deactivated due to alkali corrosion. *New J. Chem.* 44 (16): 6394–6401.

48 Xu, H., Yan, J., Lu, X. et al. (2020). Comparison of titanosilicates with different topologies as liquid-phase oxidation catalysts. *Catal. Today* 347: 48–55.

49 Zhang, S., Wei, H., Gao, X. et al. (2020). A cleaner ammoximation of cyclohexanone over Ti-MWW in dimethyl carbonate. *Catal. Commun.* 139: 105968.

50 Zhang, L., Zhu, X., Wang, X., and Shi, C. (2021). The synthesis of pure and uniform nanosized TS-1 crystals with a high titanium content and a high space–time yield. *Inorg. Chem. Front.* 8 (24): 5260–5269.

51 Wang, T., Yang, X., Ma, Y. et al. (2022). Insights into the role of titanium sites in cyclohexanone ammoximation over titanium silicalite-1. *Mol. Catal.* 524: 112298.

52 Liu, T., Wang, L., Wan, H., and Guan, G. (2014). A magnetically recyclable TS-1 for ammoximation of cyclohexanone. *Catal. Commun.* 49: 20–24.

53 Zhang, Y., Wang, Y., Bu, Y. et al. (2005). Reaction mechanism of the ammoximation of ketones catalyzed by TS-1. *React. Kinet. Catal. Lett.* 87 (1): 25–32.

54 Xue, Y., Zuo, G., Wen, Y. et al. (2019). Seed-assisted synthesis of TS-1 crystals containing Al with high catalytic performances in cyclohexanone ammoximation. *RSC Adv.* 9 (5): 2386–2394.

55 Xue, Y., Wen, Y., Wei, H. et al. (2015). Hollow TS-1 mesocrystals: hydrothermal construction and high catalytic performances in cyclohexanone ammoximation. *RSC Adv.* 5 (64): 51563–51569.

56 Lin, J., Xin, F., Yang, L., and Zhuang, Z. (2014). Synthesis, characterization of hierarchical TS-1 and its catalytic performance for cyclohexanone ammoximation. *Catal. Commun.* 45: 104–108.

57 Zhong, Z., Li, D., Liu, X. et al. (2009). The fouling mechanism of ceramic membranes used for recovering TS-1 catalysts. *Chin. J. Chem. Eng.* 17 (1): 53–57.

58 Wang, Y., Wang, S., Zhang, T. et al. (2017). Effect of extra-framework titanium in TS-1 on the ammoximation of cyclohexanone. *Trans. Tianjin Univ.* 23 (3): 230–236.

59 Yang, L., Xin, F., Lin, J. et al. (2014). Continuous heterogeneous cyclohexanone ammoximation reaction using a monolithic TS-1/cordierite catalyst. *RSC Adv.* 4 (52): 27259–27266.

60 Yip, A.C.K. and Hu, X. (2011). Formulation of reaction kinetics for cyclohexanone ammoximation catalyzed by a clay-based titanium silicalite-1 composite in a semibatch process. *Ind. Eng. Chem. Res.* 50 (24): 13703–13710.

61 Liang, X., Mi, Z., Wang, Y. et al. (2004). An integrated process of H_2O_2 production through isopropanol oxidation and cyclohexanone ammoximation. *Chem. Eng. Technol.* 27 (2): 176–180.

62 Niu, C., Liu, M., Gao, X. et al. (2019). Fabrication of microspheres by nano-TS-1 crystals via a spray-forming process: from screening to scale-up. *ACS Omega* 4 (2): 4397–4404.

63 Yip, A.C.K., Lam, F.L.Y., and Hu, X. (2009). A heterostructured titanium silicalite-1 catalytic composite for cyclohexanone ammoximation. *Microporous Mesoporous Mater.* 120 (3): 368–374.

64 Lewis, R.J., Ueura, K., Liu, X. et al. (2022). Highly efficient catalytic production of oximes from ketones using in situ–generated H_2O_2. *Science* 376 (6593): 615–620.

65 Zhong, Z., Liu, X., Chen, R. et al. (2009). Adding microsized silica particles to the catalysis/ultrafiltration system: catalyst dissolution inhibition and flux enhancement. *Ind. Eng. Chem. Res.* 48 (10): 4933–4938.

66 Lewis, R.J., Ueura, K., Fukuta, Y. et al. (2022). Cyclohexanone ammoximation *via in situ* H_2O_2 production using TS-1 supported catalysts. *Green Chem.* 24: 9496–9507.

67 Xu, Y., Jin, F., Wu, G. et al. (2022). Effect of surface polarity of titanium-loaded ERB-1 catalyst on epoxidation and ammoximation. *Catal. Lett.* https://doi.org/10.1007/s10562-022-03954-z.

68 Sun, R., Xin, F., and Yang, L. (2015). Modeling of monolithic catalyst washcoated with TS-1 and reactor for continuous cyclohexanone ammoximation reaction. *Ind. Eng. Chem. Res.* 54 (7): 2254–2258.

69 Lin, J., Yang, T., Lin, C., and Sun, J. (2018). Hierarchical MFI zeolite synthesized via regulating the kinetic of dissolution-recrystallization and their catalytic properties. *Catal. Commun.* 115: 82–86.

70 Dal Pozzo, L., Fornasari, G., and Monti, T. (2002). TS-1, catalytic mechanism in cyclohexanone oxime production. *Catal. Commun.* 3 (8): 369–375.

71 Deng, X., Wang, Y., Shen, L. et al. (2013). Low-cost synthesis of titanium silicalite-1 (TS-1) with highly catalytic oxidation performance through a controlled hydrolysis process. *Ind. Eng. Chem. Res.* 52 (3): 1190–1196.

72 Wei, H., Zhang, N., Zhao, T. et al. (2015). Controlling the microstructure of MFI zeolites with $Mg(OH)_2$ nanocrystals to improve their catalytic performances. *RSC Adv.* 5 (5): 3642–3647.

73 Liu, X., Zhang, Y., Wang, Z. et al. (2018). A catalytic hydrocracking approach for zeolite detemplation at mild condition. *Chem. Eng. J.* 346: 600–605.

74 Tatsumi, T. and Jappar, N. (1996). Ammoximation of cyclic ketones on TS-1 and amorphous SiO2–TiO2. *J. Catal.* 161 (2): 570–576.

75 Jorda, E., Tuel, A., Teissier, R., and Kervennal, J. (1997). An original and very interesting precursor to the synthesis of titanium containing silicalite-1. *Zeolites* 19 (4): 238–245.

76 Sooknoi, T. and Chitranuwatkul, V. (2005). Ammoximation of cyclohexanone in acetic acid using titanium silicalite-1 catalyst: activity and reaction pathway. *J. Mol. Catal. A Chem.* 236 (1–2): 220–226.

77 Park, S.-E. and Choi, K.-M. (2008). Green catalysis by microwave synthesized nanostructured materials. *J. Phys. Chem. Solids* 69 (5–6): 1501–1504.

78 Lv, G., Wang, T., Chen, Y. et al. (2022). Hydrophobized hollow TS-1 zeolite as pickering interfacial catalyst for selective oxidation reactions. *Colloids Surf. A Physicochem. Eng. Asp.* 633: 127842.

79 Chen, R., Mao, H., Zhang, X. et al. (2014). A dual-membrane airlift reactor for cyclohexanone ammoximation over titanium silicalite-1. *Ind. Eng. Chem. Res.* 53 (15): 6372–6379.

80 Xu, L., Lv, G., Li, H. et al. (2015). Surface-modified TS-1 with enhanced activity for cyclohexanone ammoximation in a pickering emulsion and increased stability in hot aqueous ammonia. *RSC Adv.* 5 (77): 62652–62658.

81 Saxena, S., Basak, J., Hardia, N. et al. (2007). Ammoximation of cyclohexanone over nanoporous TS-1 using UHP as an oxidant. *Chem. Eng. J.* 132 (1–3): 61–66.

82 Hu, Y., Dong, C., Wang, T., and Luo, G. (2018). Cyclohexanone ammoximation over TS-1 catalyst without organic solvent in a microreaction system. *Chem. Eng. Sci.* 187: 60–66.

83 Liang, X., Mi, Z., Wang, Y. et al. (2004). Process integration of H_2O_2 generation and the ammoximation of cyclohexanone. *J. Chem. Technol. Biotechnol.* 79 (6): 658–662.

84 Liu, T., Meng, X., Wang, Y. et al. (2004). Integrated process of H_2O_2 generation through anthraquinone hydrogenation–oxidation cycles and the ammoximation of cyclohexanone. *Ind. Eng. Chem. Res.* 43 (1): 166–172.

85 Zhang, F., Wang, W., Hu, S. et al. (2017). A nanosized aluminoborate (PKU-5) with Cr-centered octahedral framework: solid-phase synthesis, characterizations and catalytic ammoximation of cyclohexanone to cyclohexanone azine. *Appl. Catal. A Gen.* 531: 60–68.

86 Liu, G., Wu, J., and Luo, H. (2012). Ammoximation of cyclohexanone to cyclohexanone oxime catalyzed by titanium silicalite-1 zeolite in three-phase system. *Chin. J. Chem. Eng.* 20 (5): 889–894.

87 Song, F., Liu, Y., Wang, L. et al. (2007). Highly selective synthesis of methyl ethyl ketone oxime through ammoximation over Ti-MWW. *Appl. Catal. A Gen.* 327 (1): 22–31.

88 Ding, J., Xu, L., Xu, H. et al. (2014). Highly efficient synthesis of methyl ethyl ketone oxime through ammoximation over Ti-MOR catalyst: highly efficient synthesis of methyl ethyl ketone oxime through ammoximation over Ti-MOR catalyst. *Chin. J. Catal.* 34 (1): 243–250.

89 Zhuo, Z., Wu, L., Wang, L. et al. (2014). Lewis acidic strength controlled highly selective synthesis of oxime via liquid-phase ammoximation over titanosilicates. *RSC Adv.* 4 (99): 55685–55688.

90 Chu, Q., Wang, P., He, G. et al. (2017). Continuous three-phase 2-butanone ammoximation process via spray forming TS-1 microspheres in a highly efficient jet loop reactor. *Chem. Eng. J.* 325: 169–175.

91 Wang, W., Fu, Y., Guo, Y. et al. (2018). Preparation of lamellar-stacked TS-1 and its catalytic performance for the ammoximation of butanone with H_2O_2. *J. Mater. Sci.* 53 (6): 4034–4045.

92 Yang, D., Wang, H., Wang, W. et al. (2019). Nickel-modified TS-1 catalyzed the ammoximation of methyl ethyl ketone. *Catalysts* 9 (12): 1027.

93 Zhang, F., Shang, H., Jin, D. et al. (2017). High efficient synthesis of methyl ethyl ketone oxime from ammoximation of methyl ethyl ketone over TS-1 in a ceramic membrane reactor. *Chem. Eng. Process. Process Intensif.* 116: 1–8.

94 Wang, L., Zhang, J., and Mi, Z. (2006). Ammoximation of methyl ethyl ketone with H_2O_2 and ammonia over TS-1. *J. Chem. Technol. Biotechnol.* 81 (4): 710–712.

95 Yu, H., Shang, F., Chu, Q. et al. (2020). Cleaner and atomic economy production of hydroxylamine hydrochloride under solvent-free conditions through process intensification. *J. Clean. Prod.* 269: 122187.

96 Chu, Q., He, G., Xi, Y. et al. (2018). Green synthesis of low-carbon chain nitroalkanes via a novel tandem reaction of ketones catalyzed by TS-1. *Catal. Commun.* 108: 46–50.

97 Ding, J., Xu, L., Yu, Y. et al. (2013). Clean synthesis of acetaldehyde oxime through ammoximation on titanosilicate catalysts. *Catal. Sci. Technol.* 3 (10): 2587.

98 Jin, H., Meng, C., He, G. et al. (2018). Green synthesis of acetaldehyde oxime using ammonia oxidation in the TS-1/H_2O_2 system. *React. Kinet. Mech. Catal.* 125 (2): 1113–1125.

99 Meng, C., Yang, S., He, G. et al. (2016). The reaction mechanism of acetaldehyde ammoximation to its oxime in the TS-1/H_2O_2 system. *Catalysts* 6 (7): 109.

100 Ding, J. and Wu, P. (2014). Selective synthesis of dimethyl ketone oxime through ammoximation over Ti-MOR catalyst. *Appl. Catal. A Gen.* 488: 86–95.

101 Zhang, Z., Chu, Q., Sun, Y. et al. (2022). Green synthesis of 2-nitropropane via ammoximation-oxidation over organic base modified TS-1 catalysts. *ChemistryS-elect* 7 (35): e202202475.

102 Li, Z., Chen, R., Jin, W., and Xing, W. (2010). Catalytic mechanism and reaction pathway of acetone ammoximation to acetone oxime over TS-1. *Korean J. Chem. Eng.* 27 (5): 1423–1427.

103 Liang, X., Mi, Z., Wang, Y. et al. (2004). Synthesis of acetone oxime through acetone ammoximation over TS-1. *React. Kinet. Catal. Lett.* 82 (2): 333–337.

104 Chen, R., Bu, Z., Li, Z. et al. (2010). Scouring-ball effect of microsized silica particles on operation stability of the membrane reactor for acetone ammoximation over TS-1. *Chem. Eng. J.* 156 (2): 418–422.

105 Li, Z., Chen, R., Xing, W. et al. (2010). Continuous acetone ammoximation over TS-1 in a tubular membrane reactor. *Ind. Eng. Chem. Res.* 49 (14): 6309–6316.

106 Wang, Y., Zhang, S., Zhao, Y., and Lin, M. (2014). The mechanism of catalyst deactivation and by-product formation in acetone ammoximation catalyzed by hollow titanium silicalite. *J. Mol. Catal. A Chem.* 385: 1–6.

107 Zhang, S., Zhao, H., Wang, X. et al. (2019). Formation and transformation of nitroso compounds of ketoxime isomers in ketone ammoximation catalyzed by hollow titanium silicalite. *React. Kinet. Mech. Catal.* 127 (2): 787–801.

108 Zhang, S., Zhao, Y., and Zhang, H. (2012). Synthesis of 2,3-dimethyl-2,3-dinitrobutane by oxidative coupling of acetone oxime. *Chin. J. Org. Chem.* 32 (4): 786.

109 Zhang, S., Deng, X., Shen, L., and Liu, Y. (2013). Ammoximation of citronellal with H_2O_2 catalyzed by TS-1: ammoximation of citronellal with H_2O_2 catalyzed by TS-1. *Chin. J. Catal.* 33 (4): 723–729.

110 Xu, L., Li, N., Peng, H., and Wu, P. (2013). Clean synthesis of amides over bifunctional catalysts of rhodium-loaded titanosilicates. *ChemCatChem* 5 (8): 2462–2470.

111 Peng, H., Xu, L., Wu, H. et al. (2013). One-pot synthesis of benzamide over a robust tandem catalyst based on center radially fibrous silica encapsulated TS-1. *Chem. Commun.* 49 (26): 2709.

112 Lu, X., Guan, Y., Xu, H. et al. (2017). Clean synthesis of furfural oxime through liquid-phase ammoximation of furfural over titanosilicate catalysts. *Green Chem.* 19 (20): 4871–4878.

113 Xu, Z., Jin, X., Zhou, T. et al. (2022). Preparation of aldoxime through direct ammoximation using titanium silicalite-1 catalyst. *Chin. J. Chem. Eng.* 47: 11–17.

114 Xu, H., Zhang, Y., Wu, H. et al. (2011). Postsynthesis of mesoporous MOR-type titanosilicate and its unique catalytic properties in liquid-phase oxidations. *J. Catal.* 281 (2): 263–272.

115 Yang, Y., Ding, J., Wang, B. et al. (2014). Influences of fluorine implantation on catalytic performance and porosity of MOR-type titanosilicate. *J. Catal.* 320: 160–169.

116 Yang, Y., Ding, J., Xu, C. et al. (2015). An insight into crystal morphology-dependent catalytic properties of MOR-type titanosilicate in liquid-phase selective oxidation. *J. Catal.* 325: 101–110.

117 Xu, H., Wang, Y., Peng, R. et al. (2022). Synthesis of micro-mesoporous Ti-MOR/silica composite spheres in oil-in-water microemulsion system. *Chem. Res. Chin. Univ.* 38 (1): 192–199.

118 Yoo, J.W., Lee, C.W., Chang, J. et al. (2000). Characterization and catalytic properties of Ti-ZSM-5 prepared by chemical vapor deposition. *Catal. Lett.* 66 (3): 169–173.

119 Zhao, S., Xie, W., Yang, J. et al. (2011). An investigation into cyclohexanone ammoximation over Ti-MWW in a continuous slurry reactor. *Appl. Catal. A Gen.* 394 (1–2): 1–8.

120 Sun, B., Guo, Y., Xu, L. et al. (2012). Synthesis of zeolite/mesoporous silica composite microspheres by microemulsion method. *Acta Chim. Sin.* 70 (23): 2419.

121 Xu, L., Peng, H., Zhang, K. et al. (2013). Core–Shell-structured titanosilicate as a robust catalyst for cyclohexanone ammoximation. *ACS Catal.* 3 (1): 103–110.

122 Tang, Z., Yu, Y., Li, L. et al. (2019). Regulation of framework titanium siting in MWW-type titanosilicates by boron and aluminum incorporation and its effect on the catalytic performance. *Microporous Mesoporous Mater.* 287: 93–100.

123 Wen, Y., Zhang, F., Wu, K. et al. (2021). Clean synthesis of Ti-MWW by utilizing the recycled mother liquor. *ACS Omega* 6 (46): 31196–31202.

124 Xu, H., Guan, Y., Lu, X. et al. (2020). Relation of selective oxidation catalytic performance to microenvironment of Ti^{IV} active site based on isotopic labeling. *ACS Catal.* 10 (8): 4813–4819.

125 Song, F., Liu, Y., Wu, H. et al. (2005). Highly effective synthesis of cyclohexanone oxime over a novel Titanosilicate Ti-MWW. *Chem. Lett.* 34 (10): 1436–1437.

126 Song, F., Liu, Y., Wang, L. et al. (2006). A novel titanosilicate with MWW structure: highly effective liquid-phase ammoximation of cyclohexanone. *Chin. J. Catal.* 27: 562–566.

127 Wang, L., Liu, Y., Xie, W. et al. (2008). Improving the hydrophobicity and oxidation activity of Ti–MWW by reversible structural rearrangement. *J. Phys. Chem. C* 112 (15): 6132–6138.

128 Yang, J., Yao, M., Zhao, S. et al. (2010). Reaction behavior for synthesis of methyl ethyl ketone oxime catalyzed by titanosilicate/H_2O_2 systems: reaction behavior for synthesis of methyl ethyl ketone oxime catalyzed by titanosilicate/H_2O_2 systems. *Chin. J. Catal.* 31 (1): 95–99.

129 Zhao, S., Xie, W., Liu, Y., and Wu, P. (2011). Methyl ethyl ketone ammoximation over Ti-MWW in a continuous slurry reactor: methyl ethyl ketone ammoximation over Ti-MWW in a continuous slurry reactor. *Chin. J. Catal.* 32 (1): 179–183.

130 Xu, H., Tian, W., Xu, L. et al. (2020). Crossed intergrowth triggered TS-2 microsphere: formation mechanism, modification and catalytic performance. *Chin. J. Catal.* 41 (7): 1109–1117.

131 Li, H., Gao, P., Xue, X. et al. (2018). Fluorine modified TAPO-5: synthesis and catalytic performance in cyclohexanone ammoximation. *Chem. Res. Chin. Univ.* 34 (2): 247–253.

132 Bhaumik, A., Samanta, S., and Mal, N.K. (2004). Highly active disordered extra large pore titanium silicate. *Microporous Mesoporous Mater.* 68: 29–35.

133 Li, Z., Zhou, J., Sui, Z., and Zhou, X. (2017). Synthesis of titanium-containing organic-inorganic hybrid periodic mesoporous organosilicas and its application in ammoximation of cyclohexanone. *Petrochem. Technol.* 46 (9): 1099–1104.

134 Reddy, J.S., Sivasanker, S., and Ratnasamy, P. (1991). Ammoximation of cyclohexanone over a titanium silicate molecular sieve, TS-2. *J. Mol. Catal.* 69 (3): 383–392.

135 Tvaruzkova, Z., Petras, M., Habersberger, K., and Jiru, P. (1992). Surface complexes of cyclohexanone and aqueous solution of NH3 on Ti-silicalite in liquid phase. *Catal. Lett.* 13 (1–2): 117–121.

136 Ichihashi, H. and Sato, H. (2001). The development of new heterogeneous catalytic processes for the production of ε-caprolactam. *Appl. Catal. A Gen.* 221 (1–2): 359–366.

137 Haruta, M., Uphade, B.S., Tsubota, S., and Miyamoto, A. (1998). Selective oxidation of propylene over gold deposited on titanium-based oxides. *Res. Chem. Intermed.* 24 (3): 329–336.

138 Hayashi, T., Tanaka, K., and Haruta, M. (1998). Selective vapor-phase epoxidation of propylene over Au/TiO_2 catalysts in the presence of oxygen and hydrogen. *J. Catal.* 178 (2): 566–575.

139 Uphade, B.S., Yamada, Y., Akita, T. et al. (2001). Synthesis and characterization of Ti-MCM-41 and vapor-phase epoxidation of propylene using H_2 and O_2 over Au/Ti-MCM-41. *Appl. Catal. A Gen.* 215 (1–2): 137–148.

140 Laufer, W. and Hoelderich, W.F. (2001). Direct oxidation of propylene and other olefins on precious metal containing Ti-catalysts. *Appl. Catal. A Gen.* 213 (2): 163–171.

141 Taylor, B., Lauterbach, J., and Delgass, W.N. (2007). The effect of mesoporous scale defects on the activity of Au/TS-1 for the epoxidation of propylene. *Catal. Today* 123 (1–4): 50–58.

142 Ren, Y., Xu, L., Zhang, L. et al. (2011). Selective epoxidation of propylene to propylene oxide with H_2 and O_2 over Au/Ti-MWW catalysts. *Pure Appl. Chem.* 84 (3): 561–578.

143 Hiroako, A., Tomonori, K., and Makoto, Y. (2008). *US Patent, 7999125B2.*

144 Lewis, R.J., Ueura, K., Liu, X. et al. (2023). Selective ammoximation of ketones via in situ H_2O_2 synthesis. *ACS Catal.* 13 (3): 1934–1945.

145 Yin, C., Ni, R., Bao, X., and Chen, Y. (2015). Synthesis of hierarchical porous silicalite-1 and its catalytic performance in Beckmann rearrangement. *Microporous Mesoporous Mater.* 202: 133–137.

146 You, K., Mao, L., Yin, D. et al. (2008). Beckmann rearrangement of cyclohexanone oxime to ε-caprolactam catalyzed by sulfonic acid resin in DMSO. *Catal. Commun.* 9 (6): 1521–1526.

147 Wang, B., Gu, Y., Luo, C. et al. (2004). Sulfamic acid as a cost-effective and recyclable catalyst for liquid Beckmann rearrangement, a green process to produce amides from ketoximes without waste. *Tetrahedron Lett.* 45 (17): 3369–3372.

148 Owston, N.A., Parker, A.J., and Williams, J.M.J. (2007). Iridium-catalyzed conversion of alcohols into amides via oximes. *Org. Lett.* 9 (1): 73–75.

149 Owston, N.A., Parker, A.J., and Williams, J.M.J. (2007). Highly efficient ruthenium-catalyzed oxime to amide rearrangement. *Org. Lett.* 9 (18): 3599–3601.

150 Sharghi, H. and Hosseini, M. (2002). Solvent-free and one-step Beckmann rearrangement of ketones and aldehydes by zinc oxide. *Synthesis* 2002 (08): 1057–1060.

151 Park, S., Choi, Y., Han, H. et al. (2003). Rh-catalyzed one-pot and practical transformation of aldoximes to amides. *Chem. Commun.* 15: 1936.

152 García-Álvarez, R., Díaz-Álvarez, A.E., Borge, J. et al. (2012). Ruthenium-catalyzed rearrangement of aldoximes to primary amides in water. *Organometallics* 31 (17): 6482–6490.

153 García-Álvarez, R., Francos, J., Crochet, P., and Cadierno, V. (2011). Ibuprofenamide: a convenient method of synthesis by catalytic hydration of 2-(4-isobutylphenyl)propionitrile in pure aqueous medium. *Tetrahedron Lett.* 52 (32): 4218–4220.

154 Allen, C.L., Lawrence, R., Emmett, L., and Williams, J.M.J. (2011). Mechanistic studies into metal-catalyzed aldoxime to amide rearrangements. *Adv. Synth. Catal.* 353 (18): 3262–3268.

155 Ali, M.A. and Punniyamurthy, T. (2010). Palladium-catalyzed one-pot conversion of aldehydes to amides. *Adv. Synth. Catal.* 352 (2–3): 288–292.

156 Fujiwara, H., Ogasawara, Y., Kotani, M. et al. (2008). Supported rhodium hydroxide catalyst: preparation, characterization, and scope of the synthesis of primary amides from aldoximes or aldehydes. *Chem. Asian J.* 3 (8–9): 1715–1721.

157 Peng, H., Wang, D., Xu, L., and Wu, P. (2013). One-pot synthesis of primary amides on bifunctional Rh(OH)x/TS-1@KCC-1 catalysts. *Chin. J. Catal.* 34 (11): 2057–2065.

11

Titanosilicate-Based Alkene Epoxidation Catalysis

Changjiu Xia[1] *and Xiang Feng*[2]

[1] *SINOPEC, Research Institute of Petroleum Processing, State Key Laboratory of Catalytic Materials and Reaction Engineering, No.18 Xueyuan road, Beijing, 100083, China*
[2] *China University of Petroleum, State Key Laboratory of Heavy Oil Processing, College of Chemical Engineering, Department of Chemical Engineering, No. 66 West Changjiang Road, Qingdao, 266580, China*

11.1 Introduction

Epoxides compounds, including ethylene oxide (EO), propylene oxide (PO), and epichlorohydrin (ECH) in major, serve as important versatile chemical intermediates and organic building blocks in industry [1–4], owing to the existence of highly reactive three-member epoxy moiety for polymerization and addition reactions, as shown in Scheme 11.1. Consequently, epoxide compounds are widely applied to manufacture a large number of useful products, i.e. polymers, polyurethanes, polyesters, oxygenated organic solvents, surfactants, lubricating oils, textiles, and cosmetics [5, 6]. In 2022, the annual global consumption of EO, PO, and ECH had been over 42, 13, and 3.5 Mt, respectively. Obviously, they are closely connected to industrial production and human daily life, and thus their demand always goes increasing along with social development and life quality improvement with an annual growth rate over 5% as predicted [7].

With respect to the production of epoxides, the chlorohydrin process is the most conventional method, which employs hazardous chlorine (Cl_2) and alkali (such as NaOH and CaO) as raw materials [8, 9]. Although this process is very mild to be carried out, it suffers from several harsh pollution and corrosion issues, particularly with the generation of a large amount of chlorine-containing liquid and solid wastes, thus against the principles of green and sustainable chemistry. To overcome these drawbacks, several specific routes for various epoxides were developed to make epoxide production more environmentally friendly, as shown in Scheme 11.2. Some key information on the process and characteristics are summarized as follows:

(1) EO can be produced via aerobic selective oxidation of ethylene over $Ag/\alpha\text{-}Al_2O_3$ catalysts, and at present over 90% of EO product in the world is manufactured by this process [10–12]. Unfortunately, some undesired deep oxidation reactions,

Micro-Mesoporous Metallosilicates: Synthesis, Characterization, and Catalytic Applications, First Edition. Edited by Peng Wu and Hao Xu.

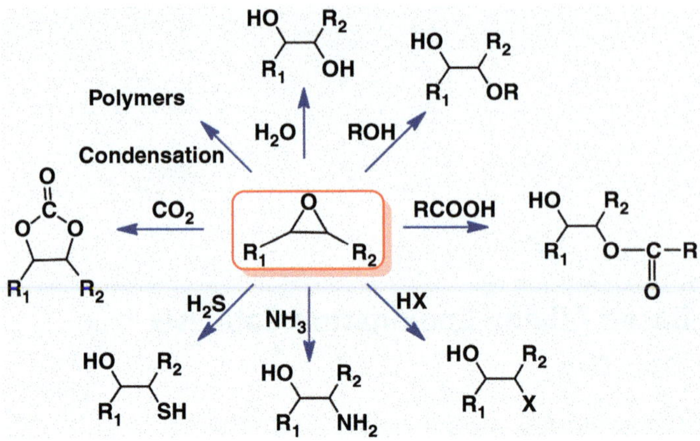

Scheme 11.1 Wide application of epoxides to synthesize various chemicals.

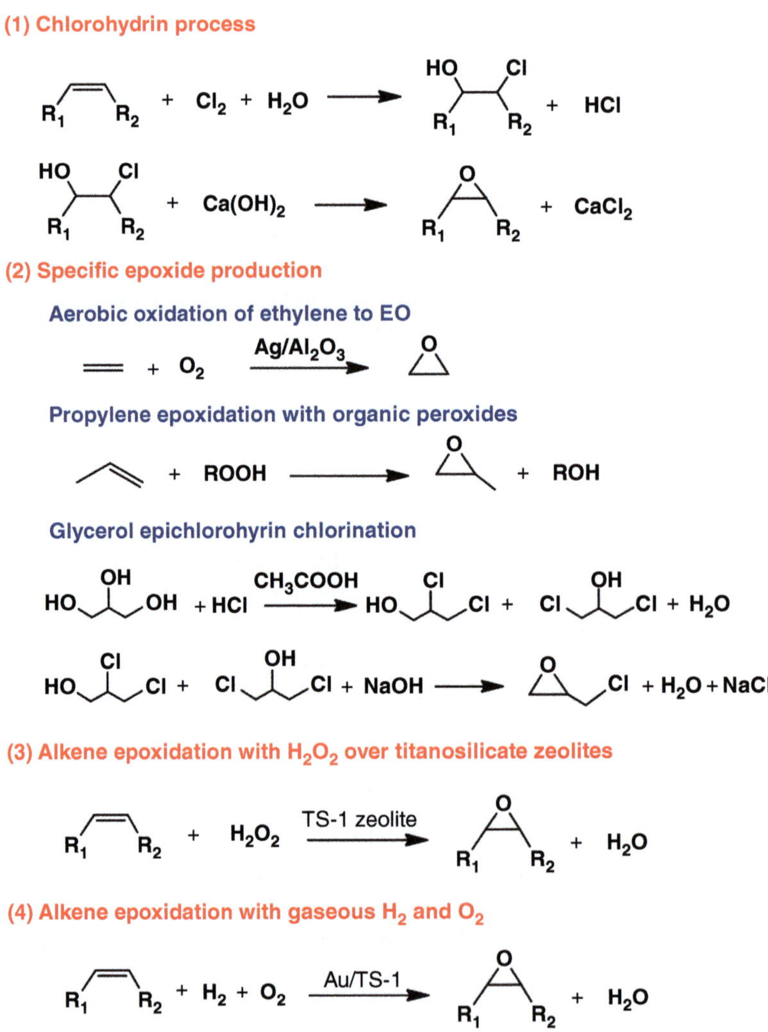

Scheme 11.2 Various epoxide manufacture routes.

with CO_2 and H_2O as final products, significantly reduce the yield of EO. Consequently, to obtain high EO selectivity, ethylene conversion should be maintained at a low level, hence with a large amount of ethylene needed for separating and recycling. However, this method works poorly for other olefins due to the inertness of allyl groups and activation of neighboring allyl H atoms via p-π conjugated effects, therefore with aldehyde, CO_2, and H_2O as main products [13–15].

(2) With respect to PO production, various hydroperoxides (such as *tert*-butyl hydroperoxide, cumene hydroperoxide, and ethylbenzene hydroperoxide) are applied as the oxidant for olefin epoxidation under different kinds of homogeneous or heterogeneous transition metal catalysts with ultra-wide substrate scope. Meanwhile, alcohol and derivates are generated as a co-product. Hence, this process is generally referred to as co-oxidation method. Obviously, this method is much cleaner and less corrosive than the chlorohydrin process, with far less waste formation [16–18]. However, this method is still of remarkable drawbacks, i.e. long reaction processing, high capital investment, and low conversion to maintain high hydroperoxide and PO selectivity. Importantly, the economic feature of this process is usually restricted by the price and market of co-products [19–22].

(3) ECH is also a versatile chemical intermediate with a variety of applications in industry. Except for the traditional Chlorohydrin route, ECH can be synthesized from glycerol through two basic steps [23–25]. Firstly, glycerin hydrochlorination by HCl gas in the liquid phase gives birth to a mixture of 1,3-dichlorohydrin (1,3-DCH) and 2,3-dichlorohydrin (2,3-DCH) with liquid acid catalysts. Then, ECH is produced via the dehydrochlorination of 1,3-DCH and 2,3-DCH isomers with alkaline compounds, quite similar to the second step of the chlorohydrin route. However, hazardous and corrosive HCl and strong alkalis are still employed in this route.

Although great progress has been achieved for specific epoxide production, they are still far away from being considered as ideal choices at industrial scale, owing to these incredible intrinsic technical drawbacks [26–28]. Fortunately enough, the discovery and evolution of titanium silicate zeolites of different topologies provide desirable opportunities for exploring an alternative environmentally friendly epoxidation approach using hydrogen peroxide (H_2O_2) as the oxidant agent under very mild conditions, with water as the sole side-product. The first case on this topic lines in the propylene epoxidation catalyzed by TS-1 zeolite of MFI topology (processing a two-dimensional channel system with diameters of 0.53×0.56 and 0.51×0.51 nm) in 1983, which was also initially reported by EniChem. After that, several Ti-substituted zeolites (i.e. Ti-MWW, TS-2, Ti-Beta, and Ti-ZSM-48) were prepared, and the epoxidation of other olefins [such as ethylene, allylic chloride (ALC), cyclohexene] was also widely evaluated [29–31]. Up to now, considerable efforts have been devoted to understanding the fundamental issues of catalyst design and epoxidation reaction processing, severing as a hot topic in heterogeneous catalysis. Notably, propylene epoxidation with H_2O_2 to PO [referred as to the hydrogen peroxide propene oxide (HPPO) process in general] has been industrialized and run smoothly by a few international large chemical companies, involving Dow/BASF, Evonik/Uhde, and SINOPEC [32–34]. Meanwhile, other olefin epoxidation processes have also achieved good progresses. For example, ALC epoxidation to ECH is ultra-close to being commercialized at SINOPEC in China. The detailed information on the epoxidation of various olefins will be summarized and discussed in following sections.

Furthermore, inspired by the successful development of the HPPO process, the in situ generation of H_2O_2 from gaseous H_2 and O_2 over noble metal nanoparticles (such as Au, Pd, and Au–Pd) is combined with olefin epoxidation under one-pot environment. In a typical procedure, the gaseous mixture of propylene, H_2, and O_2 directly passes through the fixed Au/TS-1 catalyst bed, in which H_2O_2 and PO are formed under the effect of Au nanoparticles and framework Ti species, respectively. In principle, the major advantage of this route is to avoid the use of industrial H_2O_2 synthesis units based on the anthraquinone method, thus with lower capital investment and operation cost. However, it is very hard to be carried out at a large scale, owing to the harsh intrinsic safety issues and poor product selectivity [35–37]. The main epoxidation reaction over Ti active sites plays a key role in improving PO yield; yet, this route will be further introduced in Chapter 14.

Above all, titanosilicate zeolite catalyzed olefin epoxidation shows considerable benefits for environmentally friendly epoxide manufacture. In this chapter, to provide an overview of this topic, we try our best to review the key scientific issues in epoxidation from a fundamental understanding of active sites and reaction pathways to industrial technology evolution via technology and engineering innovations.

11.2 Reaction Chemistry of Alkene Epoxidation Catalyzed by Titanosilicate Zeolites

As for the coordination structure, geometric location, and amount of Ti atoms in zeolite framework, enormous papers and reviews involving various characterization methods (i.e. X-ray diffraction, infrared and Raman vibrational spectroscopies, UV–Visible spectroscopy, UV-Raman and X-ray photoelectron spectroscopy and density functional theory (DFT) calculation) and reaction dynamics were positively reported, especially in the period from 1985 to 2000 [38–40]. Recently, X-ray absorption near edge structure (XANES), extended X-ray absorption fine structure (EXAFS) together with solid-state ^{17}O NMR were demonstrated to be effective in characterizing atomic-scale coordination mode of Ti sites and related active sites. According to these systemical works, the typical coordination mode of active Ti sites for olefin epoxidation includes the isolated tetrahedrally coordinated framework Ti species (TiO_4), pentahedrally coordinated framework Ti species (TiO_5), and mononuclear TiO_6 species. Moreover, the dinuclear Ti sites are presumed to be more active for the propylene epoxidation than the TiO_4 species [41]. In contrast, extra-framework Ti species (TiO_2) and octahedrally coordinated Ti are considered as inert active sites for epoxidation reactions, or even causing inefficient decomposition of H_2O_2 molecules to some extend.

The catalytic process of alkene epoxidation is assigned to the variation of the coordination sphere of isolated tetrahedral framework Ti sites via the generation of Lewis acid–base adducts with H_2O_2 and olefins. It is ascribed to the donating-accepting of electron pairs between the highest occupied molecular orbital (HOMO) of H_2O_2 molecules and the lowest unoccupied molecular orbital (LUMO) of framework Ti species, thus forming the highly reactive Ti-OOH intermediates. It is widely accepted that two kinds of framework Ti sites in TS-1 zeolite, including fully tetrahedrally bonded Ti sites (non-defect, with closed coordination) and tetrahedrally bonded Ti sites terminated with H atoms adjacent to Si vacancies (defect, with open coordination), are involved [42–44]. As highlighted by Lamberti et al., lattice strain of the Ti isomorphous substitution would cause the generation of neighboring framework defects, and thus open coordinated Ti sites are always nearby abundant of defects, even if they are randomly distributed at low concentrations (Figure 11.1). Generally, open-type defective framework Ti sites are considered as the active centers for epoxidation catalysis.

Besides the TiO_4, the TiO_5 could also exist in TS-1. In the liquid-phase reaction with H_2O_2 oxidant, the TiO_4 may further react with the alcohol solvent and H_2O_2 to form the TiO_5 by five-membered ring formation mechanism (5MR) [30, 45, 46]. In the five-membered ring formation mechanism, the Ti atom in TS-1 coordinates with the O atom and H atom in the alcohol solvent together with the two O atoms in the H_2O_2 to form the typical five-membered ring, then the O atom from the H_2O_2 is transferred to the olefin and subsequently the epoxidation product is synthesized by the transfer of active O species. Therefore, the TiO_5 is also considered as the active sites with higher performance to enhance the selective olefin epoxidation. Moreover, Yu et al. constructed the novel mononuclear TiO_6 species in the TS-1 (Si/Ti = 80) by the one-step rapid synthesis with the assistance of active seeds and microwave irradiation. This TS-1 contains both mononuclear TiO_6 species and isolated TiO_4 species. The TS-1 containing mononuclear TiO_6 species with open Ti sites showed better catalytic activity (conversion, 28.0%) in 1-hexene epoxidation than that of the traditional TS-1 (conversion, 18.8%) with closed Ti sites. Recently, Copéret et al. proposed an intriguing viewpoint that the dinuclear Ti sites are the active sites for

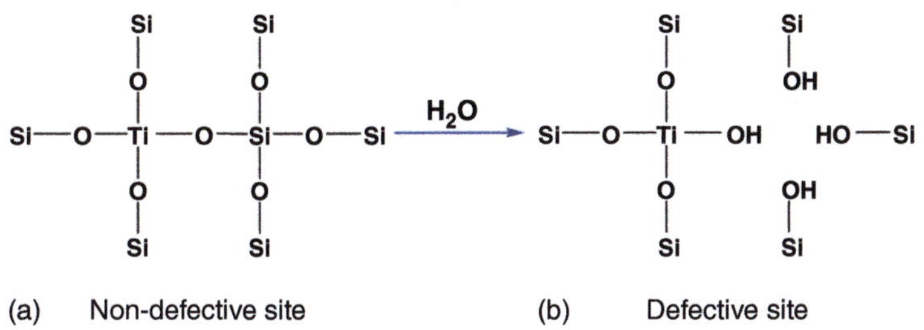

Figure 11.1 Schematic view of a fully tetrahedrally bonded Ti ion substituted for a Si ion at one of the lattice positions of TS-1 (a) and the same Ti site located near a silicon vacancy terminated with hydrogen atoms forming a silanol nest (b).

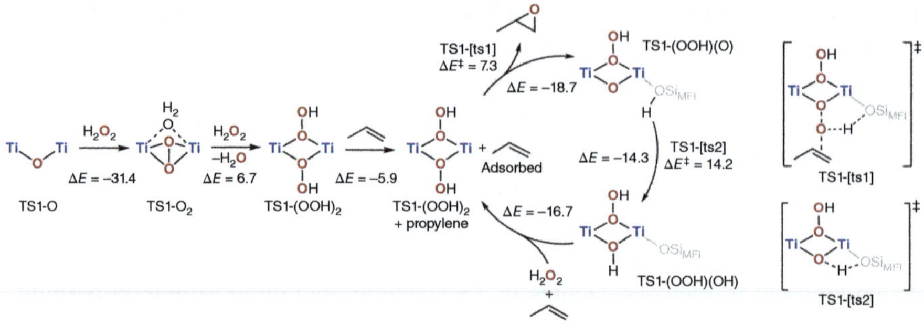

Figure 11.2 Catalytic mechanism of HPPO reaction via dinuclear Ti sites. Source: Reproduced from [47]/with permission of Springer Nature.

the propylene epoxidation instead of the TiO_4 [47]. The dinuclear Ti sites existed extensively in the Ti-content range of 0.5–1.9 wt% whether or not the TS-1 contains the TiO_2 species. The contact between the dinuclear Ti sites and H_2O_2 boosts the formation of bridging peroxo species for further epoxidation step as proved by the solid-state ^{17}O NMR spectra. The DFT calculations demonstrated that the cooperation between the adjacent Ti sites enables a low-energy epoxidation route via the oxygen-transfer transition state mechanism (Figure 11.2).

In conclusion, it requires a well-organized process to control the Ti coordination modes at the atomic scale for different reactions in order to fundamentally enhance the reaction performance. Furthermore, with respect to the alkene epoxidation mechanism over the H_2O_2/TS-1 system, several activation and transformation

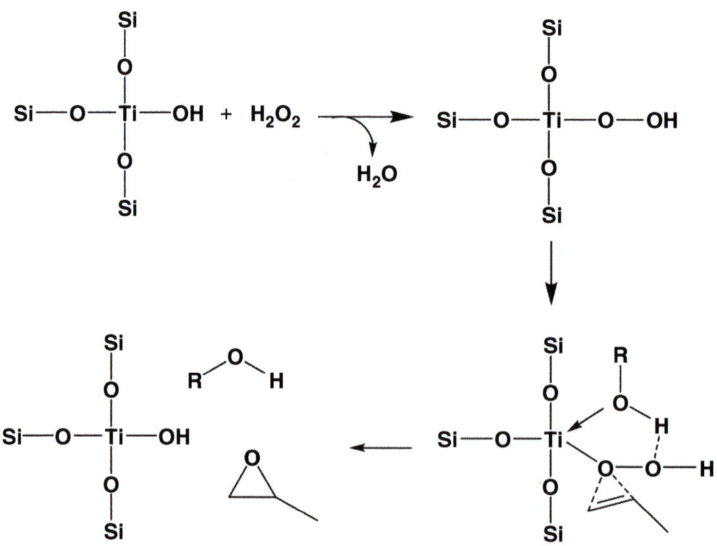

Figure 11.3 Suggested pathway for epoxide formation via a "five-member" bidentate Ti^{IV} (η_2-OOH) complex. Source: Reproduced from [49]/with permission of American Chemical Society.

models with different reactive oxygen intermediates have been proposed by employing consistent DFT methods. By taking Gibbs free energy into account, it points out that the formation of highly reactive Ti-OOH species over open coordinated framework Ti defects, which is the rate-determining step in epoxidation processing, is more favorable for alkene epoxidation due to the stable distorted octahedral coordination of active Ti species [48, 49]. In addition, the open structure of the defect sites allows the adsorption and coordination of H_2O_2, water, and other solvent molecules, while defective Ti sites are quite flexible (Figure 11.3).

11.3 Typical Alkene Epoxidation Cases

11.3.1 Propylene Epoxidation for PO Production

It is well accepted that the HPPO process is a very simple and clean approach for PO production under ultra-mild conditions, with H_2O as the only byproduct, which was initially reported by Clerici and Romano at EniChem in 1987 [50]. PO can be directly synthesized via propylene abstraction one single oxygen atom from an H_2O_2 molecule, rather than the generation of allylic products although the high reactivity of allylic hydrogen. Hence, all byproducts are caused by the side ring-opening reactions of PO product over acid sites from framework defects, involving hydrolysis, esterification, and hydrated condensation, as shown in Figure 11.4. Hence, the introduction of some alkaline agents and catalyst modification by loading metal oxides can efficiently inhibit acidic active sites, and thus improve PO selectivity.

Additionally, compared to other solvents, methanol is widely accepted as the preferred solvent for the HPPO process with more promising catalytic performance, owing to the high dissolution of propylene and participation in the formation of reactive chemical intermediates. Consequently, methanol is the most general solvent in both academic and industrial fields. To achieve a good propylene dissolution in the liquid phase, the reaction pressure is usually controlled around 2.0–2.5 MPa at a temperature of 20–60 °C [51–53]. Based on abundant experimental data, it is claimed that high reaction temperature corresponds to high H_2O_2 conversion and reaction rate, but significantly increases the side ring-opening reactions of PO at the same

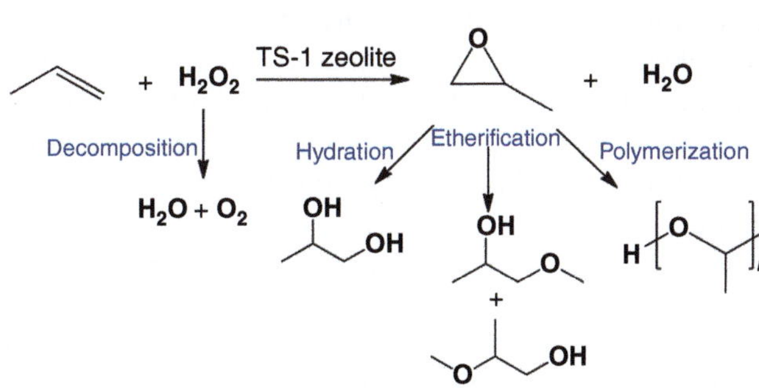

Figure 11.4 Reaction network of HPPO reaction.

time. Therefore, low temperature and fast heat exchange are usually preferential needed for high catalytic performance. In principle, stirred slurry bed reactor, in which titanosilicate catalysts are in the form of solid power, can facially control reaction temperature and make reagent mixture homogeneously in a short time, yet with unavoidable remixing of PO within mixture flows and wide residence time distribution. Hence, to achieve high PO selectivity, the packed/fixed bed reactor with pellet-shaped catalyst is also generally employed in literatures, owing to the narrow and controllable residence time for lowering side reactions; thus, this type of reactor has been put into industrialization. For this issue, the shape and size of the catalyst and liquid–solid mass transport properties are key factors.

To simultaneously achieve high conversion and PO selectivity, HPPO reaction is very difficult to be carried out, especially at large or even industrial scale. That is because the major HPPO reaction is extremely exothermic (over 220 kcal/mol PO formation), thus accelerating a variety of side reactions, such as PO hydrolysis, solvolysis, and dehydrated condensation, or H_2O_2 decomposition, due to the higher chemical reactivity of PO and H_2O_2 molecules within narrow confined microporous channels of titanosilicate zeolites. Notably, the concentration distribution of propylene and PO inside zeolite channels is always not uniform, and hence, side reactions may be promoted in some hot areas of high condensation and temperature. Herein, a successful HPPO catalyst developed by SINOPEC is valuable to be described in detail. In accordance with the deep understanding of the HPPO reaction network, the surface silicon-rich hollow titanium silicate (SR-HTS) zeolite was designed and manufactured by combing controllable recrystallization and post-synthesis methods at high temperatures under hydrothermal conditions (Figure 11.5). In the first step, the hydrolysis, nucleation, and crystal growth processes of Si and Ti sources are altered, with faster incorporation of Ti species than Si species into the framework matrix, thus giving birth to the Si-rich surface. Then, multiple hollow cavities can be created via an in situ dissolution–recrystallization process using organic structure-directing agents. Meanwhile, this post-treatment process also increases the amount of framework Ti species due to the dissolution of unstable framework defects and re-condensation between Si-OH and Ti-OH groups.

The unique benefits of this novel SR-HTS zeolite material are contributed by the following four aspects: (i) providing much more framework Ti active sites for high epoxidation performance; (ii) reducing the amount of framework defect and shortening mass diffusion pathway, thus lowering side reactions catalyzed by acidic sites; (iii) synergistic effects between high reactivity and great mass transfer; (iv) inhibiting H_2O_2 decomposition via minimizing the contact of H_2O_2 molecules to framework Ti sites. As a result, SR-HTS zeolite shows much higher H_2O_2 conversion and PO selectivity than the reference TS-1 zeolite. Similarly, the core–shell structures were also reported in literature [54].

In the case of industrial application, the deactivation and regeneration of HPPO catalyst [9] are another important issues that need to be considered, which deeply influences operation stability and the technical economy of this process. As initially detected with the GC–MS method by G.F. Thiele at Degussa AG, the side products are generally produced from the ring opening of PO with nucleophiles probably catalyzed by acidic silanols and crystalline defects. The deactivation is caused by

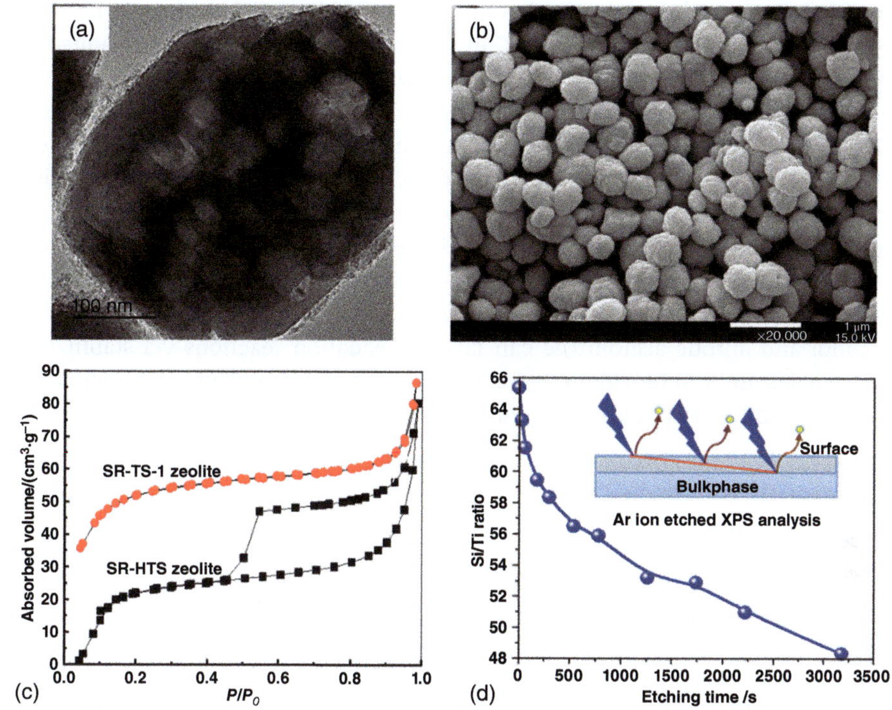

Figure 11.5 Characterization of SR-HTS zeolite for HPPO process: (a) TEM image of SR-HTS. (b) SEM image of SR-HTS zeolite; (c) Low temperature N_2 adsorption- desorption curves of SR-HTS and SR-TS-1 zeolites; (d) Ar ion etched XPS spectrum of SR-HTS zeolite.

the formation of PO oligomers within zeolite microporous channels via continuous polymerization of PO and side products. Obviously, those oligomers are difficult to diffuse outside micropores, thus inhibiting the accessibility of framework Ti species. With respect to the deactivated catalyst, calcination at a high temperature of 550 °C can completely refresh catalytic activity; yet, it is hard to be carried out in a fixed bed reactor due to difficulties in catalyst loading and unloading at a large scale. Compared to thermal treatment, chemical in situ refluxing regeneration by synergistically employing aqueous H_2O_2 solution and methanol solvent under mild conditions also provide an efficient and facial alternative approach to make the catalyst fully recovered as a fresh one, which powerfully overcomes the above-mentioned limitations and thus favors for smooth operation in the industry. The mechanism for this treatment lines in the oxidative cleavage of PO oligomers by H_2O_2 solution within zeolite channels, and then small organic fragments can be extracted by methanol and water.

11.3.2 Propylene Chloride Epoxidation

ECH is widely used for the industrial production of epoxy resins, surfactants, pharmaceuticals, and plasticizers. Over 95% of ECH is industrially manufactured via the chlorohydrin route with hazardous Cl_2 and $Ca(OH)_2$ as reagents, which brings serious operation and pollution problems. Luckily, ECH can be synthesized via the liquid phase epoxidation of ALC in a continuous slurry reactor with aqueous H_2O_2 solution by TS-1, Ti-MOR, and Ti-MWW zeolites [55–57], as shown in Figure 11.6. It is demonstrated that both TS-1 and Ti-MWW show greater catalytic activity in ALC epoxidation than Ti-MOR, which may be attributed to the side ECH solvolysis reaction catalyzed by weak acid sites, increasing the selectivity of 3-chloro-1,2-propanediol and corresponding ethers, as shown in Figure 11.7. Additionally, solvents play significant roles in determining conversion and selectivity. To improve the yield of ECH in the TS-1 catalyzed system, a mixed solvent of methanol and acetonitrile was introduced in this process [58]. That is because protic methanol and aprotic acetonitrile can favor epoxidation reactions via stabilizing intermediates and inhibit ring-opening side reactions, respectively, with increased ALC conversion and ECH selectivity at the same time.

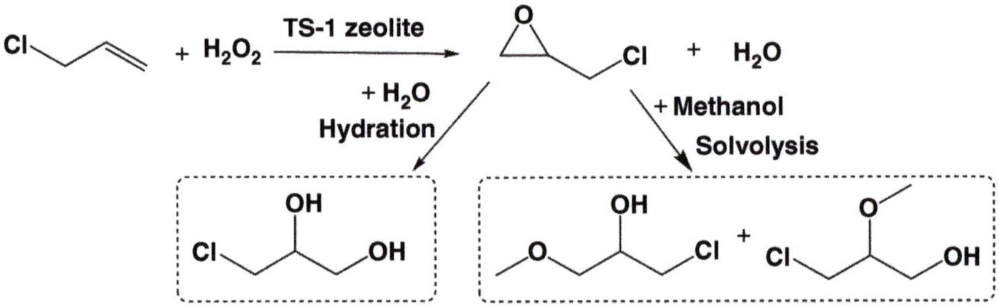

Figure 11.6 Reaction routes of titanosilicate zeolite catalyzed ALC epoxidation. Source: Reproduced from [57]/with permission of Royal Society of Chemistry.

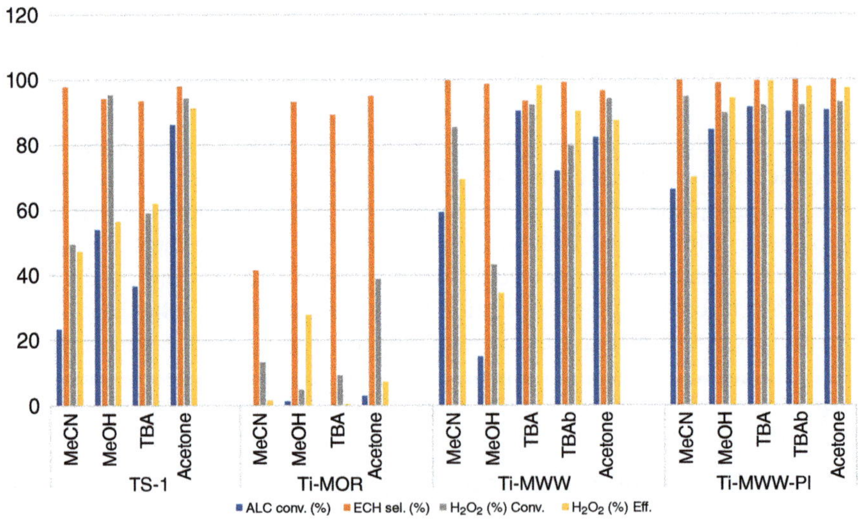

Figure 11.7 Catalytic performance of different titanosilicate zeolites adapted from literatures.

However, the selectivity of ECH is still far away from expectation at a large scale, owing to the complex side reactions under harsh reaction conditions, i.e. AC polymerization, ring-opening of ECH, and deep oxidation, which are ultra-hard to be carried out at pilot and even large scale. Hence, this route has not been commercially operated in industry, and great efforts on technological issues and engineering are greatly required [59–61].

On the other hand, to avoid the restriction of the rate-limiting epoxidation step, a novel and efficient ALC chlorohydrination route to produce dichloropropanol (DCP) with H_2O_2 and HCl was developed by using hollow titanium silicate (HTS) zeolite as the catalyst at SINOPEC [62–64]. It is well demonstrated that the Lewis acidic framework Ti sites and Brönsted acid HCl synergistically promote the ALC epoxidation and ring opening of ECH, giving birth to 1,3-DCP as the main product. This is quite different from that of the traditional chlorohydrination process, majorly with the generation of Cl_2, HOCl, and 2,3-DCP. It is found that the addition reaction of ECH with HCl is much easier than ALC epoxidation; hence, the secondary ring-opening reaction is of great help for accelerating the former epoxidation reaction via altering reaction balance. Moreover, as shown in Figure 11.8, HTS zeolite shows extremely excellent stability for this reaction even using HCl solution with a very low pH value.

Although a further DCP saponification reaction with bases is still needed for ECH generation, this novel ALC chlorohydrination process shows greater advantages than the Cl_2-based route, due to a higher DCP concentration of close to 35% in aqueous solution (around over seven times than that of the traditional route), and better DCP yield ascribing to the fast reaction of ECH with HCl, and thus these high operations and atom utilization efficiency remarkably offset the cost of H_2O_2 oxidant and energy consumption. As a consequence, this process also provides a desirable alternative method for environmentally friendly ECH manufacture under mild conditions.

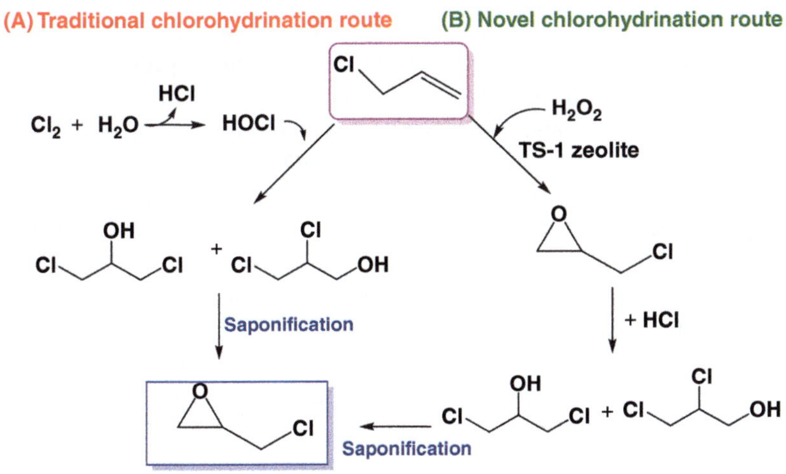

Figure 11.8 Reaction network of traditional (A) and novel ALC chlorohydrination route (B).

11.3.3 Ethylene Epoxidation to EO, EG, and Ethers

EO is the simplest epoxide and the third largest ethylene derivate, just smaller than two important polymers, including polyethylene and polyvinyl chloride, with a market demand of 3.05 Mt in 2021 in China and an average increasing rate over 5% for the future five years. Especially, ethylene glycol (EG) and derivated ethers are the largest downstream product of EO via further hydrolysis or solvolysis reactions, which are majorly used in the manufacture of polyester, surfactant, solvents, and antifreeze. Thus, EO, EG, and corresponding ethers can be attributed to the main product for ethylene epoxidation. Based on the economic and environmental assessment by the Center for Environmentally Beneficial Catalysis (CEBC), aqueous H_2O_2 solution as oxidant at mild conditions shows greater advantages than vapor-phase aerobic oxidation of ethylene, due to higher product selectivity and H_2O_2 utilization efficiency, although with great costs of oxidant and noble metal catalysts [65–67]. Therefore, when using cheap alternative heterogeneous catalysts, this H_2O_2 oxidation process is much more desirable and completive. From the viewpoint of industrial applications, heterogeneous catalysts are easy to be separated and recycled, hence avoiding the pollution and discharge of homogeneous catalysts. On the other hand, CO_2 emission is almost zero for this EO production process, agreeing well with the carbon neutrality goals worldwide.

As for heterogeneous ethylene epoxidation catalysis, many kinds of transition metals can be employed as the active component, i.e. Nb and W loaded mesoporous silica catalysts exhibit excellent catalytic performance in ethylene epoxidation. However, they suffer metal dissolution and leaching in high-concentration H_2O_2 solution, ascribing to the generation of soluble metal species [68–70]. Luckily, similarly to the HPPO process, titanium silicate zeolites also act as the dream catalyst for ethylene epoxidation due to the high stability of active sites and long lifetime. Yet, under similar reaction conditions, propylene is much more reactive for epoxidation than ethylene, and therefore, process optimization is greatly necessary in a wide range of reaction parameters, such as H_2O_2 concentration, pressure, temperature, and residence time in both batch and fixed bed reactors [71]. As summarized in Table 11.1, TS-1, Ti-MWW, Ti-MOR, Ti-Beta, and Sn-Ti-MWW zeolites exhibit specific catalytic activity and EO selectivity in various solvents under board experiment conditions. It is observed that TS-1 displays higher catalytic performance using MeOH as solvent, while the aprotic CH_3CN solvent is in favor of high EO selectivity for Ti-MWW, Ti-Beta, and Ti-MOR, owing to the restriction of EO hydrolysis and solvolysis. Although using CH_3CN solvent, TS-1 is of highest EO selectivity, but the total yield is only 5.2%. It suggests that the chemical state of Ti species and the wettability of microporous channel closely influence the efficiency of solvents. In addition, H_2O is considered as dream solvent for ethylene epoxidation, owing to the absence of organic solvent refining with high energy cost [74]. Although the total selectivity of EO and EG is around 90%, the conversion and efficiency of H_2O_2 are relatively low,

Table 11.1 Catalytic performance of various titanosilicate zeolites for ethylene epoxidation.

No.	Catalyst	Solvent	Total product yield (%)	Product distribution (%)				H_2O_2 efficiency (%)		References
				EO	EG	ME	Others	Conv.	Eff.	[72]
1	TS-1[a)]	MeOH	87.0	15.2	1.7	82.3	0.8	88.0	98.9	
2		MeCN	5.2	99.9	0.1	—	—	5.2	99.2	
3		H2O	19.8	20.3	69.7	0	10.0	44.1	45.0	
4	Ti-MWW[a)]	MeCN	44.3	98.8	1.2	—	—	54.4	81.4	
5		MeOH	7.0	24.1	—	74.2	1.7	7.5	93.2	
6		H2O	48.0	72.3	13.7	—	14.0	86.7	55.4	
7	Ti-MOR	MeCN	15.9	100.0	—	—	—	20.4	78.3	
8	Ti-Beta	MeCN	1.8	94.5	5.5	—	—	5.9	30.9	
9	Sn-Ti-MWW	H_2O	90.0	2.0	96.0	—	2.0	95.0	70.0	[73]
10	Ti-MWW	H_2O	81.0	62.0	36.0	—	2.0	95.0	65.0	

a) Reaction conditions: cat., 150 mg; ethylene, 2.5 MPa; H_2O_2, 10 mmol; solvent, 10 g; temp., 40 °C; time, 1.5 hours.

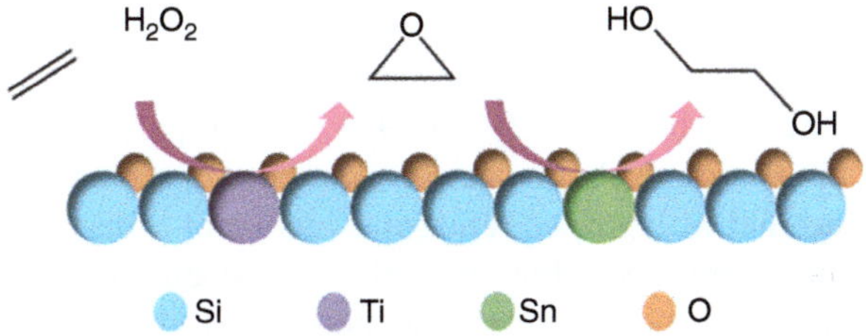

Figure 11.9 Synergistic effects between framework Ti and Sn species for ethylene to EG.

which may ascribe to poor solubility of ethylene and high H_2O_2 decomposition in aqueous solution. As a result, novel reaction processes and mass-diffusion enhanced reactors are promisingly required for further development of this green approach.

Very recently, to intensify ethylene transformation to EO and EG, a novel catalytic material Sn-Ti-MWW zeolite was reported by taking advantages of the synergistic effects between framework Ti and Sn species at atomic scale [73]. As shown in Figure 11.9, framework Ti and Sn sites can significantly promote ethylene epoxidation to EO and hydrolysis of EO to EG, respectively, due to framework Sn species is of stronger Lewis acidity. Hence, the chemical balance of epoxidation reaction is remarkably enhanced by fast consumption of EO, and therefore, Sn-Ti-MWW possesses higher total product yield and H_2O_2 efficiency at the same time.

11.4 Industrial Propylene Epoxidation Techniques and Processes

Up to the end of 2022, the HPPO process for PO production is the only commercialized olefin epoxidation case catalyzed by titanosilicate zeolite in industry, which was successfully explored by giant international petrochemical companies, including BASF/Dow, Evonik (formerly Degussa)/Uhde, and SINOPEC, respectively. Now, over 10% of the PO in global is produced by the HPPO process, and many new HPPO plants have been built and are under construction or evaluation in many countries, especially in China. Consequently, it is well estimated that the HPPO process represents for the development orientation of sustainable PO production in future, significantly overcoming the drawbacks of traditional routes.

In 2006, the first set of the joint HPPO plant (300 kt/a PO) was constructed and run successfully by BASF/Dow using H_2O_2 as oxidizing agent supplied by Solvay (230 kt/a H_2O_2), in Antwerp, Belgium. Owing to its high efficiency and environmental benefits, this process received a US Presidential Green Chemistry Challenge Award in 2010. The simplified BASF/Dow process flow sheet is illustrated in Figure 11.10. In the propylene epoxidation unit, the mixture of propylene, H_2O_2, and methanol is injected into the fixed bed reactor with shaped TS-1 zeolite as the catalyst, with reaction temperature at around 40 °C and pressure of 2 MPa.

An optimal performance in terms of PO yield is obtained when the reaction of PO synthesis is carried out in two steps with intermediate separation of the PO produced. The output mixture from the first reactor consists of methanol, water, PO, by-products, unreacted propene, and H_2O_2. The achieved yield is 85% with a PO selectivity based on H_2O_2 of 95%. The output is depressurized into a column operating at atmospheric pressure at a bottom temperature of 69 °C. PO is distilled off together with propene and some methanol, while the bottom product is fed to a second reactor operating under the same conditions as the first reactor, together with fresh propene. In the second step, the H_2O_2 conversion is 96%, and the PO selectivity is 96%. The overall H_2O_2 conversion is 99.4%, and the overall PO selectivity is 95–96%; the PO yield based on H_2O_2 is 94–95%. When the reaction is carried out in a single reactor, the H_2O_2 conversion is similar, that is, 98.4%, but the PO selectivity based on H_2O_2 is 80.3%, with a PO yield of 79%. Overall, the higher propene-to-H_2O_2 ratio in each reactor leads to a lower propene conversion, but the selectivity for PO

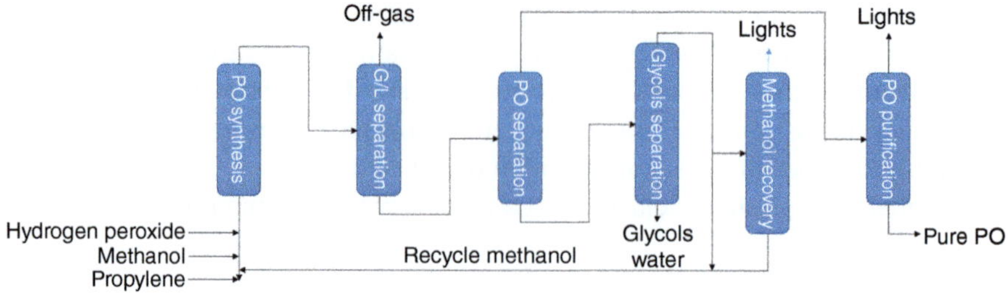

Figure 11.10 Simplified flow sheet of the BASF/Dow process for PO production.

with respect to H_2O_2 is improved. After the epoxidation reactor, a first column separates the lights, mainly including propene, propane, nitrogen (used as ballast to avoid the formation of flammable mixtures in the epoxidation reactor), and oxygen, the latter deriving from the decomposition of H_2O_2. Specifically, the off-gas stream is compressed with cooling (35 °C) and fed to a stripping column under pressure to obtain an olefin-containing bottom stream (potentially also containing propane) recycled to the epoxidation, and an off-gas stream with a low hydrocarbon content, containing nitrogen, oxygen, and further volatile by-products. The following is an example of the stream outlet composition from the epoxidation reactor with the subsequent separation of the light-boiling compounds: unreacted propene 0.01 mass%, formaldehyde 0.01%, acetaldehyde 0.03%, PO 9.45%, methanol 71.97%, water 17.54%, glycol ethers 0.43%, propene glycol 0.05%, and heavy boilers remainder.

PO is separated from methanol by extractive distillation, using water (or in another embodiment, propene glycol) as the extractor. PO is distilled overhead from the extractive column as the top stream, while the bottom stream contains methanol and water. The energy integration of the column is one important issue of the patents; the vapors of the top stream are compressed, and the condensation heat is returned to the vaporizer employed in the extractive distillation column.

Evonik developed an HPPO process with Uhde, opening a plant in Ulsan, South Korea (100 000 ta PO), licensed to SKC in 2008. And finally, an HPPO plant operated by Dow/SCG began production in Thailand in 2011 (390 000 ta PO) with a companion H O plant (330 000 ta) nearby.

As shown in Figure 11.11, in the Degussa/Uhde process [19], the reactor works at 60 °C. The raw PO is purified by extractive distillation using propene glycol that effectively removes impurities such as acetaldehyde. The final purity of PO is 99.97 wt%. The main by-product is propene glycol. The catalyst deactivates due to the formation of PO oligomers and is regenerated by calcination or treatment with H_2O_2 solutions. The high selectivity is reached owing to the moderate reaction temperature; the continuous removal from circulation of substances that can

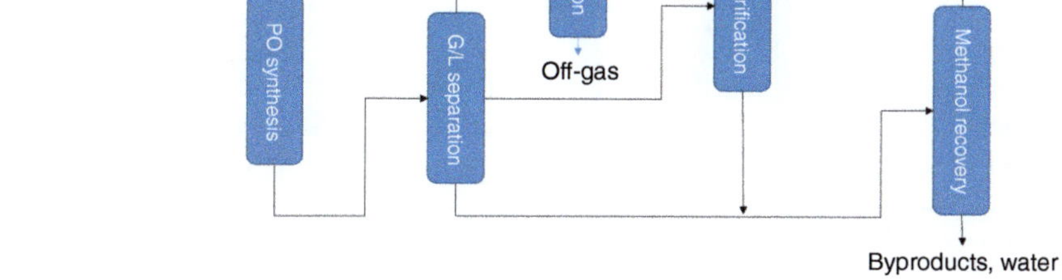

Figure 11.11 Simplified flow sheet of the Degussa/Uhde process.

deactivate the catalyst brings about a considerable improvement in the performance and catalyst longevity.

Almost at the same time, SINOPEC developed a titanium silicalite catalyst with hollow cavities and high framework Ti content, which obtained several domestic and internationally issued patents. The related HPPO process is operated in a fixed bed reactor, which achieved 95.1% selectivity for PO, 98.8% conversion of H_2O_2, and 1218 hours catalyst lifetime under laboratory conditions. This process has passed a review by the Sinopec technology department. The Pilot-scale plant of HPPO was built in October 2009 and commissioned in October 2010. From a huge amount of data in the epoxidation reaction, 98.3% conversion of H_2O_2, 96.1% selectivity for PO, and 99.3% product purity were achieved.

SINOPEC took the lead in initiating and developing of process design package in 2011. As shown in Figure 11.12, critical routes including the PO separation process, the propylene recovery process, and cyclic methanol hydrogenation rectification process were successively determined from April to December 2011. The HPPO process (100 000 t/year) design package was finished in 2012, and the detailed design was finished in 2013. The 100 000 tons industrial model plant accomplished a mid-term delivery on 25 June 2014. The commissioning production of PO products with 99.95% purity was carried out in December 2014. In April 2018, the plant accomplish 72 hours full-load calibration, and the consumption of material and energy meet the design standard. In November 2018, this plant has passed a review and evaluation by Sinopec's head office. This process possesses significant innovation with intellectual property rights, and the overall process has reached an international level. Particularly, the technology and consumption of materials for catalyst have reached an advanced international level. It is suggested to promote the application of catalyst technology.

Up to the June of 2023, SINOPEC has contracted to transfer the processing to North Huajin Chemical Industries Group Co. Ltd., Maoming Chemical Company, and Zhejiang Yuanjin new materials Co. Ltd., and two sets of 300 000 t/year HPPO plants are under design and construction, which has the following characteristics:

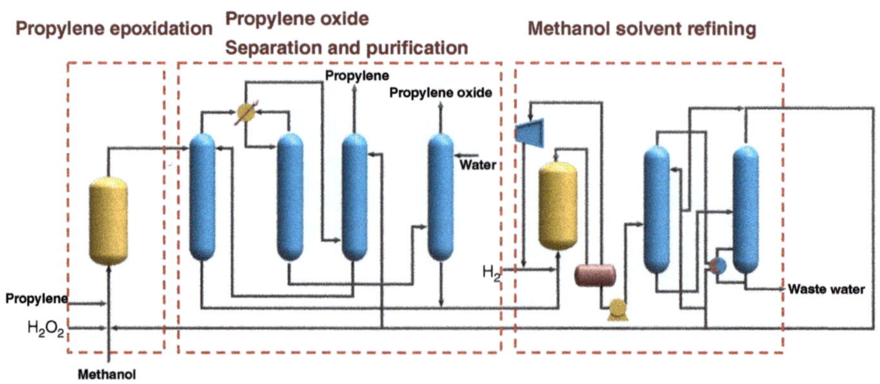

Figure 11.12 Simplified flow sheet of the HPPO process by SINOPEC.

1. The HPO-1 catalyst based on SR-HTS zeolite was employed for propylene epoxidation, with high conversion, high product selectivity, good stability, and lifespan.
2. The reaction conditions are mild, with the reaction temperature at 25–60 °C. High PO yield and low consumption of process can be achieved with few side reactions. With its high purity, superior quality compare to Chinese national standard, good color, and odorless characteristics, it could be a promising crude material for high-end polyether production.
3. Low energy consumption and thermal utilization are realized with double-effect rectification and thermal gathering of methanol.
4. Low-temperature separation of PO, the combined technique of chemical dehybridization, and extraction rectification resulted in high product quality.
5. The plant can be easily operated with smooth reaction control and stable long-cycle operation.
6. No corrosive raw materials and agents are employed in the process.
7. The in situ resurrection condition of the catalyst device is moderated without special requirements for equipment and operation.
8. The essential safety of the whole process of the plant was guaranteed due to the priority of safety given during design.
9. The production condition is mild, clean, and environmentally friendly with a small amount of wastewater.

11.5 Conclusion and Outlook

Epoxides serve as important raw materials and precursors for many chemical industries. We review the key scientific issues in epoxidation from a fundamental understanding of active sites and reaction pathways to industrial technology evolution via technology and engineering innovations. Nowadays, the most promising technology of propene epoxidation is the HPPO process in which the propene oxide is produced by oxidizing propene with hydrogen peroxide using TS-1 catalyst. The process is environmental friendly because only water is formed as a byproduct of the main reaction. TS-1 catalysts have attracted extensive attention of researchers due to its unique catalytic performance and excellent hydrothermal stability in the HPPO process. Although there are great progresses in the synthesis of TS-1 catalysts, there are still challenges. The way to construct stable Ti active sites with high coordination and the method to neutralize unfavorable strong acidity need to be further explored. In addition, inspired by the successful development of the HPPO process, the in situ generation of H_2O_2 from gaseous H_2 and O_2 over Au nanoparticles located on the TS-1 catalysts is strongly attractive in the future.

Acknowledgments

The authors would like to acknowledge the financial supports from the National Key Basic Research Development Plan “973” Project (No. 2006CB202508), the

SINOPEC Project (No. 411058 and 413025), and the National Natural Science Foundation (No. 21808244, 22278441, and 21978325), National natural science foundation of china-outstanding youngth foundation (No. 22322814).

References

1 Nijhuis, T.A., Makkee, M., Moulijn, J.A., and Weckhuysen, B.M. (2006). The production of propene oxide: catalytic processes and recent developments. *Ind. Eng. Chem. Res.* 45: 3447–3459.

2 Rebsdat, S. and Mayer, D. (2000). Ethylene glycol. In: *Ullmann's Encyclopedia of Industrial Chemistry*, vol. 33, 531–546. Wiley-VCH Verlag GmbH & Co.

3 Kwon, O., Ayla, E.Z., Potts, D.S., and Flaherty, D.W. (2023). Effects of solvent–pore interaction on rates and barriers for vapor phase alkene epoxidation with gaseous H_2O_2 in Ti-BEA catalysts. *ACS Catal.* 13: 6430–6444.

4 Neri, C., Anfossi, B., Esposito, A., and Buonomo, F. (1989). *Process for the Epoxidation of Olefinic Compounds.* Google Patents.

5 Weissermel, K. and Arpe, H.-J. (2008). *Industrial Organic Chemistry.* John Wiley & Sons.

6 Romano, U. and Ricci, M. (2013). Industrial applications. In: *Liquid Phase Oxidation via Heterogeneous Catalysis: Organic Synthesis and Industrial Applications*, 451–506. Wiley-VCH Verlag GmbH & Co.

7 Khatib, S.J. and Oyama, S. (2015). Direct oxidation of propylene to propylene oxide with molecular oxygen: a review. *Catal. Rev.* 57: 306–344.

8 Cavani, F. and Teles, J.H. (2009). Sustainability in catalytic oxidation: an alternative approach or a structural evolution? *ChemSusChem* 2: 508–534.

9 Thiele, G. and Roland, E. (1997). Propylene epoxidation with hydrogen peroxide and titanium silicalite catalyst: activity, deactivation and regeneration of the catalyst. *J. Mol. Catal. A Chem.* 117: 351–356.

10 Rebsdat, S. and Mayer, D. (2000). Ethylene oxide. In: *Ullmann's Encyclopedia of Industrial Chemistry*. Wiley-VCH Verlag GmbH & Co.

11 Jankowiak, J.T. and Barteau, M.A. (2005). Ethylene epoxidation over silver and copper-silver bimetallic catalysts: II. Cs and Cl promotion. *J. Catal.* 236: 379–386.

12 Christopher, P. and Linic, S. (2008). Engineering selectivity in heterogeneous catalysis: Ag nanowires as selective ethylene epoxidation catalysts. *J. Am. Chem. Soc.* 130: 11264–11265.

13 Montemore, M.M., van Spronsen, M.A., Madix, R.J., and Friend, C.M. (2017). O_2 activation by metal surfaces: implications for bonding and reactivity on heterogeneous catalysts. *Chem. Rev.* 118: 2816–2862.

14 Özbek, M. and Van Santen, R. (2013). The mechanism of ethylene epoxidation catalysis. *Catal. Lett.* 143: 131–141.

15 Carbonio, E.A., Rocha, T.C., Klyushin, A.Y. et al. (2018). Are multiple oxygen species selective in ethylene epoxidation on silver? *Chem. Sci.* 9: 990–998.

16 Liu, X., Wang, X., Guo, X., and Li, G. (2004). Effect of solvent on the propylene epoxidation over TS-1 catalyst. *Catal. Today* 93: 505–509.

17 Leow, W.R., Lum, Y., Ozden, A. et al. (2020). Chloride-mediated selective electrosynthesis of ethylene and propylene oxides at high current density. *Science* 368: 1228–1233.

18 Teržan, J., Huš, M., Likozar, B., and Djinović, P. (2020). Propylene epoxidation using molecular oxygen over copper- and silver-based catalysts: a review. *ACS Catal.* 10: 13415–13436.

19 Torres, D., Lopez, N., Illas, F., and Lambert, R.M. (2007). Low-basicity oxygen atoms: a key in the search for propylene epoxidation catalysts. *Angew. Chem. Int. Ed.* 46: 2055–2058.

20 Pell, M. and Korchak, E.I. (1969). *Epoxidation Using Ethylbenzene Hydroperoxide with Alkali or Adsorbent Treatment of Recycle Ethylbenzene*. Google Patents.

21 Lewandowski, G. and Milchert, E. (2001). Parameters of epoxidation of cyclohexene by tert-butyl hydroperoxide. *Ind. Eng. Chem. Res.* 40: 2402–2408.

22 Seo, T. and Tsuji, J. (2003). *Process for Producing Propylene Oxide*. Google Patents.

23 Almena, A. and Martín, M. (2016). Technoeconomic analysis of the production of epichlorohydrin from glycerol. *Ind. Eng. Chem. Res.* 55: 3226–3238.

24 Wang, M., Zhou, J., Mao, G., and Zheng, X. (2012). Synthesis of TS-1 from an inorganic reactant system and its catalytic properties for allyl chloride epoxidation. *Ind. Eng. Chem. Res.* 51: 12730–12738.

25 Wang, S.-J., Wong, D.S.-H., Lim, I.J.Q. et al. (2018). Design and control of a novel plant-wide process for epichlorohydrin synthesis by reacting allyl chloride with hydrogen peroxide. *Ind. Eng. Chem. Res.* 57: 6926–6936.

26 Wang, Q., Mi, Z., Wang, Y., and Wang, L. (2005). Epoxidation of allyl choride with molecular oxygen and 2-ethyl-anthrahydroquinone catalyzed by TS-1. *J. Mol. Catal. A Chem.* 229: 71–75.

27 Notari, B. (1988). Synthesis and catalytic properties of titanium containing zeolites. In: *Studies in Surface Science and Catalysis*, vol. 37, 413–425. Elsevier.

28 Wang, X., Zhang, X., Wang, Y. et al. (2011). Preparation and performance of TS-1/SiO_2 egg-shell catalysts. *Chem. Eng. J.* 175: 408–416.

29 Dong, J., Zhu, H., Xiang, Y. et al. (2016). Toward a unified identification of Ti location in the MFI framework of high-Ti-loaded TS-1: combined EXAFS, XANES, and DFT study. *J. Phys. Chem. C* 120: 20114–20124.

30 Wu, L., Deng, X., Zhao, S. et al. (2016). Synthesis of a highly active oxidation catalyst with improved distribution of titanium coordination states. *Chem. Commun.* 52: 8679–8682.

31 Li, T., Zuo, Y., Guo, Y. et al. (2020). Highly stable TS-1 extrudates for 1-butene epoxidation through improving the heat conductivity. *Catal. Sci. Technol.* 10: 6152–6160.

32 Russo, V., Tesser, R., Santacesaria, E., and Di Serio, M. (2013). Chemical and technical aspects of propene oxide production via hydrogen peroxide (HPPO process). *Ind. Eng. Chem. Res.* 52: 1168–1178.

33 Li, W., Wu, G., Hu, W. et al. (2022). Direct propylene epoxidation with molecular oxygen over cobalt-containing zeolites. *J. Am. Chem. Soc.* 144: 4260–4268.

34 Corma, A. and Garcia, H. (2003). Lewis acids: from conventional homogeneous to green homogeneous and heterogeneous catalysis. *Chem. Rev.* 103: 4307–4366.

35 Sinha, A., Seelan, S., Okumura, M. et al. (2005). Three-dimensional mesoporous titanosilicates prepared by modified sol-gel method: ideal gold catalyst supports for enhanced propene epoxidation. *J. Phys. Chem. B* 109: 3956–3965.

36 Song, W., Zuo, Y., Xiong, G. et al. (2014). Transformation of SiO_2 in titanium silicalite-1/SiO_2 extrudates during tetrapropylammonium hydroxide treatment and improvement of catalytic properties for propylene epoxidation. *Chem. Eng. J.* 253: 464–471.

37 Clerici, M., Bellussi, G., and Romano, U. (1991). Synthesis of propylene oxide from propylene and hydrogen peroxide catalyzed by titanium silicalite. *J. Catal.* 129: 159–167.

38 Nguyen, H.K., Sankar, G., and Catlow, R.A. (2017). Reactivities study of titanium sites in titanosilicate frameworks by in situ XANES. *J. Porous. Mater.* 24: 421–428.

39 Xiong, G., Cao, Y., Guo, Z. et al. (2016). The roles of different titanium species in TS-1 zeolite in propylene epoxidation studied by in situ UV Raman spectroscopy. *Phys. Chem. Chem. Phys.* 18: 190–196.

40 Xu, J., Wang, Y., Feng, W. et al. (2014). Effect of triethylamine treatment of titanium silicalite-1 on propylene epoxidation. *Front. Chem. Sci. Eng.* 8: 478–487.

41 Bordiga, S., Damin, A., Bonino, F. et al. (2002). The structure of the peroxo species in the TS-1 catalyst as investigated by resonant Raman spectroscopy. *Angew. Chem. Int. Ed.* 41: 4734–4737.

42 Ricchiardi, G., Damin, A., Bordiga, S. et al. (2001). Vibrational structure of titanium silicate catalysts. A spectroscopic and theoretical study. *J. Am. Chem. Soc.* 123: 11409–11419.

43 Lin, W. and Frei, H. (2002). Photochemical and FT-IR probing of the active site of hydrogen peroxide in Ti silicalite sieve. *J. Am. Chem. Soc.* 124: 9292–9298.

44 Gale, J.D. (2006). A periodic density functional study of the location of titanium within TS-1. *Solid State Sci.* 8: 234–240.

45 Wang, Y., Lin, M., and Tuel, A. (2007). Hollow TS-1 crystals formed via a dissolution–recrystallization process. *Microporous Mesoporous Mater.* 102: 80–85.

46 Zhuang, J., Yan, Z., Liu, X. et al. (2002). NMR study on the acidity of TS-1 zeolite. *Catal. Lett.* 83: 87–91.

47 Gordon, C.P., Engler, H., Tragl, A.S. et al. (2020). Efficient epoxidation over dinuclear sites in titanium silicalite-1. *Nature* 586: 708–713.

48 Bellussi, G., Carati, A., Clerici, M.G. et al. (1992). Reactions of titanium silicalite with protic molecules and hydrogen peroxide. *J. Catal.* 133: 220–230.

49 Sinclair, P.E. and Catlow, C.R.A. (1999). Quantum chemical study of the mechanism of partial oxidation reactivity in titanosilicate catalysts: active site formation, oxygen transfer, and catalyst deactivation. *J. Phys. Chem. B* 103: 1084–1095.

50 Romano, U., Esposito, A., Maspero, F. et al. (1990). Selective oxidation with Ti-silicalite. In: *Studies in Surface Science and Catalysis*, vol. 55, 33–41. Elsevier.

51 Wu, L., Zhao, S., Lin, L. et al. (2016). In-depth understanding of acid catalysis of solvolysis of propene oxide over titanosilicates and titanosilicate/H_2O_2 systems. *J. Catal.* 337: 248–259.

52 Liu, X., Liu, J., Xia, Y. et al. (2019). Catalytic performance of TS-1 in oxidative cleavage of 1-alkenes with H_2O_2. *Catal. Commun.* 126: 40–43.

53 Panyaburapa, W., Nanok, T., and Limtrakul, J. (2007). Epoxidation reaction of unsaturated hydrocarbons with H_2O_2 over defect TS-1 investigated by ONIOM method: formation of active sites and reaction mechanisms. *J. Phys. Chem. C* 111: 3433–3441.

54 Song, Z., Yuan, J., Cai, Z. et al. (2020). Engineering three-layer core-shell S-1/TS-1@dendritic-SiO_2 supported Au catalysts towards improved performance for propene epoxidation with H_2 and O_2. *Green Energy Environ.* 5: 473–483.

55 Clerici, M.G. and Ingallina, P. (1993). Epoxidation of lower olefins with hydrogen peroxide and titanium silicalite. *J. Catal.* 140: 71–83.

56 Wang, L., Liu, Y., Xie, W. et al. (2007). Highly efficient and selective production of epichlorohydrin through epoxidation of allyl chloride with hydrogen peroxide over Ti-MWW catalysts. *J. Catal.* 246: 205–214.

57 Ding, L., Yin, J., Tong, W. et al. (2021). Selective synthesis of epichlorohydrin via liquid-phase allyl chloride epoxidation over a modified Ti-MWW zeolite in a continuous slurry bed reactor. *New J. Chem.* 45: 331–342.

58 Pandey, R.K. and Kumar, R. (2007). Eco-friendly synthesis of epichlorohydrin catalyzed by titanium silicate (TS-1) molecular sieve and hydrogen peroxide. *Catal. Commun.* 8: 379–382.

59 Singh, G.S., Mollet, K., D'hooghe, M., and De Kimpe, N. (2013). Epihalohydrins in organic synthesis. *Chem. Rev.* 113: 1441–1498.

60 Heywood, D.L. and Phillips, B. (1958). The reaction of epichlorohydrin with secondary amines. *J. Am. Chem. Soc.* 80: 1257–1259.

61 Lu, Y., Wang, R., Zhang, J. et al. (2015). Evaluation of an improved epichlorohydrin synthesis from dichloropropanol using a microchemical system. *Chin. J. Chem. Eng.* 23: 1123–1130.

62 Peng, X., Xia, C., Lin, M. et al. (2017). Chlorohydrination of allyl chloride with HCl and H_2O_2 catalyzed by hollow titanium silicate zeolite to produce dichloropropanol. *Green Chem.* 19: 1221–1225.

63 Peng, X., Xia, C., Lin, M. et al. (2017). A safer and greener chlorohydrination of allyl chloride with H_2O_2 and HCl over hollow titanium silicate zeolite. *Appl. Catal. A Gen.* 543: 17–25.

64 Xia, C., Peng, X., Lin, M. et al. (2017). Understanding the pathways of improved chlorohydrination of allyl chloride with HCl and H_2O_2 catalyzed by titanium-incorporated zeolites. *Mol. Catal.* 442: 89–96.

65 Ghanta, M., Fahey, D.R., Busch, D.H., and Subramaniam, B. (2013). Comparative economic and environmental assessments of H_2O_2-based and tertiary butyl hydroperoxide-based propylene oxide technologies. *ACS Sustain. Chem. Eng.* 1: 268–277.

66 Bregante, D.T., Tan, J.Z., Schultz, R.L. et al. (2020). Catalytic consequences of oxidant, alkene, and pore structures on alkene epoxidations within titanium silicates. *ACS Catal.* 10: 10169–10184.

67 Khouw, C.B., Dartt, C.B., Labinger, J.A., and Davis, M.E. (1994). Studies on the catalytic-oxidation of alkanes and alkenes by titanium silicates. *J. Catal.* 149: 195–205.

68 Bai, R., Navarro, M.T., Song, Y. et al. (2020). Titanosilicate zeolite precursors for highly efficient oxidation reactions. *Chem. Sci.* 11: 12341–12349.

69 Yan, W., Ramanathan, A., Ghanta, M., and Subramaniam, B. (2014). Towards highly selective ethylene epoxidation catalysts using hydrogen peroxide and tungsten-or niobium-incorporated mesoporous silicate (KIT-6). *Catal. Sci. Technol.* 4: 4433–4439.

70 Lee, H.-J., Ghanta, M., Busch, D.H., and Subramaniam, B. (2010). Toward a CO_2-free ethylene oxide process: homogeneous ethylene oxide in gas-expanded liquids. *Chem. Eng. Sci.* 65: 128–134.

71 Alvear, M., Fortunato, M.E., Russo, V. et al. (2021). Continuous liquid-phase epoxidation of ethylene with hydrogen peroxide on a titanium-silicate catalyst. *Ind. Eng. Chem. Res.* 60: 9429–9436.

72 Lu, X., Zhou, W.-J., Wu, H. et al. (2016). Selective synthesis of ethylene oxide through liquid-phase epoxidation of ethylene with titanosilicate/H_2O_2 catalytic systems. *Appl. Catal. A Gen.* 515: 51–59.

73 Xu, W., Wang, X., Hou, W. et al. (2022). Synergetic effects of Sn and Ti incorporated in MWW zeolites on promoting the oxidative hydration of ethylene with H_2O_2 to ethylene glycol. *J. Catal.* 413: 554–564.

74 Yu, Y., Fang, N., Chen, Z. et al. (2022). Greening oxidation catalysis: water as a solvent for efficient alkene epoxidation over a titanosilicate/H_2O_2 system. *ACS Sustain. Chem. Eng.* 10: 11641–11654.

12

Propylene Epoxidation with Cumene Hydroperoxide/Titanosilicates

Le Xu and Hailang Liu

Nanjing Tech University, State Key Laboratory of Materials-Oriented Chemical Engineering, College of Chemical Engineering, South Puzhu Rd. 30, Nanjing 211816, China

12.1 Introduction

Propylene oxide (PO), one of the major commodities stemmed from propylene, is a key basic chemical intermediate. The annual worldwide production capacity of PO has reached over 10 million tons in 2019 and is expected to grow rapidly in next few years, particularly in Asia [1]. Its main application is for the manufacture of polyether polyols (approximately 70% of PO world production), which are primarily consumed in the production of polyurethane plastics and rigid/flexible foams. The second main usage of PO (approximately 17%) is for the synthesis of propylene glycols (in the forms of mono-, di-, tri-, and higher), which are used as raw materials for the production of unsaturated polyester resins for textile, construction, cosmetics, detergents, and automotive brake fluids. Additionally, a rapid growing market for PO is for producing propylene glycol ethers, which are increasingly replacing ethylene glycol ethers as solvent for paintings, coatings, inks, etc. (Figure 12.1) [2].

The continuous increasing needs stimulate the development of eco-efficient industrial production routes of PO, which is currently produced from propylene via various processes as summarized in Table 12.1. Generally, three generations of PO production technology are currently employed in industry, including the traditional chlorohydrin route, co-production routes with co-producing PO and alcohol or styrene monomer (PO/*t*-butanol and PO/styrene monomers process), as well as environmental-friendly routes without co-product including cumene hydroperoxide to propylene oxide (CMHPPO) and hydrogen peroxide-based propylene oxide (HPPO) processes, also called PO-only processes [3]. Since 1970s, chlorohydrin route has been the main technology for producing PO, while the market share of PO/*t*-butanol (PO/TBA) and PO/styrene monomers (PO/SM) routes started to grow in 1980s. The third generation of PO production technology, CMHPPO and HPPO processes, was implemented after 2000. Nowadays, approximately 30–40% of the world PO production is still implemented by using chlorohydrin technology, while the rest is via other oxidation routes (Figure 12.2). The past decade has witnessed

Micro-Mesoporous Metallosilicates: Synthesis, Characterization, and Catalytic Applications, First Edition. Edited by Peng Wu and Hao Xu.

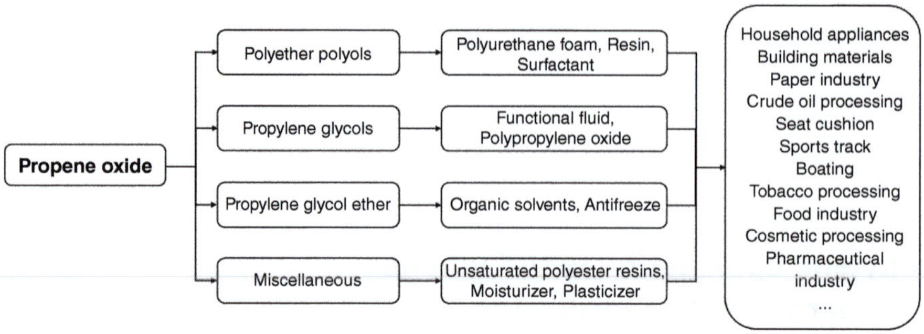

Figure 12.1 Schematic representation of the downstream usage of propylene oxide.

Table 12.1 Summary of PO industrial production technology and producers.

Chlorohydrin route	Co-production route		PO-only route	
	PO/SM	PO/TBA	CMHPPO	HPPO
The Dow Chemical Company	Lyondell-Bayer Chemical Company	Lyondell Chemical Company	Sumitomo Chemical	BASF SE – The Dow Chemical Co.
Asahi Glass Co. Ltd	Shell Chemicals Ltd.	Huntsman International LLC.	Petro Rablgh	Evonik-Degussa
Tokuyama Corporation	Nihon Oxirane Co., Ltd.		S-oil Corp.	Krupp-Uhde
	Sinopec		PTT Global Chemical Public Co., Ltd	Seoul-based Korean Company
	BASF		Sinopec	Sinopec
	Repsol			

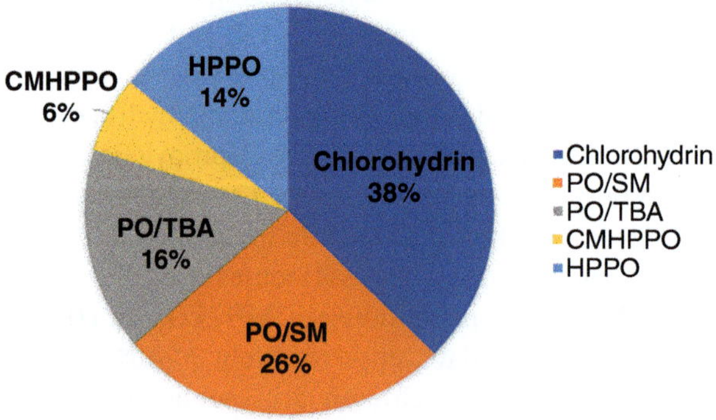

Figure 12.2 Pie chart of the world market shares for different PO production technologies. Source: Adapted from Ref. [1b].

the rapid growing of the latter approaches in market share. The major producers include Dow, BASF, Lyondell, Shell, Repsol, Sumitomo, Huntsman, Asahi, and Bayer, and they adopt different approaches as listed in Table 12.1.

12.2 Traditional Route for PO Production (Chlorohydrin Process)

As a traditional route for conversion of propylene to PO, chlorohydrin route involves the reaction of propylene with aqueous chlorine solution following stoichiometry: $H_3CCH{=}CH_2 + Cl_2 + Ca(OH)_2 \rightarrow CH_3CHCH_2O + H_2O + CaCl_2$. In detail, in the first step (Figure 12.3a), propylene is reacted with chlorine, resulting in a mixture containing two propylene chlorohydrin isomers (1-chloro-2-propanol and 2-chloro-1-propanol), while 1,2-dichloropropane is produced as by-product (up to 10% yield based on propylene). The subsequent dehydrochlorination of propylene chlorohydrin isomers uses an inorganic base solution (usually calcium hydroxide), resulting in the desirable product PO. However, the employed reaction protocol leads to one critical drawback of this route, which needs 1.4 tons of chlorine and 1.0 ton of calcium hydroxide, ultimately generating a large amount (approximately 40 tons) of waste solution containing approximately 1.9 tons of co-product $CaCl_2$ per ton of the desirable product PO [4]. Additionally, another drawback is that limited applications can be found for the by-product 1,2-dichloropropane. The disposal of two hazardous substances leads to severe environment problem, which is the main limitation of chlorohydrin process. Nowadays, although chlorohydrin route still contributes to approximately 30–40% of PO production capacity, no new chlorohydrin plant is expected to be built in the future due to increasing environment restrictions. Besides, current operated plants are needed to optimize to satisfy the environment protection requirement. This can explain the gradual decrease of this technology in the market share for PO production in recent decades.

12.3 Co-production Route for PO Production (PO/TBA and PO/SM Processes)

The second generation of PO-production technology involves the oxidation of propylene with an organic hydroperoxide following the stoichiometry: $H_3CCH{=}CH_2 +$ ROOH (oxidant) $\rightarrow CH_3CHCH_2O +$ ROH (co-product). The process is industrially practiced with two organic hydroperoxides (*tert*-butyl hydroperoxide for PO/TBA and ethylbenzene hydroperoxide for PO/SM), while corresponding alcohols are produced as co-products. Therefore, these routes are also called the co-production approaches.

- **The Halcon process (PO/TBA)**

Tert-butyl hydroperoxide (*t*-BuOOH, denoted as TBHP) is used as the oxidant in PO/TBA process. TBHP is derived from the oxidation of isobutane by oxygen or air with the isobutane conversion of 35–50% (Figure 12.3b). However, a relatively

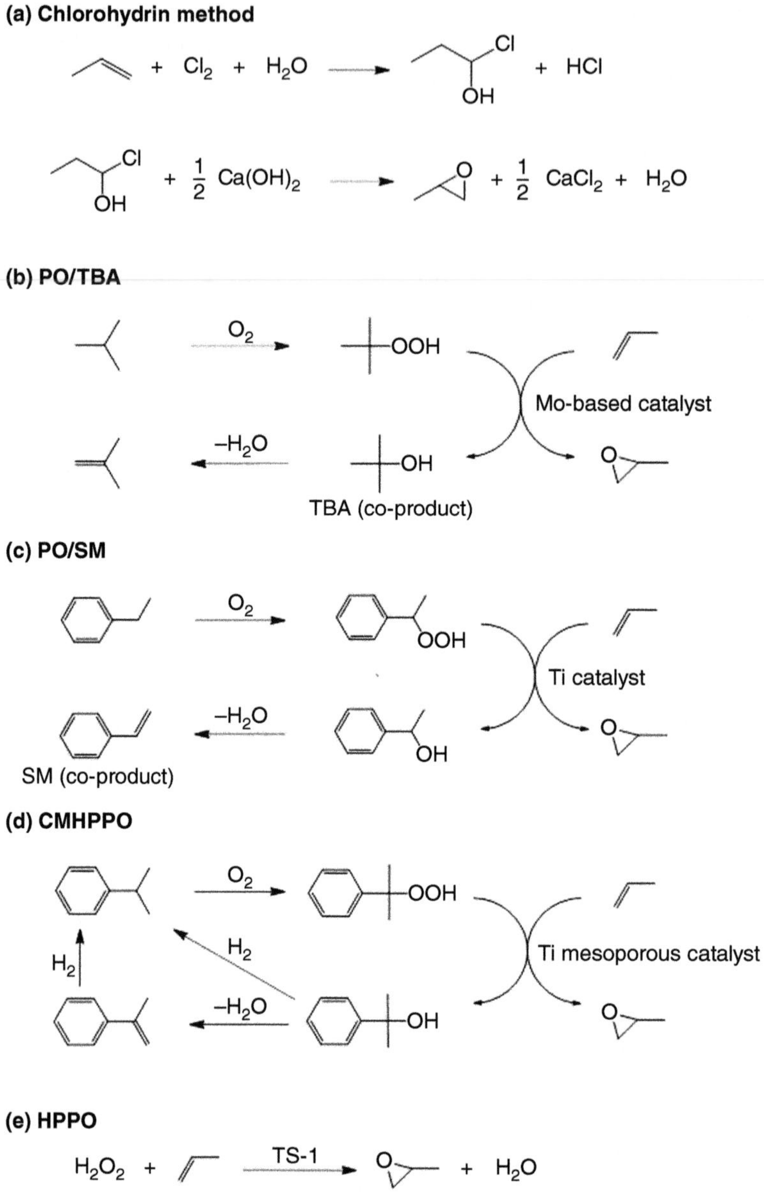

Figure 12.3 Simplified production steps for different PO production technologies. (a) Chlorohydrin, (b) PO/TBA, (c) PO/SM, (d) CMHPPO, and (e) HPPO process.

low TBHP selectivity of 50–60% was achieved due to its decomposition to *t*-butanol (TBA). In the following epoxidation step, the obtained intermediate TBHP is then used as an oxygen carrier and to convert propylene to PO, yielding co-product TBA simultaneously (Figure 12.3b). The catalyst that Halcon/ARCO used for epoxidation is a homogeneous molybdenum complex catalyst, which gives a TBHP conversion of 95% and PO selectivity over 95%. After epoxidation, the co-product alcohol TBA is subjected to dehydration to yield isobutene (Figure 12.3b), which then reacts with methanol to produce methyl *tert*-butyl ether (MTBE), an octane booster for lead-free gasoline. Although the PO/TBA route is efficient for propylene epoxidation, economic limitations that relate to co-product exist, since approximately 2.1 tons of MTBE is produced per ton of targeted product PO. It should be noted that MTBE with toxic property is reported to have a potential risk to some aquatic organisms for their reproduction [5]. The other disadvantage is that the instability of TBHP leads to the reduction of process efficiency and, on the other side, increases the energy consumption and cost for the purification and separation in PO/TBA process [4].

- **Ethylbenzene hydroperoxide process (PO/SM)**

The other type of organic oxidant, ethylbenzene hydroperoxide (EBHP), is derived from the oxidation of ethylbenzene by air (Figure 12.3c) [4]. EBHP can oxidize propylene to produce PO, while 1-phenylethanol is produced as the co-product. Shell used a heterogeneous titanium-containing catalyst for propylene epoxidation with EBHP, resulting in an EBHP conversion greater than 95% and a PO selectivity over 95%. With an acid catalyst, the co-product 1-phenylethanol can be dehydrated to give styrene monomers (SM), which are useful raw chemicals for manufacture of polystyrene and several copolymers. However, the oligomerization of SM may also occur to produce heavy polymers as side-products, which would potentially induce catalyst deactivation. Approximately 2.5 tons of SM is produced per ton of PO. Besides, similar to TBHP, organic EBHP is also not stable, resulting in a relatively low efficient process for EBHP generation. Owing to its instability, the obtained EBHP solution is normally used at low concentrations between 10% and 15%, which needs larger reactor volume to operate.

Although co-production routes (PO/TBA and PO/SM) have several advantages over the traditional chlorohydrin route in terms of environmental concerns, they both suffer from drawbacks relating to co-products (TBA and SM, respectively). The economic concern that derived from the balance of market and prices for PO and co-products is the main limitation for both PO/TBA and PO/SM routes. In addition, as mentioned before, the instability is another common drawback for both organic oxidants (TBHP and EBHP), limiting the efficiency of PO/TBA and PO/SM processes. Furthermore, building new co-production plants requires large capital investments, and thus, the market shares of PO/TBA and PO/SM technology are expected to decline over time in the future.

12.4 PO-Only Production Routes (HPPO and CMHPPO Routes)

Considering the economic issues, new environmental-friendly PO-only process without the burden of co-products is highly desirable. Under this circumstance, CMHPPO route and hydrogen peroxide to propylene oxide (HPPO) route, commercialized by Sumitomo Chemical Co. in 2003 and BASF-Dow in 2009, respectively, bring PO-production technology to the third stage. Different innovative pathways were developed to build new PO-only production routes in CMHPPO and HPPO processes.

In HPPO process, the diluent hydrogen peroxide aqueous solution is used as the oxidant, which can efficiently oxidize propylene to PO over a microporous titanosilicate TS-1 under very mild conditions (Figure 12.3e). In principle, HPPO route only produces water as a by-product, making this technology attractive and promising in the respective of green and sustainability. However, the high manufacture cost, high shipping standard, and low stability of hydrogen peroxide cause the economic and safety concerns for HPPO process. Therefore, in practice, an integrated plant involving the manufacturing of hydrogen peroxide and propylene epoxidation at one site is a potential alternative for building a safe and economic feasible HPPO process. In addition, the TS-1/H_2O_2 catalytic system prefers protic solvent of methanol, leading to by-products of methyl propyl ethers in HPPO process. Nowadays, several new plants using HPPO technology have been built and operated. The first was the plant implemented by BASF and Dow in Antwerp, Belgium, accompanied with a large plant for hydrogen peroxide production using the anthraquinone technology developed by Solvay near BASF-Dow HPPO plant. Additionally, new HPPO process was also developed by the cooperation between Degussa-Evonik and Krupp-Uhde, which was started at Ulsan by Seoul-based Korean Company (SKC) in 2009. Compared with the traditional chlorohydrin and co-production routes, the benefits of HPPO process can be summarized from two aspects. First, from the perspective of economic, HPPO technology requires significantly less capital to build, eliminates the need for additional infrastructure or markets for co-products, and requires simple raw material integration. Regarding environmental benefits, HPPO technology reduces waste water by 70–80% and energy consumption by 35%, compared with traditional technologies.

Different from HPPO process, the other new PO-only technology, CMHPPO route for PO production first developed by Sumitomo Chemical Co., uses a recyclable organic oxidant cumene hydroperoxide (CHP), which can be regenerated after propylene epoxidation, thus avoiding the burden of expensive oxidant. Typically, the closed cycle of CHP oxidant in the whole CMHPPO process involves three steps (Figure 12.3d). (1) The oxidation of cumene to generate CHP, (2) propylene epoxidation with CHP to produce PO and co-product cumene alcohol (CA), and (3) after separation and purification to obtain the target product PO, CA is subjected to hydrogenation, yielding the starting source cumene. Although the first two steps of CMHPPO process are similar to those in both of PO/TBA and PO/SM processes, the most outstanding feature of CMHPPO process is that the co-product CA obtained after epoxidation can be recycled via hydrogenation. Therefore, the oxidant regeneration process makes CMHPPO route unique compared with PO/TBA and PO/SM routes since both suffer from possible economic issues of co-products. Additionally, the adopted oxidant CHP is much more stable than TBHP and EBHP, and thus, the production efficiency of organic oxidant is greatly enhanced. The details of CMHPPO process for PO production will be discussed in the following part of this chapter.

12.5 Catalyst Design for PO-Only Routes

Considering that HPPO and CMHPPO processes are implemented without co-products in principle, both can be considered as the PO-only routes. Compared with the traditional chlorohydrin route, they are green and sustainable in terms of environment and economic considerations. It is noticeable that the industrial advancement is always fueled by the design of advanced technology based on novel functional materials, which can efficiently catalyze chemical reactions proceeding in a new energy and economic feasible pathway. The development of PO-only process is a good demonstration of catalyst-driven industrial revolution. New catalysts for propylene epoxidation under desirable conditions play the key role in the development of new PO production processes, especially for HPPO and CMHPPO.

In HPPO route, the catalyst is the microporous titanosilicate TS-1 with ZSM-Five (MFI) topology, which is a well-known heterogeneous zeolite catalyst for selective oxidations of a variety of alkenes [6]. The TS-1 catalyst was discovered and patented by Taramasso and co-workers in EniChem in 1980s [7]. The hydrophobic microporous 10-ring channel systems of TS-1 zeolite allow both reactants and oxidant, propylene and hydrogen peroxide, to access the Ti active sites in zeolite framework, making epoxidation proceeding highly efficiently under mild conditions (Figure 12.4). In addition, the combination of Ti cations that locating at the tetrahedral sites in zeolite framework makes the catalyst a potential to be active for shape-selective oxidation reactions, similar to the predominant role of its analogous aluminosilicate Zeolite Socony Mobil-5 (ZSM-5) in acid-catalyzed reactions [8]. Based on TS-1 zeolite, EniChem has successfully industrialized a new phenol hydroxylation process in 1986 using hydrogen peroxide as oxidant, replacing the traditional Brichima process for the manufacture of hydroquinone and catechol. Additionally, EniChem also applied TS-1 zeolite in cyclohexanone ammoximation for the manufacture of caprolactam using ammonia and hydrogen peroxide [6a, 9]. Dow acquired the related technology based on TS-1 zeolite via purchasing EniChem's polyurethane business in 2001. Afterwards, Dow and BASF have started to cooperate to develop the industrial feasible process for PO production since 2003. Combining with Solvay's hydrogen peroxide production technology using anthraquinone approach, HPPO process was successfully realized from a dream reaction to the industrialized process in 2009. The application of TS-1 zeolite in HPPO process for PO production has been regarded as the milestone in the history of zeolite catalysis. Dow and BASF received the 2010 US Presidential Green Chemistry Challenge Award for their jointed developed HPPO route that remarkably upgrades the production of the key chemical intermediate PO [10].

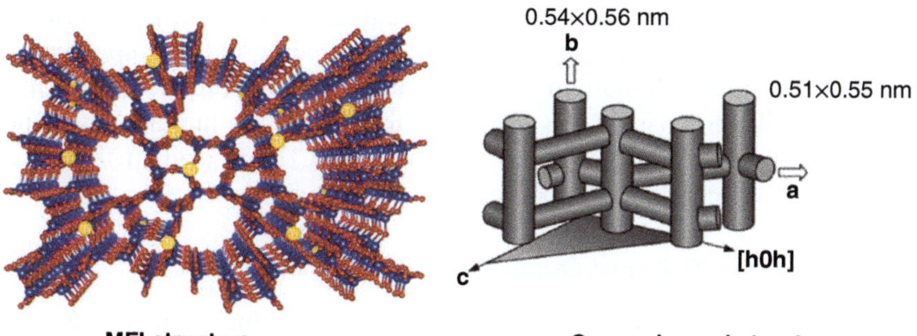

Figure 12.4 Description of TS-1 zeolite structure and intersected 10-ring channels.

On the other hand, although isolated tetrahedrally coordinated Ti sites in MFI titanosilicate exhibit spectacular catalytic properties for selective oxidations, the application of the microporous titanosilicate is mainly limited to small molecules due to their small 10-ring pore openings (5–6 Å in diameter). Therefore, TS-1 zeolite catalyst is not an ideal candidate for the oxidation technologies that based on bulky organic oxidants (such as TBHP, EBHP, and CHP), which normally have much larger kinetic diameters in comparison to 10-ring pores of MFI topology, inducing severe diffusion problems inside zeolite channels. The catalytic testing of alkene epoxidation using organic TBHP as oxidant over TS-1 zeolite has confirmed that it is not an efficient catalyst [11]. Therefore, a more open catalytic environment is required for bulky oxidant to access the Ti^{4+} active sites within catalyst channels.

Thus, the design synthesis of crystalline titanosilicates with hierarchical pore architecture or mesoporous catalysts possessing large-pore channel systems in mesoscale are two attractive pathways to solve the severe diffusion limitation problem in selective oxidation with bulky oxidant. In this aspect, the discovery of mesoporous silica-based materials in early 1990s opened a new avenue for many catalytic oxidation reactions involving bulky substrates [12]. The mesoporous channels of this new class of materials, often having a channel opening larger than 2 nm, provide a molecular highway for the diffusion of reactants and oxidants. Afterward, various Ti-doped mesoporous silicates [13], such as Ti-MCM-41, Ti-HMS, Ti-SBA-15, and Ti-MCM-48, have been synthesized and showed promising activity in catalyzing alkenes epoxidation using bulky organic oxidants, such as TBHP, EBHP, and CHP.

In addition to the mass transfer efficiency relating to the size of pore opening, many other factors can also have a great impact on the catalytic behaviors of catalysts for PO production with organic CHP as oxidant. For instance, it is well-known that the coordination environment of Ti sites is also vital to the catalyst design for alkene selective oxidations. The direct hydrothermal synthesis or post-synthesis of Ti-containing catalyst often leads to a complicated Ti site coordination. However, in many cases, the octahedral Ti sites and anatase TiO_2 were often found. The octahedral and extraframework anatase TiO_2 showed extremely low catalytic activity and sometimes could induce the unproductive decomposition of oxidant. Thus, the catalysts containing various Ti coordination states normally showed different catalytic behaviors in alkenes epoxidation, highlighting the importance of Ti sites coordination on the catalytic process. In addition, the mesoporous Ti-containing catalysts suffered the drawbacks related to its inaccessibility because a part of Ti species were incorporated inside the silica walls rather than on pore surface. Besides, the surface properties of catalyst, such as acidity and hydrophobicity, were found to be another important factor that influences the product selectivity. Therefore, the following part of this chapter will focus on the rational design of Ti-containing catalyst for propylene epoxidation with CHP as oxidant from the perspective of above-mentioned properties.

12.5.1 Mesoporous Ti-Doped Catalysts for CMHPPO Process

Since the discovery of mesoporous materials, this new kind of materials has attracted extensive research interests, particularly in the catalysis field. Its large mesoscale channel system is particularly suitable for catalytic reaction that limited by slow mass transfer efficiency. Nevertheless, although it possesses a large pore opening, the mesoporous Ti-doped catalysts often exhibit low activities for alkene epoxidation reactions with hydrogen peroxide as oxidant [14]. The hydrophilic nature of Ti-isomorphous substituted mesoporous silica catalysts is postulated to be one of the major reasons for their low activities with aqueous hydrogen peroxide. The mesoporous materials often contain a relatively high concentration of uncondensed hydroxyl groups (Si-OH) on catalyst surface, which possibly reduces the affinity of organic oxidant and alkene to the active sites on catalyst surface, and prevents the fast desorption of product and ultimately leads to deactivation. Therefore, critical issues that relate to the hydrophilic nature of catalyst surface were needed to be addressed for applying Ti-doped mesoporous materials in the propylene epoxidation with bulky CHP.

Sumitomo Chemical Co. first reported a Ti-grafted mesoporous silica catalyst for propylene epoxidation with CHP in 1998. Subsequently, based on the catalyst they developed, Sumitomo designed a new three-step industrial process that not only involves the propylene epoxidation but also includes the generation and recycling process of CHP oxidant [15]. The successful application of mesoporous Ti-doped silicate in CMHPPO process is another milestone in the development of PO production technology. In the following part, the key aspects of catalyst design for propylene epoxidation by using CHP as oxidant will be discussed, including research works in the literature and information collected from reports and patents from Sumitomo Chemical Company. The exceptional activity of Sumitomo Ti catalyst has been attributed to the following aspects.

- **Mesoporous channel system**

The Sumitomo Ti catalyst possesses the mesoporous channel system (pore size in the range of 1–20 nm according to Sumitomo report) and exhibits an external surface area much higher than that of the conventional TS-1 [1b, c], allowing bulky oxidant CHP to easily diffuse to the accessible Ti sites exposed on the surface of mesopore channels.

- **Tetrahedral Ti sites**

The Ti active sites in Sumitomo catalyst were post-grafted onto the silica-based support, resulting in a highly dispersed Ti sites that mainly located on the catalyst surface and existed only in the tetrahedral coordination state, which was confirmed by XAFS spectroscopy. The sole presence of tetrahedral Ti sites maximizes the amount of ideal active sites for propylene epoxidation.

- **Surface hydrophobicity and acidity**

The Sumitomo Ti catalyst surface was modified by post-silylation technique, leading to a hydrophobic surface that was demonstrated by the low water adsorption result. The amount of adsorbed water on Sumitomo Ti catalyst was found to be around 70% of water adsorbed on the conventional TS-1 [1b, c]. This unique property was found to be critical for propylene epoxidation, as the catalytic environment favors the diffusion of organic reactant and the desorption of product PO, thus avoiding the influence of side-reactions on product selectivity. The hydrophobic nature of Sumitomo Ti catalyst is in significant contrast to conventional mesoporous silica materials without surface modifications, which often had hydrophilic surface and showed low PO selectivity.

Except for Sumitomo's Ti catalyst, many researchers have been devoted to investigate the catalytic behavior of various mesoporous Ti-doped catalysts for propylene epoxidation with CHP (Table 12.2). Li and coworkers studied the preparation of

Table 12.2 Summary of Ti-containing catalysts for propylene epoxidation using CHP.

No.	Type of catalysts	Reaction conditions	PO selectivity (%)	CHP conversion (%)	PO yield (%)	Ref.
1	NTO	Continuous-flow fixed-bed micro-reactor; 5.0 ml/h CHP; 3.0 ml/h C_3H_6; $T = 373$ K; C_3H_6/CHP molar ratio of 5; pressure 4.0 MPa; total WHSV 6.2 h^{-1}	96.1	70.6	40.8	[16]
2	UNT-400		95.1	64.7	49.0	
3	UNT-600		95.2	59.7	42.9	
4	NT-400		95.6	56.1	40.1	
5	NT-600		94.2	52.7	34.6	
6	Ti-SBA-15EC	Autoclave, pressure 2.4 MPa; 10 ml CHP, 20 ml *tert*-butyl alcohol; reaction time = 4 h; catalyst, 0.3 g; $T = 353$ K	96.4	87.3	84.2	[18]
7	Ti-SBA-15C		92.3	84.9	78.3	
8	Ti-SBA-15		49.5	27.2	13.4	
9	Ti/SBA-15		50.6	55.1	27.9	
10	Ti-SBA-15-st	Autoclave, pressure 2.4 MPa; 10 ml CHP, 20 ml *tert*-butyl alcohol; reaction time = 4 h; catalyst, 0.3 g; $T = 353$ K	84.3	80.1	67.4	[19]
11	Ti-SBA-15-nd		84.9	81.9	69.5	
12	Ti-SBA-15-rd		92.3	84.9	78.4	
13	Ti-SBA-15-th		85.7	74.6	63.9	
14	Ti-SBA-15-rd-(4)		95.2	86.9	82.7	
15	Ti-SBA-15-rd-(6)		96.2	90.3	86.9	
16	Ti-SBA-15-rd-(12)		94.9	87.7	83.2	

Table 12.2 (Continued)

No.	Type of catalysts	Reaction conditions	PO selectivity (%)	CHP conversion (%)	PO yield (%)	Ref.
17	Ti-HMS	Autoclave; pressure, 2.2 MPa; reaction time, 2 h; catalyst, 0.3 g; $T = 353$ K	71.4	62.9	—	[20]
18	SZ-Ti-HMS		96.2	98.2	—	
19	CS-Ti-HMS		85.6	89.4	—	
20	DCS-Ti-HMS		79.9	94.1	—	
21	CTS-1	Propylene, 0.5 MPa; CHP, 1 mmol; cumene, 10 ml Reaction time = 2 h; catalyst, 0.05 g; $T = 373$ K	99.3	0.8	0.80	[27]
22	Ti-Beta		99.1	2.0	1.98	
23	Ti-MWW		99.5	0.8	1.79	
24	LTS-1		99.7	20.0	19.94	
25	Ti-MCM-41		99.2	7.3	7.24	
26	Ti-ECNU-7	Propylene, 0.5 MPa; CHP, 1 mmol; cumene, 10 ml Reaction time = 2 h; catalyst, 0.05 g; $T = 373$ K	99.2	38.1	37.8	[28]
27	Ti-MWW		99.1	0.7	0.7	
28	TS-1		99.6	0.6	0.6	
29	LTS-1		99.7	20.0	19.9	
30	Ti-MCM-41		99.2	6.8	6.8	

Ti/SiO_2 by chemical vapor deposition (CVD) of $TiCl_4$ with nitrogen as carrier gas at 900 °C on a commercial silica gel powder, which has a surface area of 256 m^2/g and a pore volume of 1.65 cm^3/g [17]. It is well-known that $TiCl_4$ precursor reacts with the Si-OH groups on the support surface during CVD process. They found that the doped Ti contents in the resultant Ti/SiO_2 have a close relation with the deposition period, leaving different levels of residual Si-OH group after deposition. For the catalyst with a low Ti content (more Si-OH group remaining), CHP was found to be a more suitable oxidant than TBHP for achieving a high PO yield. This is attributed to the fact that CHP is a more stable oxidant and thus has a significantly lower decomposition rate than TBHP. The residual Si-OH group after Ti doping was estimated to have a promoting effect in catalyzing the decomposition of organic oxidants. A PO yield of 93% was achieved on the catalyst with the highest Ti content (around 1.12 μmol/m^2), when CHP was adopted as oxidant, while acetophenone, α,α-dimethylbenzyl alcohol, α-methylstyrene, and α-methylstyrene dimer were determined as side-products by gas chromatography mass spectrometry (GCMS). This study highlights that CVD is a useful deposition method to dope Ti species and that the hydrophilic nature due to the remaining Si—OH is critical to catalyst behaviors. However, this study lacks of details regarding the mesopore size and the coordination status of grafted Ti species.

Lu and coworkers prepared the Ti-SBA-15 catalyst by CVD with $TiCl_4$ using nitrogen as carrier gas [18]. They found that the doped Ti content can be controlled by using different approaches of removing surfactant from as-synthesized Ti-SBA-15. The SBA-15E that prepared by extracting surfactant with ethanol solution has a relative higher concentration of free Si—OH groups than that in the Ti-SBA-15C obtained after direct calcination. Therefore, with more Si—OH groups on support surface, a higher Ti content (4.62 wt.%) was found after Ti sites deposition at 900 °C for 1.5 hours on Ti-SBA-15EC, while 4.14 wt.% of Ti was loaded on direct calcined Ti-SBA-15C at the same condition. The obtained Ti-SBA-15EC catalyst has a BET surface area of 346 m^2/g, pore volume of 0.39 cm^3/g, and mesopore size of 6.4 nm. It was demonstrated by UV–visible spectra that the doped Ti species were distributed in multiple states, including tetrahedral, octahedral, and the extraframework anatase TiO_2, since the amounts of surface Si—OH groups could not allow all Ti sites to highly dispersed on catalyst surface. With Ti-SBA-15EC as catalyst, 87.3% CHP conversion and 96.4% PO selectivity were observed, when the propylene epoxidation was carried out at 80 °C for four hours (Table 12.2). The Ti-SBA-15C showed a slightly lower CHP conversion (84.9%) and PO selectivity (92.3%). In addition, the hydrothermally synthesized Ti-SBA-15 with co-hydrolysis of silicon and titanium source and the impregnated Ti/SBA-15 exhibited significantly lower catalytic activity, giving CHP conversion of 27.2% and PO selectivity of 49.5% for Ti-SBA-15 and CHP conversion of 55.1% and PO selectivity of 50.6% for Ti/SBA-15, respectively (Table 12.2). This difference between catalysts prepared by various approaches (CVD, direct synthesis, impregnation) highlighted the advantage of CVD as a more suitable route to achieve an efficient catalyst with high Ti loading. However, CVD condition should still be carefully controlled to improve the coordination state of doped Ti species. Additionally, Ti-SBA-15EC showed a good recyclability as the activity barely decreased after used repeatedly six times.

A number of recent reports have described the advantages of silylation/organic group functionalization of the surface of Ti-MCM-41 materials, thus making more hydrophobic surface to improve the activity. Ti-HMS catalyst was also reported to be more active after surface hydrophobization by silylation with organosilane bearing methyl groups. Furthermore, Li and coworkers found the selection of silylated agents had a great impact on the catalytic activity of modified catalysts [20]. They compared the catalytic activities of the catalysts after modification with different organosilanes, such as hexamethyldisilazane (HMDSZ), trimethylchlorosilane (TMCS), and dimethyldichlorosilane (DMDCS). The catalytic results indicated that the Cl-functionalized silylating agents (TMCS and DMDCS) had a negative impact on the catalytic activity because it potentially induced the deconstruction of Ti coordination environment during modification, thus leading to the decreased activity. The highest CHP conversion of 98.2% and PO selectivity of 96.2% was observed over the silylated Ti-HMS with HMDSZ, while the unsilylated catalyst with hydrophilic surface exhibited the lowest CHP conversion of 62.9% and PO selectivity of 71.4% (Table 12.2). Meanwhile, owing to the partially deconstructed Ti environment, the catalysts that modified with TMCS and DMDCS showed a CHP conversion of 89.4% and PO selectivity of 85.6% and a CHP conversion of 84.1% and PO selectivity of 79.9%, respectively (Table 12.2).

The incorporation of nitrogen atoms into zeolite framework to replace the framework oxygen atoms can reduce the catalyst acidity caused by framework oxygen atoms. Ni and coworkers reported that the modification of Ti-MCM-41 catalyst by ammonia effectively reduced the acid strength of catalyst and thus suppressed the non-productive decomposition of the oxidant CHP [21]. They first synthesized Ti-MCM-41 via sol-gel process, during which titanium was introduced by the co-hydrolysis of tetrabutyl titanate (TBOT) with tetraethyl orthosilicate (TEOS). Then, the as-synthesized Ti-MCM-41 material containing cetyltrimethylammonium bromide (CTAB) surfactant inside mesopore channels was subjected to the ammonia treatment at 400 °C for four hours. After treatment, the resultant N-doped Ti-MCM-41 showed a decreased surface area of 692.3 m^2/g, while the mesopore size shrank to 3.21 nm. Element analysis indicated that nitrogen content in the N-doped Ti-MCM-41 was around 0.8 wt.%. The catalysis test of propylene epoxidation using CHP indicated that the CHP utilization efficiency increased from 66.1% on the unmodified Ti-MCM-41 to 80.4% on the N-doped Ti-MCM-41 (Table 12.2). This result further implied that the free Si—OH groups not only endow the catalyst surface hydrophilic property but also provide the acidity, both of which have negative influence on the catalytic behaviors.

12.5.2 Hierarchical Titanosilicates for CMHPPO Process

Although the mesoporous silica-based materials possessing large mesopore channels are promising for propylene epoxidation with bulky organic CHP, its amorphous nature leads to a high concentration of uncondensed free hydroxyl groups on catalyst surface, which brings new issues relating to hydrophilic and acidity. Furthermore, the doped Ti species often existed in various coordination states, which are hardly avoided, especially during CVD synthesis. To solve these drawbacks, the introduction of mesoporosity into the conventional microporous zeolite materials has becoming an attractive alternative pathway to prepare catalysts containing mesopores. Zeolite catalysts containing both microporosity and mesoporosity have been widely used for industrial process. For instance, the well-known ultra-stable Y (USY) zeolite that used in fluid catalytic cracking (FCC) reaction for oil refining process is prepared by the steam treatment of aluminum-rich Y zeolite at elevated temperature to selectively remove Al atoms from Y zeolite framework, ultimately resulting in mesoporosity in USY zeolite [22]. Inspired by the pioneering work based on crystalline zeolite possessing mesoporosity, various approaches have been developed to introduce mesoporosity into zeolite catalyst, such as soft/hard templating, acid/base etching, delamination, and many others [23]. The microporous channel system in this kind of material provides a rigid confinement for shape-selective catalysis, while the mesoscale channels allow the fast diffusion of reactants and products during catalysis. Furthermore, the combination of this hierarchical porosity and redox active sites in zeolite framework makes the selective oxidation of propylene possible by using bulky oxidant, such as CHP.

In 2009, Ryoo and coworkers reported a novel protocol of designing the Gemini-type organic structure-directing agent (OSDA) for the direct synthesis of hierarchical MFI zeolite, which contains only 2.5 nm thick MFI nanosheets that pillared by the long-tail of OSDA [24]. The hydrophilic part of OSDA plays the structure-directing role for MFI nanosheets formation during synthesis, while the hydrophobic long-chain alkyl part of OSDA restricts the MFI crystal growth along the *b*-axis, leading to the alternative stacking of inorganic MFI nanosheets and organic moieties in unit-cell scale. The interlayer spacing between adjacent MFI nanosheets was found to be around 4 nm. Owing to such large interlayer spacing, the removal of this bifunctional OSDA by calcination leads to a hierarchical zeolite containing intracrystalline mesopores due to the random stacking of MFI nanosheets. Based on this methodology, Ryoo group found that the obtained MFI nanosheets had outstanding activity in various acid-catalyzed reactions [25], such as alkylation, cracking, rearrangement, and isomerization. Detailed characterization and catalytic study showed that the hierarchical pore architecture exposed a large amount of catalytic active sites at or near external surface that is much more accessible, compared with conventional ZSM-5 zeolite. Furthermore, they successfully incorporated heteroatoms (including Ti, Sn) into MFI nanosheet framework and found that the resultant hierarchical MFI zeolites exhibited an extremely high activity in selective oxidation reactions, especially those involving bulky reactants [26].

Inspired by the synthesis of ZSM-5 nanosheets with hierarchical architecture, Wu and coworkers reported the hydrothermal synthesis of the multilayer-structured MFI titanosilicate (LTS-1) by using a similar Gemini-type OSDA [27]. The as-synthesized LTS-1 zeolite material is constructed by alternative stacking of interlayer OSDA surfactant and MFI nanosheet with single unit-cell thickness. Further calcination removes OSDA molecules, yielding a hierarchical LTS-1 zeolite with intracrystalline mesopores of 3–4 nm (Figure 12.5). The resultant LTS-1 zeolite has an external surface area of 357 m^2/g, which is much higher than conventional TS-1 zeolite (73 m^2/g). Besides, the Ti sites in LTS-1 are in tetrahedral coordination in zeolite framework, which exhibited a typical absorption band around 210 nm in UV–visible spectrum. Additionally, solid-state ^{29}Si nuclear magnetic resonance spectroscopy (NMR) and Fourier transform infrared spectroscopy (FTIR) spectra of LTS-1 indicates that it has a relatively hydrophilic surface, which was probably due to the presence of terminal silanol groups (Si—OH) on the surface. Although possessing hydrophilic property, the LTS-1 zeolite exhibited a promising catalytic behavior in propylene epoxidation using CHP as oxidant. Under the same reaction conditions, the conventional TS-1 only showed propylene conversion of 0.8% and PO selectivity of 0.8% (Table 12.2), probably due to its small pore opening. Other

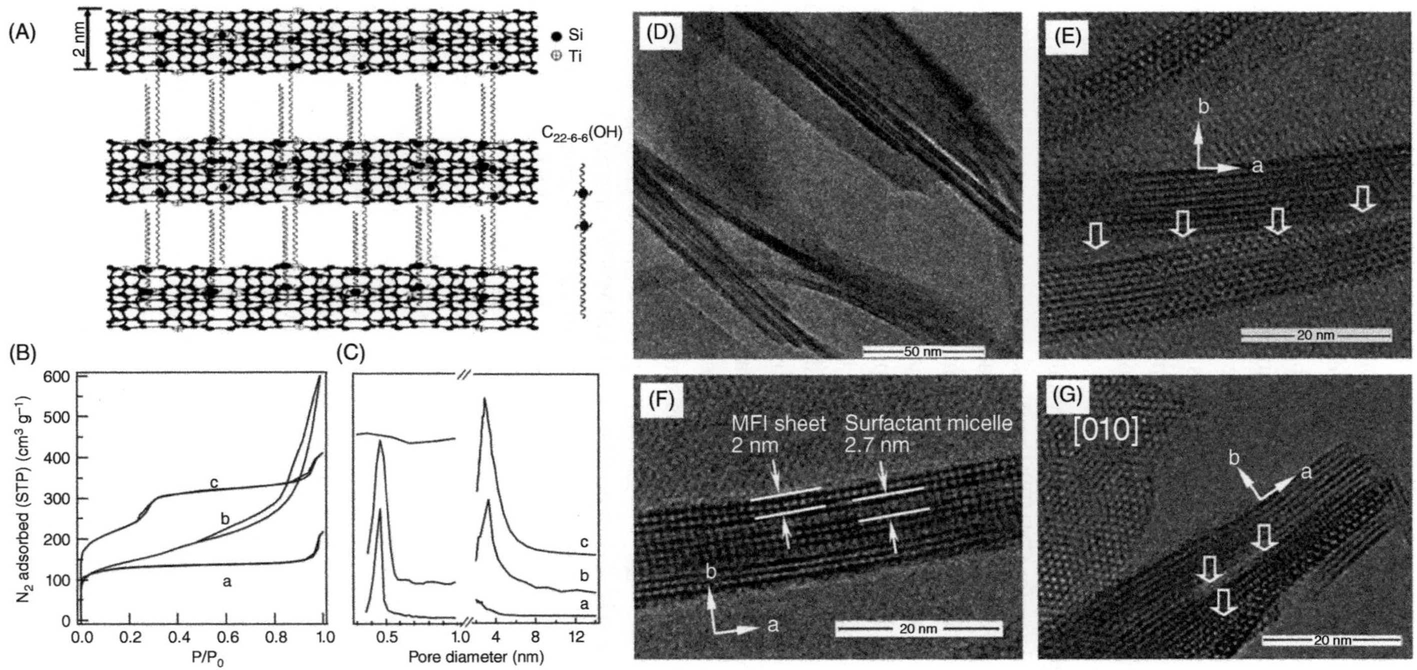

Figure 12.5 Characterization of hierarchical LTS-1 zeolite catalyst. (A) Multilayered MFI titanosilicate directed by bifunctional SDA. N_2 adsorption isotherms at 77 K (B) and pore size distribution (C) of CTS-1 (a), LTS-1 (b), and Ti-MCM41 (c). The isotherm is offset vertically by 100 $cm^3\ g^{-1}$ for Ti-MCM-41. TEM images of LTS-1 as-synthesized at $C_{22-6-6}(OH)_2/Si = 0.04$ (D and F) and after calcination (E and G). The arrows indicate the presence of mesopores. Source: Reproduced from Ref. [27] with permission from Elsevier.

typical microporous titanosilicates, Ti-Beta and Ti-MWW, were also not efficient, both showing very low propylene conversions. In contrast, the hierarchical LTS-1 zeolite showed significantly higher conversion of propylene (20.0%) and PO yield (19.94%). Turnover frequency (TOF) achieved on LTS-1 zeolite was four times higher that of TS-1. This result was reasonably explained by that the bulky CHP could only access the Ti sites on the external surface of microporous TS-1. Owing to the hierarchical LTS-1 zeolite structure, more Ti sites were exposed to the zeolitic layer surface, allowing bulky CHP to access. Compared with LTS-1 with solely tetrahedral Ti sites and lower concentration of Si—OH groups, a much lower catalytic activity was achieved on the mesoporous Ti-MCM-41 catalyst, which gave a medium propylene conversion of 7.3% and PO yield of 7.24% (Table 12.2). This study provides an example that a crystalline zeolite catalyst with more open framework is promising for CMHPPO process.

In addition to MFI zeolite nanosheets, a similar multilayered titanosilicate containing MWW-type nanosheets was prepared by direct hydrothermal synthesis [28]. In contrast to the multilamellar MFI nanosheets, the multilamellar MWW layers were successfully pillared by the simple surfactant CTAB micelles during hydrothermal synthesis. The synthesis used the small imidazolium-based OSDA to direct the crystallization of MWW nanosheets with the assistance of zeolite seed, while the simple single-head surfactant CTAB was adopted to pillar the neighboring MWW layers. This interlayer distance between adjacent MWW zeolitic layers in the as-synthesized Ti-ECNU-7 material was found to be 2.4 nm, which was significantly larger than that of the as-synthesized Ti-MWW (often between 2 and 3 Å). Owing to the large interlayer spacing, the removal of CTAB surfactant by subsequent calcination leads to a novel hierarchical MWW-structured titanosilicate due to the disorder stacking of MWW layers (Figure 12.6). The calcined Ti-ECNU-7 zeolite possessed a hierarchical porous structure, showing a high external surface area of 211 m^2/g and a high concentration of accessible Ti sites (surface Si/Ti ratio of 68 determined by XPS) on the external surface of catalyst, making it highly suitable for propylene epoxidation with bulky CHP. In comparison with other catalysts, including microporous TS-1, Ti-MWW, hierarchical LTS-1, and mesoporous Ti-MCM-41, the hierarchical Ti-ECNU-7 was found to be the most active catalyst. The highest propylene conversion of 38.1% was achieved on Ti-ECNU-7 catalyst, while the hierarchical LTS-1 zeolite showed the propylene conversion of 20.0% (Table 12.2). The comparable PO selectivity around 99% was observed for both Al-ECNU-7 and LTS-1 catalysts. Additionally, Ti-ECNU-7 zeolite catalyst also exhibited a desirable resistance against Ti leaching and deactivation. This study on the synthesis and catalytic test based on Ti-ECNU-7, accompanied with previous LTS-1 work, demonstrates that the crystalline microporous titanosilicates with highly opened porosity are good candidates for propylene epoxidation with CMHP as oxidant, as the Ti coordination environment is more easy to control during zeolite synthesis and less free Si—OH compared with mesoporous catalysts with amorphous silica walls.

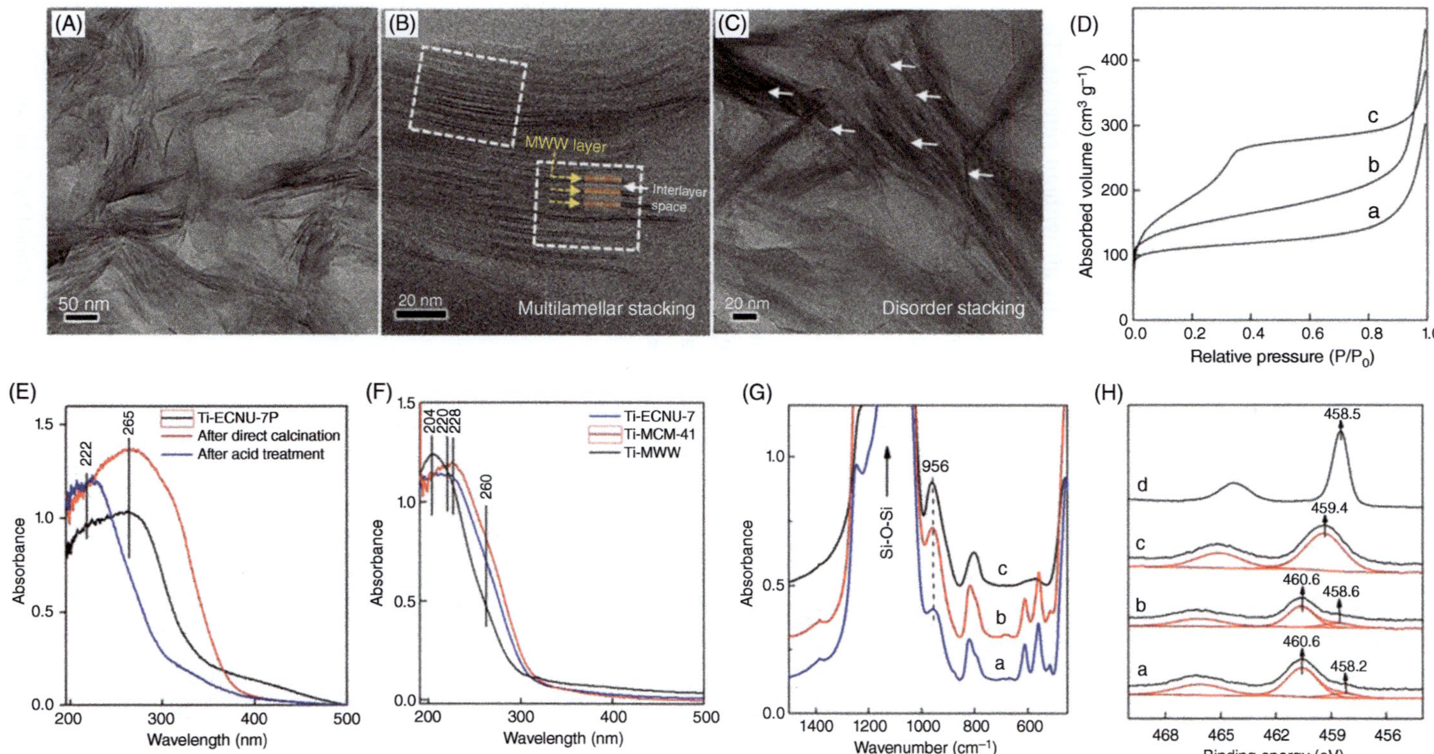

Figure 12.6 Characterization of hierarchical Ti-ECNU-7 zeolite catalyst. TEM images of (A and B) the mesostructured multilamellar Ti-ECNU-7P and (C) the calcined hierarchical Ti-ECNU-7. (D) N_2 adsorption isotherms at 77 K of (a) Ti-MWW, (b) Ti-ECNU-7 and (c) Ti-MCM-41. (E, F) UV-visible spectra of different samples. (G) IR spectra of (a) Ti-ECNU-7, (b) Ti-MWW and (c) Ti-MCM-41. (H) XPS spectra of (a) Ti-ECNU-7, (b) Ti-MWW, (c) Ti-MCM-41 and (d) anatase TiO_2. Source: Reproduced from Ref. [28] with permission from Royal Society of Chemistry.

12.6 Industrial CMHPPO Process

The commercial CMHPPO process was implemented by Sumitomo Chemical. This is also the first industrial PO-only production process. The first plant was operated in Chiba, Japan, with an annual PO production capacity of 0.2 million tons. The second factory was implemented in 2009 in Saudi Arabia with the cooperation with Saudi Arabian Oil Company. As mentioned above, three steps were combined, including the generation of CHP, propylene epoxidation, and regeneration of CHP. The simplified flow-sheet of the CMHPPO process is shown in Figure 12.7 [29]. The detailed characteristics and main advantages will be discussed as following.

- **Generation of CHP**

CHP was produced via the auto-oxidation of cumene by air without catalyst. The oxidation reaction was performed between 90 °C and 130 °C at 0–1 MPa, giving rise to a high CHP yield and over 90% selectivity. In comparison with other organic oxidants (such as TBHP and EBHP), the resultant oxidant CHP is ultra-stable, and thus its production is more selective and it can be used at much higher concentrations (75–85%). The cumene loss is minimal in the whole process, requiring less capital investment and facilitates to recover cumene.

- **Propylene epoxidation**

Sumitomo used a Ti-doped mesoporous material in a designed fix-bed reactor to catalyze the reaction between propylene and CHP. In addition, Sumitomo used cumene as solvent during catalytic process, which largely reduces the possibility of the hydrolysis of PO to propylene glycol. The propylene epoxidation was carried out

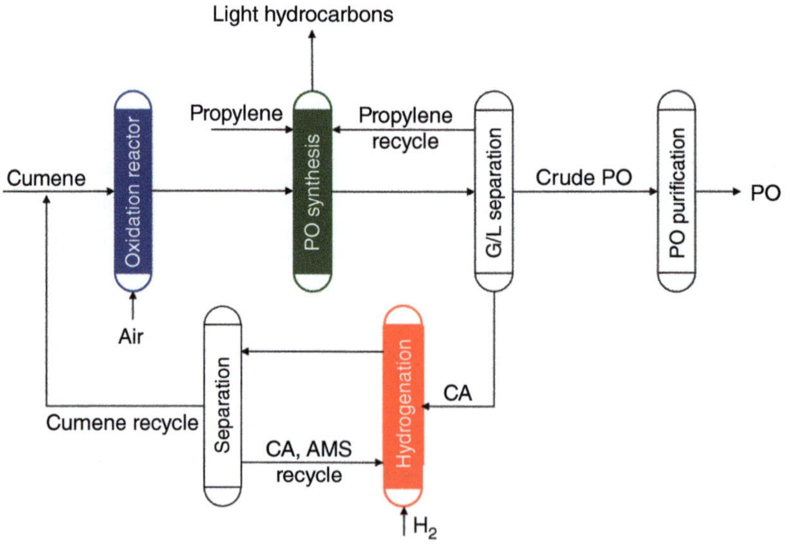

Figure 12.7 Simplified flow-sheet of the CMHPPO process developed by Sumitomo. Source: Adapted from Ref. [29].

at the propylene-to-CHP ratio of 10 : 1 and at the temperature of 60 °C. Because of the excellent catalytic performance of epoxidation catalyst and the use of cumene as solvent, the whole CMHPPO process was reported to have an extremely high PO yield. Overall, the propylene conversion can reach over 95%, while the PO selectivity based on CHP can achieve 95%.

- **Regeneration of CHP**

The co-product α-cumyl alcohol (CA) was regenerated by subsequent hydrogenation process that performed by reacting the mixed α-cumyl alcohol/cumene with hydrogen. This hydrogenation was carried out quantitatively, yielding cumene and water. After separation and purification, the cumene was recycled to the first CHP generation step. Much lower energy consumption for separation and purification was needed due to the efficient heat recovery. It was estimated that Sumitomo CMHPPO process reduced about 40% energy consumption, even compared with HPPO process. These characteristics allow CMHPPO process to be a cost-effective technology for PO production.

12.7 Conclusions and Outlooks

The PO production technology has been evolved from the traditional chlorohydrin process and co-production processes (PO/TBA and PO/SM). Nowadays, HPPO and CMHPPO routes have been successfully implemented. Since both processes are PO-only routes, they are promising for an increasing market share in PO production industry in the future. In particular, the CMHPPO technology is the only PO production industrial process that can generate and recycle oxidant. The key step during the development of CMHPPO production technology is the discovery of new heteroatoms-doped catalysts, which exhibit large mesopore opening, the desirable Ti active sites in tetrahedral coordination and the enhanced hydrophobicity of catalyst surface. In addition to mesoporous catalysts, the hierarchical zeolitic catalysts with crystalline frameworks have been proved to be active in PO production with CHP because of easy control over Ti coordination and surface hydrophobicity. This kind of material would also be promising in the future. The catalytic result is the consequence of a complicated process that relate to diffusion, adsorption, catalysis, and desorption steps. The successful application of Ti-doped mesoporous catalyst is a demonstration of how to design a material that can induce complicated chemical reaction proceeding in a new energy-feasible pathway. It is worthy to note that the oxidant CHP may partially decompose and/or convert to byproducts, leading to incomplete recovery of cumene. The existence of byproducts and other polar organic molecules would also affect the catalytic activity of catalysts.

In addition to Sumitomo, Sinopec also successfully used Ti-HMS catalyst in a similar PO process with CHP as oxidant. Researchers at Sinopec found that impurities in feedstocks would be one of the main factors that affected catalyst performance. They showed that organic acids, phenol, acetone, and water that existed in CHP solution brought negative effects on the catalyst. The impurities not only decreased the stability of the catalyst but also decreased activity and selectivity to the desirable product PO. The performance was greatly improved after removing the polar impurities by proper treating processes. Therefore, in a real industrial operation, the purification and other factors should also be taken into considerations to achieve ideal process for PO production [30].

References

1 (a) Tullo, A.H. and Short, P.L. (2006). *Chem. Eng. News* 84 (41): 22. (b) Hidaka, J., Shimizu, I., and Takemoto, L. (2019). *Sumitomo Kagaku* 5: 1. (c) Tsuji, J., Yamamoto, J., Ishino, M., and Oku, N. (2006). *Sumitomo Kagaku* 1: 1.

2 (a) http://www.chemicalsafetyfacts.org/propylene; (b) https://www.shell.com/business-customers/chemicals/our-products/propylene-oxide.html; (c) https://www.epa.gov/sites/default/files/2016-09/documents/propylene-oxide.pdf; (d) https://www.lyondellbasell.com/globalassets/documents/polymers-technical-literature/poproductsafetybulletin.pdf. All web sites were accessed on September 2022.

3 (a) Nijhuis, T.A., Makkee, M., Moulijn, J.A., and Weckhuysen, B.M. (2006). *Ind. Eng. Chem. Res.* 45: 3447. (b) Lin, H.B., Chen, H.B., He, S. et al. (2014). *Chem. Ind. Eng. Prog.* 33 (8): 2192. (c) Buijink, J.K.F., van Vlaanderen, J.J.M., Crocker, M., and Niele, F.G.M. (2004). *Catal. Today* 93: 199. (d) Ghanta, M., Fahey, D.R., Busch, D.H., and Subramaniam, B. (2013). *ACS Sustain. Chem. Eng.* 1 (2): 268. (e) Tsuji, J., Ishino, M. and Uchida, K. (2000) *US Patent 6,160,137 (assigned to Sumitomo Chem).*

4 Cavani, F. and Gaffney, A.M. (2009). *Sustainable Industrial Chemistry* (ed. F. Cavani, G. Centi, S. Perathoner, and F. Trifiró), 323. Hoboken: John Wiley & Sons Publishing.

5 Yan, Z.F., Chen, J.K., Zhao, X. et al. (2019). *Environ. Eng.* 37 (6): 170.

6 (a) Russo, V., Tesser, R., Santacesaria, E., and Serio, M. (2013). *Ind. Eng. Chem. Res.* 52: 1168. (b) Clerici, M.G. and Ingallina, P. (1993). *J. Catal.* 140: 71. (c) Lane, B.S. and Burgess, K. (2003). *Chem. Rev.* 103: 2457. (d) Perego, C., Carati, A., Ingallina, P. et al. (2001). *Appl. Catal. A Gen.* 221 (1–2): 63.

7 (a) Taramasso, M., Perego, G. and Notari, B. (1983) *US Patent 4,410,501 (assigned to SnamProgetti)*; (b) Clerici, M.G., Bellussi, G., and Romano, U. (1991). *J. Catal.* 129: 159. (c) Neri, C., Anfossi, B., Esposito, A. and Buonomo, F. (1986) *Eur Patent 100,119 (assigned to EniChem)*; (d) Paparatto, G., Forlin, A. and Tegon, P. (2000) *Eur Patent 1,072,599 (assigned to EniChem)*; (e) Clerici, M.G. and Ingallina, P. (1993) *US Patent 5,221,795 (assigned to Enichem).*

8 (a) Jia, Y.J., Liu, Z.C., and Gao, H.X. (2013). *Chem. React. Eng. Technol.* 28 (6): 519. (b) Sankar, G., Thomas, J.M., and Catlow, C.R.A. (2000). *Top. Catal.* 10 (3–4): 255.

9 (a) Tonti, S., Raffia, P. and Gervasutti, V. (1993), *US Patent 5,227,525 (assigned to Enichem)*; (b) Tatsumi, T., Yuasa, K., and Tominaga, H. (1992). *J. Chem. Soc. Chem. Commun.* 1446. (c) Notari, B. (1993). *Catal. Today* 18: 16. (d) Kitamura,

M., Shimazu, Y. and Yako, M. (2001) *US Patent 6,265,574 (assigned to Sumitomo Chemical Company)*; (e) Ratnasamy, O., Srinivas, D., and Knözinger, H. (2004). *Adv. Catal.* 48: 1.

10 (a) Tullo, A.H. (2004). *Chem. Eng. News* 82 (36): 15. (b) Campos-Martin, J.M., Blanco-Brieva, G., and Fierro, J.L.G. (2006). *Angew. Chem. Int. Ed.* 45: 6962. (c) Russo, V., Tesser, R., Santacesaria, E., and Di Serio, M. (2013). *Ind. Eng. Chem. Res.* 52 (3): 1168. (d) Laufer, W., Meieres, R., and Holderich, W. (1999). *J. Mol. Catal. A Chem.* 141 (1–3): 215.

11 (a) Wu, P., Tatsumi, T., Komatsu, T., and Yashima, T. (2001). *J. Catal.* 202 (2): 245. (b) Serrano, D.P., Sanz, R., Pizarro, P. et al. (2014). *Appl. Catal. B* 146: 35–42. SI. (c) Chang, C.C., Lee, J.F., and Cheng, S. (2017). *J. Mater. Chem. A* 5 (30): 15676.

12 (a) Kresge, C.T., Leonowicz, M.E., Roth, W.J. et al. (1992). *Nature* 359: 710. (b) Beck, J.S., Vartuli, J.C., Roth, W.J. et al. (1992). *J. Am. Chem. Soc.* 114 (27): 10834. (c) Tanev, P.T., Chibwe, M., and Pinnavaia, T.J. (1994). *Nature* 368: 321. (d) Zhao, D.Y., Feng, J.L., Huo, Q.S. et al. (1998). *Science* 279 (5350): 548. (e) Wan, Y. and Zhao (2007). *Chem. Rev.* 107 (7): 2821.

13 (a) Blasco, T., Corma, A., Navarro, M.T., and Pariente, J.P. (1995). *J. Catal.* 156 (1): 65. (b) Morey, M.S., O'Brien, S., Schwaez, S., and Stuck, G.D. (2000). *Chem. Mater.* 12 (4): 898. (c) Anpo, M., Yamashita, H., Ikeue, K. et al. (1998). *Catal. Today* 44 (1–4): 327. (d) Zhang, W.H., Froba, M., Wang, J.L. et al. (1996). *J. Am. Chem. Soc.* 118 (39): 9164. (e) Wu, P. and Tatsumi, T. (2002). *Chem. Mater.* 14 (4): 1657.

14 (a) Koyano, K.A. and Tatsumi, T. (1997). *Micropor. Mater.* 10: 259. (b) Brutchey, R.L., Ruddy, D.A., Andersen, L.K., and Tilley, T.D. (2005). *Langmuir* 21 (21): 9576. (c) Fraile, J.M., Garcia, J.I., Mayoral, J.A., and Vispe, E. (2001). *J. Catal.* 204 (1): 146. (d) Igarashi, N., Kidani, S., Ahemaito, R. et al. (2005). *Microporous Mesoporous Mater.* 81 (1–3): 97.

15 (a) Tsuji, J., Ishino, M. and Uchida, K. (2000) *US Patent 6,160,137 (assigned to Sumitomo Chem)*; (b) Oku, N. and Seo, T. (2003) *US Patent 6,646,138 (assigned to Sumitomo Chem)*; (c) Tsuji, J. and Ishino, M. 2006 *US Patent Appl. 2006/293531*; (d) Tsuji, J. 2005 *US Patent Appl. 2005/085647*; (e) Tsuji, J. and Ishino, M. 2004 *US Patent Appl. 2004/254386.*

16 Liu, J., Ni, X.L., and Hu, Y.K. (2015). *React. Kinet. Mech. Catal.* 114: 685.

17 Li, K.-T., Lin, P.-H., and Lin, S.-W. (2006). *Appl. Catal. A Gen.* 301: 59.

18 Yao, J.M., Zhan, W.C., Liu, X.H. et al. (2012). *Microporous Mesoporous Mater.* 148: 131.

19 Zhan, W.C., Yao, J.M., Xiao, Z. et al. (2014). *Microporous Mesoporous Mater.* 183: 150.

20 Li, X.-F., Gao, H.-X., Jin, G.-J. et al. (2008). *J. Mol. Struct.* 872: 10.

21 Ni, X.L., Liu, J., Wang, A.J., and Hu, Y.K. (2016). *Petrochem. Technol.* 45 (9): 1050.

22 (a) Agostini, G., Lamnerti, C., Palin, L. et al. (2010). *J. Am. Chem. Soc.* 132 (2): 667. (b) Corma, A., Huber, G.W., Sauvanaud, L., and O'Connor, P. (2007). *J. Catal.* 247 (2): 307. (c) Verboekend, D., Vile, G., and Perez-Ramirez, J. (2012).

Adv. Funct. Mater. 22 (5): 916. (d) Peters, A.W. and Wu, C.C. (1995). *Catal. Lett.* 30 (1–4): 171.

23 (a) Thomas, A., Goettmann, F., and Antonietti, M. (2008). *Chem. Mater.* 20 (3): 738. (b) Na, K. and Somorjai, G.A. (2015). *Catal. Lett.* 145 (1): 193. (c) Liu, J.H., Cui, L., Wang, L. et al. (2014). *Mater. Lett.* 132: 78. (d) Yang, X.Y., Leonard, A., Lemaire, A. et al. (2011). *Chem. Commun.* 47 (10): 2763. (e) Zhang, K. and Ostraat, M.L. (2016). *Catal. Today* 264: 3. (f) Pérez-Ramirez, J., Christensen, C.H., Egeblad, K. et al. (2008). *Chem. Soc. Rev.* 37: 2530. (g) Fan, W., Snyder, M.A., Kumar, S. et al. (2008). *Nat. Mater.* 7: 984. (h) Hartmann, M., Machoke, A.G., and Schwieger, W. (2016). *Chem. Soc. Rev.* 45: 3313. (i) Chen, L.-H., Sun, M.-H., Wang, Z. et al. (2020). *Chem. Rev.* 120: 11194.

24 Choi, M., Na, K., Kim, J. et al. (2009). *Nature* 461 (7261): 246.

25 (a) Jo, C., Ryoo, R., Zilkova, N. et al. (2013). *Catal. Sci. Technol.* 3 (8): 2119. (b) Seo, Y., Cho, K., Jung, Y., and Ryoo, R. (2013). *ACS Catal.* 3 (4): 713. (c) Verheyen, E., Jo, C., Kurttepeli, M. et al. (2013). *J. Catal.* 300: 70. (d) Kim, J., Park, W., and Ryoo, R. (2011). *J. Catal.* 1 (4): 337. (e) Kim, J., Kim, W., Seo, Y. et al. (2013). *J. Catal.* 301: 187.

26 (a) Kim, J., Chun, J., and Ryoo, R. (2015). *Chem. Commun.* 51 (66): 13102. (b) Na, K., Jo, C., Kun, J., and Ryoo, R. (2011). *ACS Catal.* 1 (8): 901. (c) Jo, C., Park, W., and Ryoo, R. (2017). *Microporous Mesoporous Mater.* 239: 19.

27 Wang, J.G., Xu, L., Zhang, K. et al. (2012). *J. Catal.* 288: 16.

28 Ji, X.Y., Xu, L., Du, X. et al. (2017). *Catal. Sci. Technol.* 7: 2874.

29 (a) Goto, S., Shinohara, K. and Katao, M. (2006) *US Patent 7,030,254 (assigned to Sumitomo Chem)*; (b) Ito, K. and Katao, M. 2003 *Jap Patent 2003/261551 (assigned to Sumitomo Chem)*; (c) Ito, K. and Katao, M. 2003 *Jap Patent 2003/160519 (assigned to Sumitomo Chem)*; (d) Oku, N., Nakayama, T. and Shinohara, K. (2005) *Eur Patent 1,498,414 (assigned to Sumitomo Chem)*; (e) Goto, S., Shinohara, K. and Katao, M. (2004) *Eur Patent 1,420,014 (assigned to Sumitomo Chem)*; (f) Tsuji, J. and Yamamoto, Y. (2003) *US Patent 6,639,085 (assigned to Sumitomo Chem).*

30 Zhou, L.J., Jin, G.S., and Li, Y.S. (2013). *Polyurethane Ind.* 28 (6): 18.

13

Hydroxylation of Benzene and Phenol on Zeolite Catalysts

Shiying Li, Shanhui Zhu, Sen Wang, Mei Dong, and Weibin Fan

Institute of Coal Chemistry, Chinese Academy of Sciences, State Key Laboratory of Coal Conversion, Institute of Coal Chemistry, Chinese Academy of Sciences, 27 South Taoyuan Road, Taiyuan, 030001, China

13.1 Introduction

Phenol is an industrially important chemical for production of many compounds such as bisphenol A, phenolic resins, caprolactam, alkylphenols, and adipic acid [1]. Dihydroxybenzenes (DHB), which can be processed to produce pesticides, perfumes, pharmaceuticals, photosensitive material, etc., is another indispensable chemical in industry fields [2]. Phenol is mainly obtained via well-established technology through oxidation of isopropylbenzene with co-generation of acetone on acid catalysts such as $AlCl_3$ and H_2SO_4. However, the procedures are complex and the used liquid acid catalyst is unfriendly to environments. Additionally, the raw materials of propene are relatively expensive. Thus, it is necessary to develop a simple one-step process by direct oxidation of benzene to phenol. Subsequently, the vanadium, molybdenum, and tungsten supported on oxides are used for catalyzing benzene to phenol. The vanadia catalyst provided a 71.5% selectivity toward phenol at 10.5% benzene conversion [1]. However, its low selectivity toward phenol limits the application in industry due to rather intensive side reactions, i.e. the complete oxidation of benzene.

In order to realize the high selectivity to phenol and thermal stability, the zeolites such as ferrialuminosilicalite (Fe-ZSM-5) are applied as catalyst in benzene oxidation to phenol [3]. The path, in which benzene is directly oxidized to phenol with nitrous oxide (N_2O) as oxidizing agent via the reaction: $C_6H_6 + N_2O \rightarrow C_6H_5OH + N_2$, has attracted much attention due to the high selectivity to phenol (~100%) upon Fe-ZSM-5 zeolite catalyst [4–6]. The formation of surface oxygen species on Fe active sites, which is called "α-oxygen" species, is responsible for the conversion of benzene to phenol. The types of active Fe species and the proposed reaction mechanisms are discussed in this context. It is concluded that the active Fe species can capture the atomic oxygen of N_2O to form the "α-O" species during N_2O decomposition processes. This type of O species subsequently transfers themselves into the benzene molecules to produce phenol molecules.

Micro-Mesoporous Metallosilicates: Synthesis, Characterization, and Catalytic Applications, First Edition. Edited by Peng Wu and Hao Xu.

DHB could be produced by hydrolysis of chlorophenol and oxidation of aminobenzene. However, this process suffers from some problems such as complicated reaction condition, environmental pollution, and high cost. Phenol hydroxylation is an energy-saving, green, and promising method for DHB production [2, 7, 8]. DHBs (including hydroquinone (HQ) and catechol (CAT), the two most widely used isomers) are usually obtained simultaneously by liquid-phase oxidation of phenol with hydrogen peroxide using strong organic acid catalysts, Fenton's reagent, titanosilicates (TS-1), etc. Moreover, on an industrial scale, the phenol hydroxylation process (installation by Enichem Company) is conducted in liquid phase at 100 °C, in which the molar ratio of phenol/hydrogen peroxide is 3 : 1, with the methanol concentration and the TS-1 content of 30 wt% and 3 wt%, respectively. The obtained molar ratio of CAT to HQ is 1 : 1. The total selectivity of the main products in relation to phenol consumed amounts to 90–94 mol% with the phenol conversion of 25–30 mol% [9, 10]. Furthermore, the ratio of CAT to HQ is highly dependent on the types of catalysts. The Fe-containing catalysts usually give equal amounts of CAT and HQ products, whereas the CAT and HQ are produced on the external surface and inside of Ti-containing zeolites due to the space-confined effect in hydroxylation of benzene.

In this context, the benzene oxidation and phenol hydroxylation reactions are summarized. The effects of reaction conditions on benzene oxidation and corresponding reaction mechanism are illustrated on Fe-containing zeolite catalysts in gas phase. Moreover, benzene and phenol hydroxylation reactions in liquid phase are discussed in detail upon titanosilicate catalysts. Zeolites supported other transition metals are also listed. The recent progresses on hydroxylation of benzene/phenol are provided. This will guide the catalyst designing for researchers in both fundamental chemistry and industry fields.

13.2 Hydroxylation of Benzene to Phenol

13.2.1 Gas-Phase Reactions

13.2.1.1 Heteroatomic Molecular Sieves as Catalysts

Traditionally, the conversion of benzene to phenol usually proceeds by means of the so-called cumene method, which consists of three steps as shown in Figure 13.1. Alkylation of benzene with propene to produce cumene (step 1), oxidation of cumene to form the carboxylic acid (step 2), and decomposition of cumene hydroperoxide (step 3) occur in liquid phase in the presence of acid catalysts. This process is complex and consumes large amounts of C_3H_6 and unrecoverable acid catalyst. Therefore, it is of demand that the introduction of a hydroxyl group via one-step direct oxidation of benzene to phenol.

ZSM-5 zeolites containing Ti, V, Cr, Mn, Co, Ni, Cu, Zn, Pd, or Pt, obtained through direct synthesis or ion exchange, are found almost inactive in oxidation of benzene to phenol with N_2O [11, 12]. Notably, the Cu/ZSM-5 prepared by impregnation method shows certain catalytic activity in benzene oxidation reaction with O_2 and H_2O at

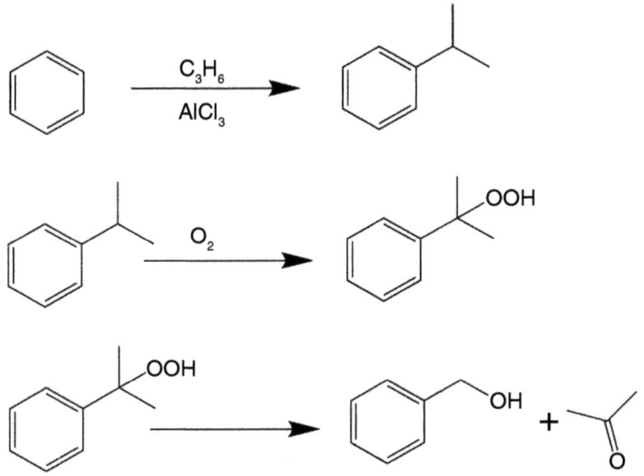

Figure 13.1 Scheme of benzene to phenol conversion via the cumene route.

Figure 13.2 Scheme of hydroxylation of benzene to phenol.

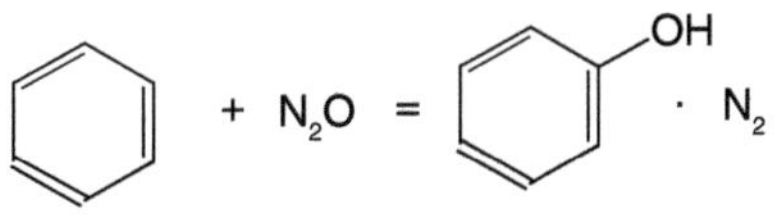

500 °C. In this case, the phenol selectivity is about 15%–30%, accompanied by plenty of CO_2 and CO [13].

In order to improve the phenol selectivity, in 1982, a new reaction was discovered by Iwamoto and co-workers: benzene oxidation by dinitrogen monoxide: $C_6H_6 + N_2O \rightarrow C_6H_5OH + N_2$ (Figure 13.2) using Fe-containing ZSM-5 [14]. It not only gives a desired one-step ecologically safe and commercially feasible process but also can decrease the consumed amount of dinitrogen monoxide due to scarcely happening of side reactions [15–18]. ZSM-5 with 0.1 wt% Fe_2O_3 exhibited 100% selectivity to phenol at about 10% conversion of benzene at 300 °C. With increase of reaction temperature to 350 °C, the conversion of benzene reached up to 25% with slightly decrease of phenol selectivity to 95% [11]. According to the kinetic study, the stoichiometric ratio of feed gas has a significant influence on the selectivity to phenol [19]. The stoichiometric excess of N_2O (i.e. N_2O/benzene molar ratio > 1) caused a relatively high benzene conversion (>20%) and a low selectivity of N_2O for phenol (~40%). In contrast, with a stoichiometric excess of benzene, it gave the low conversion of benzene and increased N_2O selectivity for phenol (~90%).

Panov and co-workers also found that ZSM-5 zeolites with different Fe content were active for benzene to phenol oxidation with nitrous oxide [12, 20, 21]. When Fe active sites are passivated with sulfide groups to form the $[HO\text{-}Fe\text{-}(SO_4)\text{-}Fe\text{-}OH]^{2+}$, phenol cannot be formed [22], indicating Fe metal played an essential role for benzene hydroxylation. Panov's group also evidenced that the α-oxygen that formed in binuclear Fe(II) sites was active in benzene oxidation, implying that the Fe(II) sites were responsible for hydroxylation of benzene to phenol.

Notably, the Fe supported on the zeolite is usually pre-treated with calcination or steaming at elevated temperatures before reaction. It leads to the dealumination of zeolite lattice and expulsion of isomorphously substituted Fe(III) [20]. This process was also accompanied by partial Fe(III) auto-reduction resulting in formation of extraframework Fe(II) species, which can interact with N_2O to produce active oxygen $(O)_{Fe}$ [23]. There are three types of Fe species in zeolite after calcination and/or steaming: (1) mononuclear Fe, catalyzing the phenol formation; (2) dinuclear oxygen-bridged Fe oxo-ions, catalyzing NO_x reduction; and (3) Fe oxide nanoparticles, activating benzene combustion [15, 17, 18, 22]. The state of Fe species can be tuned by changing its content. It was found that the turnover frequency (TOF) value was the highest at a phenol yield near 70% with H-MFI catalyst synthesized by hydrothermal method with a Fe/Al ratio of 0.014 and followed by the treatment of steaming at 650 °C [22]. With increasing Fe/Al ratio to 1.2, the TOF value of this sample decreased by more than 3 orders of magnitude. Similarly, the FeZSM-5 with the Fe content ranging from 0.015 to 2.1 wt% was obtained under hydrothermal treatment and then was activated by steaming and/or calcinations in He at 1050 °C for the formation of activated Fe species before reactions [18]. It is found that the degree of nuclearity of the Fe(II) species (mono-, bi-, oligonuclear species or nanoparticles) increased with iron content and activation temperature/time (Figure 13.3). At extremely low Fe concentration (0.58 wt%), the phenol productivity is ~0.29 mmol h^{-1} g^{-1} due to the mononuclear Fe(II)-O-Al complex and/or oligonuclear species containing at least two oxo-bridged Fe(II) sites formed via treating with He at high temperature of 1050 °C. When the Fe content increased to 2.1 wt%, the productivity to phenol decreases to ~0.22 mmol h^{-1} g^{-1}, despite the almost fourfold higher total Fe content after steaming at 550 °C. This was attributed to the formation of some inactive agglomerates of Fe(III) oxide during the steaming process, which can cause benzene to be totally oxidized to produce CO_2 and H_2O [24].

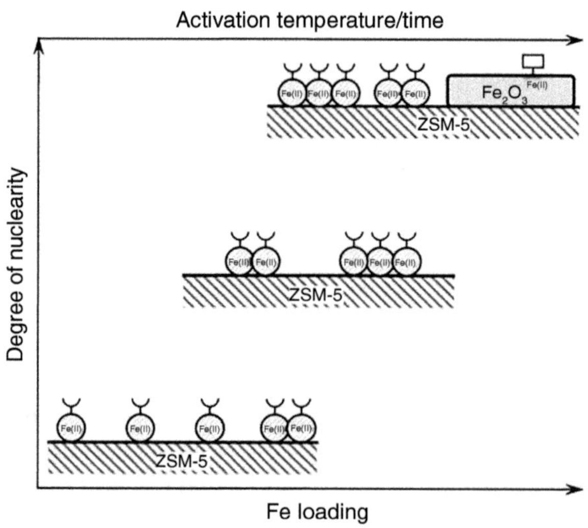

Figure 13.3 Scheme of evolution of Fe(II) states in Fe-ZSM-5 with the increase of Fe amount. Source: Reproduced from Ref. [18]/with permission of Elsevier.

13.2.1.2 Fe-Containing Molecular Sieves as Catalysts

13.2.1.2.1 Effect of Pore Structure Zeolite framework with different pore structures has a significant effect on the performance in hydroxylation of benzene reaction. Panov and co-workers investigated the catalytic activities on benzene oxidation using various zeolite as catalysts. According to the results, $AlPO_4$ and Fe-ZSM-5 give benzene conversion of 4.5% and 8% and high selectivity to phenol [1]. Furthermore, Burch et al. found that ZSM-5 zeolite showed high conversion of benzene (27.2%) and selectivity of phenol (98%) [25]. However, the amorphous silica-alumina exhibits no catalytic activity in benzene oxidation reaction despite comparable acidity, indicating that the pore structure is essential for benzene conversion. The constrained space in the zeolite can promote the reacted molecules to collide with each other and/or with active sites, and the electrostatic field in the zeolite channels can facilitate the activation of benzene and N_2O. However, when the EU-1 with similar acidities and pore size to ZSM-5 was used as the catalyst, the conversion of benzene (0.75%) and the selectivity to phenol (86.2%) are lower than that of ZSM-5, owing to relatively high deactivated rates originating from the coke deposition at the aperture of EU-1 with only one-dimensional channel structure [26]. The Y zeolite with supercages also showed low activity as a result of the formation of large polyaromatics in the supercages [27]. In addition, the probe experiment of oxidation of 1,3,5-trimethylbenzene (TMB) with kinetic diameter of 8 Å over FeZSM-5A_s, which is obtained through steaming treatment of FeZSM-5 at 500 °C for 5 h, is performed to investigate the locations of active sites [24]. When applying TMB as a substrate in the reaction, no product was observed, revealing the active sites for the benzene oxidation with nitrous oxide located inside the pores, instead of outer surface.

The large coking species, such as phenol condensation and grafted phenolate species, can lead to the deactivation of catalysts, because benzene oxidation occurs in the micropores of zeolite. In order to improve the catalytic lifetime of Fe-ZSM-5, the hierarchical porous structures are introduced [6, 28, 29]. The hierarchical Fe-ZSM-5 zeolite can be prepared via alkaline post-treatment method [3, 30]. It shows a much higher catalytic activity and stability compared to the parent Fe-ZSM-5 catalysts due to the elimination of internal mass-transfer limitation. Hensen et al. recently synthesized a hierarchical Fe-ZSM-5 by adding $[(CH_3O)_3SiC_3H_6N(CH_3)_2C_{18}H_{37}]$ as silica source [31]. It is found that the initial reaction rate and the stability are improved for hierarchical Fe-ZSM-5 (Table 13.1). After the reaction of 5 h, the reaction rate of the hierarchical Fe-ZSM-5 (R_{phenol} = 5.3 mmol g^{-1} h^{-1}) is almost 3.5 times higher than that of conventional Fe-ZSM-5 (R_{phenol} = 1.5 mmol g^{-1} h^{-1}) because the mesoporous matrix structures promote the accessibility to the Fe active sites and slow the pore blocking by carbonaceous deposits. Similar results are obtained for the hierarchical Fe-ZSM-5 zeolites synthesized with different methods, including carbon black template, a second organosilane source, and desilication. Furthermore, a stable Fe/ZSM-5 nanosheet zeolite was synthesized with di- and triquaternary ammonium structure directing agents for benzene oxidation [16, 28]. Nanosheet Fe/ZSM-5 zeolite with a *b*-axis thickness of ~3 nm and Fe content of 0.24 wt% after steaming treatment exhibited higher activity, compared to bulk Fe/ZSM-5 zeolite that are

Table 13.1 Physicochemical properties of conventional and mesoporous Fe-ZSM-5 and their reaction results for benzene hydroxylation.

Catalysts	Fe content	BET/$m^2\ g^{-1}$	Conv.	R_{phenol}/ $mmol\ g^{-1}\ h^{-1}$	Reference
Meso-Fe/ZSM-5-org[a)]	0.54 wt%	371	24[g)]	5.3[g)]	[31]
Meso-Fe/ZSM-5-C1[b)]	0.44 wt%	345	16[g)]	2.4[g)]	
Meso-Fe/ZSM-5-des[c)]	0.58 wt%	355	10[g)]	2.8[g)]	
Meso-Fe/ZSM-5-nano[d)]	0.54 wt%	489	19[g)]	3.9[g)]	
Fe-ZSM-5	0.49 wt%	356	6[g)]	1.5[g)]	
Fe/ZSM-5(TPA,360)-st	Si/Fe = 180	356	4.9[h)]	0.6[h)]	[16]
Fe/ZSM-5(2 N,360)-st[e)]		483	23.4[h)]	6.0[h)]	
MZ(0.18)-L (Silylated)[f)]	0.93 wt%	–	38.6[i)]	16.7[i)]	[29]
MZ(0.18)-L[f)]	0.93 wt%	765	42.7[i)]	10.2[i)]	
HZSM-5 (steamed)	0.02 wt%	409	28.2[i)]	8.1[i)]	

a) Meso-Fe/ZSM-5-org was prepared by adding $[(CH_3O)_3SiC_3H_6N(CH_3)_2C_{18}H_{37}]$ as silica source.
b) Meso-Fe/ZSM-5-C1 was synthesized by using carbon black as mesopore generating template.
c) Meso-Fe/ZSM-5-des was obtained by treating with 0.2 M NaOH.
d) Meso-Fe/ZSM-5-nano was attained by adding phenylaminopropyl-trimethoxysilane as a second organosilane.
e) Fe/ZSM-5(2 N,360)-st was synthesized with di-quanternary ammonium as template.
f) MZ(0.18)-L was obtained using octadecyldimethyl(3-trimethoxysilylpropyl)ammonium chloride as mesopore template; MZ(0.18)-L (Silylated) was generated via pre-treating with methoxytrimethylsilane vapors at 150 °C.
g) Reaction conditions: 1.01 vol% C_6H_6 and 4.01 vol% N_2O in He, 100 mg catalyst, T = 350 °C.
h) 1.01 vol% C_6H_6 and 4.01 vol% N_2O in He, 100 mg catalyst, T = 350 °C.
i) 1 vol% C_6H_6 and 4 vol% N_2O in He, 100 mg catalyst, T = 400 °C.

prepared using tetrapropylammonium hydroxide (TPAOH) as template. The phenol productivity is remained at 6.0 mmol g^{-1} h^{-1}, which is 10 times higher than that of bulk Fe/ZSM-5 even after 24 h (Table 13.1). The amounts of deposited coke were 149 and 53 mg g^{-1} for nanosheet and bulk Fe/ZSM-5, respectively, indicating higher coke tolerance for zeolite with mesopores. When the coke deposits were removed by calcination, the initial activity can be restored. In this paper, the authors pointed out that the framework Fe^{3+} irons are inactive for benzene oxidation consistent with the abovementioned results and only isolated Fe^{2+} centers in the extraframework position are the active sites for catalyzing the benzene oxidation. However, the oligomeric complexes such as $[Fe(\mu\text{-}O)Fe]^{2+}$ and the ferryl cations $[FeO]^+$ can stably adsorb phenolate species, resulting in the deactivation and blockage of the micropores. Silylation of the mesoporous catalyst for improving the hydrophobicity and reducing the amount of Lewis acid sites can further lead to the increase by one time of initial phenol formation rate, compared to the microporous

H-ZSM-5 zeolite (Table 13.1), owing to accelerating the desorption of phenol product [29].

13.2.1.2.2 Promotion for Formation of Active Fe Species by Extraframework Metal Species

According to abovementioned Fe species, the formation of three types of Fe sites: (1) mononuclear Fe, catalyzing phenol formation; (2) dinuclear, oxygen-bridged Fe oxo-ions, catalyzing NO*x* reduction; and (3) Fe oxide nanoparticles, catalyzing deep oxidation, is dependent on the zeolite acidity. The heteroatoms in the zeolite, such as Al and Ga, not only promote the formation of Fe-acid oxo species, such as Fe-O-Al and Fe-Ga-O, but also suppress Fe species agglomeration into oxide particles that can be served as active sites for benzene combustion.

The strong Al-Lewis sites are essential for benzene hydroxylation to phenol together with the α-sites, which are primarily linked to Fe species [32–34]. Motz et al. observed an enhancement of hydroxylation activity with benzene conversion increased from 22% to ~30% at 350 °C in the presence of extraframework Al species [35]. Hensen et al. took an insight into the role of Al in zeolites. They synthesized three types of MFI zeolite, including [Fe]MFI, [Al]MFI, and [Fe, Al]MFI under hydrothermal condition [36]. For [Fe]MFI and [Al]MFI samples, most of the Fe and Al atoms are incorporated into the zeolite framework. After calcination, a large amount of framework iron is removed to form the Fe^{3+} clusters for both [Fe]MFI and [Fe, Al]MFI samples due to the iron is less stable in the framework positions than aluminum atom. To remove the Al atoms from the framework, the severe steam activation step is performed. Before and after the treatment with steaming, the [Al]MFI exhibited a negligible nitrous oxide conversion up to 800 K, indicating that Brønsted acid sites and Lewis acidic extraframework Al species cannot activate N_2O at moderate temperatures. Similarly, the [Fe]MFI sample also produces negligible amounts of phenol, whereas both [Fe, Al]MFI and [Fe, Al]MFI(st) (st stands for sample with steaming treatment) display high rates of phenol formation. Particularly, the latter one has a higher rate of phenol formation than the former one probably because the dispersed or mononuclear Fe species can be stabilized by extraframework Al species that were generated by steaming, denoted as Fe-O-Al, which further interacted with defect sites of zeolites. It was also verified that the migration of Al from framework made the Fe—O bond retard, thus leading to the formation of active Fe-O-Al sites for benzene oxidation under the steam-activated treatment [37]. This sample gave the conversion of benzene with 7% after 1 h and the selectivity to phenol close to 100%.

The Fe/Ga-ZSM-5 catalysts with different Ga contents are prepared by solid-state ion-exchange method. It was found that large amount of Fe(II) sites were formed because extra-framework Ga facilitated the reduction of oligo-nuclear Fe(III) sites, thus leading to the formation of Fe(II)-Ga-O species [38]. With the increase of Ga amount, Fe/Ga-ZSM-5(1.61)-HT showed a high activity of direct N_2O decomposition due to high concentrations of Fe(II) sites (5.0×10^{19} sites/g) and α-sites (2.8×10^{19} sites/g). This suggested that extra-framework Ga species promote the formation of much more amounts of Fe(II) sites and α-sites.

ZSM-5 zeolites having different acid strength were prepared through the incorporation of different heteroatoms, such as B, Al, Ga, and Ti, and were applied as support of Fe. The Al-ZSM-5 and Ga-ZSM-5 exhibit high activity even in the presence of 0.01–0.03 wt% Fe. However, much higher amount of Fe with 10–100 times for B-ZSM-5 and Ti-ZSM-5 was needed for obtaining the same activity [39]. This is because the acid sites of zeolite can trap the Fe atoms excluded from the zeolite crystal framework under steaming conditions, forming Fe-containing α-sites. It is implied that Fe species and moderate acidity are essential for benzene hydroxylation. Furthermore, the B atom in the zeolite was easier to migrate from framework to the extra-framework position, compared to Al atom during heat treatment with a flow of helium at 700 °C, thus creating much more defect sites [40]. These sites can interact with the iron ions for enhancing the phenol productivity. The maximum phenol productivity is up to 3.116 mmoles h^{-1} g^{-1} with Si/B and Si/Al ratios of 191 and >700, respectively. The phenol productivity decreases by 23% when increasing Si/B ratio to 232 at the same Si/Al ratio.

13.2.1.2.3 Catalytic Mechanism In the past, the researchers believed that the Brønsted acid plays an important role in the oxidation of benzene [11, 14]. The N_2O molecules can interact with a Brønsted acid center as follows:

$$N_2O + H - O - \text{surface} = N_2OH^+ + O^- - \text{surface}$$

Then, the N_2OH^+ species can decompose to produce N_2 and OH^+: the electrophile attacks the electron-rich aromatic center. Another hypothesis is that the benzene ring itself can be activated, causing the formation of carbonium ion, which is then attacked by the nitrous oxide molecule. However, this proposed reaction mechanism is overturned with the evidence of no catalytic activity upon ZSM-5 only with acid sites [36].

Recently, a new pathway of direct conversion of benzene to phenol is disclosed. It is proposed as follows: (1) the composition of nitrous oxide at the active site; (2) the formation of the benzene complex at the iron active site; (3) the hydrogen abstraction from benzene; (4) the recombination between the OH and C_6H_5 ligands; and (5) the release of product phenol [41]. The radical nature of α-oxygen that significantly affects the mechanistic concepts of oxidation catalysis process has sparked an intense discussion on the role of O^- radicals in selective oxidation procedures. The iron is the acknowledged active sites through interacting with O atom of N_2O molecules, resulting in the formation of α-O species. In order to investigate the active site for catalyzing the conversion of benzene to phenol, three types of extraframework Fe(II) sites were synthesized in zeolite through steaming and/or calculation method: (1) Fe(II) sites in mononuclear species, (2) oligonuclear species with at least two oxygen-bridged Fe(II) sites, and (3) Fe(II) sites with Fe_2O_3 nanoparticles. It was found that $(O)_{Fe}$ can be formed at only mono- and oligonuclear Fe(II) locations and served as active sites in benzene hydroxylation reactions [18]. The dinuclear Fe species can capture the atomic oxygen of N_2O to form the α-O species, accompanied by N_2O decomposition. Then, this type of O species can transfer themselves into the benzene molecules. Furthermore, a linear relationship

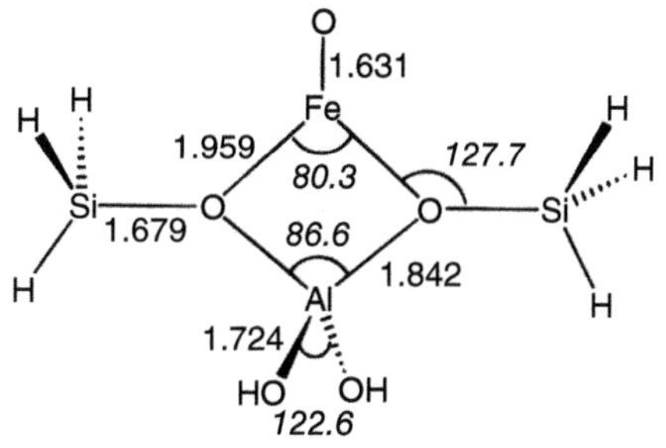

Figure 13.4 A proposed structure of "α-oxygen" model. Source: Reproduced from Ref. [4]/with permission of Elsevier.

between the reaction rate and the concentration of independently identified O^- radicals (α sites), i.e. α-O species, is obtained.

Yoshizawa et al. proposed a Fe cluster that possesses a structure of "α-oxygen" in the quartet state form (Figure 13.4) [4]. The iron-oxo species locate on the AlO_4 surface site. The oxo species can be used as oxygen source in the oxidation of benzene reactions. The benzene molecules interact with iron species, resulting in the formation of reactant complex with binding energy of 10.2 kcal/mol (Figure 13.5). The H atom was abstracted from benzene molecule via the production of a four-centered transition state (TS1) with C—H and O—H bond length of 0.1298 nm and 0.1364 nm, respectively, thus giving a hydroxy intermediate. The energy barrier height for this process is about 15.4 kcal/mol. Next, it can be transformed into the phenol through forming a three-centered transition state (Fe, C, O) (TS2), involving the process of a phenyl migration form the iron atom to the oxygen atom. The broken Fe—C bond and formed C—O bond are calculated to be 0.1999 and 0.1730 nm, respectively. The energy barrier for TS2 is 31.1 kcal/mol. The desorption energy of phenol is 20.2 kcal/mol. In summary, the overall endothermic reaction is 8.6 kcal/mol on the quarter surface.

Different from above reaction mechanism in a concerted manner, Ryder et al. put the whole cluster into the zeolite framework, resulting in the formation of

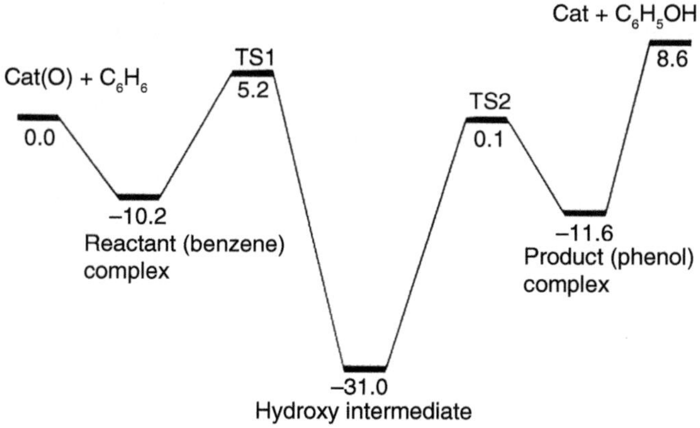

Figure 13.5 Potential energy diagram for benzene hydroxylation on a model "α-oxygen" of Fe-ZSM-5 zeolite. Relative energy unit is in kcal/mol. Source: Reproduced from Ref. [4]/with permission of Elsevier.

physically realistic constrained cluster [42]. It was found that the potential energy surface of sextet coordinate state is below those of the double and quartet state for Fe. It is well-known that the dissociation of N_2O on the Fe species could occur. Then, the benzene oxidation reaction process probably proceeded over $[FeO_2]^+$ or $[FeO]^+$ species as candidate active centers. There are two pathways proposed according to the benzene oxidation reaction. The first one is as follows:

(1) $[FeO]^+ + N_2O(g) \rightarrow [FeO]^+(N_2O)$
(2) $[FeO]^+(N_2O) \rightarrow [FeO_2]^+ + N_2(g)$
(3) $[FeO_2]^+ + N_2O(g) \rightarrow [FeO_2]^+(N_2O)$
(4) $[FeO_2]^+(N_2O) \rightarrow [FeO]^+(O_2) + N_2(g)$
(5) $[FeO]^+(O_2) \rightarrow [FeO]^+ + O_2(g)$
(6) $[FeO_2]^+ + C_6H_6(g) \rightarrow [FeO_2]^+(C_6H_6)$
(7) $[FeO_2]^+(C_6H_6) \rightarrow [FeO]^+(C_6H_5OH)$
(8) $[FeO]^+(C_6H_5OH) \rightarrow [FeO]^+ + C_6H_5OH(g)$

The first five steps illustrate a representation of the N_2O decomposition mechanism. Step 6 involves the reversible adsorption of benzene, whereas step 7 gives the oxidation of benzene to phenol. The desorption of phenol is described by step 8. The energy barrier for every step 6–8 was obtained by calculation. The different value of benzene adsorption energy (ΔE_{ads}) was −6.6 kcal/mol between structures (a) and (b). The activation energy for benzene oxidation (E_{act}) was defined as the difference of energy between transition state (c) and adsorbed one (b). This value, corrected by zero-point energy, was 41.9 kcal/mol. The phenol desorption energy (ΔE_{des}) between structures (d) and (e) was +20.2 kcal/mol. The C* and O* represented the adsorption state in the reactions. It was found that the distance between the two O atoms in $[Fe(O)_2]^+$ was 0.135 nm. This is a typical distance of peroxide species. The bond length of O—O* bond in $[Fe(O)_2]^+$ was increased to 0.268 nm and the O* began to form C—O* (0.141 nm) and O*—H (0.144 nm) bonds with benzene simultaneously, corresponding to the phenol formation process. A further lengthening of the O—O* bond to 0.271 nm and shortening of the O*—H bond to 0.097 nm implied the formation of the phenol molecule. Once phenol molecule desorbed, the Fe—O bond length in $[FeO]^+$ decreased to 0.165 nm. In conclusion, the benzene molecule adsorbed on $[FeO_2]^+$ species to form phenoxy and hydroxyl groups. Then, the proton was migrated to the phenoxy group, leading to the formation of phenol. In addition, the reaction rate of benzene oxidation was higher than that of N_2O dissociation. Thus, the benzene oxidation was preferred over $[FeO_2]^+$.

A second possible pathway for oxidation of benzene on $[FeO]^+$ species. It proceeded as follows:

(9) $[FeO]^+ + C_6H_6(g) \rightarrow [FeO]^+(C_6H_6)$
(10) $[FeO]^+(C_6H_6) \rightarrow [Fe]^+(C_6H_5OH)$
(11) $[Fe]^+(C_6H_5OH) \rightarrow [Fe]^+ + C_6H_5OH(g)$

Similarly, the benzene adsorption energy in this route was −5.4 kcal/mol. However, the activation energy for benzene oxidation, corrected by zero-point energy, is 55.3 kcal/mol. The desorption of phenol is +4.5 kcal/mol. Correspondingly, the distance of Fe—O* bond in $[FeO]^+$ increased from 0.165 to 0.195 nm because O* can inserted into C*—H bond of benzene, which increased from 0.109 to 0.126 nm. Compared to the former pathway, the reaction rate of N_2O dissociation is higher than that of benzene oxidation. Thus, it is concluded that the benzene oxidation reaction is much hard to proceeded over $[FeO]^+$ owing to competitive N_2O dissociation reaction and higher energy barriers.

13.2.1.2.4 Deactivation by Deposition of Coke Species Surface acidity plays a major role in determining catalyst lifetime because it causes deactivation by coking [15, 43, 44]. It is found that the simple or condensed polyphenols and products derived from further condensation are produced on the zeolite surface acid sites. These species block the pore mouth and channel resulting in the limitation of the reactant access to the active sites, thus leading to catalyst deactivation. For example, Fe/ZSM-5 zeolites with different Fe/Al ratios of 0.05, 30, 0.27 were synthesized through ion-exchange method, corresponding to surface acid amount of 224, 204, 193 μmol/g, respectively [43]. It was found that the benzene conversion was decreased with increasing acid sites on the surface because of undesired reactions involving the phenol condensation on acid sites, leading to the deactivation of catalysts.

13.2.1.3 Other Methods for Benzene Hydroxylation to Phenol

Besides, the benzene hydroxylation can proceed using Pd membrane reactors with H_2 and O_2 gas mixtures [45, 46]. The hydrogen dissociated on the palladium layer reacts with oxygen to form active oxygen species such as HOO· and HO·. Then, these oxygen species reacted with benzene to produce phenol. In 2002, Niwa et al. achieved phenol formation selectivity of 80% to 97% and benzene conversion of 2% to 16% below 250 °C by one-step benzene conversion to phenol using Pd membranes [47]. By this way, Liu et al. investigated the different membrane reactors and feed mode, etc. (Figure 13.6). The palladium layer penetrated the porous α-alumina support by electroless plating technique resulting in the formation of Pd capillary membrane with a thickness of 4–5 μm. It was found that narrow flow channel and large surface area-to-volume ratios of Pd membrane favored the formation of more effective area of Pd interface and provided more reacted opportunities. The benzene conversion and phenol yield reached 19.6% and 18.1% at optimum

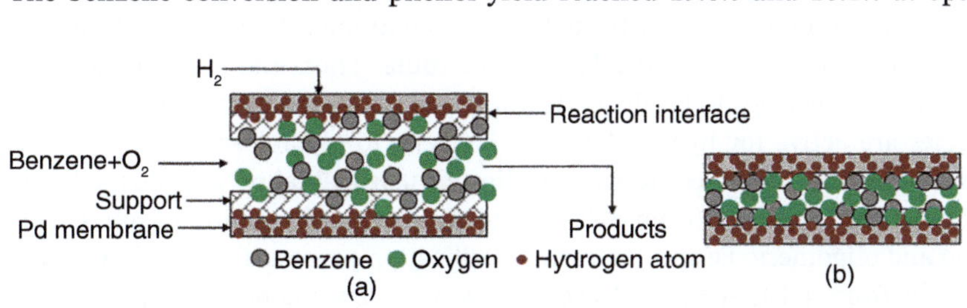

Figure 13.6 Reactants diffusion diagram on (a) large Pd membrane reactor and (b) Pd capillary membrane reactor. Source: Reproduced from Ref. [48]/with permission of Royal Society of Chemistry.

H_2/O_2 ratio of 4 at 200 °C [48]. However, the Pd membrane is easily poisoned by hydrocarbons or other impurities at high temperature. In order to improve the catalytic efficiency, the bifunctional titanium silicalite (TS) zeolite modified by Pd membrane microreactor was synthesized [49]. The TS-1 zeolite was grown on the tube support and treated with TPAOH for forming the mesopores in the zeolite. Then, the Pd membrane was covered on this mesoporous TS-1 zeolite. TS-1 can serve as intermediate support layer that intercepts the contaminants. This membrane reactor shows phenol selectivity of 95% and benzene conversion of ca. 5.4%. Similarly, the TS-1 zeolite with particle size of 180 nm was coated on the Pd membrane surface using alkoxysilane linkers via forming covalent bonds [50]. This membrane exhibited a maximum benzene conversion of 23.4% and phenol yield of 22.3% at 200 °C and 100 kPa. Simultaneously, it also displayed good stability by preventing the hydrocarbon molecules from contacting the Pd surface.

In addition, the one-step benzene conversion to phenol can be realized using other reaction system, such as manganese oxides supported on USY zeolite and Fe-CN/TS-1 as catalysts under visible light irradiation [51, 52]. Compared to sole Fe-CN and TS-1, Fe-CN/TS-1 showed a superior performance with up to 10.0% phenol yield (based on benzene) and 18.4% phenol selectivity (based on H_2O_2) in photocatalytic selective oxidation of benzene to phenol using H_2O_2 as the benign oxidant under ambient conditions [52]. Pd/MCM-41 synthesized by impregnation-reduction method exhibited benzene conversion of 63% and phenol selectivity of 98% using 300 W mercury lamp as light source with H_2O_2 as oxidation [53].

13.2.2 Liquid-Phase Reactions

13.2.2.1 Fe-Containing Molecular Sieves as Catalysts

According to the former section, the benzene hydroxylation reaction usually proceeds under high temperature and needs to consume expensive N_2O as oxidant. However, compared to gas-phase reaction of benzene hydroxylation, liquid-phase transformation from benzene to phenol is not only under mild reaction conditions (60–100 °C) but also using H_2O_2 as an ideal alternative oxidant. Various kinds of metal-containing zeolites, including iron, titanium, copper, and vanadium, are served as catalysts (Table 13.2) [54, 55, 57, 62]. Among them, Fe-ZSM-5 is an important one for catalyzing this reaction. Different from the gas-phase mechanism, the H_2O_2 molecules are firstly activated on Fe-containing catalyst for the formation of radicals (HO·) followed by the Fenton-like route. Then, the HO· radicals react with benzene molecules to produce phenol. In addition, the isolated or oligomeric Fe species are active for hydroxylation of benzene to phenol with H_2O_2 [56–61]. When the Fe-silicalite-1 zeolite was desilicated using alkaline aqueous solution, the phenol yield increases from 5.7% to 7.6% due to the formation of much more isolated and oligomeric Fe species during desilication process along with increased Fe content from 4.0 to 4.5 wt%. Nevertheless, very low phenol yield (0.1%–0.2%) was achieved over silicalite-1 supported Fe by impregnation method, in which iron oxide clusters and bulk iron oxides existed after calcination [57]. The Al species provide the location site for Fe species and make them highly disperse

Table 13.2 Comparison of catalytic performances of different catalysts for benzene hydroxylation using H_2O_2 as the oxidant.

Catalysts	Metal loading	Benzene Conv. (%)	Phenol yield (%)	Phenol Sel. (%)	Solvent	Temp. (°C)	Reference
TS-1-0.2	Ti (2.80 wt%)	57.5	27.7	48.2	H_2O	70	[54]
CuAPO-5	Cu (1.85 wt%)	5.9	—	68	CH_3CN	60	[55]
VS-1	V (1.0 wt%)	—	11	>90	HOAc	70	[56]
ALT-FSD15	Fe (1.3 wt%)	—	7.6	92	CH_3CN	60	[57]
[Fe, Al]-MFI	Fe (1.04 wt%)	2.8	>2.74	~100	Acetonitrile	60	[58]
TS-1	—	1.8	<1.44	~78	Acetonitrile	60	[58]
TS-1B	TiO_2 (0.9 wt%)	8.6	8.08	91	Sulfolane	100	[59]
FeZ-2-72 h	Si/Fe = 42.7	12.1	—	97	Acetonitrile	60	[60]
Fe-ZSM-5(50)-S	Fe (1.71 wt%)	25.5	22.95	~90	Acetonitrile	60	[61]

in zeolite, but are not responsible for the benzene hydroxylation reactions. The Fe-MFI zeolite synthesized with Fe-containing zeolite seed sol showed benzene conversion of 12.1% and phenol selectivity of 97% at 60 °C owing to formation of much higher amount of iron centers [60]. When the mesopores were introduced into the Fe-ZSM-5 zeolite by steam-assisted conversion method, this material gave a benzene conversion of 25.5% with a phenol selectivity of ~90% and a good stability by alleviating the diffusion limitation [61].

13.2.2.2 Microporous Crystalline Titanosilicates as Catalysts

In addition to Fe-containing zeolites, titanosilicates are found to be another type of active catalyst for oxidation process with hydrogen peroxide, such as alkene epoxidation, alkane oxidation, alcohol oxidation, cyclohexanone ammoximation, and phenol hydroxylation. Bianchi et al. used titanium silicalite-1 (TS-1) as the catalyst, showing 4.5% of benzene conversion with phenol selectivity of 43% in acetonitrile [59]. In order to improve the catalytic activity, much higher amount of Ti was incorporated into the zeolite framework through adding the $(NH_4)_2CO_3$ into the synthesis gel for slowing down the crystallization process [63]. The benzene conversion and phenol yield of TS-1 without assistance of $(NH_4)_2CO_3$ are only 46.9% and 23.1% at 70 °C. In contrast, they are increased to 57.5% and 27.7%, respectively, for TS-1 with the addition of $(NH_4)_2CO_3$ (Table 13.2) [54]. Furthermore, when TS-1 is modified with NH_4HF_2 and H_2O_2, the benzene conversion and phenol selectivity using sulfolane as solvent were 8.6% and 94%, respectively, due to selective removal of titanium atoms from the framework without destroying the structure of zeolite [64].

Although TS-1 has been widely applied into the liquid-phase benzene hydroxylation process, it often suffers from the poor selectivity to phenol. The phenol is reactive toward oxidation for producing the over-oxygenated byproducts such as bisphenol and tar. The phenol selectivity in the reaction is strongly affected by the nature of solvent [57, 59, 61, 65]. Generally, the organic agents, including acetone,

Table 13.3 Effect of the solvent on catalytic performance for benzene hydroxylation.

Entry	Solvent[a)]	Benzene conv. (%)	H_2O_2 conv. (%)	Selectivity to phenol based on benzene[b)]	Selectivity to phenol based on H_2O_2[c)]
1	Acetone	4.2	90	41	19
2	Acetonitrile	4.6	93	43	21
3	*tert*-Butanol	4.1	90	44	20
4	Sulfolane	7.8	97	83	67
5	Sulfolane/TS-1B[d)]	8.6	97	94	83

a) Reaction conditions: benzene /solvent = 2 : 8 w/w, H_2O_2/benzene molar ratio = 0.1, TS-1/benzene = 0.1 : 1 w/w, temperature = 100 °C.
b) Moles of phenol/mol of reacted benzene × 100.
c) Moles of phenol/mol of reacted H_2O_2 × 100.
d) Reaction carried out using the TS-1B (TS-1 modified with NH_4HF_2 and H_2O_2).
Source: Adapted from [59].

acetonitrile, alcohols, and acid, are used as solvents. As shown in Table 13.3, the selectivity based on benzene was lower than 50% in acetone, acetonitrile, and tert-butanol owing to the formation of deoxygenated products and tars [64]. In addition, even worse results were obtained using methanol and 2-propanol as solvents. Negligible phenol was obtained with nitrogen-containing solvents, including dimethylformamide and N-methylpyrrolidone. When the tetramethylene sulfone (sulfolane) was used as a co-solvent, the benzene conversion is 8% and the selectivity to phenol is higher than 80%. All the by-products (CAT, HQ, 1,4-benzoquinone, and tars) were decreased. The sulfolane, which is a solvent with high dipole moment (4.81 debye) and dielectric constant (43.26), can manufacture complexes with phenolic compounds (Figure 13.7) that hardly diffuse into the zeolites with steric hindrance, thus suppressing the further phenol oxidation [64, 66]. Similarly, Michael and co-workers not only certificated that the sulfolane (3.5 mL in a 5 mL-total-volume reaction) as co-solvent made a decrease in phenol oxidation of ~100-fold in benzene oxidation but also obtained analogous results when 1 mmol acetonitrile was added into the reaction system for Fe-based catalysts [58]. The acetonitrile might react with hydrogen peroxide to form a peroxycarboximidic acid, which is slower to react with a catalyst than pure hydrogen peroxide. It is needed to point out that the inorganic salts can also interact with H_2O_2 molecule through acid–base reactions or hydrogen bond in solution for preventing the phenol oxidation. It was found that addition of 1 mmol $(NH_4)H_2PO_4$ or $K_2H_2PO_4$ to the

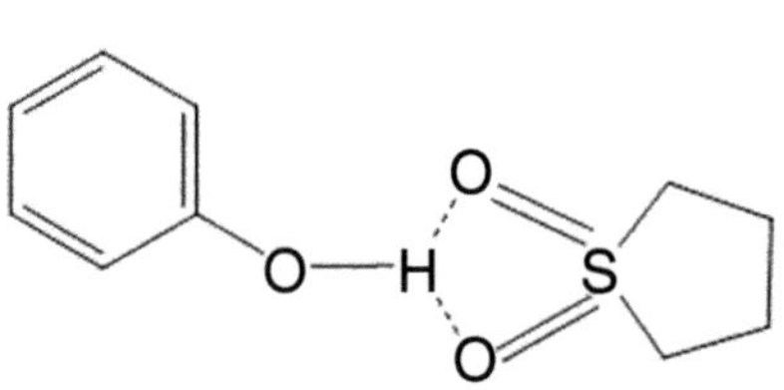

Figure 13.7 Proposed complexes with phenolic and sulfolane compounds. Source: Adapted from Ref. [64].

system of phenol oxidation reaction led to a decreased phenol conversion to less than 25% of that in the absence of an additive.

13.2.2.3 Catalytic Recyclability of Fe- and Ti-Containing Zeolites

Reusability of the catalyst is a crucial factor for a heterogeneous catalyst. In liquid-phase reaction, the catalyst is recycled to be used again through filtrating, washing, drying, and further calcination at high temperature after each run. The stability of TS-1-0.2 in Table 13.2 was investigated by reusing the catalyst for five times [54]. The yield of phenol and conversion of benzene decreased slightly after the first run and then remains almost the same. Meanwhile, the selectivity of phenol and the conversion of hydrogen peroxide showed no change. After five runs, the yield of phenol was still about 27%, indicating that this material exhibited good activity and excellent stability. Different from Ti-based catalyst, the Fe catalysts suffer from Fe leaching, resulting in declined catalytic activity. The benzene conversion decreases to 8.3% after five cycles on FeZ-2 in Table 13.2, compared to 12.1% on fresh one [60]. However, the phenol selectivity (>95%) is stable in each cycle. It is probably due to the leaching of Fe species, verified by the increase of Si/Fe molar ratio from 42.7 (fresh one) to 57.2 after the fifth run. Xiao et al. found similar results with decreasing Fe content from 4.0 to 3.3 wt% [57]. They also investigated the difference of leaching Fe species. When the filtrated liquid after reaction is re-reacted again without adding any other catalyst, 0.5% phenol yield and 0.4% HQ yield were observed, implying that the existence of active Fe species. The same result was obtained with large amount of Fe^{3+} originating from $Fe(NO_3)_3$ $9H_2O$ replace of Fe species in filtered liquid. It suggests that the activity of Fe species in filtered liquid is higher than Fe^{3+} ions. In addition, it was found that the active α-Fe_2O_3 phase in gas reaction shows no activity in the liquid-phase oxidation reaction.

13.3 Hydroxylation of Phenol to DHB

The HQ and CAT are two important isomers of DHB, which have been widely used as photographic developers, rubber and fat antioxidants, polymerization inhibitors, antiseptic and reducing agents, and pharmaceutical intermediates. The phenol hydroxylation reaction as an environmental-friendly route for producing diphenol is of industrial importance. DHB can be obtained through phenol hydroxylation using either homogeneous catalysts or heterogeneous catalysts at mild conditions with temperature of 50–100 °C. The use of metallic salts as homogeneous catalysts would induce additional pollution. With consideration of environmental problems based on the homogeneous reaction, the heterogeneous catalysts, particularly molecular sieve-based catalysts, are more attractive for researchers in the hydroxylation of phenol.

In general, trace amount of *p*-benzoquinone and CAT can be detected with only H_2O_2 in the absence of any catalyst [67]. At the same time, the higher selectivity to *p*-benzoquinone is obtained. Its proposed mechanism is shown in Figure 13.8.

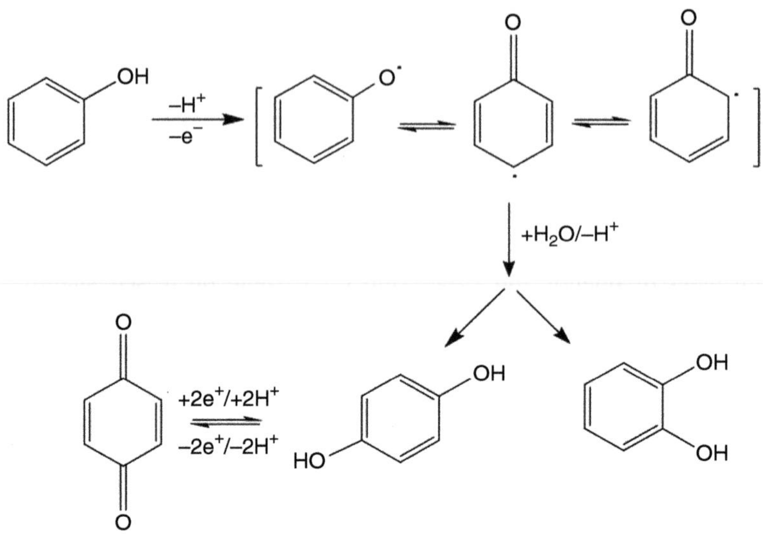

Figure 13.8 Scheme of phenol oxidation reaction process and major product without adding catalysts. Source: Reproduced from Ref. [67]/with permission of Springer Nature.

The phenolate anion is formed when losing a proton H^+, which promotes resonant stabilization to hydroxylate phenol at the *p*- and *o*-positions. Finally, the over-oxidation product of *p*-benzoquinone is observed.

13.3.1 Fe-Containing Molecular Sieves as Catalysts

13.3.1.1 Effect of Reaction Conditions

The reaction conditions had a considerable influence on the catalytic performance as reported by Choi et al. [68]. The induction period became more pronounced at lower reaction temperature using Fe-MCM-41 as catalysts. With the increase of temperature, the reaction time of induction period can be decreased. When the reaction temperature is below 10 °C, the phenol is less solubilized in solvent. The optimal reaction temperature is in the range of 60–90 °C. Nevertheless, the higher temperature will facilitate the decomposition of H_2O_2, leading to much higher amount of gaseous product. Besides, the conversion of phenol is highly dependent on solvents. The phenol hydroxylation (less than ca. 4%) hardly occurs with acetone or ethanol as a solvent because the hydroxyl radicals cannot be formed in these solvents. It is well-known that radicals are more stable in the polar solvents [69]. Therefore, the acetonitrile and water can be used. In this case, both phenol and H_2O_2 are easily dissolved, and active hydroxyl radicals are effectively formed upon Fe sites in water. The trend of phenol conversion was with the same trend of polarity of solvents: water > acetonitrile > acetone and ethanol. The amounts of Fe and H_2O_2 are related to the formation of hydroxyl radicals. It is simply concluded that the higher phenol conversion is obtained with relatively higher contents of Fe and H_2O_2, which represent plenty of active sites and radicals, respectively.

Furthermore, excessive Fe and H_2O_2 will cause the formation of Fe oxidation particles and competitive decomposition of H_2O_2 into H_2O and O_2 or probably deep oxidation of DHB. It was evidenced by the work with Adam and co-workers [70]. They synthesized zeolite L nanocrystals modified with Fe as catalyst for phenol hydroxylation reaction. It was found that the phenol conversion was increased from 64.61% to 93.40% when phenol to H_2O_2 molar ratio changed from 1 : 1 to 1 : 3. In contrast, the conversion decreased slightly to 90.66% at a molar ratio of 1 : 4.

13.3.1.2 Catalytic Recyclability

The catalyst reusability, which is concerningly responsible by Fe state, is an important evaluation index for phenol hydroxylation reaction. The Fe can be incorporated into the mesoporous material to form Fe^{3+} species. For example, Fe-MCM-48 synthesized by hydrothermal method exhibited phenol conversion of 43.6% and diphenol (CAT and HQ) selectivity of 97.7% with phenol/H_2O_2 ratio of 1 using water as solvent at 60 °C [71]. In addition, this material used in the third run gave phenol conversion about 70% of that in the first run due to the poor thermal stability for framework Fe atoms during removing the by-products through calcination. In order to improve the Fe species stability and catalyst reusability, the Fe atoms are introduced into the zeolite framework such ZSM-5, Beta, Y, and L. According to the previous researches in 1996, it was found that the phenol activity upon ZSM-5 decreased slowly after successive batch tests for four times at 70 °C and pH 3.5 [72]. Moreover, the activity of Fe-ZSM-5 can be restored again by calcination at 450 °C in air for 10 h. Similarly, the Fe/KL with Fe^{3+} in zeolite L lattice was synthesized under hydrothermal conditions [70]. It showed phenol conversion of 93.40% in 30 min in the presence of acetic acid and high selectivity to CAT and HQ of 77.47% and 22.53%, respectively, under the optimized condition (phenol/H_2O_2 = 1/3, 70 °C) because the formation of peracetic acid can serve as a better oxidizing agent. The catalytic activity decreases by 6.83% after consecutive runs using this catalyst, which was recovered via rinsing with distilled water and acetone (Figure 13.9). It suggested that Fe^{3+} substituted in zeolite frameworks can effectively alleviate the formation of Fe clusters through leaching from zeolite framework. This result was consolidated by Fe incorporated into the aluminophosphate framework, like Fe-APO-5, as catalyst for phenol hydroxylation [73, 74].

13.3.1.3 Improving Catalytic Performance by Developing Hierarchically Porous Structure

The introduction of hierarchical structure can eliminate the diffusion limitation and enhance the catalytic performance in the phenol hydroxylation due to the kinetic diameter of biphenyl close to the pore size of ZSM-5 zeolite. The hierarchical structures can be constructed through various methods: (1) mixing with mesoporous silica materials [75]; (2) adding the zeolite seeds or Fenton's reagent [60, 76]; and (3) introducing the soft-templates with hydrophobic alkyl chain groups or gemini-type cationic surfactants [77, 78]. For example, the single-unit-cell nanowires (NWs) of Fe-NW-ZSM-5 zeolite were prepared using diquaternary ammonium-type surfactant of $C_{12}H_{25}$-$N^+(CH_3)_2$-C_6H_{12}-$N^+(C_3H_7)_3$ ($C_{12\text{-}6}$) [78].

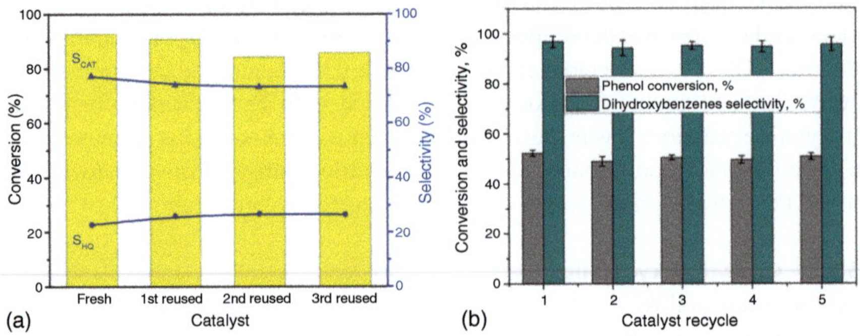

Figure 13.9 (a) Reusability and selectivities of CAT and HQ on Fe/KL under optimized reaction conditions: water 10 mL with three drops of acetic acid, phenol/H_2O_2 molar ratio 1 : 2, 70 °C, reaction time of 30 min. Source: Reproduced from Ref. [70]/with permission of Elsevier. (b) Stability of Fe-ZSM-5 zeolite in the hydroxylation of phenol reaction. Reaction conditions: water 10 mL, phenol/H_2O_2 molar ratio 1 : 1, catalyst 100 mg, 65 °C, reaction time of 4 h. Source: Reproduced from Ref. [75]/with permission of John Wiley & Sons.

The specific surface area and pore volume of this sample with three-dimensionally intergrown nanowire morphology were 645 $m^2\,g^{-1}$ and 1.02 $m^3\,g^{-1}$, respectively. It showed a higher catalytic activity with phenol conversion of 26.1% and DHB selectivity of 98.9% even at room temperature when the molar ratio of phenol to H_2O_2 was 3 : 1 (Table 13.4). In contrast, conventional Fe-ZSM-5 nanosheet (defined as Fe-NS-ZSM-5) gave phenol conversion of 23.9% and DHB selectivity of 93.2% at the same reaction condition (Table 13.4). This was due to the enhancement of diffusion properties for single-unit-cell nanowires. It was also found that increased amount of H_2O_2 can improve the phenol conversion to 69.3% but cause the selectivity of DHB decrease to 80.8% as a result of the formation of large amount of hydroxyl radicals, thus leading to further oxidation, which is in agreement with aforementioned results. In addition, this catalyst exhibited great stability and maintained good catalytic performance after five cycles even through calcination at 550 °C for 4 h after washing with ethanol. The hollow-structured Fe-ZSM-5 zeolite synthesized with gemini-type cationic surfactants under hydrothermal treatment exhibited phenol conversion of 52.2% and DHB selectivity of 96.8% at 65 °C (Figure 13.9), with the phenol/H_2O_2 ratio of 1 : 1 and water as the solvent [77]. Simultaneously, it showed great stability and the activity remained almost unchanged after five cycles due to accelerated mass transfer properties and accessibility to active sites.

13.3.1.4 Catalytic Mechanism

When the catalysts are introduced into the reaction system, the distribution of products, including *p*-benzoquinone, CAT, HQ, and [1,1′-biphenyl]-2,5′-diol, can be controlled. With acid catalysts, the phenoxy radicals are formed through resonance stabilization, resulting in the formation of benzoquinone via a consecutive reaction of CAT/HQ [81]. H-BEA, having larger pore size compared to MFI, can produce few amounts of condensed product, such as biphenyl as depicted in Figure 13.10, as a result of the sufficient space to access acid sites. In summary, the acid catalysts

Table 13.4 Catalytic results of Fe-substituted ZSM-5 for phenol hydroxylation.

Catalyst[a)]	Si/Fe[b)]	Phenol/H_2O_2[c)]	X_{Ph} (%)	$S^{d)}$ (%) CAT	HQ	S_{DHB} (%)
NS-ZSM-5[e)]	–	3 : 1	0	0	0	0
NW-ZSM-5[f)]	–	3 : 1	0	0	0	0
Fe-NS-ZSM-5	131	3 : 1	23.9	56.5	36.7	93.2
Fe-NW-ZSM-5	131	3 : 1	26.1	59.7	39.2	98.9
Fe-NW-ZSM-5	131	2 : 1	46.5	52.8	38.8	91.6
Fe-NW-ZSM-5	131	1 : 1	69.3	38.3	42.5	80.8
TS-1 (1-3)[g)]	50[f)]	3 : 1	23	52	46	98
FeCoNaβ[h)]	1.12[h)]	3 : 1	21.4	49.4	16.7	66.1

a) Reaction conditions: water, 10 mL; catalyst, 100 mg; molar ratio of phenol to H_2O_2, 3 : 1; reaction temperature, 25 °C; reaction time, 12 h.
b) Molar ratio of Si/Fe.
c) Molar ratio of phenol to H_2O_2.
d) CAT and HQ represent CAT and HQ, respectively.
e) NS represents nanosheet morphology.
f) NW represents nanowire morphology.
g) Ref. [79], reaction conditions: reaction temperature, 30 °C; molar ratio of phenol to H_2O_2, 3 : 1; 2 g of phenol; 60 g of water; 0.2 g of catalyst. Molar ratio of Si/Ti.
h) Ref. [80], reaction conditions: reaction temperature, 30 °C; molar ratio of phenol to H_2O_2, 3 : 1; 2 g of phenol; 60 g of water; 0.2 g of catalyst. Fe content by mass (wt%).
Source: Reproduced from Ref. [78]/with permission of Royal Society of Chemistry.

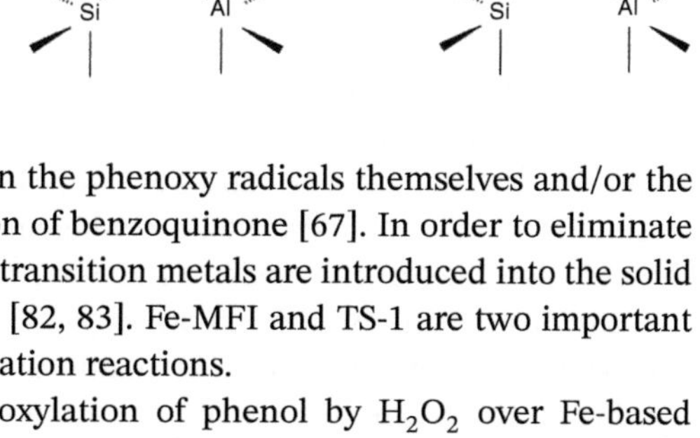

Figure 13.10 Phenoxy radicals on H-BEA surface. Source: Reproduced from Ref. [67]/with permission of Springer Nature.

usually enhance the reaction between the phenoxy radicals themselves and/or the consecutive reaction for the formation of benzoquinone [67]. In order to eliminate the over-oxidation reactions, various transition metals are introduced into the solid acid catalysts to form redox catalysts [82, 83]. Fe-MFI and TS-1 are two important types of catalysts for phenol hydroxylation reactions.

It is acknowledged that the hydroxylation of phenol by H_2O_2 over Fe-based catalysts is followed by a radical mechanism [67, 82]. In the presence of Fe^{2+} ions, the Fenton reaction proceeds resulting in the decomposition of hydrogen peroxide to produce the hydroxyl radicals [84, 85]. Then, these hydroxyl radicals react with phenol to attain the CAT and HQ molecules (Figure 13.11) accompanied

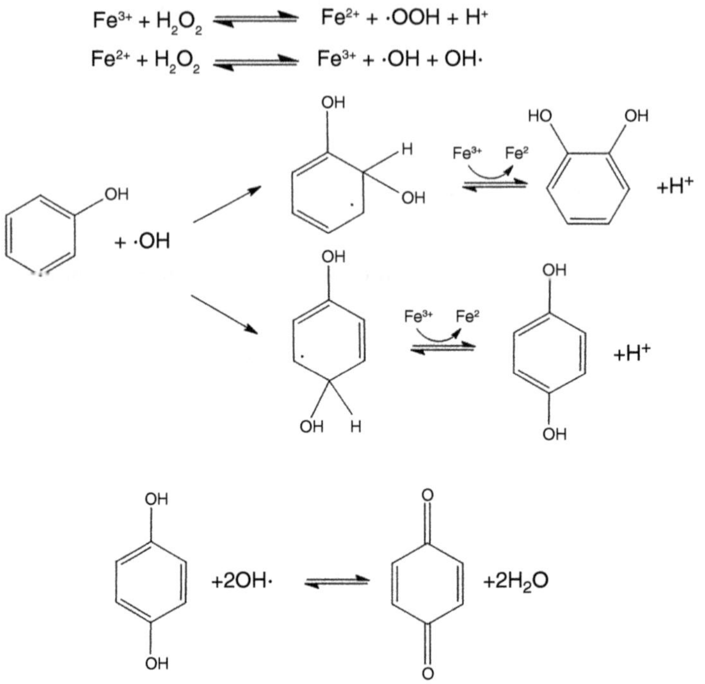

Figure 13.11 Scheme of proposed phenol hydroxylation reaction pathway followed by Fenton reaction involving H_2O_2 and Fe^{2+}/Fe^{3+}. Source: Reproduced from Ref. [67]/with permission of Springer Nature.

by the reduction from Fe^{3+} to Fe^{2+}. Tetrahedral Fe^{3+} incorporated into the ZSM-5 framework was also reported as active site for phenol hydroxylation to highly selectively generate DHB due to redox transition of Fe(III)/Fe(II) pairs [71]. The superoxide radical (·OOH) and Fe^{2+} are primarily created by the reaction of H_2O_2 with Fe^{3+}. The superoxide radical and Fe^{2+} consequently react to produce hydroxyl radical for the hydroxylation processes via a Fenton-like reaction [86]. Then, the reaction cycle can be repeated.

Loiha et al. synthesized Fe^{n+}-ZSM-5 through ion exchange method, including Fe^{2+}, Fe^{3+}, and equivalent molar of Fe^{2+}/Fe^{3+}. For comparison, bulk-iron oxides of α-Fe_2O_3 and α-FeOOH were synthesized via hydrothermal method [87]. It was found that the isolated Fe^{2+} and Fe^{3+} in ZSM-5 framework gave no catalytic conversion in the phenol hydroxylation, whereas both of bulk α-Fe_2O_3 and α-FeOOH showed high phenol conversion of about 75% at 90 min with phenol/H_2O_2 ratio of 1/2 in 60 mL water solution. For Fe^{2+}/Fe^{3+}-HZSM-5, the phenol conversion and CAT/HQ ratio were about 80% at 90 min and 2.4, respectively, and the turnover number (TON) was up to 1800 at 150 min. It is proposed that the iron-oxo (Fe—O) and iron-dioxo (O—Fe—O) can serve as the active sites. As shown in Figure 13.12, one way is that the iron-oxo species interact with hydrogen peroxide (H_2O_2) to produce a FeOOH intermediate. This species can further react with H_2O_2 to form hydroxyl radical. Then, the phenol reacts with FeOOH intermediate in the ortho

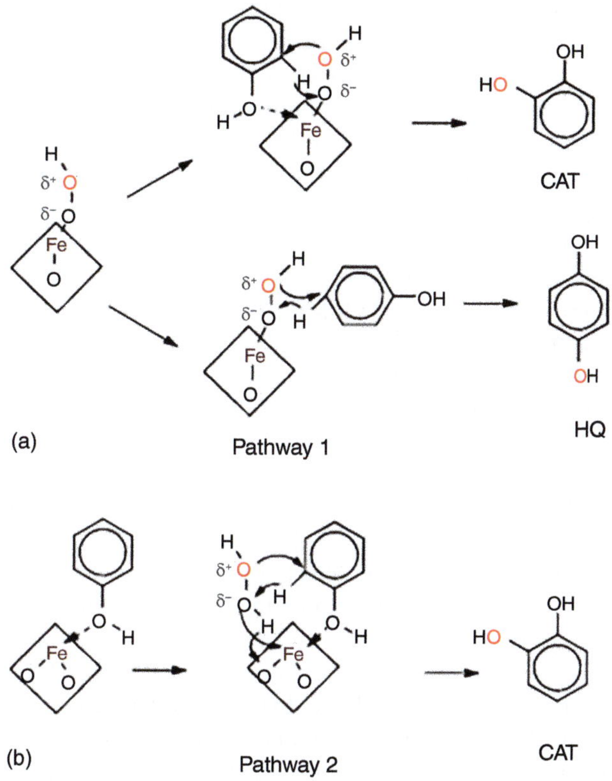

Figure 13.12 Scheme of proposed reaction mechanism for phenol hydroxylation on (a) iron-oxo and (b) iron-dioxo species. Source: Reproduced from Ref. [87]/with permission of Elsevier.

and para positions to obtain CAT and HQ, respectively (Figure 13.12 Pathway 1). Another way is that phenol molecules adsorb on the iron-dioxo species to form the phenoxo-iron complex and subsequently are oxidized to produce CAT by activated H_2O_2. The higher CAT/HQ ratio is attributed to large amount of iron-dioxo species with longer Fe—O bond distance that can promote the reaction pathway followed by Pathway 2 in Figure 13.12.

13.3.2 Titanosilicate Molecular Sieves as Catalysts

13.3.2.1 Effect of Pore Structure

TS-1 zeolite, which was first discovered by Taramasso et al. in 1983, can be used as an active catalyst in the epoxidation of propylene and higher α-olefins, ammoximation of cyclohexanone, and hydroxylation of phenol. Different from the Fe-MFI zeolite, this material provides chemoselective control of product formation due to space confined effect. Various aspects of phenol hydroxylation, such as the reaction mechanism, contribution of the external surface, effect of solvents, and diffusion of the reactants and products, have been deeply investigated. It is generally concluded that the HQ is preferentially formed on the internal pore channel of zeolite stemming from the confined effect, while the CAT is usually produced on the external surface. In 1991, Tuel et al. found that CAT was the major product and the reaction hardly proceeded over uncalcined TS-1 due to the diffusion limitation of phenol into the zeolite channels [9]. In contrast, the calcined TS-1 at 500 °C exhibited initial HQ selectivity of 60% and up to 80% at complete total H_2O_2 conversion with methanol as a solvent. Therefore, it is hypothesized that two different stages are distinguished: (1) phenol is firstly oxidized on the active sites of the external surface to produce CAT for a few minutes. This is attributed to a rapid poisoning of these sites via the formation of tar; (2) after a few minutes, oxidation process takes place inside the channels of the zeolite for the formation of HQ.

Wu and co-workers reported that the Ti-UZM-35 synthesized through modifying the highly dealuminated UZM-35 with $TiCl_4$ vapor showed excellent activity in hydroxylation phenol reaction [88]. They compared five types of zeolites, involving Ti-UZM-35, Ti-MCM-68, TS-1, Ti-MWW, and Ti-MOR, with tetrahedral Ti coordination state. Ti-UZM-35 and Ti-MCM-68, which have the same MSE framework topology, exhibited higher activity than other titanosilicalites due to their larger pore sizes. TS-1 gives lower total yield and the HQ selectivity than those of Ti-MSE zeolites, possibly attributed to the geometric constraints in 10-MR pore channels. Additionally, Ti-MWW and Ti-MOR produce extremely low total yield of 0.4% and 3%, respectively. It is probably because the Ti sites confined in MWW and MOR frameworks were not suitable for catalyzing the phenol hydroxylation reaction.

The micropores of TS-1 limit the diffusion of reactants to some extent because the active Ti species are primarily located inside zeolite. To solve this problem, the hierarchical pores are introduced into the zeolite structures through the alkaline treatment using TPAOH and/or NaOH. Treatment with alkaline solution of NaOH can selectively dissolve framework silica, resulting in the formation of mesopores in zeolite and increase of titanium content. According to the works by Guo group, the micrometer-sized umTS1 treated with only TPAOH solution gave a phenol conversion of 17.1% but a lower selectivity of DHB due to the hollow structures, compared to a phenol conversion of 1.6% for pristine umTS1 [89]. The catalytic performance can be further improved by successive NaOH post-treatment as a result of improvement of diffusivity and further exposing much higher amount of titanium sites. Similarly, they also synthesized mesoporous TS-1 through treatment with different alkaline solution such as LiOH, NaOH, and KOH [90]. Treating with cesium hydroxide caused MFI framework partially collapse. All desilicated TS-1 samples exhibit phenol conversion (18.4%) and DHB yield (17.9%) being two times as that of untreated samples (8.8%) with acetone as solvent (Table 13.5). The hierarchical TS-1 zeolite obtained using hard carbon templates gave significantly enhanced catalytic activity owing to rapid diffusion rate in the mesopores [8]. In order to conveniently collect the catalysts, Xiao's group synthesized bulky TS-1 zeolite (B-TS-1) with particle size of 20 μm by hydrothermal method and subsequently being collected via filtrating [91]. The B-TS-1 possessed rich mesopores (pore volume: $0.55\ cm^3\ g^{-1}$) due to the aggregation of small TS-1 crystals. It showed phenol conversion of 25%, comparable to that of 28% for conventional TS-1 zeolite,

Table 13.5 Catalytic performance of parent TS-1-NC and various alkaline treated TS-1 samples.[a)]

Catalyst	Si/Ti	X (%)	S_R (%)	CT/HQ	Product distribution (%)			
					CT	HQ	BQ	OCHD
TS1-NC	46.5	8.8	99.89	1.92	65.68	31.21	0	0.11
dLi-TS1-NC	36.0	17.0	99.37	1.62	61.44	37.93	0.02	0.61
dNa-TS1-NC	40.6	18.0	99.65	1.57	60.88	38.77	0.01	0.34
dK-TS1-NC	42.1	18.4	99.52	1.56	60.65	38.88	0.02	0.46
dCs-TS1-NC	34.6	18.4	99.05	1.06	50.97	48.08	0.67	0.28

a) Reaction conditions: phenol: H_2O_2 = 3 : 1, solvent: acetone, temperature: 60 °C, reaction time: 6 h.

Source: Reproduced from Ref. [90]/with permission of Springer Nature.

and high stability. The phenol conversion of recycled B-TS-1 zeolite remained 24% similar to that of fresh catalysts. Besides, when Sn was incorporated into the zeolite framework, the crystal size of TS-1 became smaller with higher acidity [92]. The acid–base sites (MgO-Al_2O_3 binary oxide) were also introduced into the micropores of hollow titanium silicalite (HTS, Ti/Si = 25) zeolite [93]. These materials exhibited good performance and high *p*-DHB selectivity in phenol dihydroxylation reaction.

13.3.2.2 Effect of Solvent

The solvent can alter the product distribution in hydroxylation of phenol through interacting with zeolite surface and/or dissolve the over-oxidation species. In the above case, replacing methanol with acetone, that cannot be oxidized by H_2O_2, can effectively dissolve the tar species to make the zeolite external surface clean, resulting in the exposed active sites again. Thus, selectivities to CAT and HQ remained at 53% and 47% (vs. 35% and 65% in the methanol as solvent) during the whole reaction process [9]. It implied that acetone prevented the rapid deposition of the tar species and irreversible poisoning of the surface, thus changing the product distributions.

However, the reaction mechanism is ambiguous and not discussed in detail before. Wilkenhöner and co-workers further pay much attention to the influences of pore size and different solvents (including water, acetone, and methanol) on the phenol hydroxylation reaction mechanisms [94]. The phenol (k_p) and hydrogen peroxide (k_d) consumption rate constant are summarized. In addition, their rate constants on the internal and external surfaces are also calculated. It was found that the catalytic activity of TS-1 (expressed as k_p) in methanol and HQ/CAT ratio were significantly higher than that obtained using acetone as solvent. It indicated that the reactions in acetone occurred on the external surface to a larger extent. This was in agreement with Tuel's results. With water as the solvent, the highest catalytic activity, which was 10 times higher than those in methanol and acetone, was obtained. According to the amount of tar formed, the trend is followed by water > methanol > acetone. The decrease of tar amount is due to tar components or precursors are more soluble in the acetone than in the water.

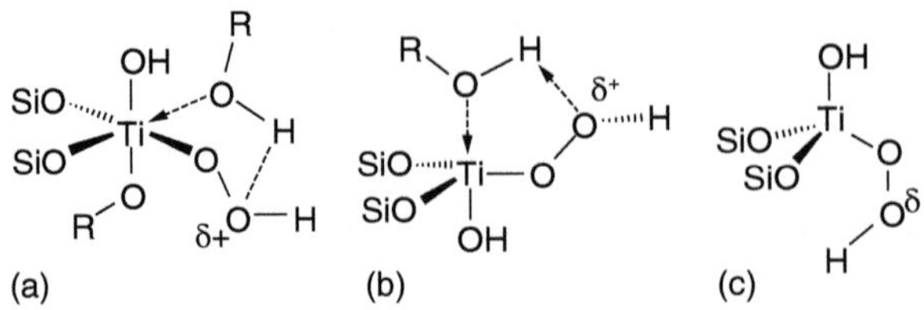

Figure 13.13 Scheme of proposed configurations of hydroperoxo-titanium active sites: (a) hexa-coordinate octahedral, (b) penta-coordinate trigonal bipyramidal, and (c) tetra-coordinate tetrahedral. Source: Reproduced from Ref. [94]/with permission of Elsevier.

The mechanism is related to the protonic properties of solvents. Figure 13.13 shows the possible configurations of Ti without (species c) and with coordination of 1 or 2 protic solvent molecules (species a and b). Generally, the protic solvents coordinate with Ti in the zeolite framework for the formation of coordination sphere to 5 or 6, along with the increased average Ti—O bond length. For example, a five-membered ring with hydrogen bonds between methanol and the peroxo group at Ti site is proposed as the active intermediate complex. This expanding structure causes the Ti species size increased, leading to a geometric constraint for an approaching phenol molecule. Therefore, the phenolic OH group will be away from the bulky Ti species resulting in the formation of HQ. In addition, the interaction between the protic molecules and peroxo groups leads to the formation of active sites, and simultaneously destabilizing H-bonding with the phenol molecules in the ortho-positions. Besides, there are two reaction pathways for the formation of CAT in non-protic solvent, such as acetone. One way is that the phenol molecules play a similar role as the protic solvent. The Ti peroxo species (Figure 13.13: species 3) can interact with phenol to yield CAT through forming penta-coordinated (trigonal bipyramidal) Ti sites. Another pathway is that six-membered transition state is formed between the Ti peroxo species and phenol resulting in the formation of CAT. In summary, HQ is preferably formed inside the internal channel under limiting geometric conditions in protic solvent. In contrast, the CAT is predominantly produced on the outer surfaces due to less restricted geometric Ti sites. Furthermore, the reaction mechanisms influenced by solvents are proved by Yin and co-workers. The phenol hydroxylation proceeded with various solvents such as MeOH (methanol), EtOH (ethanol), iso-PrOH (iso-propanol), t-BuOH (t-butanol), water, and MeCN (methyl cyanide) [88]. The highest total yield and *p*-selectivity of 48.9% and 92.7% were obtained in the EtOH over Ti-UZM-35 with Si/Ti ratio of 70. MeOH gave the similar results with EtOH as a solvent. When EtOH was replaced by high-carbon alcohols like *iso*-PrOH and *t*-BuOH, they produced much lower total yield with *p*-selectivity of ca. 92% due to large-sized alcohols inside the pore channels causing more severe constrains for phenol molecules. When water was used as solvent, lower para-selectivity was obtained compared to EtOH, because for hydrophobic Ti-UZM-35, it is difficult for water to bring the phenol molecules diffusing into the channels, leading to the reaction occurred on the external surface. Additionally, the Ti-UZM-35 in MeCN showed very high but lower total yield and *p*-selectivity than that in EtOH because the strong coordination between the —CN and Ti sites weakened catalytic activity of Ti sites.

13.3.2.3 Effect of Reaction Conditions

The moderate reaction conditions are also important for improving the catalytic activities in phenol hydroxylation. It is well-known that the elevation of reaction temperature and prolongation of reaction time can enhance the activity. The Ti-UZM-35 showed the increase of total product yield from 41.8% at 60 °C to 48.9% at 70 °C, whereas the *p*-selectivity was decreased from 95.1% at 60 °C to 90.3% at 90 °C [88]. This is probably because HQ is prone to form tar in the presence of H_2O_2 at high temperature. In addition, the color of reaction solution changed from brown to dark, indicating that much more tars were formed at relatively high temperatures. Another problem is that high temperature causes an increased H_2O_2 decomposition rate and further results in lower H_2O_2 utilization [95].

The molar ratio of phenol to H_2O_2 influences the reaction pathway. At low amount of phenol in the system, excessive H_2O_2 can lead to the over-oxidation for the formation of benzoquinone. However, the increase of phenol proportion results in higher total product but lower *p*-selectivity. In addition, the H_2O_2 feeding style has a great influence on the catalytic activity. The slow addition of H_2O_2 into the reaction system can not only suppress the over oxidation but also increase the selectivity to desired product as well as the utilization of efficiency of H_2O_2. It is needed to point out that the HQ selectivity is almost not influenced by the reaction conditions, such as phenol/H_2O_2 ratio, and reaction time except the catalyst amount. The selectivity of HQ was slightly increased from 50% to 54 mol% with the content of Ti-MWW catalyst increased from 3 to 15 wt% under phenol/H_2O_2 ratio of 3 at 100 °C for 120 min as reported by Wróblewska et al. [96]. Simultaneously, higher catalyst content promotes the ineffective decomposition of hydrogen peroxide to water and oxygen, as well as oxidation reaction of DHB to *p*-benzoquinone and subsequently to tars (mainly from pyroCAT).

Wang and co-workers investigated the chemical kinetics of hydroxylation of phenol using TS-1/diatomite as the catalyst [95]. With the inlet phenol concentration varied ranging from 1.989 to 4.973 mol L^{-1} at different reaction temperatures, the conversion of phenol remained unchanged. It was found that both the conversion of phenol and the selectivity of DHB increased with the reaction temperature increased. Moreover, the conversion of H_2O_2 and DHB selectivity greatly increases with the increase of the phenol concentration, and the rate of phenol hydroxylation was near first order with respect to the concentration of phenol. Similarly, the effects of the H_2O_2 concentration and reaction temperature on the hydroxylation of phenol are deeply studied. With the increase of H_2O_2 concentration and temperature, the rate of the hydroxylation of phenol increased, despite that the increased degree gradually decreased. It suggested that the reaction rate was between zero order and first order with respect to the concentration of hydrogen peroxide. In other words, it is about zero order at higher H_2O_2 concentration and about first order at lower

H_2O_2 concentration. The proposed kinetic equations of hydroxylation of phenol are listed as follows:

Hydroxylation of phenol: $r_i = k_0 e^{-(E/RT)} C_p{}^a C_{H2O2}{}^b = 7.18 \times 103\ e^{-(42.0/RT)}\ C_p{}^{1.06}\ C_{H2O2}{}^{0.34}$

13.3.2.4 Stability of Ti Species

As we all known, the Ti species states have a significant influence on the reaction pathway. Generally, the framework Ti is the active site for phenol hydroxylation reaction. In the past, all the mesoporous materials, such as amorphous titanosilicates [97], Ti-SBA-12 [7], Ti-MCM-41, and Ti-MTS-9 [98], showed low catalytic activity in phenol hydroxylation. For examples, Ti-MCM-41 showed phenol conversion of only 2.5% [98]. In addition, Ti in mesoporous material was easy to migrate from the framework to external surface resulting in its deactivation. In order to improve the thermal stability of Ti species, an effective way is the incorporation of titanium atoms into zeolite framework. Thangaraj and co-workers found that about eight titanium atoms can exist in a TS-1 unit cell [99]. There are many methods to synthesize TS-1, including (1) adding the small molecules like $(NH_4)_2CO_3$ and hexanediamine (C_6DN) into the synthesis gel [100, 101], (2) post-treated by chemical vapor of $TiCl_4$ [102], (3) wetness impregnation method [103], and (4) introduction of non-ionic surfactants in the synthesis system [104]. The critical factor to increase the content of framework Ti is to match the crystallization rate well with the hydrolysis rate of silicon source and titanium source. Fan et al. reported that Ti-rich TS-1 could be synthesized by adding the $(NH_4)_2CO_3$ into the synthesis gel [63]. It was found that the $(NH_4)_2CO_3$, as crystallization-mediating agent, not only decreased the pH of gel but also slowed down the crystallization process to harmonize the crystallization rate for incorporation of titanium into the framework of TS-1. This material showed good catalytic activity in oxidation of 1-hexene and 2-hexanol reactions. Furthermore, Han and co-workers investigated that effect of $(NH_4)_2CO_3$ on the synthesis of TS-1 zeolite. When the molar ratio of $(NH_4)_2CO_3$/Si was 0.3, the Si/Ti ratio reached 34.43 [100]. This Ti-rich TS-1 was further treated with TPAOH to form hierarchical Ti-rich TS-1 zeolite. It showed phenol conversion of 27.45%, DHB selectivity of 99.50%, and H_2O_2 utilization of 92.54% at 60 °C with phenol/H_2O of 3 : 1 for 6 h, compared to 17.58%, 91.01%, and 65.93%, respectively, for conventional TS-1 with Si/Ti ratio of 42.82. Similarly, glycine and starch can also decrease the pH value of synthesis gel, leading to high content of Ti in zeolite framework [105, 106]. The Ti content in the ERB-1 framework was increased by a sequential treatment of ERB-1 with deboronation and atom-planting method [102]. The mechanism was proposed that the silanol hydroxyl nests were firstly formed during the deboronation process and subsequently consumed by the dehydrochlorination of $TiCl_4$ for improving the Ti content. This material gave an excellent performance in the epoxidation of 1-hexene with H_2O_2 reaction. The TS-1 zeolite post-treated with NH_4HF_2 and H_2O_2 at 80 °C exhibited a high catalytic activity in the benzene hydroxylation due to the increase of Ti content in zeolite [59]. In addition, the increase of Ti species can also change *p*- and *o*-DHB ratios in the products [101]. The Si/Ti ratio of 76 was obtained with C_6DN to SiO_2 ratio of 0.3, corresponding to Ti atom number of 5.367 ± 2/u.c., although its synthesis ratio was 100 in the gel. This material exhibits a *p*/*o* ratio up to 2.5, being higher than that of conventional TS-1 sample ($p/o = 1.9$) with Si/Ti ratio of 88.

13.3.3 Transition Metal Complexes Encapsulated in Zeolite Y as Catalysts

Apart from the above catalysts, the encapsulations of Cu(II), Co(II), Zn(II), Ni(II), Cr(III), Bi(III), Fe(III), and oxovanadium(IV) coordinated by complicated ligands in large-pore Y zeolite are used to prepare catalysts for the phenol hydroxylation reaction to improve the CAT selectivity by weakening shape selectivity. When the various metals (including Cu(II), Ni(II), Zn(II), Cr(III), Bi(III), Fe(III), and V(IV)O) supported on the zeolite Y were synthesized through ion-exchange method without the coordination with ligands, the highest phenol conversion (36%) and CAT/HQ ratio (> 10) were obtained using OV-Y catalyst with 100% H_2O_2 efficiency [107]. The highest TOF value of 1226 h^{-1} per catalytic site after 6 h reaction time was achieved over Zn-Y catalyst. However, the NaY zeolite with Si/Al ratio of 10 without any metal showed a very poor activity. In order to improve the catalytic activity, the oxovanadium tetraaza complexes were introduced into the Y zeolite [108]. The synthesis processes are listed as followed: adsorption of [bis(diamine)VO(IV)] (diamine = 1,2-diaminoethane, 1.3-diaminopropane, 1,2-diaminobenzene, 1,3-diaminobenzene); $[VO(N\text{-}N_2)]^{2+}$-NaY; in the nanopores of the zeolite-Y; subsequently in situ condensation of oxovanadium(IV) precursor complex with ethylcinnamate to form $[VO([4]aneN_4)]^{2+}$-NaY, $[VO([16]aneN_4)]^{2+}$-NaY, $[VO(Bzo_2[14]aneN_4)]^{2+}$-NaY, and $[VO(Bzo_2[14]aneN_4)]^{2+}$-NaY. The $[VO(Bzo_2[14]aneN_4)]^{2+}$-NaY showed a higher phenol conversion of 43.2% with H_2O_2/phenol ratio of 1 in CH_3CN at 80 °C for 6 h, compared to the traditional ion-exchanged samples. Similarly, the OV-Y obtained by ion-exchange method was mixed with various salen molecule ligands, such as N,N′-bis(salicylidene)ethane-1,2-diamine (H_2salen), N,N′-bis(salicylidene)propane-1,3-diamine (H_2sal-1,3-pn), N,N′-bis(salicylidene) propane-1,2-diamine (H_2sal-1,2-pn), and N,N′-bis(salicylidene)diethylenetriamine (H_2saldien) to form the structures with encapsulation of oxovanadium complexes [109]. These materials showed obviously improved TOF values, in which the [VO(salen)]-Y gave phenol conversion of 32.6% and CAT selectivity of 92.7% with highest TOF value of 1940 h^{-1} in MeCN at 80 °C for 6 h. This is attributed to peroxo-coordinated intermediate that can be effectively formed for transferring oxygen to the substrate, further producing much more CAT than HQ. In addition, the Y-zeolite encapsulated Cu(II) and Co(II) coordinated with 1,2-bis(2-hydroxybenzamido) ethane (H_2hybe), bis(picolinato) ($[Pic]_2$), or 2-phenylbenzimidazole (PhBzlH) catalysts also exhibited high catalytic activity and CAT selectivity in phenol hydroxylation [110–112]. It is proposed that H_2O_2 can interact with transition metal complexes with an electron transferring from the occupied metal *d*-orbital to the π^*-orbital of H_2O_2; meanwhile, the bis(picolinato) Cu(II) and Co(II) that are 15- and 17-electron species, respectively, provide electron back-donation through forming $[M(Pic)_2]\cdot H_2O_2$, thus showing great catalytic activity.

13.4 Summary

In this chapter, the benzene and phenol hydroxylation reactions are systematically summarized. The phenol is obtained through gas-phase benzene hydroxylation reaction upon Fe-containing zeolite at 300–450 °C. The different proposed reaction mechanisms are discussed. Therein, the reaction mechanism is proposed as followed: (1) the decomposition of N_2O to form active "α-oxygen" sites on Fe species; (2) the formation of benzene complex at iron active site; (3) the hydrogen abstraction from benzene; (4) the recombination between OH and C_6H_5 ligands; (5) the release of product phenol. Oligomer Fe species and/or dinuclear O-Fe(II)-O rather than framework Fe are proved as active sites. To improve the activity, various post-treatment methods, like steaming treatment, are developed to create higher amount of active Fe species. The effects of reaction conditions on the catalytic activities are also discussed. In addition, the hierarchical structures lead to improvement of the activity and catalytic lifetime on Fe-containing zeolites.

The liquid-phase hydroxylation of benzene to phenol and phenol to bisphenol are performed using Fe- or Ti-zeolite catalyst with the assistance of H_2O_2. Compared to gas-phase reaction, it not only leads to high selectivity to phenol but also proceeds under mild conditions. The reaction mechanism is highly dependent on the types of active sites. Fe species reacts with H_2O_2 followed by Fenton reaction resulting in decomposition of hydrogen peroxide, thus producing the hydroxyl radicals. These radicals can react with benzene to generate the phenol.

Similarly, the phenol molecules are hydroxylated by this way to produce equal amounts of CAT and HQ on Fe-based catalysts. However, the biphenol distribution upon Ti-based catalysts is considerably related to the solvents in the reaction system. The protic solvents coordinated with Ti in the zeolite framework to form the coordination sphere of 5 or 6 as active intermediate complex. This expanding structure causes the active Ti species size increased, leading to a geometric restriction for an approaching phenol molecule, resulting in the formation of HQ. For aprotic solvents, the phenol molecules play a similar role of the protic solvent. The Ti peroxo species can interact with phenol to yield CAT through forming penta-coordinated (trigonal bipyramidal) Ti sites. The six-membered-ring transition state is formed between the Ti peroxo species and phenol, resulting in the formation of CAT. It is concluded that HQ is preferably formed inside the internal channel under limiting geometric conditions in protic solvent, whereas the CAT is preferably produced on the outer surfaces over less-restricted geometric Ti sites. Therefore, the CAT/HQ ratio can be adjusted through controlling the location of Ti active sites. In addition, it is found that the high selectivity to CAT can be obtained using Y zeolite encapsulated with VO or Cu species coordinated with complicated complex as catalysts.

The high selectivity to a specific product is one goal for researchers although phenol hydroxylation reaction has been industrially used in large scale. The catalysts should be delicately designed according to the existed theory and result. The reaction mechanism also needs to be deeply investigated with advanced characterization techniques and theoretic calculations. It would provide not only a change in industry technology but also economic benefit.

With the development of the recent progresses in benzene and phenol hydroxylation to phenol and DHB, respectively, the high selectivity to phenol in gas-phase reaction on Fe-containing zeolite and adjustable product distribution of CAT and HQ on Ti-containing zeolite are almost realized via changing the acid properties, designing the catalyst structure and tuning the active sites and so on. It is of a great challenge for feasibility of industrial implementation in large scale of oxidation hydroxylation of benzene and phenol in an environmental and commercial way. Fe- and Ti-containing zeolites provide a possibility to achieve these reaction processes, as well as to expand them to be applied in other oxidation reactions.

Acknowledgments

The authors are grateful to the financial supports of the National Key R&D Program of China (2020YFB0606402), the National Natural Science Foundation of China (21991090, 21991092, 22272195, U1910203, and U22A20431), the Natural Science Foundation of Shanxi Province of China (202203021224009), the Youth Innovation Promotion Association CAS (2021172), and project funded by the China Postdoctoral Science Foundation (2022TQ0351).

References

1 Panov, G.I., Kharitonov, A.S., and Sobolev, V.I. (1993). Oxidative hydroxylation using dinitrogen monoxide—a possible route for organic-synthesis over zeolites. *Appl. Catal. A-Gen.* 98: 1–20.

2 Zhu, M., Liu, Y., Yao, Y. et al. (2018). Preparation and catalytic performance of Ti-MWW zeolite membrane for phenol hydroxylation. *Microporous Mesoporous Mater.* 268: 84–87.

3 Zhang, F., Chen, X., Zhuang, J. et al. (2011). Direct oxidation of benzene to phenol by N_2O over meso-Fe-ZSM-5 catalysts obtained via alkaline post-treatment. *Catal. Sci. Technol.* 1: 1250–1255.

4 Yoshizawa, K., Shiota, Y., Yumura, T., and Yamabe, T. (2000). Direct methane-methanol and benzene-phenol conversions on Fe-ZSM-5 zeolite: theoretical predictions on the reaction pathways and energetics. *J. Phys. Chem. B* 104: 734–740.

5 Ivanov, D.P., Pirutko, L.V., and Panov, G.I. (2014). Effect of steaming on the catalytic performance of ZSM-5 zeolite in the selective oxidation of phenol by nitrous oxide. *J. Catal.* 311: 424–432.

6 Li, L., Meng, Q., Wen, J. et al. (2016). Improved performance of hierarchical Fe-ZSM-5 in the direct oxidation of benzene to phenol by N_2O. *Microporous Mesoporous Mater.* 227: 252–257.

7 Kumar, A. and Srinivas, D. (2013). Hydroxylation of phenol with hydrogen peroxide catalyzed by Ti-SBA-12 and Ti-SBA-16. *J. Mol. Catal. A-Chem.* 368: 112–118.

8 Xin, H., Zhao, J., Xu, S. et al. (2010). Enhanced catalytic oxidation by hierarchically structured TS-1 zeolite. *J. Phys. Chem. C* 114: 6553–6559.

9 Tuel, A., Moussakhouzami, S., Bentaarit, Y., and Naccache, C. (1991). Hydroxylation of phenol over TS-1 surface and solvent effects. *J. Mol. Catal.* 68: 45–52.

10 Thangaraj, A., Kumar, R., and Ratnasamy, P. (1991). Catalytic properties of crystalline titanium silicalites: 2. Hydroxylation of phenol with hydrogen-peroxide over TS-1 zeolites. *J. Catal.* 131: 294–297.

11 Suzuki, E., Nakashiro, K., and Ono, Y. (1988). Hydroxylation of benzene with dinitrogen monoxide over H-ZSM-5 zeolite. *Chem. Lett.* 953–956.

12 Panov, G.I., Sheveleva, G.A., Kharitonov, A.S. et al. (1992). Oxidation of benzene to phenol by nitrous-oxide over Fe-ZSM-5 zeolites. *Appl. Catal. A Gen.* 82: 31–36.

13 Shibata, Y., Hamada, R., Ueda, T. et al. (2005). Gas-phase catalytic oxidation of benzene to phenol over Cu-impregnated HZSM-5 catalysts. *Ind. Eng. Chem. Res.* 44: 8765–8772.

14 Iwamoto, M., Hirata, J., Matsukami, K., and Kagawa, S. (1983). Catalytic-oxidation by oxide radical ions: 1. One-step hydroxylation of benzene to phenol over group-5 and group-6 oxides supported on silica-gel. *J. Phys. Chem.* 87: 903–905.

15 Meloni, D., Monaci, R., Solinas, V. et al. (2003). Activity and deactivation of Fe-MFI catalysts for benzene hydroxylation to phenol by N_2O. *J. Catal.* 214: 169–178.

16 Meng, L., Zhu, X., and Hensen, E.J.M. (2017). Stable Fe/ZSM-5 nanosheet zeolite catalysts for the oxidation of benzene to phenol. *ACS Catal.* 7: 2709–2719.

17 Zhu, Q., van Teeffelen, R.M., van Santen, R.A., and Hensen, E.J.M. (2004). Effect of high-temperature treatment on Fe/ZSM-5 prepared by chemical vapor deposition of $FeCl_3$: II. Nitrous oxide decomposition, selective oxidation of benzene to phenol, and selective reduction of nitric oxide by isobutane. *J. Catal.* 221: 575–583.

18 Yuranov, I., Bulushev, D.A., Renken, A., and Kiwi-Minsker, L. (2004). Benzene hydroxylation over FeZSM-5 catalysts: which Fe sites are active? *J. Catal.* 227: 138–147.

19 Ivanov, A.A., Chernyavsky, V.S., Gross, M.J. et al. (2003). Kinetics of benzene to phenol oxidation over Fe-ZSM-5 catalyst. *Appl. Catal. A-Gen.* 249: 327–343.

20 Panov, G.I., Sobolev, V.I., and Kharitonov, A.S. (1990). The role of iron in N_2O decomposition on ZSM-5 zeolite and reactivity of the surface oxygen formed. *J. Mol. Catal.* 61: 85–97.

21 Kharitonov, A.S., Sheveleva, G.A., Panov, G.I. et al. (1993). Ferrisilicate analogs of ZSM-5 zeolite as catalysts for one-step oxidation of benzene to phenol. *Appl. Catal. A-Gen.* 98: 33–43.

22 Jia, J.F., Pillai, K.S., and Sachtler, W.M.H. (2004). One-step oxidation of benzene to phenol with nitrous oxide over Fe/MFI catalysts. *J. Catal.* 221: 119–126.

23 Dubkov, K.A., Ovanesyan, N.S., Shteinman, A.A. et al. (2002). Evolution of iron states and formation of alpha-sites upon activation of FeZSM-5 zeolites. *J. Catal.* 207: 341–352.

24 Ribera, A., Arends, I., de Vries, S. et al. (2000). Preparation, characterization, and performance of FeZSM-5 for the selective oxidation of benzene to phenol with N_2O. *J. Catal.* 195: 287–297.

25 Burch, R. and Howitt, C. (1992). Direct partial oxidation of benzene to phenol on zeolite catalysts. *Appl. Catal. A-Gen.* 86: 139–146.

26 Burch, R. and Howitt, C. (1993). Factors affecting the deactivation of various zeolites used as catalysts for the direct partial oxidation of benzene to phenol. *Appl. Catal. A-Gen.* 106: 167–183.

27 Burch, R. and Howitt, C. (1993). Investigation of zeolite catalysts for the direct partial oxidation of benzene to phenol. *Appl. Catal. A-Gen.* 103: 135–162.

28 Koekkoek, A.J.J., Kim, W., Degirmenci, V. et al. (2013). Catalytic performance of sheet-like Fe/ZSM-5 zeolites for the selective oxidation of benzene with nitrous oxide. *J. Catal.* 299: 81–89.

29 Rana, B.S., Singh, B., Kumar, R. et al. (2010). Hierarchical mesoporous Fe/ZSM-5 with tunable porosity for selective hydroxylation of benzene to phenol. *J. Mater. Chem.* 20: 8575–8581.

30 Gopalakrishnan, S., Zampieri, A., and Schwieger, W. (2008). Mesoporous ZSM-5 zeolites via alkali treatment for the direct hydroxylation of benzene to phenol with N_2O. *J. Catal.* 260: 193–197.

31 Koekkoek, A.J.J., Xin, H., Yang, Q. et al. (2011). Hierarchically structured Fe/ZSM-5 as catalysts for the oxidation of benzene to phenol. *Microporous Mesoporous Mater.* 145: 172–181.

32 Kustov, L.M., Tarasov, A.L., Bogdan, V.I. et al. (2000). Selective oxidation of aromatic compounds on zeolites using N_2O as a mild oxidant – a new approach to design active sites. *Catal. Today* 61: 123–128.

33 Zholobenko, V.L., Senchenya, I.N., Kustov, L.M., and Kazanskii, V.B. (1991). Complexation and decomposition of N_2O on bronsted acid sites in high-silicon zeolites – spectral and quantum chemical study. *Kinet. Catal.* 32: 132–137.

34 Notte, P.P. (2000). The AlphOx^{TM} process or the one-step hydroxylation of benzene into phenol by nitrous oxide: understanding and tuning the ZSM-5 catalyst activities. *Top. Catal.* 13: 387–394.

35 Motz, J.L., Heinichen, H., and Holderich, W.F. (1998). Direct hydroxylation of aromatics to their corresponding phenols catalysed by H-[Al] ZSM-5 zeolite. *J. Mol. Catal. A-Chem.* 136: 175–184.

36 Hensen, E.J.M., Zhu, Q., Janssen, R.A.J. et al. (2005). Selective oxidation of benzene to phenol with nitrous oxide over MFI zeolites. 1. On the role of iron and aluminum. *J. Catal.* 233: 123–135.

37 Hensen, E.J.M., Zhu, Q., and van Santen, R.A. (2003). Extraframework Fe-Al-O species occluded in MFI zeolite as the active species in the oxidation of benzene to phenol with nitrous oxide. *J. Catal.* 220: 260–264.

38 Xia, H., Sun, K., Fan, F. et al. (2008). Effect of extra-framework gallium on the structure of iron species in Fe/ZSM-5. *J. Catal.* 259: 269–275.

39 Pirutko, L.V., Chernyavsky, V.S., Uriarte, A.K., and Panov, G.I. (2002). Oxidation of benzene to phenol by nitrous oxide—activity of iron in zeolite matrices of various composition. *Appl. Catal. A-Gen.* 227: 143–157.

40 Centi, G., Perathoner, S., Arrigo, R. et al. (2006). Characterization and reactivity of Fe-Al,B)MFI catalysts for benzene hydroxylation with N_2O. *Appl. Catal. A-Gen.* 307: 30–41.

41 Yoshizawa, K., Shiota, Y., and Kamachi, T. (2003). Mechanistic proposals for direct benzene hydroxylation over Fe-ZSM-5 zeolite. *J. Phys. Chem. B* 107: 11404–11410.

42 Ryder, J.A., Chakraborty, A.K., and Bell, A.T. (2003). Density functional theory study of benzene oxidation over Fe-ZSM-5. *J. Catal.* 220: 84–91.

43 Selli, E., Rossetti, I., Meloni, D. et al. (2004). Effect of surface acidity on the behaviour of Fe-MFI catalysts for benzene hydroxylation to phenol. *Appl. Catal. A-Gen.* 262: 131–136.

44 Waclaw, A., Nowinska, K., and Schwieger, W. (2004). Benzene to phenol oxidation over iron exchanged zeolite ZSM-5. *Appl. Catal. A-Gen.* 270: 151–156.

45 Sayyar, M.H. and Wakeman, R.J. (2008). Comparing two new routes for benzene hydroxylation. *Chem. Eng. Res. Des.* 86: 517–526.

46 Guo, Y., Zhang, X., Zou, H. et al. (2009). Pd-silicalite-1 composite membrane for direct hydroxylation of benzene. *Chem. Commun.* 5898–5900.

47 Niwa, S., Eswaramoorthy, M., Nair, J. et al. (2002). A one-step conversion of benzene to phenol with a palladium membrane. *Science* 295: 105–107.

48 Wang, X., Tan, X., Meng, B. et al. (2013). One-step hydroxylation of benzene to phenol via a Pd capillary membrane microreactor. *Catal. Sci. Technol.* 3: 2380–2391.

49 Wang, X., Guo, Y., Zhang, X. et al. (2010). Catalytic properties of benzene hydroxylation by TS-1 film reactor and Pd-TS-1 composite membrane reactor. *Catal. Today* 156: 288–294.

50 Wang, X., Zou, Y., Meng, B. et al. (2018). Catalytic palladium membrane reactors for one-step benzene hydroxylation to phenol. *J. Membr. Sci.* 563: 864–872.

51 Einaga, H., Teraoka, Y., and Ogata, A. (2013). Catalytic oxidation of benzene by ozone over manganese oxides supported on USY zeolite. *J. Catal.* 305: 227–237.

52 Ye, X., Cui, Y., Qiu, X., and Wang, X. (2014). Selective oxidation of benzene to phenol by Fe-CN/TS-1 catalysts under visible light irradiation. *Appl. Catal. B-Environ.* 152: 383–389.

53 Yi, R., Bai, S., and Bao, Z. (2016). Preparation of Pd/MCM-41 and its photocatalytic performance for benzene hydroxylation. *J. Mol. Catal. (China)* 30: 583–593.

54 Shen, X., Wang, J., Liu, M. et al. (2019). Preparation of the hierarchical Ti-rich TS-1 via tritonx-100-assisted synthetic strategy for the direct oxidation of benzene. *Catal. Lett.* 149: 2586–2596.

55 Qi, X., Li, J., Ji, T. et al. (2009). Catalytic benzene hydroxylation over copper-substituted aluminophosphate molecular sieves (CuAPO-11). *Microporous Mesoporous Mater.* 122: 36–41.

56 Guo, B., Zhu, L., Hu, X. et al. (2011). Nature of vanadium species on vanadium silicalite-1 zeolite and their stability in hydroxylation reaction of benzene to phenol. *Catal. Sci. Technol.* 1: 1060–1067.

57 Xiao, P., Wang, Y., Kondo, J.N., and Yokoi, T. (2019). Consequences of Fe speciation in MFI zeolites for hydroxylation of benzene to phenol with H_2O_2. *Appl. Catal. A-Gen.* 579: 159–167.

58 Neidig, M.L. and Hirsekorn, K.F. (2011). Insight into contributions to phenol selectivity in the solution oxidation of benzene to phenol with H_2O_2. *Catal. Commun.* 12: 480–484.

59 Bianchi, D., Balducci, L., Bortolo, R. et al. (2007). Oxidation of benzene to phenol with hydrogen peroxide catalyzed by a modified titanium silicalite (TS-1B). *Adv. Synth. Catal.* 349: 979–986.

60 Yue, M., Jiang, X., Zhang, H. et al. (2020). Quasi-solid-phase synthesis of Fe-MFI zeolites by using Fe-containing zeolite seed sol for hydroxylation of benzene with H_2O_2. *Microporous Mesoporous Mater.* 294–301.

61 Shahami, M., Dooley, K.M., and Shantz, D.F. (2018). Steam-assisted crystallized Fe-ZSM-5 materials and their unprecedented activity in benzene hydroxylation to phenol using hydrogen peroxide. *J. Catal.* 368: 354–364.

62 Dimitrova, R. and Spassova, M. (2007). Hydroxylation of benzene and phenol in presence of vanadium grafted Beta and ZSM-5 zeolites. *Catal. Commun.* 8: 693–696.

63 Fan, W., Duan, R.-G., Yokoi, T. et al. (2008). Synthesis, crystallization mechanism, and catalytic properties of titanium-rich TS-1 free of extraframework titanium species. *J. Am. Chem. Soc.* 130: 10150–10164.

64 Balducci, L., Bianchi, D., Bortolo, R. et al. (2003). Direct oxidation of benzene to phenol with hydrogen peroxide over a modified titanium silicalite. *Angew. Chem. Int. Ed.* 42: 4937–4940.

65 Barbera, D., Cavani, F., D'Alessandro, T. et al. (2010). The control of selectivity in benzene hydroxylation catalyzed by TS-1: the solvent effect and the role of crystallite size. *J. Catal.* 275: 158–169.

66 Bianchi, D., Bortolo, R., Tassinari, R. et al. (2000). A novel iron-based catalyst for the biphasic oxidation of benzene to phenol with hydrogen peroxide. *Angew. Chem. Int. Ed.* 39: 4321–4323.

67 Li, J.P.H., Kennedy, E., and Stockenhuber, M. (2014). Oxidative coupling and hydroxylation of phenol over transition metal and acidic zeolites: insights into catalyst function. *Catal. Lett.* 144: 9–15.

68 Choi, J.S., Yoon, S.S., Jang, S.H., and Ahn, W.S. (2006). Phenol hydroxylation using Fe-MCM-41 catalysts. *Catal. Today* 111: 280–287.

69 Mal, N.K. and Ramaswamy, A.V. (1996). Hydroxylation of phenol over Sn-silicalite-1 molecular sieve: solvent effects. *J. Mol. Catal. A-Chem.* 105: 149–158.

70 Adam, F., Wong, J.-T., and Ng, E.-P. (2013). Fast catalytic oxidation of phenol over iron modified zeolite L nanocrystals. *Chem. Eng. J.* 214: 63–67.

71 Zhao, W., Luo, Y.F., Deng, P., and Li, Q.Z. (2001). Synthesis of Fe-MCM-48 and its catalytic performance in phenol hydroxylation. *Catal. Lett.* 73: 199–202.

72 Fajerwerg, K. and Debellefontaine, H. (1996). Wet oxidation of phenol by hydrogen peroxide using heterogeneous catalysis Fe-ZSM-5: a promising catalyst. *Appl. Catal. B-Environ.* 10: L229–L235.

73 Zhao, X., Sun, Z., Zhu, Z. et al. (2013). Evaluation of iron-containing aluminophosphate molecular sieve catalysts prepared by different methods for phenol hydroxylation. *Catal. Lett.* 143: 657–665.

74 Zhao, X., Zhang, X., Hao, Z. et al. (2018). Synthesis of FeAPO-5 molecular sieves with high iron contents via improved ionothermal method and their catalytic performances in phenol hydroxylation. *J. Porous. Mater.* 25: 1007–1016.

75 Shen, Y., Wang, F., Liu, W., and Zhang, X. (2018). The preparation of Fe^{3+} ion-exchanged mesopore containing ZSM-5 molecular sieves and its high catalytic activity in the hydroxylation of phenol. *J. Porous. Mater.* 25: 1587–1595.

76 Han, Z., Zhang, F., and Zhao, X. (2019). Green energy-efficient synthesis of Fe-ZSM-5 zeolite and its application for hydroxylation of phenol. *Microporous Mesoporous Mater.* 290–297.

77 Shen, Y., Wang, F., Han, Z., and Zhang, X. (2018). Gemini-type cationic surfactant-directed synthesis of hollow ZSM-5 zeolite with intracrystalline mesopores and its application in the hydroxylation of phenol. *J. Chem. Technol. Biotechnol.* 93: 1347–1358.

78 Shen, Y., Li, H., Zhang, X. et al. (2020). The diquaternary ammonium surfactant-directed synthesis of single-unit-cell nanowires of ZSM-5 zeolite. *Nanoscale* 12: 5824–5828.

79 Zuo, Y., Wang, X., and Guo, X. (2011). Synthesis of titanium silicalite-1 with small crystal size by using mother liquid of titanium silicalite-1 as seed. *Ind. Eng. Chem. Res.* 50: 8485–8491.

80 Wang, J., Park, J.N., Wei, X.Y., and Lee, C.W. (2003). Room-temperature heterogeneous hydroxylation of phenol with hydrogen peroxide over Fe^{2+}, Co^{2+} ion-exchanged Na beta zeolite. *Chem. Commun.* 628–629.

81 Rege, S.U., Yang, R.T., and Cain, C.A. (1998). Desorption by ultrasound: phenol on activated carbon and polymeric resin. *AICHE J.* 44: 1519–1528.

82 Villa, A.L., Caro, C.A., and de Correa, C.M. (2005). Cu- and Fe-ZSM-5 as catalysts for phenol hydroxylation. *J. Mol. Catal. A-Chem.* 228: 233–240.

83 Park, J.N., Wang, J., Choi, K.Y. et al. (2006). Hydroxylation of phenol with H_2O_2 over Fe^{2+} and/or Co^{2+} ion-exchanged NaY catalyst in the fixed-bed flow reactor. *J. Mol. Catal. A-Chem.* 247: 73–79.

84 Mohamed, M.M. and Eissa, N.A. (2003). Characterization of intrazeolitic Fe^{3+} prepared by chemical vapor deposition of $[(C_5H_5)Fe(CO)_2]_2$ inside NaY and FSM-16 zeolites and their catalytic activities towards phenol hydroxylation. *Mater. Res. Bull.* 38: 1993–2007.

85 Liu, C.B., Shan, Y.J., Yang, X.G. et al. (1997). Iron(II)-8-quinolinol/MCM-41-catalyzed phenol hydroxylation and reaction mechanism. *J. Catal.* 168: 35–41.

86 Lin, S.S. and Gurol, M.D. (1998). Catalytic decomposition of hydrogen peroxide on iron oxide: kinetics, mechanism, and implications. *Environ. Sci. Technol.* 32: 1417–1423.

87 Buttha, S., Youngme, S., Wittayakun, J., and Loiha, S. (2018). Formation of iron active species on HZSM-5 catalysts by varying iron precursors for phenol hydroxylation. *Mol. Catal.* 461: 26–33.

88 Yin, J., Lu, X., Yan, J. et al. (2020). Postsynthesis of Ti-UZM-35 titanosilicate as efficient catalyst for phenol hydroxylation reaction. *Microporous Mesoporous Mater.* 305–313.

89 Tsai, S.-T., Chao, P.-Y., Tsai, T.-C. et al. (2009). Effects of pore structure of post-treated TS-1 on phenol hydroxylation. *Catal. Today* 148: 174–178.

90 Chao, P.-Y., Tsai, S.-T., Tsai, T.-C. et al. (2009). Phenol hydroxylation over alkaline treated TS-1 catalysts. *Top. Catal.* 52: 185–192.

91 Shan, Z., Lu, Z., Wang, L. et al. (2010). Stable bulky particles formed by TS-1 zeolite nanocrystals in the presence of H_2O_2. *ChemCatChem* 2: 407–412.

92 Klaewkla, R., Kulprathipanja, S., Rangsunvigit, P. et al. (2003). Phenol hydroxylation using Ti- and Sn-containing silicalites. *Chem. Commun.* 1500–1501.

93 Xia, C., Long, L., Zhu, B. et al. (2016). Enhancing the selectivity of Para-dihydroxybenzene in hollow titanium silicalite zeolite catalyzed phenol hydroxylation by introducing acid-base sites. *Catal. Commun.* 80: 49–52.

94 Wilkenhoner, U., Langhendries, G., van Laar, F. et al. (2001). Influence of pore and crystal size of crystalline titanosilicates on phenol hydroxylation in different solvents. *J. Catal.* 203: 201–212.

95 Liu, H., Lu, G.Z., Guo, Y.L. et al. (2006). Chemical kinetics of hydroxylation of phenol catalyzed by TS-1/diatomite in fixed-bed reactor. *Chem. Eng. J.* 116: 179–186.

96 Wroblewska, A. (2013). Water as the solvent for the process of phenol hydroxylation over the Ti-MWW catalyst. *React. Kinet. Mech. Catal.* 108: 491–505.

97 Keshavaraja, A., Ramaswamy, V., Soni, H.S. et al. (1995). Synthesis, characterization, and catalytic properties of micro-mesoporous, amorphous titanosilicate catalysts. *J. Catal.* 157: 501–511.

98 Meng, X.J., Li, D.F., Yang, X.Y. et al. (2003). Synthesis, characterization, and catalytic activity of mesostructured titanosilicates assembled from polymer surfactants with preformed titanosilicate precursors in strongly acidic media. *J. Phys. Chem. B* 107: 8972–8980.

99 Thangaraj, A., Eapen, M.J., Sivasanker, S., and Ratnasamy, P. (1992). Studies on the synthesis of titanium silicalite, TS-1. *Zeolites* 12: 943–950.

100 Han, Z., Shen, Y., Qin, X. et al. (2019). Synthesis of hierarchical titanium-rich titanium silicalite-1 zeolites and the highly efficient catalytic performance for hydroxylation of phenol. *Chemistryselect* 4: 1618–1626.

101 Tuel, A. (1996). Crystallization of titanium silicalite-1 (TS-1) from gels containing hexanediamine and tetrapropylammonium bromide. *Zeolites* 16: 108–117.

102 Yan, M., Jin, F., Ding, Y. et al. (2019). Synthesis of titanium-incorporated MWW zeolite by sequential deboronation and atom-planting treatment of ERB-1 as an epoxidation catalyst. *Ind. Eng. Chem. Res.* 58: 4764–4773.

103 Camblor, M.A., Constantini, M., Corma, A. et al. (1995). A new highly efficient method for the synthesis of Ti-Beta zeolite oxidation catalyst. *Appl. Catal. A-Gen.* 133: L185–L189.

104 Liu, H., Lu, G.Z., Guo, Y.L., and Guo, Y. (2005). Synthesis of TS-1 using amorphous SiO_2 and its catalytic properties for hydroxylation of phenol in fixed-bed reactor. *Appl. Catal. A-Gen.* 293: 153–161.

105 Zhang, T., Zuo, Y., Liu, M. et al. (2016). Synthesis of titanium silicalite-1 with high catalytic performance for 1-butene epoxidation by eliminating the extraframework Ti. *ACS Omega* 1: 1034–1040.

106 Guo, Q., Feng, Z., Li, G. et al. (2013). Finding the "missing components" during the synthesis of TS-1 zeolite by UV resonance raman spectroscopy. *J. Phys. Chem. C* 117: 2844–2848.

107 Abbo, H.S. and Titinchi, S.J.J. (2009). Di-, tri- and tetra-valent ion-exchanged NaY zeolite: active heterogeneous catalysts for hydroxylation of benzene and phenol. *Appl. Catal. A-Gen.* 356: 167–171.

108 Salavati-Niasari, M. (2009). Synthesis, characterization and liquid-phase hydroxylation of phenol with hydrogen peroxide over host (nanopores of zeolite-Y)/guest (oxovanadium(IV) complexes of tetraaza macrocyclic ligands) nanocatalyst. *Inorg. Chim. Acta* 362: 2159–2166.

109 Maurya, M.R., Kumar, M., Titinchi, S.J.J. et al. (2003). Oxovanadium(IV) Schiff base complexes encapsulated in zeolite-Y as catalysts for the liquid-phase hydroxylation of phenol. *Catal. Lett.* 86: 97–105.

110 Nethravathi, B.P., Reddy, K.R., and Mahendra, K.N. (2014). Catalytic activity of supported solid catalysts for phenol hydroxylation. *J. Porous. Mater.* 21: 285–291.

111 Maurya, M.R., Titinchi, S.J.J., and Chand, S. (2003). Liquid-phase catalytic hydroxylation of phenol using Cu(II), Ni(II) and Zn(II) complexes of amidate ligand encapsulated in zeolite-Y as catalysts. *Catal. Lett.* 89: 219–227.

112 Bania, K.K. and Deka, R.C. (2013). Zeolite-Y encapsulated metal picolinato complexes as catalyst for oxidation of phenol with hydrogen peroxide. *J. Phys. Chem. C* 117: 11663–11678.

14

Bifunctional Titanosilicate Systems for the Gas-Phase Catalytic Propylene Epoxidation with Hydrogen and Oxygen

Gang Wang, Qianhong Wang, Zhihua Zhang, Xuezhi Duan, and Xinggui Zhou**

East China University of Science and Technology, State Key Laboratory of Chemical Engineering, School of Chemical Engineering, No. 130 Meilong Road, Shanghai 200237, China

14.1 Introduction

Propylene oxide (PO), as one of the largest propylene derivatives, is an important chemical raw material and widely used in the production of polyether polyols, surfactants, propylene glycol, dimethyl carbonate, etc. [1]. With the rapid increase in the demand for polyurethane and nonionic surfactants, the global annual production of PO has continued to increase. The market research report shows that the global PO market is expected to expand with an average annual growth rate of 5.1% from 2020 to 2027 and mainly concentrated in Asian countries [2].

Nowadays, the industrial production of PO is mainly achieved by three routes including chlorohydrin process (CHPO), organic hydroperoxide processes that co-produce tertiary butyl alcohol (PO/TBA) and styrene monomer (PO/SM), cumene hydroperoxide process (CHP), and hydrogen peroxide process (HPPO) [3]. Among them, the development of CHPO has been restricted due to the considerations of environment and equipment corrosion; the organic hydroperoxide processes would produce two to four times co-products when producing PO so that the economy of these processes is greatly influenced by the fluctuation of the co-products market; although the CHP method avoids the above problems, the industrial process is complicated involving multiple reaction steps, and the operating cost is high; the HPPO process route uses H_2O_2 as the oxidant, which is a simple and green process. However, the production, transportation, and storage of H_2O_2 have to face the disadvantages of high cost and explosion risk, and it is necessary to provide the HPPO process with an in situ H_2O_2 production device. Moreover, a large amount of methanol solvent needs to be used in the reaction, and thus, the energy consumption in the separation process is high [1, 3]. Therefore, it is highly desirable to develop a simple and efficient process for the direct epoxidation of propylene to PO.

Theoretically, the direct gas-phase epoxidation of propylene with molecular oxygen as the oxidant is considered to be the most ideal catalytic oxidation for the

*Corresponding author

Micro-Mesoporous Metallosilicates: Synthesis, Characterization, and Catalytic Applications, First Edition. Edited by Peng Wu and Hao Xu.

synthesis of PO [1]. Since the direct epoxidation of ethylene with oxygen to ethylene oxide (EO) has achieved success over supported silver catalysts, researchers have tried to use silver catalysts for the direct epoxidation of propylene with oxygen. However, the products are mainly acrolein and carbon dioxide, and the selectivity to PO is usually lower than 5% [4, 5]. This is mainly because compared with ethylene, propylene has active α-C—H bonds, with lower bond energy and higher reactivity than the carbon-carbon double bonds. The active oxygen species formed by the dissociation of oxygen on the silver catalyst would preferentially react with the α-C—H bond and undergo deep oxidation to generate acrolein and carbon dioxide, and thus it is difficult to obtain the target PO product at desirable selectivity [6, 7].

Considering that TS-1 can effectively catalyze the propylene epoxidation reaction with H_2O_2 to produce PO in the HPPO process, researchers proposed to in situ generate H_2O_2 on the catalyst from H_2 and O_2 and further react with propylene on the Ti site to produce PO. Although Pd and Pt are highly active for H_2 and O_2 activation, they can also lead to the hydrogenation of propylene to propane, resulting in a low PO selectivity [8, 9]. Haruta et al. reported the direct gas-phase epoxidation of propylene with H_2 and O_2 to produce PO with high selectivity over Au-Ti catalysts for the first time in 1998 [10]. This catalytic reaction quickly aroused great interest in the academic and industrial field because it is an environmental benign, highly selective, and simple process compared to the traditional processes. Considerable efforts have been devoted aiming to enhance the catalytic performance, including employing numerous titanium-containing materials to immobilize gold nanoparticles, probing the effects of size distribution, oxidation states and location sites of Au nanoparticles, promoter effects, and so on.

14.2 Advances in the Catalytic Propylene Epoxidation with H_2 and O_2

14.2.1 Mechanistic Insights into Au-Ti Synergy

For the Au/TiO_2 catalyst, Haruta et al. found that the PO formation rate was not significantly related to the concentration of propylene but increased with the concentration of hydrogen and oxygen [10]. The results of C_3H_6-TPD showed that propylene had adsorption interaction on both gold nanoparticles and TiO_2 support. Delgass et al. carried out isotope switching experiments and found that the PO formation rate had an obvious isotopic effect, and they proposed that peroxygen species participated in the rate-determining step of propylene epoxidation [11]. Goodman et al. studied the adsorption of propylene on Au (111) and Au (100) surfaces. It was found that the desorption energy of propylene on the cleaned surface was 39.3 kJ/mol, which was slightly smaller than the value of 45.1–52.7 kJ/mol on the Ag (110) surface and significantly smaller than the value of 79.4 kJ/mol on the Pd (111) surface. Moreover, the adsorption of propylene would be strengthened on the oxygen-adsorbed gold surface [12]. DFT calculations show that propylene interacts with Au atoms at corner sites on Au clusters and electrons transfers from the highest occupied molecular orbital (HOMO) of the propylene molecule to the lowest unoccupied molecular orbital (LUMO) of Au [13]. Compbell et al. found that propylene was weakly adsorbed on the surface of Au, moderately adsorbed on the surface of TiO_2 (111), and strongly adsorbed at the interface of Au and TiO_2 (111) by using temperature programmed desorption (TPD), X-ray photoelectron spectroscopy (XPS), and low energy ion scattering (LEIS) methods [14]. Nijhuis et al. believed that propylene was mainly adsorbed on the gold sites instead of TiO_2 by infrared spectroscopy [15], and the bidentate propoxy species on the TiO_2 surface formed during the reaction was due to the adsorption and ring-opening of PO at the interface between Au particles and the carrier oxide [16]. Furthermore, they investigated the role of water in the reaction process and proposed that water can accelerate the desorption of PO from the catalyst surface, but inhibit the adsorption of propylene at the same time [17]. By employing FT-IR, TPD, and density functional theory (DFT) calculations, Morris et al. found that propylene adsorbs on the surface of TiO_2 through π bonds while adsorbs on Au particles through $\pi\sigma$ bonds, and the interaction in the former is greater than that in the latter. The adsorption of propylene on gold would elongate its double bond, possibly being an important step in its activation and epoxidation [18].

Goodman et al. used inelastic neutron scattering (INS) to study reactive oxygen species on Au/TiO_2 catalysts, demonstrating the presence of OOH radicals and H_2O_2 species during the reaction [19]. Barton et al. studied the reaction of hydrogen with oxygen to form water through experiments and theoretical calculations and found that Au/TS-1 or Au/S-1 catalysts were 60–70 times more active than Au/SiO_2 catalysts, probably due to the fact that there were more Au_{13} clusters in the mordenite framework inverted (MFI) channels [20].

Nijhuis et al. proposed the reaction mechanism on the Au/TiO_2 catalyst by means of infrared spectroscopy and kinetic studies [15, 21, 22]. They believed that H_2 reacts with O_2 on the surface of Au to generate H_2O_2, while propylene reacts with the adjacent hydroxyl groups on the surface of TiO_2 to generate bidentate propoxy species and then reacts with H_2O_2 to generate PO. The bidentate propoxy species could also undergo oxidation reaction with oxygen to generate carboxylate species and cause catalyst deactivation.

In terms of the properties and reaction mechanism of the Ti site, Delgass et al. used theoretical calculations to study the effect of the structure of the Ti site in TS-1 for the epoxidation of propylene [23]. It was found that the Ti site located near a silicon vacancy in TS-1 would more easily form the Ti—OOH intermediate species and the transition state for propylene epoxidation, contributing to a higher PO formation rate. Subsequently, Delgass et al. investigated five possible pathways for the reaction of hydrogen peroxide and propylene on TS-1 using theoretical calculations and proposed the Sinclair–Catlow mechanism and Ti/defect mechanism as two possible reaction pathways. For the former Tripodal site, the energy barriers of Ti—OOH generation and propylene epoxidation are 7.9 and 8.5 kcal/mol, respectively, which are the energy-optimal pathways among the five mechanisms, and may mainly occur on the outer surface of TS-1 zeolite. The latter Ti sites are more likely to be located in the interior of the zeolite channels, and the formation energy barriers of Ti—OOH species are comparable to those of the Sinclair-Catlow mechanism [24, 25]. Song et al. studied the step-by-step and synergistic mechanisms on Tripodal sites and

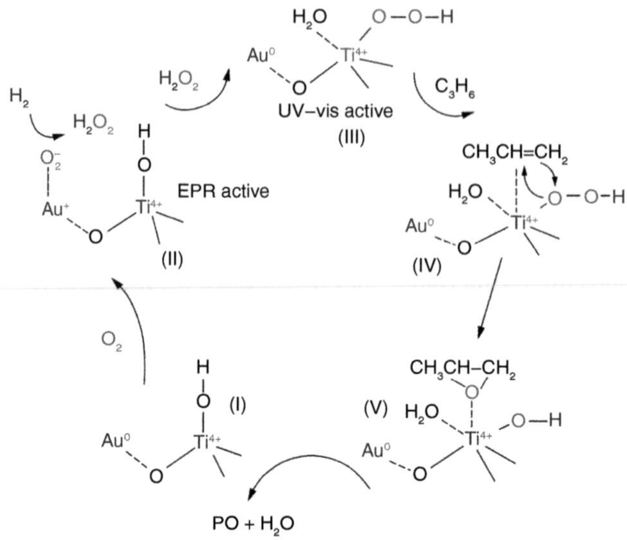

Figure 14.1 Possible sequence of steps for propylene epoxidation with H_2 and O_2 on Au-supported titanosilicates. Source: Adapted with permission from Ref. [27].

Ti/defect sites on TS-1 and found that in the step-by-step mechanism, hydrogen peroxide needs to dissociate and form Ti—OOH species with Ti sites, while in the synergistic mechanism, hydrogen peroxide directly adsorbs on Ti sites to generate Ti—OOH species [26]. By comparing the reaction energy barriers under various pathways, it was found that the formation of Ti-(η1-OOH)-5MR species via a stepwise mechanism on the Ti/defect site is the kinetically optimal pathway.

Haruta et al. employed in situ UV-Vis spectroscopy, X-ray absorption spectroscopy, and infrared spectroscopy to monitor the structural changes of Au-Ti catalysts during the reaction process and believed that the Ti—OOH species produced from the reaction of H_2O_2 and Ti—OH are the intermediate species for the propylene epoxidation [27, 28]. They proposed a mechanism for this reaction mainly involving three steps (Figure 14.1): (1) H_2O_2 formation on the gold site from O_2 and H_2; (2) diffusion and transferring of H_2O_2 to the adjacent titanium sites to form Ti—OOH species; and (3) the reaction of Ti—OOH species with adsorbed propylene to generate PO. For the reaction of O_2 and H_2 on gold sites, the formation of hydroperoxide species requires the dissociation of H_2 while O_2 cannot be dissociated; otherwise, the PO activity and H_2 utilization efficiency would decrease [29].

14.2.2 Effects of Au Particle Properties

14.2.2.1 Au Particle Sizes

This in-situ H_2O_2 formation is structure-sensitive reaction, closely depending on the gold particles. Corma [30, 31], Barrio [32], and Duan [33] carried out DFT calculations to probe the active sites and the results showed that H_2 mainly dissociated on the low-coordinated Au atoms. Feng et al. conducted kinetics analysis and model

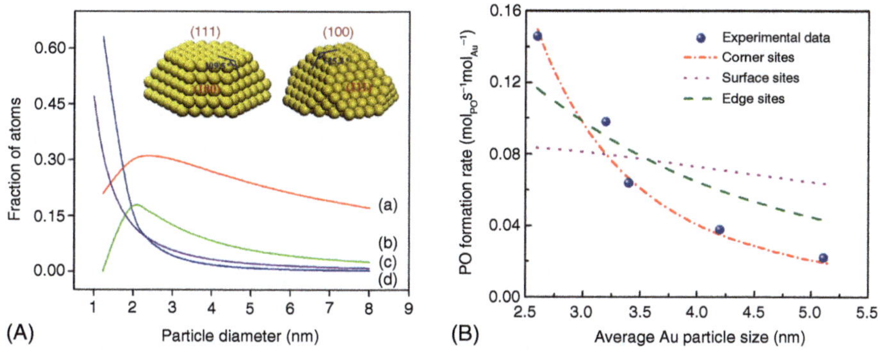

Figure 14.2 (A) Calculated fractions of Au atoms at crystal surface (a), edge (b), perimeter (c), and corner (d) sites in Au nanoparticles. (B) Calculations based on different sites and experimental data of the PO formation rate over Au/TS-1. Source: Adapted with permission from Ref. [34].

calculations and found that the corner sites of Au nanoparticles with the size of 2.5–5.5 nm were the main active sites for the formation of hydroperoxide species from H_2 and O_2, and the smaller Au nanoparticles had higher activity for PO production (Figure 14.2) [34]. Hayashi et al. found that the Au particle with the size of 2–5 nm in the Au/TiO_2 catalyst was favorable for the propylene epoxidation reaction, while those with size smaller than 2 nm would promote the hydrogenation of propylene to propane [10]. Qi et al. reported the effects of promoter and gold particle size on the selectivity of PO and propane and proposed that without promoter on the Au-Ti catalyst, the gold particles in the range of 2–5 nm could effectively catalyze the propylene epoxidation, while those clusters smaller than 2 nm and particles larger than 5 nm promoted propylene hydrogenation. For the gold clusters smaller than 2 nm, the formation of propane could be inhibited by adding a small amount of alkali metal additives [35]. Chen et al. found that introducing 1% CO into the reactants over Au/Ti-SiO_2 catalysts could inhibit the formation of propane and water, thereby improving PO selectivity and hydrogen efficiency [36]. However, Joshi et al. found that small clusters such as Au_3 clusters in the micropores of TS-1 also had high activity for this reaction through theoretical calculations [25, 37]. At the same time, Delgass et al. reported that the Au/TS-1 catalysts prepared with 0.05 and 0.01 wt.% Au loadings exhibited ultra-high activity for PO production with Au particles smaller than 2 nm [38].

Huang et al. prepared Au clusters smaller than 2 nm with solid grinding method and exhibited a higher PO formation rate than the Au particles larger than 2 nm [39]. Haruta et al. immobilized gold clusters (1–2 nm) and nanoparticles (>2 nm) on the external surfaces of TS-1 by solid grinding method, and loaded gold clusters on the external surfaces of TS-1 (1–2 nm) and in the MFI channels (<0.55 nm) by deposition precipitation method, respectively. The results showed that the former exhibited a higher PO formation rate, and they believed that gold clusters with the size of 1–2 nm outside the channels had higher activity than gold clusters with the size of less than 0.55 nm inside the channels [40, 41].

14.2.2.2 Au Oxidation States

Zwijnenburg et al. used XPS and Mössbauer spectroscopy to study the effect of the valence of Au on the performance of Au/TiO_2 catalyst [42]. The results showed that the activity of Au/TiO_2 catalyst was positively correlated with the content of zero-valent gold, and they believed that zero-valent gold had higher epoxidation activity. However, Oyama et al. used extended X-ray absorption fine structure (EXAFS) to study the valence states of gold particles with different sizes on Au/Ti-TUD catalyst [43]. The results showed that the Au nanoparticles with smaller particle size had higher content of oxidized gold. Gaudet et al. studied the effects of Au oxidation states through cyanide treatment on Au/TS-1 by EXAFS. They found that nanoscale metallic gold was active for propylene epoxidation and hydrogenation, as well as combustion of hydrogen, and nanoscale gold (+1) cyanide was active for hydrogenation with significantly reduced hydrogen combustion and absolutely no propylene oxidation [44].

14.2.2.3 Au Sites Location

Delgass et al. found that the gold nanoparticles on Au/TS-1 catalyst prepared with Cs_2CO_3 as precipitant exhibited much higher activity than that prepared with Na_2CO_3 as precipitant. The presence of Cs was proposed to be helpful to stabilize the Au clusters in the pores of TS-1 and they had higher catalytic activity [45]. Haruta et al. immobilized 1–2 nm gold clusters on the external surface of TS-1 by solid grinding method and found that the activity of this catalyst was better than that of Au/TS-1 catalyst prepared by deposition precipitation method with gold nanoparticles in the pores of TS-1. They believed that the 1–2 nm gold nanoparticles on the external surface of TS-1 could catalyze propylene epoxidation to generate PO [40]. Lu and Zhou found that the gold nanoparticles immobilized on external surfaces of the uncalcined TS-1 with size of 2–5 nm also had good catalytic activity. This indicated that the gold particles on the external surface of TS-1 can also catalyze the epoxidation of propylene with H_2 and O_2 [46, 47]. Ma et al. compared the activity of Au/TS-1, Au/TS-1-B, and Au/S-1/TS-1 catalysts, and they found that both the gold clusters located on the external surfaces of TS-1 and those in the microporous channels had high propylene epoxidation activity [48]. Feng et al. reported a strategy to manipulate Au spatial location inside micropores of TS-1 support by tuning Au hydrolysis process using different preparation temperatures. The optimum Au/TS-1 catalyst showed a high and stable PO formation rate without adding promoters (Figure 14.3) [49].

14.2.3 Effects of Materials' Properties for Immobilizing Au Particles

The chosen materials to immobilize Au particles have been found to play an important role in the formation of PO. PO could be generated when anatase TiO_2 was used as the support [50], but not on rutile and amorphous TiO_2 [51, 52]. Further research found that Au/TiO_2 catalyst faced the problems of low activity and rapid deactivation. Even though PO activity can be enhanced by increasing the reaction temperature, it resulted in a severe drop in PO selectivity [53]. This was possibly due

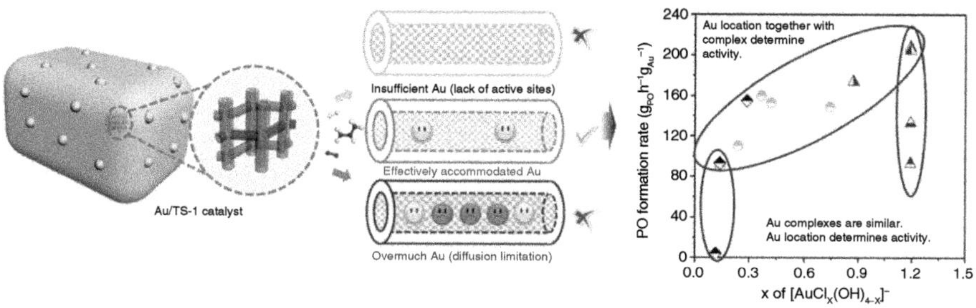

Figure 14.3 Schematic diagram of the effectively accommodated Au species in Au/TS-1 catalyst. Source: Adapted with permission from Ref. [49].

to the strong adsorption of PO on adjacent titanium sites and ring-opening to form bidentate propoxy species, which would inhibit the activity and lead to byproducts formation as well as catalyst deactivation (Figure 14.4) [16, 54]. The strategy of using Si to isolate Ti to form relatively isolated active sites was proposed, and thus many

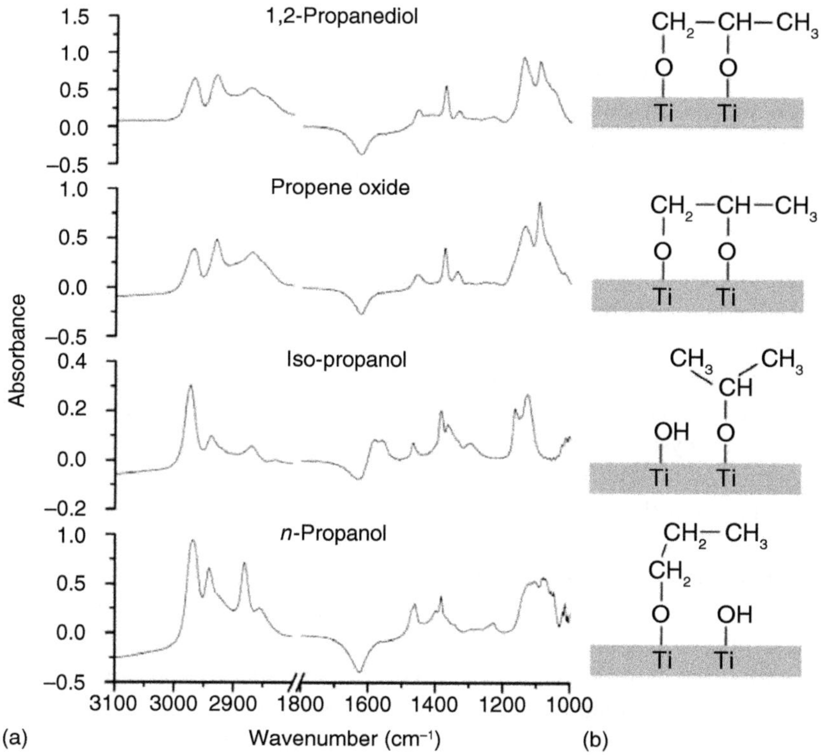

Figure 14.4 (a) Infrared spectra after adsorption and subsequent desorption of 1,2-propanediol, propene oxide, 2-propanol, and *n*-propanol on titania (323 K). (b) Most likely adsorbate species corresponding to the infrared spectra in part a. Source: Reproduced from Ref. [16]/with permission of American Chemical Society.

kinds of Ti-containing materials including Ti-SiO_2 and titanium-silicalite molecular sieve with microporous and/or mesoporous channels were developed.

14.2.3.1 Ti-SiO_2 Materials

Compared with TiO_2, using SiO_2 to isolate TiO_2 species can greatly reduce the degree of aggregation of TiO_2 species, and the stability of Au/TiO_2-SiO_2 catalyst was significantly higher than that of Au/TiO_2 catalyst [55]. This indicated that the bidentate propoxy species may be mainly formed on the agglomerated TiO_2 species. Haruta et al. found that increasing the calcination temperature benefits for improving the interaction between TiO_2 and SiO_2 and producing more isolated tetra-coordinated Ti species. The catalyst evaluation showed that Ti-SiO_2 material treated at high temperature had higher PO yield [56]. Nijhuis et al. prepared Ti-SiO_2 composite oxides with tetra-coordinated Ti species highly dispersed on the surface of SiO_2 with the Ti loading to 0.2–0.3 wt.% by grafting method, and the catalysts showed PO formation rates of 120–130 g kg_{cat}^{-1} h^{-1} [57]. Turner et al. prepared TiO_2-SiO_2 by atomic layer deposition and found that the selectivity of PO was strongly correlated to the coordination number of Ti species and Ti—O bond length. The fourfold coordinated Ti sites favored high selectivity of PO and H_2 efficiency, while the sixfold coordinated Ti sites promoted the generation of by-products such as propanal and CO_2 [58].

14.2.3.2 Mesoporous Ti-Containing Materials

Uphade et al. prepared Ti-MCM-41 with average pore diameters of 4.55 and 6.12 nm with sol-gel method. The results showed that gold nanoparticles supported on Ti-MCM-41 with larger pores had higher propylene conversion and PO selectivity probably because the larger pores benefit for the dispersion of gold and diffusion [51]. Sinha et al. found that the activity and stability of the Ti-MCM-41 supported gold catalyst prepared with hydrothermal method followed by grafting were much higher, which could be contributed by the more isolated four-coordinated Ti sites and less acidic Si—OH sites. This not only increased the number of Au-Ti active sites but also inhibited the acid catalyzed ring-opening of PO and thus prohibited coke deposition [59]. Sacaliuc et al. also reported that the performance of Ti-SBA-15 supported gold catalyst prepared by grafting method was better than that prepared by hydrothermal method [60].

Haruta et al. found that the performance of Au/Ti-MCM-48 catalyst was better than that of Au/Ti-MCM-41 catalyst. This may be because the pore system of Au/Ti-MCM-48 catalyst not only helps to transport reactants but also helps to strengthen the diffusion of product PO, thus inhibiting the ring-opening of PO and the adsorption of organic chemicals on the active site [61]. Haruta et al. used Ti-TUD-1 with three-dimensional disorder and wormlike pores as the support. The Au/Ti-TUD-1 catalyst evaluation showed that the initial conversion of propylene was up to 9.8% and the selectivity of PO was 90.3% with the promotion of $Ba(NO_3)_2$ [62]. Yuan et al. prepared Ti-HMS with wormhole structure using long-chain alkyl primary amines as template agents. The Au/Ti-HMS catalyst exhibited superior initial performance with the propylene conversion of about 9.0%, PO selectivity of 97.3%, and H_2 efficiency of 30.4% [63]. Cheng et al. prepared pillared Ti-MCM-36 with an average pore diameter of 2 nm and used it as the support for Au nanoparticles. The research results showed that the particle size of gold nanoparticles and Ti content had great impact on the performance of Au/Ti-MCM-41 catalyst. The main product on gold nanoparticles with an average particle size of 2 nm was propanal and only gold particles with the particle size of about 1 nm could selectively generate PO [64].

14.2.3.3 Microporous Ti-Containing Materials

TS-1 is a typical microporous titanium-containing material. Delgass et al. conducted systematic studies over TS-1 to optimize the preparation method of TS-1, and they found that the performance of 0.1 wt.% Au/TS-1 catalyst prepared by fine adjustment of preparation conditions was much higher than that of gold supported on other Ti-containing materials [38, 65, 66]. The Au/TS-1 catalysts prepared with Cs_2CO_3 as precipitant were found to have PO formation rates up to ~300–350 g_{PO} h^{-1} kg_{cat}^{-1}, which were ascribed to the strong interaction between Cs and Au. This could promote most of gold species to generate highly active gold clusters in the microporous channels of TS-1, thus showing extremely high activity [45].

Zhou et al. analyzed the deactivation mechanism of the Au/TS-1 catalyst and found that the diffusion of PO molecules in the microporous channels is slow, making the produced PO easy to suffer ring-opening and polymerization at the acid sites [47]. This could lead to the block of the micropores and result in the loss of Au and Ti sites. On this basis, the strategy of using uncalcined TS-1 with blocked micropores to disperse gold nanoparticle was proposed, which made the catalyst relatively stable for more than 30 hours at a PO formation rate of 125 g_{PO} h^{-1} kg_{cat}^{-1} (Figure 14.5) [47, 67]. Delgass et al. also reported the obvious promotional effect of the pore-blocking strategy on improving stability, and the PO formation rate (100 g_{PO} h^{-1} kg_{cat}^{-1}) was stable for more than 40 hours [68]. In addition, for the Au nanoparticles supported on TS-2, the pore blocking strategy also has a good promotional effect on the catalyst stability [69]. Feng et al. further carried out a lot of modification works on the structure of TS-1, including introducing mesoporous

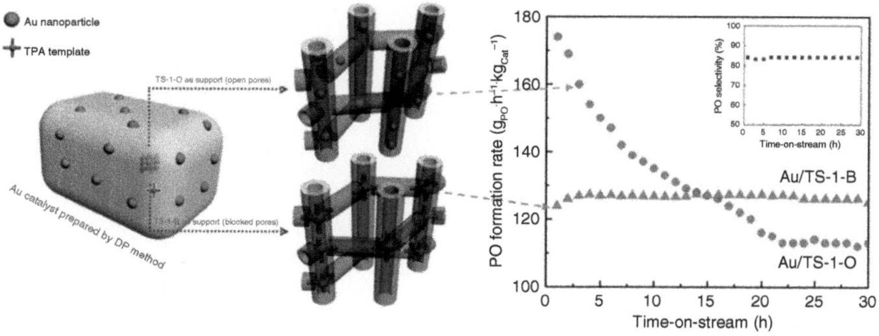

Figure 14.5 Schematic diagram of Au locations of Au/TS-1-O and Au/TS-1-B catalysts, and PO formation rate as a function of the time-on-stream over them. The inset shows the selectivity of Au/TS-1-B catalyst. Source: Adapted with permission from Ref. [47].

[70], constructing core-shell [71], and hierarchical pore [72] structures, reducing particle size [73], to enhance the catalytic performance.

14.2.3.4 Effects of Hydrophobicity

FT-IR and kinetic studies revealed that the produced PO would adsorbed on the catalyst surface by bonding with hydroxyl groups, which not only occupied the active site of propylene epoxidation and inhibited the PO activity but also promoted PO isomerization and cracking to produce by-products (Figure 14.6) [54, 55, 74, 75]. The modification of Au-Ti bifunctional catalyst with silane coupling agent to improve PO performance was first proposed by Haruta et al. in 2002 [61]. They treated Au/Ti-MCM-48 catalyst with trimethylmethoxysilane, and the stability, PO selectivity, and hydrogen efficiency were improved. Subsequently, they loaded Au nanoparticles on *N*-methyl-*N*-trimethylsilyltrifluoroacetamide (MSTFA) treated Ti-MCM-48 with impregnation method, and they immobilized Au nanoparticles on the trimethylmethoxysilane modified Ti-MCM-41 by precipitation method, respectively. It was found that trimethyl modification can improve the selectivity to PO. At the same time, the modified catalyst still had strong adsorption on water, which may lead to the reconstruction of hydrophilic surface and deactivation of the catalyst [76]. In addition, Haruta's research group proposed to co-feed a low concentration of trimethylamine (10–20 ppm) into the reaction system to modify the catalyst. At the same time, it could play a role in protecting the hydrolysis of the grafted hydrophobic groups, making the propylene conversion and PO selectivity significantly improved [77, 78].

Chowdhury et al. studied the modification effect of trimethylmethoxysilane on Au/Ti-SiO_2 catalyst. The silylation treatment was proposed to improve the hydrophobicity of the catalyst, reduce the number of acid sites, and redisperse the

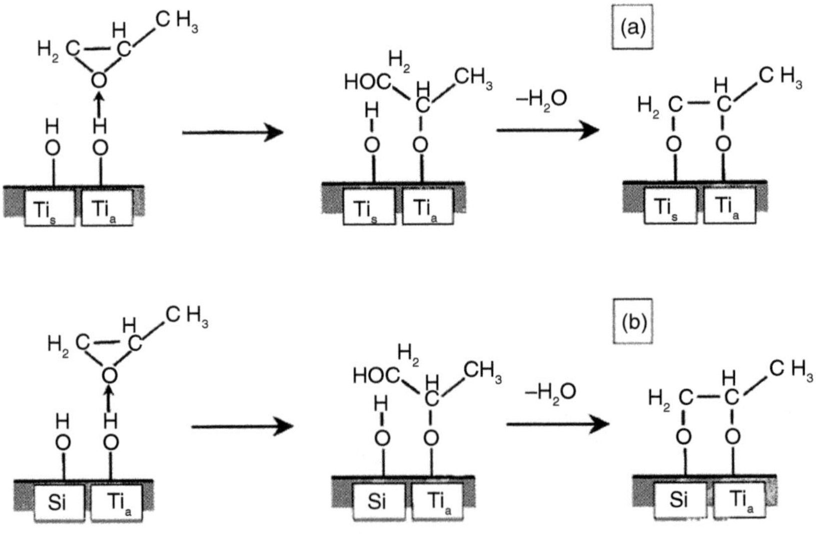

Figure 14.6 Formation of bidentate propoxy species over the TiO_2 (a) and TiO_2/SiO_2 (b) support. Source: Adapted with permission from Ref. [55].

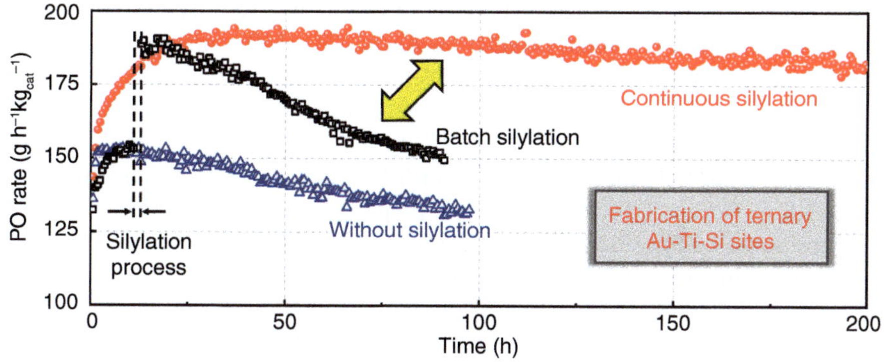

Figure 14.7 Effects of batch and continuous silylation on PO formation rate as a function of time on stream. Source: Reproduced from Ref. [81]/with permission of Elsevier.

gold particles, thereby promoting the desorption rate of PO and increasing PO activity, selectivity, and hydrogen efficiency [79]. Kanungo et al. studied the modification effect over Au/Ti-SiO_2 catalysts with different treatment methods and reagents, including hexamethyldisilazane, trimethylchlorosilane, triethoxyfluorosilane, and tetramethyldisilazane [80]. It was found that the gas-phase modification after Au loading benefits for the catalytic activity and PO selectivity, and the reagent type and treatment time had great influence on the performance.

Duan et al. investigated the silylation effects of hexamethyldisiloxane (HMDSO) and hexamethyldisilazane (HMDS) comparatively over the uncalcined TS-1 immobilized Au nanoparticles. They developed a strategy to selectively consume silanol groups and fabricate the unique Au—Ti(—OH)—Si(—O—SiR_3) moiety. The obtained catalyst imparted enhanced PO formation rate as high as $193\,g\,h^{-1}\,kg_{cat}^{-1}$ and PO selectivity of 95.7% at a H_2 efficiency of 42.5% owing to the weakened PO adsorption and increased byproduct formation barriers, and no obvious deactivation was observed in the tested 200 hours (Figure 14.7) [81].

14.2.4 Effect of Promoters

CsCl was first reported by Haruta et al. as a kind of promoter by physical mixing with Au/Ti-MCM-41 catalyst, which can significantly improve the selectivity of PO and reduce the conversion of H_2 by about 90% [82]. This indicated that the addition of CsCl is helpful to inhibit the direct oxidation reaction of H_2. Lu et al. investigated the effect of alkali metals and alkaline-earth metals on the performance of Au/TS-1 catalyst. The results showed that the addition of alkali metals such as K and Cs had little effect on the loading of Au particles but improved the catalyst activity. The addition of alkaline-earth metals such as Mg, Ca, Sr, and Ba not only improved the catalyst activity but also increased the loading of gold particles. Characterization results showed that the addition of alkali metals and alkaline earth metals did not change the acidity of the catalyst, and the change of catalytic activity was attributed to the improvement of Au particle dispersion [83].

Zwijnenburg et al. introduced Pd and Pt into Au/TiO_2-SiO_2 catalysts to prepare bimetallic catalysts. The main product on Au-Pd/TiO_2-SiO_2 catalyst was propane, which was due to the enhanced hydrogenation of propylene catalyzed by Pd nanoparticles, while the amount of water generated on Au-Pt/TiO_2/SiO_2 catalyst was significantly lower than that on Au/TiO_2-SiO_2 catalyst [9]. Jin et al. prepared core-shell Au-Pd/TiO_2 spherical catalyst with TS-1 film on its surface by hydrothermal method. Compared with the catalyst prepared by traditional physical mixing with the same composition, the encapsulated catalyst showed higher propylene conversion. This was attributed to that its core and shell components can catalyze the reaction independently. However, the PO selectivity over this catalyst is very low [84]. Feng et al. used a two-step method to prepare Au-Ag/TS-1-B catalyst. The results showed that compared with the Au/TS-1-B catalyst, the bimetallic Au-Ag/TS-1-B catalyst exhibits significantly improved the PO formation rate and hydrogen efficiency calculated by the mass of gold. In addition, it was found that the introduction of Ag reduced the particle size of gold nanoparticles and caused the XPS binding energy of Au to shift to the lower regions, while the content of adsorbed O_2 species was significantly increased. This indicated that Au got electrons from Ag, thus promoting the activation of oxygen molecules and generating more adsorbed O^{2-} species, which helped to accelerate the formation rate of hydroperoxide species on gold nanoparticles [85].

Ma et al. found that the PO formation rate on Au-Pd/TS-1 bimetallic catalyst prepared with ethylene glycol as reductant was much higher than that on Au/TS-1 and Pd/TS-1 catalysts. The formation rate of PO increased with the increase of Pd/Au molar ratio, and the selectivity of acetone also increased with the increase of Pd content. This may be due to that Pd transfers electrons to Au, which makes the electron-rich Au easier to generate hydrogen protons, thus promoting the generation of H_2O_2 [86]. Zhou et al. found that combining trace Pt addition with silylation over the uncalcined TS-1 immobilized Au catalyst could significantly enhance the propylene conversion up to 11% with the PO selectivity of 88% [87].

Liu et al. studied the influence of Ge addition on the performance of Au/TS-1 catalyst. X-ray diffraction (XRD) results showed that Ge entered the framework of TS-1, and the propylene conversion increased from 1.6% to 4% and the propylene oxide selectivity increased from 87% to 91%. According to the DFT calculation, the main reason for the enhanced catalyst activity was ascribed to the electronic effect between Ge in the framework and the active Ti sites, which reduced the activation energy of C_3H_6 epoxidation with Ti—OOH to PO [88]. Delgass et al. used Cs_2CO_3 as the precipitant in the process of preparing Au/TS-1 catalyst by deposition-precipitation method, which promoted Au to enter the micropores of TS-1 more easily. Therefore, the content of Au nanoclusters in the TS-1 microporous channel was significantly increased, and the catalytic activity was greatly improved (Figure 14.8) [45].

Lu et al. found that the particle size of gold nanoparticles on the In modified Au/TS-1 catalyst was significantly reduced. The content of oxidized gold on the catalyst was relatively higher after adding In, which made the PO formation rate on the In modified Au/TS-1 catalyst increased by 80% compared with the Au/TS-1 catalyst.

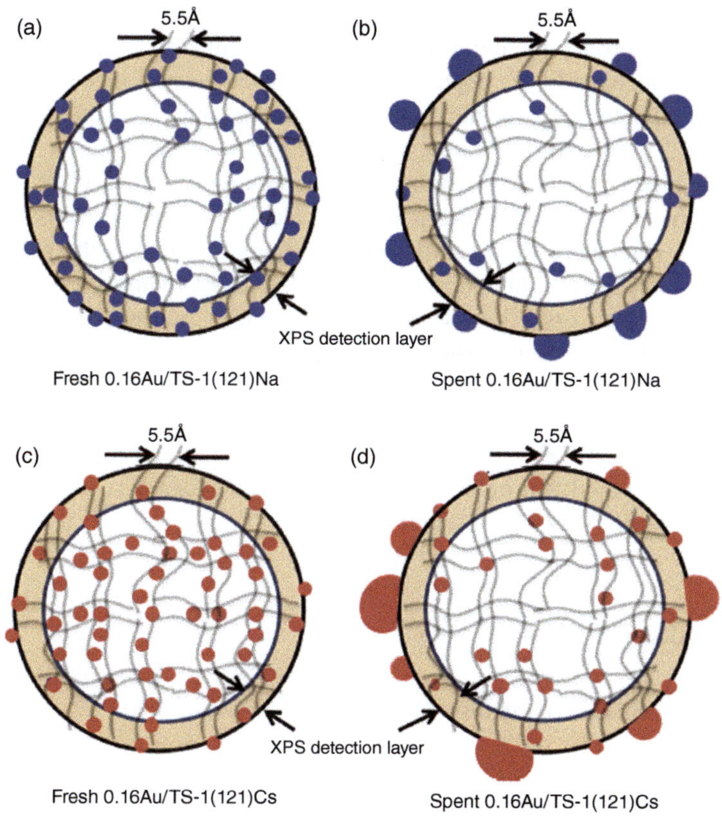

Figure 14.8 Proposed model for the gold species spatial distribution for (a) fresh 0.16Au/TS-1(121)Na, (b) spent 0.16Au/TS-1(121)Na, (c) fresh 0.16Au/TS-1(121)Cs, and (d) spent 0.16Au/TS-1(121)Cs. Source: Adapted with permission from Ref. [45].

Moreover, the addition of In also helped to improve the stability of the catalyst, which may be due to that In covered some acidic sites on TS-1, thus inhibiting the ring-opening of PO on the acidic sites to form bidentate propoxy species that led to catalyst deactivation [89].

14.3 Conclusions and Outlook

In summary, the progress of bifunctional Au-Ti catalysts for the gas-phase epoxidation of propylene with H_2 and O_2 has been reviewed, including the properties of Au nanoparticles, Ti-containing materials, and promoters. Among the Ti-containing materials, the Au nanoparticles immobilized on uncalcined TS-1 have exhibited state-of-the-art PO formation rate and stability by avoiding the deactivation caused by micropore blocking. The activity PO formation is closely related to the particle size of Au nanoparticles, and the smaller particle size is conducive to the activation of reactants and the transferring of the H_2O_2 to the surrounding Ti active sites.

In addition, the strong adsorption of PO on the hydroxyl groups would not only limit the conversion of propylene and PO formation rate but also promote its ring-opening to byproducts and coke deposition, which would accelerate catalyst deactivation. Surface silylation to consume the hydroxyl groups could be an effective way to enhance the PO desorption and improve catalytic performance.

Future studies are expected to focus on the understanding of active sites at micro-scale by multiple methods, including theoretical calculation, kinetics studies, and advanced in-situ characterization techniques. In addition, although the best catalytic performance reported in literature could be close to the requirements of industrial production in terms of PO activity, selectivity, and hydrogen efficiency, the catalyst stability is still far from industrial application. Considering that the deactivation of catalyst could be mainly caused by the agglomeration of gold nanoparticles and the coke deposition, exploring the stabilization method of small gold cluster and/or nanoparticles, the redispersion strategy of large gold particles and the coke removal method will provide an important basis for improving the catalyst stability and establishing the regeneration strategy.

Conflicts of Interest

The authors declare no conflict of interest.

References

1 Teržan, J., Huš, M., Likozar, B., and Djinović, P. (2020). Propylene epoxidation using molecular oxygen over copper- and silver-based catalysts: a review. *ACS Catal.* 10: 13415–13436.

2 Global Industry Analysts, Inc. (2022). *Propylene Oxide: Global Strategic Business Report*. Research and Markets.

3 Alexander Nijhuis, T., Makkee, M., Moulijn, J.A., and Weckhuysen, B.M. (2006). The production of propene oxide: catalytic processes and recent developments. *Ind. Eng. Chem. Res.* 45: 3447–3459.

4 Akimoto, M., Ichikawa, K., and Echigoya, E. (1982). Kinetic and adsorption studies on vapor-phase catalytic oxidation of olefins over silver. *J. Catal.* 76: 333–344.

5 Geenen, P.V., Boss, H.J., and Pott, G.T. (1982). A study of the vapor-phase epoxidation of propylene and ethylene on silver and silver-gold alloy catalysts. *J. Catal.* 77: 499–510.

6 Takei, T., Akita, T., Nakamura, I. et al. (2012). Chapter one—heterogeneous catalysis by gold. In: *Advances in Catalysis*, vol. 55 (ed. B.C. Gates and F.C. Jentoft), 1–126. Academic Press.

7 He, J., Zhai, Q., Zhang, Q. et al. (2013). Active site and reaction mechanism for the epoxidation of propylene by oxygen over CuO_x/SiO_2 catalysts with and without Cs^+ modification. *J. Catal.* 299: 53–66.

8 Jenzer, G., Mallat, T., Maciejewski, M. et al. (2001). Continuous epoxidation of propylene with oxygen and hydrogen on a Pd-Pt/TS-1 catalyst. *Appl. Catal. A Gen.* 208: 125–133.

9 Zwijnenburg, A., Saleh, M., Makkee, M., and Moulijn, J.A. (2002). Direct gas-phase epoxidation of propene over bimetallic Au catalysts. *Catal. Today* 72: 59–62.

10 Hayashi, T., Tanaka, K., and Haruta, M. (1998). Selective vapor-phase epoxidation of propylene over Au/TiO_2 catalysts in the presence of oxygen and hydrogen. *J. Catal.* 178: 566–575.

11 Stangland, E.E., Stavens, K.B., Andres, R.P., and Delgass, W.N. (2000). Characterization of gold-titania catalysts via oxidation of propylene to propylene oxide. *J. Catal.* 191: 332–347.

12 Davis, K.A. and Goodman, D.W. (2000). Propene adsorption on clean and oxygen-covered Au(111) and Au(100) surfaces. *J. Phys. Chem. B* 104: 8557–8562.

13 Chrétien, S., Gordon, M.S., and Metiu, H. (2004). Binding of propene on small gold clusters and on Au(111): simple rules for binding sites and relative binding energies. *J. Chem. Phys* 121: 3756–3766.

14 Ajo, H.M., Bondzie, V.A., and Campbell, C.T. (2002). Propene adsorption on gold particles on TiO_2(110). *Catal. Lett.* 78: 359–368.

15 Nijhuis, T.A., Visser, T., and Weckhuysen, B.M. (2005). The role of gold in gold-titania epoxidation catalysts. *Angew. Chem. Int. Ed.* 44: 1115–1118.

16 Nijhuis, T.A., Visser, T., and Weckhuysen, B.M. (2005). Mechanistic study into the direct epoxidation of propene over gold/titania catalysts. *J. Phys. Chem. B* 109: 19309–19319.

17 Nijhuis, T.A. and Weckhuysen, B.M. (2005). The role of water in the epoxidation over gold-titania catalysts. *Chem. Commun.* 48: 6002.

18 Driscoll, D.M., Tang, W., Burrows, S.P. et al. (2017). Binding sites, geometry, and energetics of propene at nanoparticulate Au/TiO_2. *J. Phys. Chem. C* 121: 1683–1689.

19 Sivadinarayana, C., Choudhary, T.V., Daemen, L.L. et al. (2004). The nature of the surface species formed on Au/TiO_2 during the reaction of H_2 and O_2: an inelastic neutron scattering study. *J. Am. Chem. Soc.* 126: 38–39.

20 Barton, D.G. and Podkolzin, S.G. (2005). Kinetic study of a direct water synthesis over silica-supported gold nanoparticles. *J. Phys. Chem. B* 109: 2262–2274.

21 Nijhuis, T.A. and Weckhuysen, B.M. (2006). The direct epoxidation of propene over gold-titania catalysts—a study into the kinetic mechanism and deactivation. *Catal. Today* 117: 84–89.

22 Nijhuis, T.A., Sacaliuc-Parvulescu, E., Govender, N.S. et al. (2009). The role of support oxygen in the epoxidation of propene over gold–titania catalysts investigated by isotopic transient kinetics. *J. Catal.* 265: 161–169.

23 Wells, D.H., Delgass, W.N., and Thomson, K.T. (2004). Evidence of defect-promoted reactivity for epoxidation of propylene in titanosilicate (TS-1) catalysts: a DFT study. *J. Am. Chem. Soc.* 126: 2956–2962.

24 Wells, D.H., Joshi, A.M., Delgass, W.N., and Thomson, K.T. (2006). A quantum chemical study of comparison of various propylene epoxidation mechanisms using H_2O_2 and TS-1 catalyst. *J. Phys. Chem. B* 110: 14627–14639.

25 Joshi, A.M., Delgass, W.N., and Thomson, K.T. (2007). Mechanistic implications of Au_n/Ti-lattice proximity for propylene epoxidation. *J. Phys. Chem. C* 111: 7841–7844.

26 Nie, X., Ji, X., Chen, Y. et al. (2017). Mechanistic investigation of propylene epoxidation with H_2O_2 over TS-1: active site formation, intermediate identification, and oxygen transfer pathway. *Mol. Catal.* 441: 150–167.

27 Bravo-Suárez, J.J., Bando, K.K., Lu, J. et al. (2008). Transient technique for identification of true reaction intermediates: hydroperoxide species in propylene epoxidation on gold/titanosilicate catalysts by X-ray absorption fine structure spectroscopy. *J. Phys. Chem. C* 112: 1115–1123.

28 Chowdhury, B., Bravo-Suárez, J.J., Mimura, N. et al. (2006). In situ UV-vis and EPR study on the formation of hydroperoxide species during direct gas phase propylene epoxidation over Au/Ti-SiO_2 catalyst. *J. Phys. Chem. B* 110: 22995–22999.

29 Ranganathan, S. and Sieber, V. (2018). Recent advances in the direct synthesis of hydrogen peroxide using chemical catalysis—a review. *Catalysts* 8: 379.

30 Boronat, M., Illas, F., and Corma, A. (2009). Active sites for H_2 adsorption and activation in Au/TiO_2 and the role of the support. *J. Phys. Chem. A* 113: 3750–3757.

31 Corma, A., Boronat, M., González, S., and Illas, F. (2007). On the activation of molecular hydrogen by gold: a theoretical approximation to the nature of potential active sites. *Chem. Commun.* 3371–3373.

32 Barrio, L., Liu, P., Rodríguez, J.A. et al. (2006). A density functional theory study of the dissociation of H_2 on gold clusters: importance of fluxionality and ensemble effects. *J. Chem. Phys.* 125: 164715.

33 Tang, Y., Zhang, Z., Lu, M. et al. (2019). Site-dependent activity and selectivity of H_2O_2 formation from H_2 and O_2 over Au-based catalysts. *Ind. Eng. Chem. Res.* 58: 15119–15126.

34 Feng, X., Duan, X., Qian, G. et al. (2014). Insights into size-dependent activity and active sites of Au nanoparticles supported on TS-1 for propene epoxidation with H_2 and O_2. *J. Catal.* 317: 99–104.

35 Qi, C., Huang, J., Bao, S. et al. (2011). Switching of reactions between hydrogenation and epoxidation of propene over Au/Ti-based oxides in the presence of H_2 and O_2. *J. Catal.* 281: 12–20.

36 Chen, J., Halin, S.J.A., Ferrandez, D.M.P. et al. (2012). Switching off propene hydrogenation in the direct epoxidation of propene over gold–titania catalysts. *J. Catal.* 285: 324–327.

37 Joshi, A.M., Delgass, W.N., and Thomson, K.T. (2006). Partial oxidation of propylene to propylene oxide over a neutral gold trimer in the gas phase: a density functional theory study. *J. Phys. Chem. B* 110: 2572–2581.

38 Taylor, B., Lauterbach, J., and Delgass, W. (2005). Gas-phase epoxidation of propylene over small gold ensembles on TS-1. *Appl. Catal. A Gen.* 291: 188–198.

39 Huang, J., Takei, T., Akita, T. et al. (2010). Gold clusters supported on alkaline treated TS-1 for highly efficient propene epoxidation with O_2 and H_2. *Appl. Catal. B* 95: 430–438.

40 Huang, J., Lima, E., Akita, T. et al. (2011). Propene epoxidation with O_2 and H_2: identification of the most active gold clusters. *J. Catal.* 278: 8–15.

41 Nguyen, V.-H., Nguyen, B.-S., Hu, C. et al. (2020). Advances in designing Au nanoparticles for catalytic epoxidation of propylene with H_2 and O_2. *Catalysts* 10: 442.

42 Zwijnenburg, A., Goossens, A., Sloof, W.G. et al. (2002). XPS and Mössbauer characterization of Au/TiO_2 propene epoxidation catalysts. *J. Phys. Chem. B* 106: 9853–9862.

43 Lu, J., Zhang, X., Bravo-Suárez, J.J. et al. (2007). Direct propylene epoxidation over barium-promoted Au/Ti-TUD catalysts with H_2 and O_2: effect of Au particle size. *J. Catal.* 250: 350–359.

44 Gaudet, J., Bando, K.K., Song, Z. et al. (2011). Effect of gold oxidation state on the epoxidation and hydrogenation of propylene on Au/TS-1. *J. Catal.* 280: 40–49.

45 Lee, W.-S., Akatay, M.C., Stach, E.A. et al. (2013). Enhanced reaction rate for gas-phase epoxidation of propylene using H_2 and O_2 by Cs promotion of Au/TS-1. *J. Catal.* 308: 98–113.

46 Lu, X., Zhao, G., and Lu, Y. (2013). Propylene epoxidation with O_2 and H_2: a high-performance Au/TS-1 catalyst prepared via a deposition-precipitation method using urea. *Catal. Sci. Technol.* 3: 2906–2909.

47 Feng, X., Duan, X., Qian, G., and Zhou, X. (2014). Au nanoparticles deposited on the external surfaces of TS-1: enhanced stability and activity for direct propylene epoxidation with H_2 and O_2. *Appl. Catal. B* 150–151: 396–401.

48 Li, Z., Zhang, J., Wang, D. et al. (2017). Confirmation of gold active sites on titanium-silicalite-1-supported nano-gold catalysts for gas-phase epoxidation of propylene. *J. Phys. Chem. C* 121: 25215–25222.

49 Feng, X., Song, Z., Liu, Y. et al. (2018). Manipulating gold spatial location on titanium silicalite-1 to enhance the catalytic performance for direct propene epoxidation with H_2 and O_2. *ACS Catal.* 8: 10649–10657.

50 Gąsior, M., Grzybowska, B., Samson, K. et al. (2004). Oxidation of CO and C_3 hydrocarbons on gold dispersed on oxide supports. *Catal. Today* 91–92: 131–135.

51 Uphade, B.S., Yamada, Y., Akita, T. et al. (2001). Synthesis and characterization of Ti-MCM-41 and vapor-phase epoxidation of propylene using H_2 and O_2 over Au/Ti-MCM-41. *Appl. Catal. A Gen.* 215: 137–148.

52 Haruta, M., Uphade, B.S., Tsubota, S., and Miyamoto, A. (1998). Selective oxidation of propylene over gold deposited on titanium-based oxides. *Res. Chem. Intermed.* 24: 329–336.

53 Nijhuis, T.A., Huizinga, B.J., Makkee, M., and Moulijn, J.A. (1999). Direct epoxidation of propene using gold dispersed on TS-1 and other titanium-containing supports. *Ind. Eng. Chem. Res.* 38: 884–891.

54 Zwijnenburg, A., Makkee, M., and Moulijn, J.A. (2004). Increasing the low propene epoxidation product yield of gold/titania-based catalysts. *Appl. Catal. A Gen.* 270: 49–56.
55 Mul, G., Zwijnenburg, A., van der Linden, B. et al. (2001). Stability and selectivity of Au/TiO_2 and Au/TiO_2/SiO_2 catalysts in propene epoxidation: an in situ FT-IR study. *J. Catal.* 201: 128–137.
56 Qi, C., Akita, T., Okumura, M., and Haruta, M. (2001). Epoxidation of propylene over gold catalysts supported on non-porous silica. *Appl. Catal. A Gen.* 218: 81–89.
57 Chen, J., Halin, S.J.A., Pidko, E.A. et al. (2013). Enhancement of catalyst performance in the direct propene epoxidation: a study into gold-titanium synergy. *ChemCatChem* 5: 467–478.
58 Lu, Z., Liu, X., Zhang, B. et al. (2019). Structure and reactivity of single site Ti catalysts for propylene epoxidation. *J. Catal.* 377: 419–428.
59 Sinha, A.K., Seelan, S., Akita, T. et al. (2003). Vapor phase propylene epoxidation over Au/Ti-MCM-41 catalysts prepared by different Ti incorporation modes. *Appl. Catal. A Gen.* 240: 243–252.
60 Sacaliuc, E., Beale, A.M., Weckhuysen, B.M., and Nijhuis, T.A. (2007). Propene epoxidation over Au/Ti-SBA-15 catalysts. *J. Catal.* 248: 235–248.
61 Uphade, B.S., Akita, T., Nakamura, T., and Haruta, M. (2002). Vapor-phase epoxidation of propene using H_2 and O_2 over Au/Ti-MCM-48. *J. Catal.* 209: 331–340.
62 Sinha, A.K., Seelan, S., Tsubota, S., and Haruta, M. (2004). A three-dimensional mesoporous titanosilicate support for gold nanoparticles: vapor-phase epoxidation of propene with high conversion. *Angew. Chem. Int. Ed.* 43: 1546–1548.
63 Yang, H., Tang, D., Lu, X., and Yuan, Y. (2009). Superior performance of gold supported on titanium-containing hexagonal mesoporous molecular sieves for gas-phase epoxidation of propylene with use of H_2 and O_2. *J. Phys. Chem. C* 113: 8186–8193.
64 Jin, F., Lin, T.H., Chang, C.C. et al. (2015). Gold supported on Ti incorporated MCM-36 as efficient catalysts in propylene epoxidation with H_2 and O_2. *RSC Adv.* 5: 61710–61718.
65 Lee, W.-S., Akatay, M.C., Stach, E.A. et al. (2012). Reproducible preparation of Au/TS-1 with high reaction rate for gas phase epoxidation of propylene. *J. Catal.* 287: 178–189.
66 Lee, W.-S., Lai, L.-C., Akatay, M.C. et al. (2012). Probing the gold active sites in Au/TS-1 for gas-phase epoxidation of propylene in the presence of hydrogen and oxygen. *J. Catal.* 296: 31–42.
67 Feng, X., Duan, X., Yang, J. et al. (2015). Au/uncalcined TS-1 catalysts for direct propene epoxidation with H_2 and O_2: effects of Si/Ti molar ratio and Au loading. *Chem. Eng. J.* 278: 234–239.
68 Lee, W.-S., Akatay, M.C., Stach, E.A. et al. (2014). Gas-phase epoxidation of propylene in the presence of H_2 and O_2 over small gold ensembles in uncalcined TS-1. *J. Catal.* 313: 104–112.

69 Zhang, Z., Zhao, X., Wang, G. et al. (2020). Uncalcined TS-2 immobilized Au nanoparticles as a bifunctional catalyst to boost direct propylene epoxidation with H_2 and O_2. *AIChE J* 66: e16815.

70 Feng, X., Sheng, N., Liu, Y. et al. (2017). Simultaneously enhanced stability and selectivity for propene epoxidation with H_2 and O_2 on Au catalysts supported on nano-crystalline mesoporous TS-1. *ACS Catal.* 7: 2668–2675.

71 Song, Z., Feng, X., Sheng, N. et al. (2018). Cost-efficient core-shell TS-1/silicalite-1 supported Au catalysts: towards enhanced stability for propene epoxidation with H_2 and O_2. *Chem. Eng. J.* 377: 119927.

72 Sheng, N., Liu, Z., Song, Z. et al. (2018). Enhanced stability for propene epoxidation with H_2 and O_2 over wormhole-like hierarchical TS-1 supported Au nanocatalyst. *Chem. Eng. J.* 377: 119954.

73 Song, Z., Feng, X., Sheng, N. et al. (2018). Propene epoxidation with H_2 and O_2 on Au/TS-1 catalyst: cost-effective synthesis of small-sized mesoporous TS-1 and its unique performance. *Catal. Today* 347: 102–109.

74 Panayotov, D., Mcentee, M., Burrows, S. et al. (2016). Infrared studies of propene and propene oxide adsorption on nanoparticulate Au/TiO_2. *Surf. Sci.* 652: 172–182.

75 Harris, J.W., Arvay, J., Mitchell, G. et al. (2018). Propylene oxide inhibits propylene epoxidation over Au/TS-1. *J. Catal.* 365: 105–114.

76 Qi, C., Akita, T., Okumura, M. et al. (2003). Effect of surface chemical properties and texture of mesoporous titanosilicates on direct vapor-phase epoxidation of propylene over Au catalysts at high reaction temperature. *Appl. Catal. A Gen.* 253: 75–89.

77 Chowdhury, B., Bravo-Suárez, J.J., Daté, M. et al. (2006). Trimethylamine as a gas-phase promoter: highly efficient epoxidation of propylene over supported gold catalysts. *Angew. Chem. Int. Ed.* 45: 412–415.

78 Chowdhury, B., Bando, K.K., Bravo-Suárez, J.J. et al. (2012). Activity of silylated titanosilicate supported gold nanoparticles towards direct propylene epoxidation reaction in the presence of trimethylamine. *J. Mol. Catal. A Chem.* 359: 21–27.

79 Kanungo, S., Keshri, K.S., Hoof, A.J.F.V. et al. (2016). Silylation enhances performance of Au/Ti-SiO_2 catalysts in direct epoxidation of propene using H_2 and O_2. *J. Catal.* 344: 434–444.

80 Kanungo, S., Keshri, K.S., Hensen, E.J. et al. (2018). Direct epoxidation of propene on silylated Au-Ti catalysts: a study on silylating procedures and effect on propane formation. *Catal. Sci. Technol.* 8: 3052–3059.

81 Wang, G., Du, W., Duan, X. et al. (2021). Mechanism-guided elaboration of ternary Au–Ti–Si sites to boost propylene oxide formation. *Chem Catal.* 1: 885–895.

82 Uphade, B., Okumura, M., Tsubota, S., and Haruta, M. (2000). Effect of physical mixing of CsCl with Au/Ti-MCM-41 on the gas-phase epoxidation of propene using H_2 and O_2: drastic depression of H_2 consumption. *Appl. Catal. A Gen.* 190: 43–50.

83 Lu, J., Zhang, X., Bravo-Suárez, J.J. et al. (2009). Effect of composition and promoters in Au/TS-1 catalysts for direct propylene epoxidation using H_2 and O_2. *Catal. Today* 147: 186–195.

84 Jin, Q., Bao, J., Sakiyama, H., and Tsubaki, N. (2011). Preparation, structure and performance of TS-1 zeolite-coated Au–Pd/TiO_2–SiO_2 capsule catalyst for propylene epoxidation with oxygen and hydrogen. *Res. Chem. Intermed.* 37: 177–184.

85 Feng, X., Yang, J., Duan, X. et al. (2018). Enhanced catalytic performance for propene epoxidation with H_2 and O_2 over bimetallic Au–Ag/uncalcined titanium silicate-1 catalysts. *ACS Catal.* 8: 7799–7808.

86 Li, Z., Gao, L., Ma, W., and Zhong, Q. (2019). Higher gold atom efficiency over Au-Pd/TS-1 alloy catalysts for the direct propylene epoxidation with H_2 and O_2. *Appl. Surf. Sci.* 497: 143749.

87 Wang, G., Du, W., Zhang, Z. et al. (2021). Combining trace Pt with surface silylation to boost Au/uncalcined TS-1 catalyzed propylene epoxidation with H_2 and O_2. *AIChE J* 68: e17416.

88 Liu, T., Hacarlioglu, P., Oyama, S.T. et al. (2009). Enhanced reactivity of direct propylene epoxidation with H_2 and O_2 over Ge-modified Au/TS-1 catalysts. *J. Catal.* 267: 202–206.

89 Lu, J.Q., Li, N., Pan, X.R. et al. (2012). Direct propylene epoxidation with H_2 and O_2 over in modified Au/TS-1 catalysts. *Appl. Catal. A Gen.* 28: 179–182.

15

Zeolites Containing Heteroatoms/Metal Nanoparticles for Catalytic Conversion of Light Alkanes

Hang Zhou, Liang Wang, and Feng-Shou Xiao

Zhejiang University, Key Lab of Biomass Chemical Engineering of Ministry of Education, College of Chemical and Biological Engineering, Yuhangtang Road 866, Hangzhou 310027, China

15.1 Introduction

The efficient utilization of fossil resources has profound significance on the economy [1]. C_1–C_4 light alkanes can be derived from a wide range of sources such as natural gas, oilfield-associated gas, coal bed gas, and refinery gas, and they have become cheap and abundant carbon resources with the exploitation of shale gas and undersea fire ice [2–4]. Direct production of value-added chemicals or important intermediates from light alkanes as raw materials is currently important research in catalysis [5–7]. However, activation of light alkanes is quite difficult due to their weak molecular polarity and the inertness of their C—H bonds. For example, commercial methane steam reforming usually requires reaction temperatures higher than 800 °C to achieve substantial conversion and productivity [8, 9]. In the past decades, various routes of light alkanes transformation into value-added products have been extensively studied such as the selective oxidation of methane [10–12] and dehydrogenation of C_2–C_4 alkanes [13–15]. For these processes, researchers have developed plentiful types of catalysts and reaction processes, some of which have achieved a high level of performance and applied in industrial routes. However, the above processes still have challenges such as insufficient stability of catalysts, rapid deactivation due to coke deposition, and low catalytic efficiency under mild conditions.

The construction of highly efficient catalysts is a central issue in improving the reaction processes [16]. Compared with generally supported metal nanoparticle catalysts for the conversion of light alkanes, zeolite materials have significant advantages including high stability, uniform micropores, accurately controllable acidity/basicity, and heteroatoms in the framework with uniform coordination [17, 18]. The zeolite-supported metal nanoparticle catalysts, such as Pt/ZSM-22 [19, 20], Ni/ZSM-5 [21, 22], and Pt/SAPO-11 [23, 24], have been successfully applied

Micro-Mesoporous Metallosilicates: Synthesis, Characterization, and Catalytic Applications, First Edition. Edited by Peng Wu and Hao Xu.

in petrochemical reactions. Compared with these supported metal nanoparticle catalysts, a trend to fix the metal nanoparticles within zeolite crystals (metal@zeolite) rather than supporting them on the external surface could enhance the selectivity and stability of the metal nanoparticle-based catalysts [25, 26]. For example, the zeolite framework effectively stabilizes the metal nanoparticles against sintering; the zeolite micropores significantly provide shape selectivity to the metal nanoparticles within zeolite crystals. In addition to the metal@zeolite catalysts, the siliceous zeolites containing heteroatoms offer conceptually new active sites for catalysis [27–29]. The heteroatoms-containing zeolites, that is, metallosilicates with tetrahedrally coordinated framework metal ions, are different from those bulk oxides, resulting in the formation of new catalysts for the conversion of light alkanes with distinctive properties, compared to the general ones.

In this chapter, we summarized the representative cases of zeolites in solving a variety of emerging problems in catalytic propane dehydrogenation (PDH, Section 15.2), ethane dehydrogenation (EDH, Section 15.3), and methane oxidation (Section 15.4). The current techniques for these reactions and the structure-performance interplay of zeolite-based catalysts relative to the previous catalysts are briefly introduced in each section. We also summarized the current challenges and discussed the trends of zeolite and metallosilicate-based catalysts in light alkane conversions (Section 15.5).

15.2 Metal Nanoclusters and Heteroatoms in Zeolite for Propane Dehydrogenation

Light olefins, such as propylene and ethylene, are the most important building blocks in the modern chemical industry. These olefins have been extensively used as upstream raw materials for synthesizing polymers and other important chemicals [13–15]. In 2020, the annual production of propylene and ethylene reached 140 000 000 and 200 000 000 tons, respectively [13]. Before the 2010s, both ethylene and propylene are mainly obtained from the cracking of naphtha [30, 31]. Recently, with the shortage of petroleum resources, developing a new route for obtaining ethylene and propylene from alternative feedstocks is highly desired. Currently, various routes for producing light olefins from coal and biomass have been developed [32–34], but with the increased production of shale gas, the dehydrogenation of light alkanes to olefins becomes the most profitable process [13–15]. Particularly, PDH has been successfully commercialized in the past years, where various techniques and catalysts have been developed such as the Catofin technology using CrO_x/Al_2O_3 catalysts operated by ABB Lummus and Oleflex technology using $PtSn/Al_2O_3$ catalysts operated by Honeywell UOP. These routes produced over 15 000 000 tons of propylene globally in 2020 to fill the gap between propylene supply and demand [35].

There are still several residual challenges remaining in catalytic PDH to propylene. For example, the reaction generally suffers from multiple side reactions such as cracking and coking. The catalyst deactivation from coke formation results in frequent regeneration in the PDH operations. For example, the CrO_x/Al_2O_3 catalysts in the commercial Catofin process were regenerated each 10–15 minutes in dehydrogenation operation [13, 36, 37]. In the commercial Oleflex process, where the alumina-supported Pt nanoparticles with SnO_x promoter (Pt-Sn/Al_2O_3) were employed as catalysts, the average operating durability of the catalyst in one cycle is more than one day due to the hydrogen protection, but the catalysts still need for calcination in air to remove the coke [13, 38]. In the regeneration process, oxychlorination treatment is required to re-disperse the sintered Pt nanoparticles [39]. The coking- and sintering-resistant catalysts for PDH are highly desired to reduce the cost and improve the production efficiency of propylene, but it is still a challenge for the traditionally supported metal/oxide nanoparticle catalysts.

15.2.1 Zeolite-Based Catalysts for Non-Oxidative PDH

In contrast to traditional supports such as alumina and amorphous silica, zeolites have ordered and uniformed microporous channels, which could serve as an ideal platform to disperse and stabilize nanometric and subnanometric particles. Corma and co-workers developed a one-pot synthesis method to fix the PtSn nanoclusters within siliceous MFI zeolite crystals (PtSn@MFI) [40]. They considered that the tetrapropylammonium hydroxide (TPAOH) templates could fill the intersectional voids of zeolites, and K^+ ions are introduced to compensate for the silanol groups. These features benefit Pt locating in the sinusoidal channels of MFI zeolites and prevent their migration under harsh conditions. After introducing atomically dispersed Sn species for modifying the electronic structures of Pt clusters, the K-PtSn@MFI catalysts with subnanometric Pt–Sn clusters fixed in zeolite sinusoidal channels were finally obtained. In their more recent works, Corma and co-workers further revealed the influence of modulating the exact interaction of Pt–Sn species located in the sinusoidal 10-MR channels of zeolites on the catalytic performances [41]. The modulation of Pt–Sn interactions was achieved via activating (and/or reducing) the catalysts under different conditions. The structural features of Pt–Sn species were recognized via in situ extended X-ray absorption fine structure (EXAFS), high-angle annular dark-field scanning transmission electron microscopy (HAADF-STEM), and CO-adsorption. They tested the catalytic performance of the catalysts for the dehydrogenation of undiluted propane under more industrially relevant conditions. The optimized catalyst showed a 1 order of magnitude lower deactivation rate in the PDH reaction, and the initial propane conversion was ~20% with propylene selectivity >97% at 550 °C under an extremely high weight hourly space velocity of ~118 h^{-1}. After the reaction for 70 hours, the propane conversion decreased slightly to ~17%, demonstrating the enhanced catalytic activity and stability of the PDH catalyst with optimized Pt–Sn interaction.

A variety of other metals such as zinc and rare earth metals have also been widely used as promoters for Pt-based PDH catalysts [42, 43]. The effects of these metal promoters were considered to be both geometric and electronic modifications. They may optimize the dispersion of Pt species, modify the electronic structure of the Pt particles, and further improve the catalytic performance of the catalyst.

For example, Ryoo and co-workers fixed PtLa-, PtCe-, and PtY-alloyed nanoparticles on the mesoporous MFI zeolite [43]. In the PDH reaction using undiluted propane, these catalysts all showed an initial conversion of up to 40%, which are close to the thermodynamic equilibrium. The PtLa catalyst (PtLa/mz-deGa) can even maintain the conversion above 25% for more than 10 days. Yu and co-workers encapsulated Pt–Zn bimetallic sub-nano clusters into siliceous MFI zeolite ($PtZn_4$@S-1-H) under the assistance of ethylenediamine ligand [42]. The $PtZn_4$@S-1-H catalyst exhibited stable performances even in a test for more than 200 hours with the absence of hydrogen in the feed gas, giving an average coke formation rate two orders lower than that on the alumina-supported catalyst under the equivalent test conditions (Figure 15.1).

Besides acting as a promoter for platinum-based catalysts, ZnO has abundant Lewis acid sites that could directly catalyze the dehydrogenation of alkanes [44]. Kondratenko and Jiang successfully synthesized binuclear ZnO_x species on Silicalite-1 by simple mechanical mixing and in situ reduction in hydrogen (Figure 15.2). The obtained catalyst exhibited superior catalytic durability and regenerability relative to the commercial K-CrO_x/Al_2O_3 catalysts in the 400 hours of PDH reaction test, and the space–time yield of propylene was improved by more than three times [44].

15.2.2 Zeolite-Based Catalysts for Oxidative PDH

Although the direct non-oxidative dehydrogenation (ODH) process of alkanes has been commercialized in the past decades, the conversion-per-pass of propane under anaerobic conditions could not be infinitely improved due to the thermodynamic limitation. Direct dehydrogenation coupled with hydrogen combustion could overcome this limitation. Besides, due to the hindered coke formation in oxidative atmospheres, rapid deactivation was effectively inhibited. However, the propylene product and the intermediates (isopropyl, propoxy, etc.) are very active in being over-oxidized into CO_2 under aerobic conditions. A wide scope of metal oxide catalysts has been developed for the oxidative PDH, but a few of them can achieve propylene selectivity higher than 50% while maintaining propane conversion higher than 20% [2, 45].

In 2016, Hermans and co-workers found that boron nitride catalysts showed very promising selectivity to olefins in propane ODH (e.g. selectivity of 79% propene and 12% ethene at 14% propane conversion) [46]. This finding has got extensive attention as it represents a big step forward in selective oxidation research. Subsequently, a variety of boron-containing compounds, such as boron carbide, silicon boride, and metal borides, were also found catalytically active and selective for oxidative PDH [47, 48]. Notably, the intrinsic active sites of both boron nitride and other boron-containing catalysts are the surface BO_xH_y species in situ formed during the reaction [49, 50]. Unfortunately, the amorphous molten BO_xH_y species is viscous, which is obviously incapable for large-scale industrial

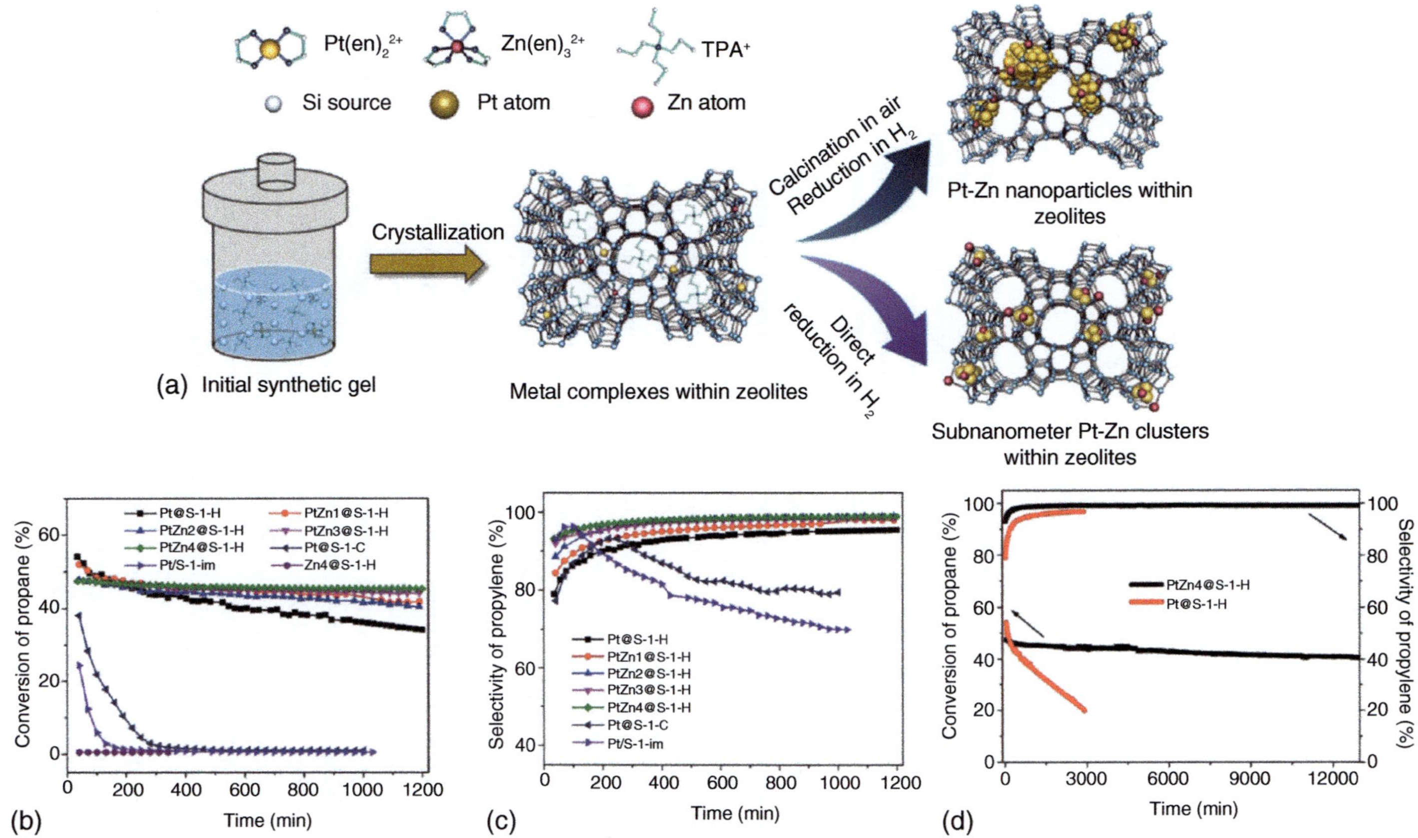

Figure 15.1 (a) Schematic representation of the synthesis of zeolite-encaged subnanometer Pt–Zn cluster catalysts. (b) The conversion of propane and (c) selectivity of propylene over various catalysts. (d) Long-term stability test for PDH conversion over PtZn4@S-1-H and Pt@S-1-H catalysts. Source: Reproduced with permission from Ref. [42]. Copyright 2020 Wiley.

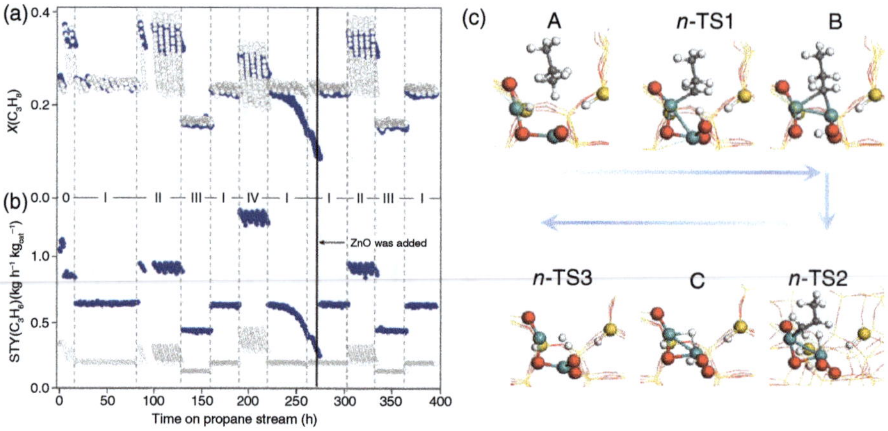

Figure 15.2 On-stream durability test profiles of (a) propane conversion and (b) propylene space–time-yield over ZnO_x–S-1 catalyst (blue) and K–CrO_x/Al_2O_3 catalysts (gray). Reaction conditions were varied for region O to IV. (c) Scheme showing the reaction mechanism of PDH on the reduced binuclear ZnO_x sites on the surface of S-1. Cyan, red, gray, yellow, and white spheres stand for Zn, O, C, Si, and H atoms, respectively. Source: Reproduced with permission from Ref. [44] Copyright 2021 Springer Nature.

devices at high temperatures. Dispersing boron species into the framework of highly hydrothermal stable inorganic materials, forming isolated boron species, seems to be an effective solution for boron leaching. Hermans and co-workers synthesized B-MWW zeolite containing solely isolated trigonal boron species in the zeolite framework [51]. Although the boron content is as high as ~10 wt%, no propane conversion was observed under conventional reaction conditions. This result is consistent with the previous view that the excellent reaction performance of boron catalysts is attributed to the general knowledge of multiple boron centers as active sites, whereas isolated boron centers were catalytically inactive in propane ODH [52].

By contrast, recent research by Xiao and co-workers indicates that isolated boron sites substituted in MFI-typed silicate zeolite (BS-1) show promising catalytic activity, durability, and selectivity to propylene in propane ODH, which is superior to the traditionally supported boron oxide catalytic material [53]. It is confirmed that the boron sites in BS-1 are modified with a double hydroxyl structure and existed in the form of —B[OH…O(H)—Si]$_2$. In propane ODH, the boron dihydroxy and the adjacent silicon hydroxyl groups simultaneously activate propane and molecular oxygen, forming a stable eight-member ring transition state, and the reaction energy barrier is significantly lower than that on the boron monohydroxy or B—O—B structure. More importantly, the reversible hydrolysis-condensation process of the B—O—H…O(H)—Si and B—O—Si structures in the reaction effectively inhibit boron-leaching from the zeolite, which efficiently improved the stability of boron-based catalysts (Figure 15.3).

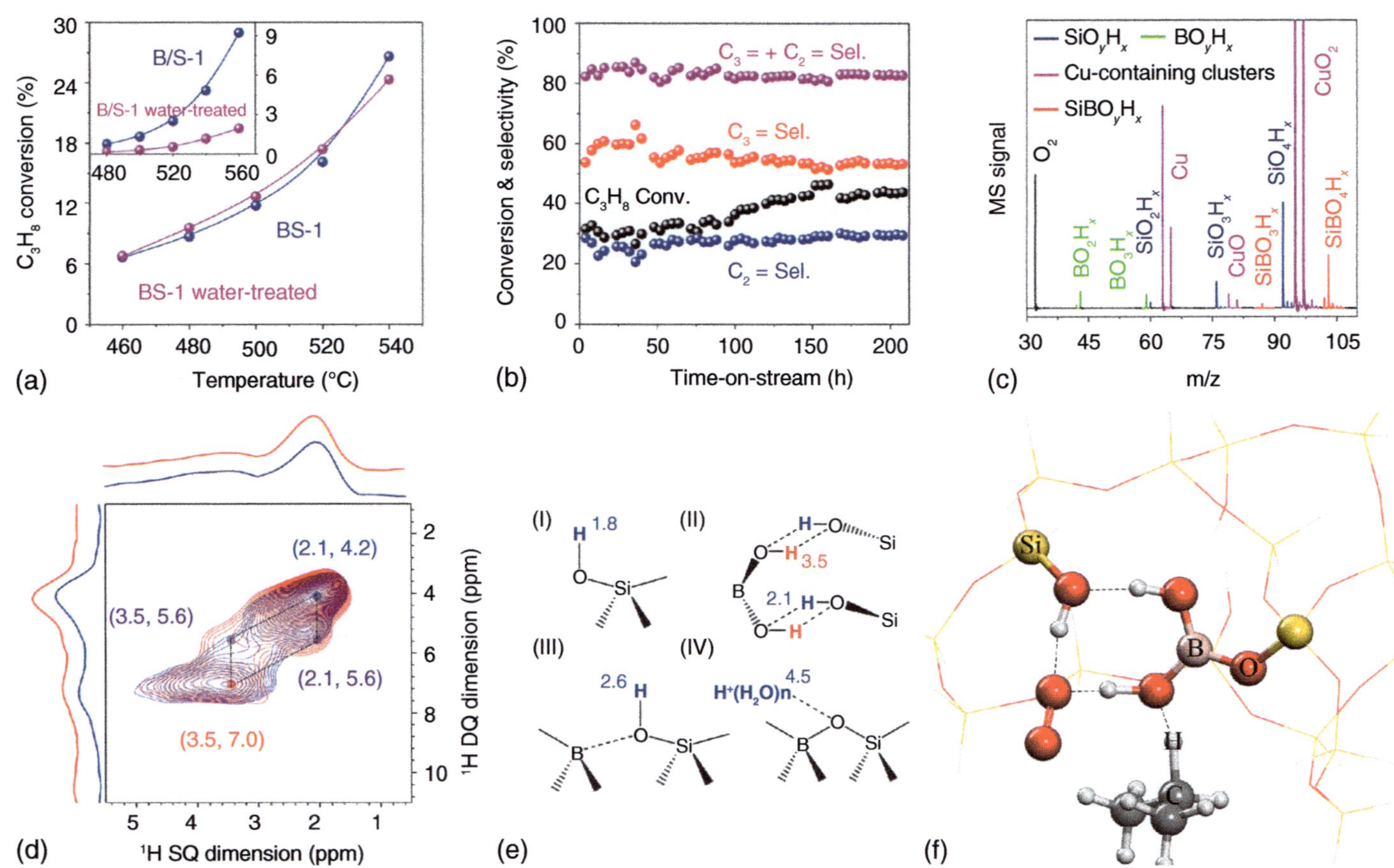

Figure 15.3 (a) Data characterizing the performances of BS-1 and B/S-1 before and after water treatment. (b) Data characterizing the durability of BS-1. (c) ToF-MS spectra of BS-1. B-, Si-, and Cu-containing species were characterized by a set of peaks because of the isotopes of these elements. The isotope effect allowed differentiating a few fragment ions with similar m/z values from each other. (d) 2D $^1H-^1H$ DQ MAS NMR of BS-1 before (blue) and after (red) catalytic propane oxidative dihydrogen reaction. (e) Different H species and their chemical shifts. (f) Transition state structure of $-B[OH\ldots O(H)-Si]_2$ reacting with oxygen and propane. Source: Reproduced with permission from Ref. [53] Copyright 2021 American Association for the Advancement of Science.

15.3 Metallosilicates for Ethane Dehydrogenation

Ethylene is mainly produced from the steam cracking of naphtha before 2010s [30, 31]. With the advancement of the shale gas revolution during the past decade, cheap and abundant ethane has gradually replaced naphtha as the main feedstock for ethylene production. Currently, commercial production of ethylene from ethane is still mainly carried out in tubular steam cracking furnaces following non-catalytic pyrolysis/dehydrogenation processes, in which ethane is dehydrogenated in the presence of steam at high temperatures of 800–1100 °C [8, 9]. This process is accompanied by a minor proportion of C—C bond cracking, reforming, and other side reactions, but the ethylene yield can still reach as high as 80%, benefiting from the high reaction temperature. By contrast, traditional steam cracking of naphtha simultaneously produces a large proportion of heavy hydrocarbons, and the yield to ethylene is typically below 35% [30, 31].

Catalytic dehydrogenation of ethane has also been studied for decades, where the catalyst could activate the C—H bond and benefit the dehydrogenation under mild conditions, compared with the steam cracking [1]. Propylene selectivity is expected to be further improved because the complex side reactions under high-temperature conditions (e.g. >800 °C) can be avoided. The catalytic kinetics and mechanisms of EDH are believed to be consistent with that of PDH, and the catalyst system for the dehydrogenation of ethane and propane is mutually universal [54]. Accordingly, the technical challenges in PDH such as rapidly coke forming and metal sintering are even more serious in EDH.

15.3.1 Metallosilicates for Non-Oxidative Dehydrogenation of Ethane

Recently, several studies revealed that isolated metal sites in zeolite crystals were potentially efficient active sites in EDHs. Xiao and co-workers synthesized iron-containing siliceous MFI zeolite (FeS-1) with the assistance of sodium ethylenediaminetetraacetate, which is an efficient, cheap, and coke-resistant Fe-based catalyst for the direct dehydrogenation of ethane, overcoming the fast deactivation of conventional supported Fe catalysts [55]. Dehydrogenation of ethane over the FeS-1 can achieve conversion close to the limit of thermodynamic equilibrium under various conditions, with negligible deactivation in a durable test of 50 hours. This catalyst has overturned the catalytic properties of conventional supported metallic Fe and FeO_x, which suffer from rapid reduction and carbonization, the resultant FeC_x would efficiently catalyze the coke formation to give poor activity and durability. Other isolated metal sites in zeolite were also reported as highly active and selective catalysts for EDH. They further developed a more efficient cobalt-containing zeolite catalyst (Co/S-1) through a facile mechanical milling method to benefit the cobalt species that were spontaneously dispersing and riveting into the siliceous zeolite crystals [56]. The isolated cobalt species within zeolite frameworks are extraordinarily active for EDH. The space–time yield of ethylene could reach as high as 13.4 $kg_{C2H4}\ kg_{cat}^{-1}\ h^{-1}$, which outperforms the non-oxidative and even oxidative EDH catalysts tested previously (Figure 15.4).

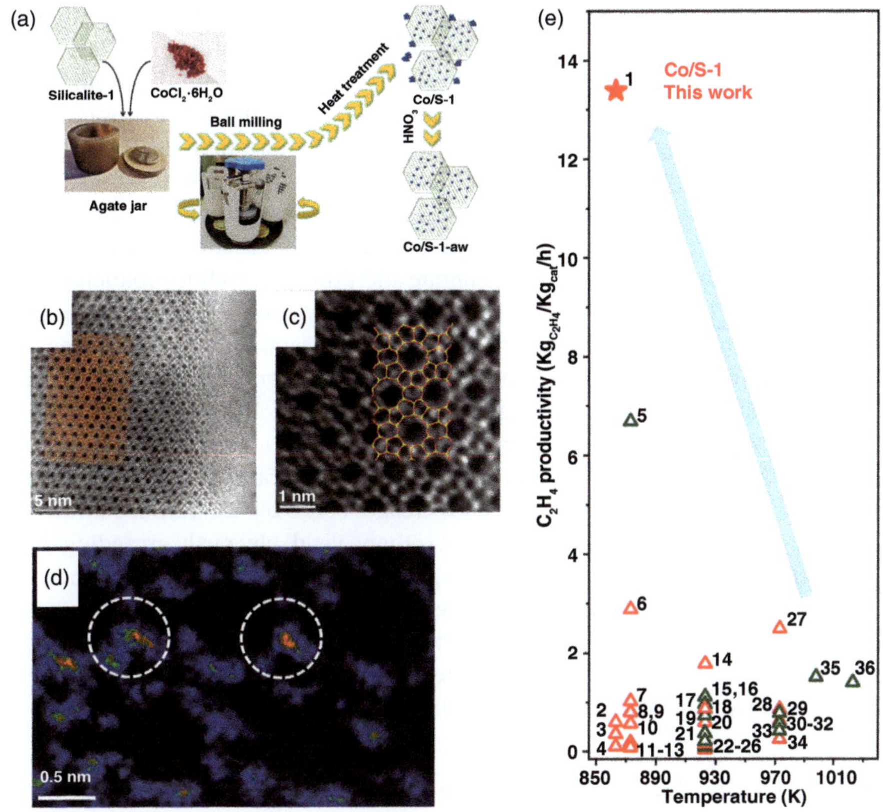

Figure 15.4 (a) Procedures for the synthesis of Co/S-1 following a mechanical method. (b) HAADF-STEM images and (c) STEM-iDPC images of Co/S-1. (d) The colored HAADF-STEM image with enhanced contrast highlights the cobalt sites in the white circles with higher Z-contrast. (e) Comparison between the performances of Co/S-1 (pentagram) with various catalysts previously tested (triangles) in EDH. The green symbols represent the data in oxidative EDH, and the red symbols represent those in non-oxidative EDH. Source: Reproduced with permission from Ref. [56] Copyright 2022 Elsevier.

The *d*-orbitals of In^{3+} are fully occupied with delocalized *s-/p*-bands as the host orbitals; therefore, the unique electronic properties of the isolated In^{3+} site may facilitate the desorption of alkenes. In the studies by Shimizu and co-workers, it was found that indium hydrides could be formed on In^{3+} exchanged CHA zeolite (In-CHA) after H_2 treatment [57]. The detailed structure of indium hydrides in CHA zeolites was confirmed by in situ FTIR and XAFS spectroscopy. In-CHA also showed a higher ethylene selectivity and stability for EDH compared with CHA zeolites exchanged with other ions such as Zn^{2+} and Ga^{3+}.

15.3.2 Metallosilicates for Oxidative Dehydrogenation of Ethane

Although the practice has proved that the commercialized catalytic dehydrogenation technologies such as Catofin and Oleflex can be applied to convert mixed alkanes including ethane, currently, very few units are used for EDH to ethylene [13, 14, 54]. An important reason is that EDH is more significantly limited by the thermodynamic equilibrium relative to PDH. For example, the equilibrium conversion of dehydrogenation of 0.1 MPa atmospheric ethane at 600 °C is only 16% [13]. To achieve a high conversion-per-pass (such as more than 70%) that meets the requirements of industrialization, a high reaction temperature is still required. Therefore, the industrialization process of catalytic dehydrogenation of ethane to ethylene is slower than that of PDH. The oxidative process could overcome such limitation. More importantly, it may generate additional heat while producing ethylene. According to the assessment, if the current ethylene production technology, such as steam cracking, was replaced by an aerobic process, nearly half of the energy consumption in ethylene production can be saved [30]. Concerning that ethylene is the most important organic chemical in the world, the technological revolution of ethylene production will have a crucial impact on the world energy market pattern.

Zhang and co-workers craft a main-group indium oxide catalyst In/HY [58], where the protons of Brønsted acid sites in the supercages of HY zeolite were substituted by $[InOH]^{2+}$. This catalyst exhibited up to 60% ethene yield, obviously exceeding those of most transition metal oxide catalysts reported previously. The isolated $[InOH]^{2+}$ sites were active sites for EDH, confirmed by both experimental and theoretical characterizations (Figure 15.5). This work may open a new door for designing a unique catalytic process of ethane to ethylene with efficiency far beyond the general non-oxidative and oxidative dehydrogenation of ethane.

15.4 Zeolite Catalysts for Selective Oxidation of Methane

Converting methane to methanol is one of the most important research in heterogeneous catalysis, as it can provide a potentially important route for methanol synthesis from cheap and abundant methane [59]. Methane to methanol can also be considered as a basic model reaction to understand the catalytic process and structure–activity relationship in fundamental catalytic research. In a typical process of natural gas chemical plant, methane to methanol is performed via a two-step process. Initially, methane is reformed to produce syngas (a mixture of CO and H_2) at temperatures higher than 800 °C [60, 61], and then the syngas is transformed to methanol at high pressure in a CO hydrogenation process [62, 63]. Although direct oxidation of methane seems more commercially attractive, at present this catalytic efficiency is still insufficient to meet the demand. It is difficult to activate methane under mild conditions due to its low polarity and high C—H bond energy (140 kcal/mol) [59]. Currently, the typical effective reaction system of methane oxidation to methanol or other oxygenates can be divided into categories in the following:

(i) Direct gas phase oxidation at high temperature

Direct oxidation of methane at high temperatures is the most common route to activate C—H bonds effectively, which has been used in commercial processes

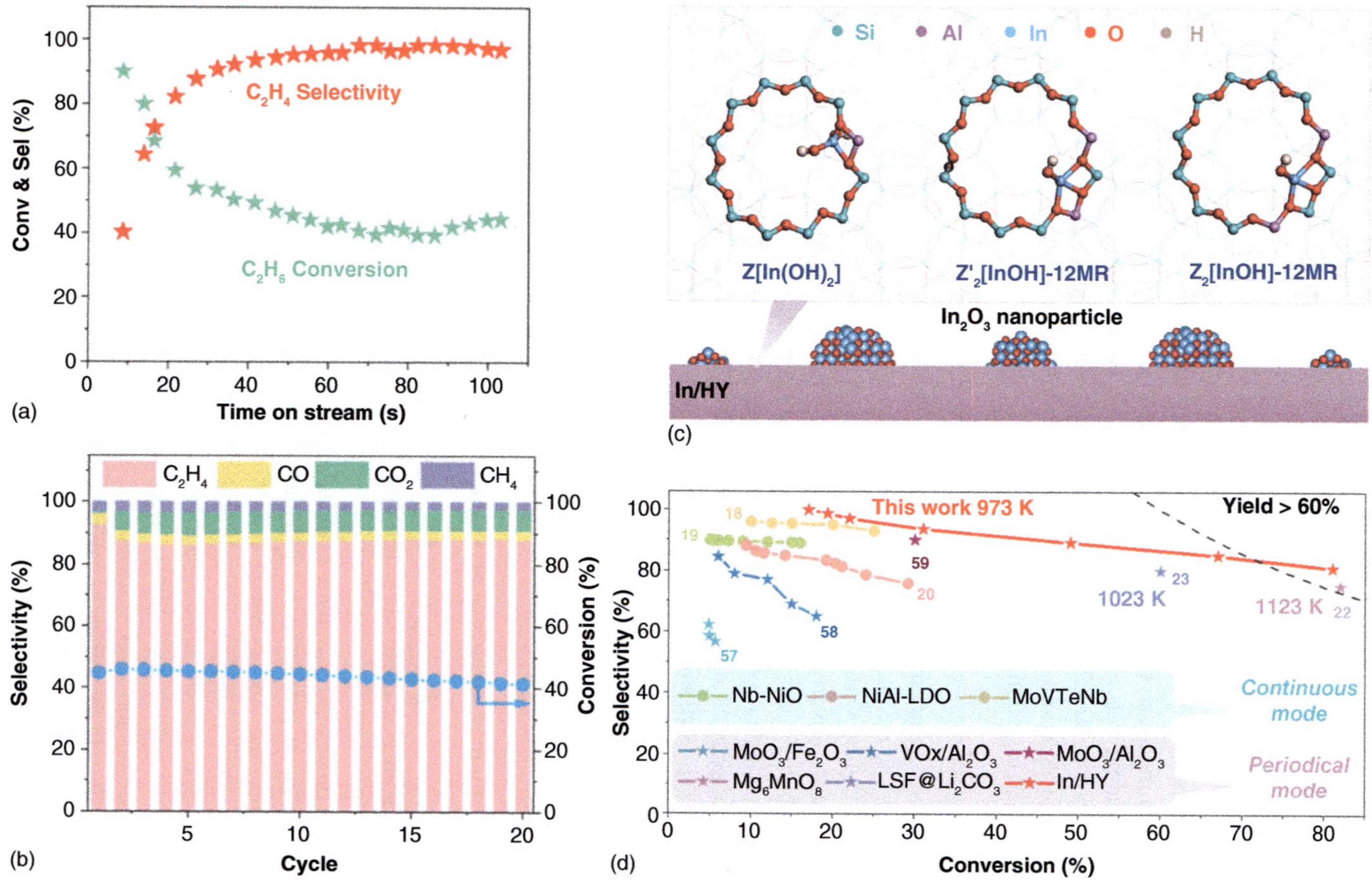

Figure 15.5 (a) Ethane conversion and ethene selectivity vs. time for 11In/HY. (b) Stability test of 11In/HY during 20 cycles. (c) Scheme of the proposed local structure of indium-modified HY. (d) Performance comparison of oxidative dehydrogenation of ethane. Continuous mode: co-feeding of C_2H_6 and O_2; periodical mode: alternate C_2H_6 and O_2 feeds. Source: Reproduced with permission from Ref. [58] Copyright 2022 American Chemical Society.

such as oxidative reforming methane to syngas. In the 1980s, MO_3/SiO_2 catalyst was used to activate methane to methoxy intermediates and further reacts with water to form methanol [64]. Herman et al. used a double-bed catalyst system consisting of SrO/La_2O_3 and 2 wt% MO_3/SiO_2 to oxidize methane to formaldehyde. At 630 °C, the methane conversion and formaldehyde productivity reached to 8.2% and 129.0 g kg_{cat}^{-1} h^{-1}, respectively [65]. Parmaliana et al. further optimized the reaction process and improved the yield of formaldehyde by timely separating the products formed [66]. Wang and co-workers used mesoporous silica-supported transition metals to directly oxidize methane and ethane with air to produce the corresponding aldehydes at 600 °C [67–69]. It is of significance that the target product can be directly obtained by the one-step process, but the yield to aldehydes was relatively low as it is easier to generate a large proportion of CO_x by deep oxidation at high temperatures.

(ii) Low-temperature oxidation in liquid phase using active oxidant

Low-temperature methane oxidation in the liquid phase has also been extensively studied in the past century. Initially, a catalyst system containing Pt and toxic Hg was widely used with fuming sulfuric acid as an oxidant. In this process, methane was initially converted to methyl sulfate, which was subsequently hydrolyzed to methanol [70]. Later, Au, Pd, and other precious metals were used in the catalyst preparation instead of toxic Hg, but the catalytic process is still complicated to operate and requires the use of highly oxidizing and corrosive fuming sulfuric acid as an oxidant [71]. Furthermore, reprocessing the refractory by-products such as SO_2 is also a severe problem.

Generally, methane activation still requires a high energy input or strong oxidants. In addition, methanol or formaldehyde is more reactive than methane, which often leads to deep oxidation forming carbon oxides.

15.4.1 Bionic Catalysis of Methane to Methanol Using Metallosilicates

Different oxidation catalysts and reaction processes have been developed to activate methane effectively while inhibiting deep oxidation. By analyzing the previous research progress, the scientific trend of methane oxidation is to develop advanced catalytic processes or catalytic materials that enable the conversion of methane to methanol using green and cheap oxidants under mild conditions.

Methane monooxygenase is a class of proteases containing single or double Fe/Cu centers [73]. The high dispersion and unique coordination structure of Fe/Cu sites enable the monooxygenase to effectively catalyze the selective oxidation of methane with high methanol selectivity. According to this structural characteristic, Groothaert et al. dispersed Cu species into the micropores of ZSM-5 and MOR zeolite via the ion-exchange method, and the resultant zeolites containing Cu ions were further activated by oxygen [72]. The obtained catalyst can catalyze the selective oxidation of methane to methanol at a low temperature of 125 °C. The high efficiency is a benefit of the zeolite pores stabilizing the bis (μ-oxo) dicopper Core (Cu_2O_2), similar to the structure of Cu centers in the enzyme (Figure 15.6). Based on the Cu-zeolite structure, Bokhoven et al. designed a stepwise reaction process [74].

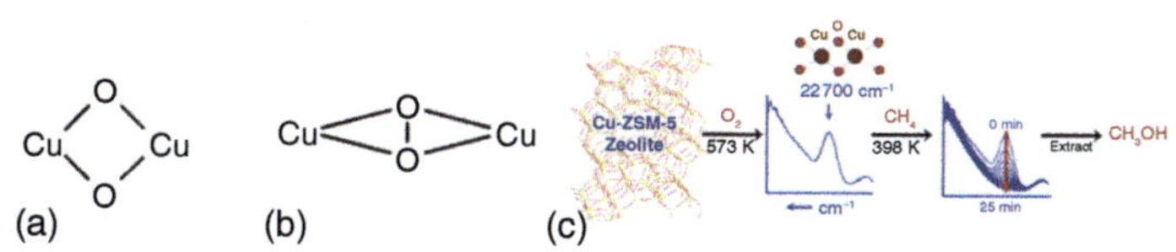

Figure 15.6 Schemes showing the structure of (a) bis (μ-oxo) dicopper core, $[Cu_2\text{-}(\mu\text{-}O)_2]^{2+}$, and (b) (μ-$\eta^2$:$\eta^2$-peroxo) dicopper center. (c) Fiber-optic UV–vis spectra of O_2-activated Cu-ZSM-5 catalyst during reaction with CH_4. Source: Reproduced with permission from Ref. [72] Copyright 2005 American Chemical Society.

First, the Cu-MOR catalyst was activated by oxygen at 450 °C, and then the reactor was cooled down to 200 °C for methane activation. The methanol product was still adsorbed in the zeolite pore. Feeding a mixture of water and acetonitrile is required to remove the main product of methanol. The main stoichiometric equation of the reaction is $CH_4 + 1/2\,O_2 \rightarrow CH_3OH$, accompanied by a minor proportion of peroxide product CO_2 in the actual process. Later, Bokhoven et al. also found that water could be used instead of oxygen in the catalyst activation step, and the reaction was shifted to an anaerobic process while by-producing hydrogen ($CH_4 + H_2O \rightarrow CH_3OH + H_2$) [75]. 0.204 mol of methanol can be produced per mole of Cu species in a whole cycle. Biomimetic catalysis provides a novel strategy for highly selective methane oxidation, which has extended further to ethylene oxidation reactions [76, 77]. However, several steps including alkane oxidation, product deoxidation, and catalyst regeneration still need to be periodically switched, and even the catalyst bed temperature needs to be changed with the reaction cycling.

Tao et al. anchored single-site Pd inside ZSM-5 zeolite to form an isolated Pd_1O_4 center and catalyzed methane oxidation to formic acid, methyl hydrogen peroxide, and methanol products at low temperature (50–95 °C) using H_2O_2 as the oxidant [78]. It was also found that formic acid formation can be significantly suppressed when CuO_x additive was added, and the selectivity of methanol can be further improved. All of these provide a good case for low-temperature methane oxidation. But converting methane using hydrogen peroxide as an oxidant is commercially unattractive due to the costly hydrogen peroxide. Using cheap and environmentally friendly hydrogen peroxide as the oxidant for methane activation can solve this issue.

15.4.2 In situ Synthesizing Hydrogen Peroxide for Low-Temperature Methane Oxidation Using Metal@zeolite

As early as 2001, Mizuno et al. developed hydrogen to assist oxygen in activating methane and produced formic acid at a low temperature of 150 °C [79]. It is also discovered that hydrogen could directly react with oxygen to produce hydrogen peroxide under the catalysis of precious metals such as Au and Pd [80]. These discoveries provide a new strategy for the in situ synthesis of hydrogen peroxide and further catalyzing selective oxidation of alkanes. Hutchings and co-workers coupled the two reactions of hydrogen peroxide synthesis and methane oxidation by directly mixing methane, hydrogen, and oxygen to produce the target product of methanol [81]. This process is simple to operate but is still far less efficient than those systems using hydrogen peroxide directly. Taking the $AuPd/TiO_2$ catalytic methane oxidation as an example, the productivity of methanol and methyl hydrogen peroxide in the above process is only 0.116 mol kg_{cat}^{-1} h^{-1}, much lower than that of directly using hydrogen peroxide (1.86 mol kg_{cat}^{-1} h^{-1}). This is due to that the oxidation rate of hydrogen peroxide is closely related to its concentration. In situ generated hydrogen peroxide (or related species, such as HOO* and HO* radicals) is a slow step, and the H_2O_2 products or intermediates will be quickly diluted into the reaction solution, resulting in a low oxidation rate. Besides, the side reactions of hydrogen peroxide decomposition ($H_2O_2 \rightarrow H_2O + 1/2O_2$) and deep hydrogenation ($H_2O_2 + H_2 \rightarrow 2H_2O$) will also affect the oxidation efficiency. Although considerable efforts have been made to optimize the performance of supported metal nanoparticles, the yield to methanol can hardly achieve higher than 3%.

As seen from the previous research progress, most of the studies mainly focused on the composition and size effects of the active sites (single atoms, nanoparticles, etc.) or the strong metal-support interactions for the step of C—H bond activation. The diffusion behavior of the molecules and intermediate involved in the reaction system, especially the hydrogen peroxide, the key intermediate molecule, has not been discussed. The research focus should be extended from the regulation of the active site to the construction of local environment of the catalyst, and the influence on the diffusion of the key intermediate should be highly regarded, which will provide a new idea for the efficient oxidation of light alkanes. Considering that the oxidation efficiency of hydrogen peroxide is closely related to its concentration, Xiao and co-workers designed a “molecular fence” around the active centers to inhibit the dilution of hydrogen peroxide and form local enrichment [82]. They encapsulated the AuPd nanoparticles inside the zeolite crystal and modified the zeolite surface with hydrophobic alkyl chains so that the hydrophilic hydrogen peroxide molecules and the related species generated could hardly diffuse out of the zeolite crystal. By contrast, reactant molecules, including methane, oxygen, and hydrogen, can easily penetrate through the fence, which significantly improves the reaction efficiency of in situ hydrogen peroxide formation and methane activation. These studies indicate that tuning the diffusion properties of the key intermediate plays an important role in the reaction (Figure 15.7).

Converting methane to other higher value-added products is also an attractive process. Flytzani-Stephanopoulos and co-workers used ZSM-5 zeolite containing isolated Rh center as the catalyst to convert methane directly to acetic acid in CO and O_2 gas mixture [83]. On this basis, Tao et al. revealed the mechanism of synergistic catalytic methane activation and C—C coupling between isolated Rh center and MFI zeolite [82]. All these provide important guidance for the low-temperature conversion of methane into value-added chemicals.

15.4.3 Direct Metal Oxidation with Oxygen by Zeolite-Supported Nanoparticle Catalysts

Direct partial oxidation of methane into value-added chemicals using molecular oxygen has attracted much attention in the academia and industry, which is one of the most challenging reactions in catalysis [84, 85]. Due to more active nature of

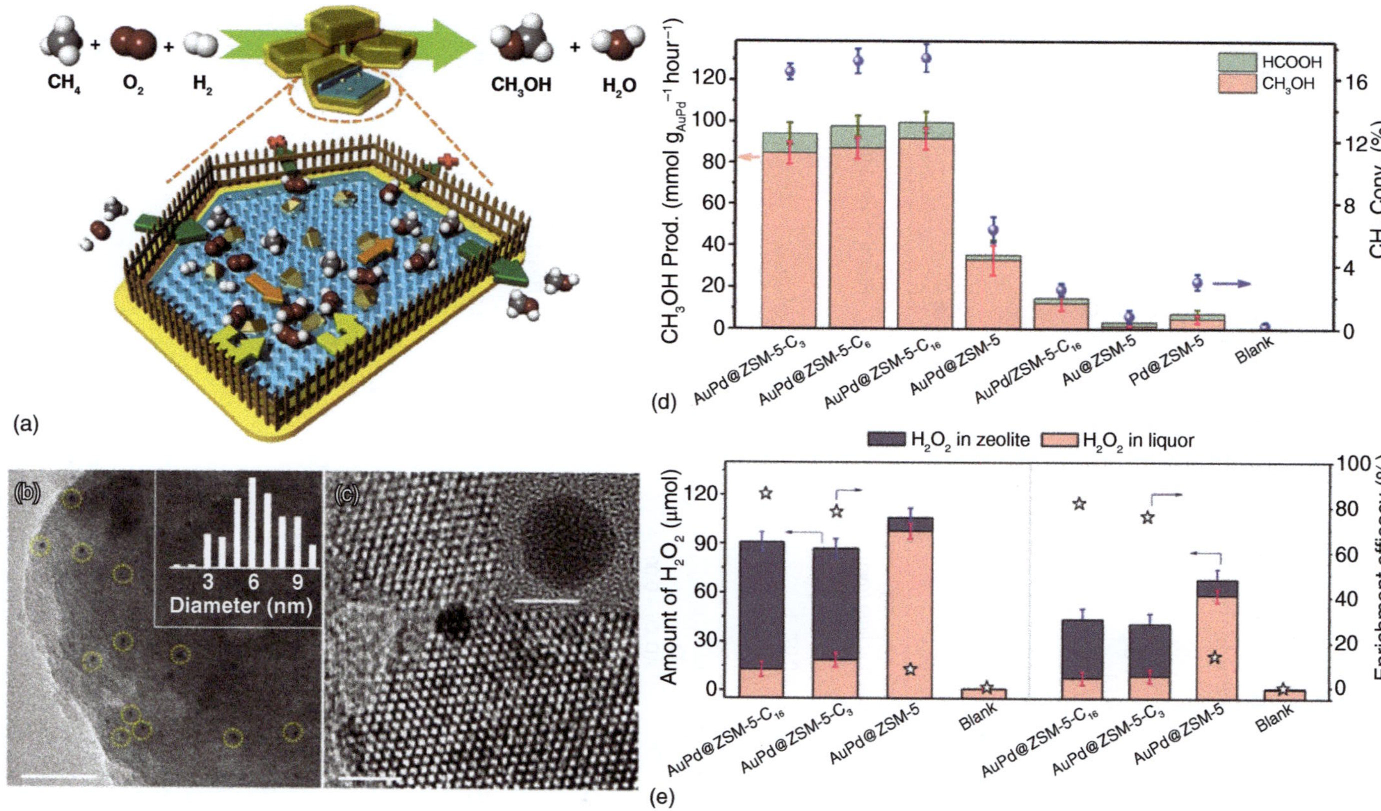

Figure 15.7 (a) Scheme showing the action principle of "molecular fence" in methane oxidation. (b,c) TEM images of AuPd@ZSM-5-C_{16} catalyst. (d) Data characterizing the oxidation of methane with H_2 and O_2 over various catalysts. (e) Data showing the amount of H_2O_2 in the (left) H_2O_2 synthesis and (right) methane oxidation process. The enrichment efficacy is the percentage of H_2O_2 in zeolite crystals to the total amount of H_2O_2 in the reactor. Source: Reproduced with permission from Ref. [82] Copyright 2020 American Association for the Advancement of Science.

the generated oxygen-containing compounds than methane, it is usually easier to over-oxidize to CO_2 and other by-products. The search for selective oxidation routes of methane with industrial prospects under mild conditions is hopeful.

Xu and Hutchings developed ZSM-5 zeolite-supported Au nanoparticles (Au-ZSM-5) for the aerobic oxidation of methane to methanol and acetic acid using molecular O_2 in the temperature range of 120–240 °C [86]. The reductant (H_2 or CO) was absent in this system (Figure 15.8). The content of reaction products

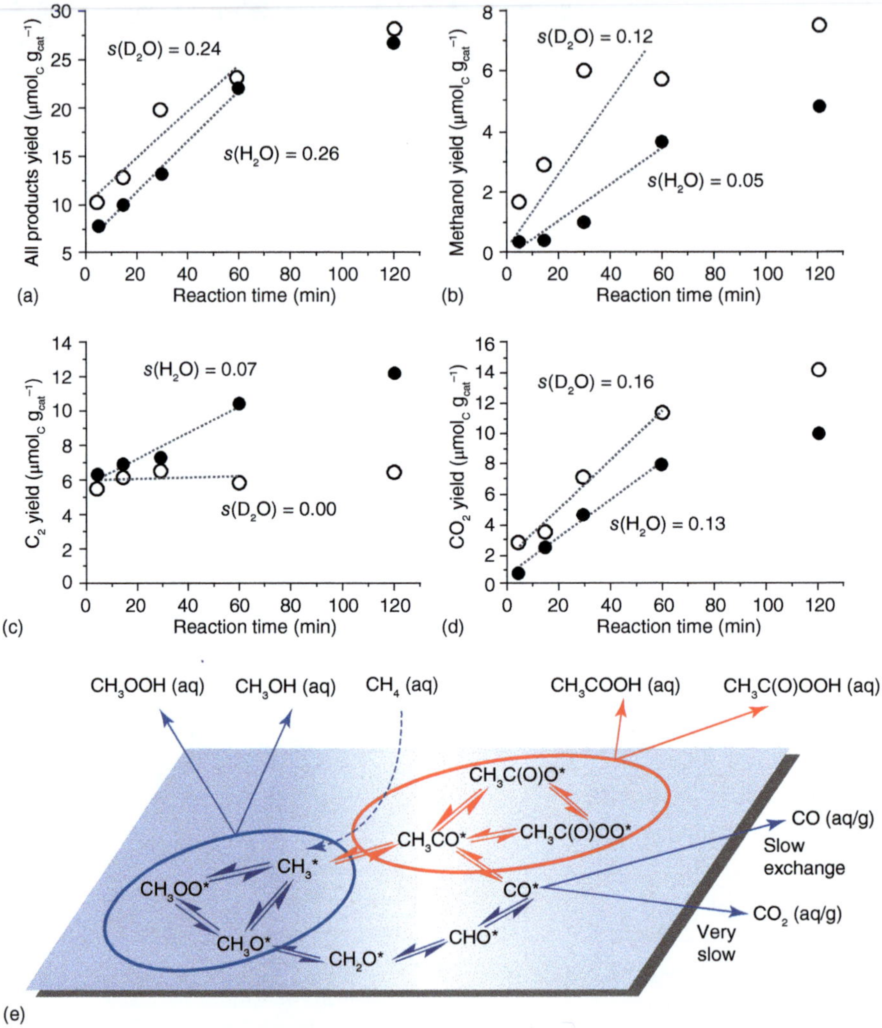

Figure 15.8 Yield of (a) all products, (b) methanol, (c) C_2 product, and (d) CO_2 on Au-ZSM-5 catalysts as a function of time. Data for H_2O solvent are shown with filled symbols and that for D_2O solvent with open symbols. (e) Schematic illustration of the proposed surface catalyzed reactions. Blue arrows and lines are used for C1 intermediates and products, while red arrows and lines are used for C_2 intermediates and products. Source: Reproduced with permission from Ref. [86] Copyright Springer Nature.

and intermediates was quantitatively analyzed using nuclear magnetic resonance (NMR) spectroscopy. It was shown that the maximum yield of oxygen-containing compounds reached 7.3 mol mol_{Au}^{-1} h^{-1} in a relatively short reaction time. Different from the conventional Cu-zeolite catalysts for the formation of C_1 products, this catalyst can also produce C_2 oxygenate products such as acetic acid. This indicates that there are different catalytic reaction mechanisms on the Au-ZSM-5 catalyst in comparison with those on conventional methane oxidation catalysts. Two-dimensional ^1H–^{13}C correlation NMR spectrum combined with $^{12}C/^{13}C$ isotope tracer techniques was used to explore the methane conversion mechanism. It was found that Au nanoparticles on the catalyst surface could promote oxygen activation and generate reactive oxygen species for methane activation. This reaction involves the formation and subsequent conversion of active species (methyl, peroxymethyl, acetyl, etc.) on solid catalytic surfaces, rather than only the oxidation reaction of methane in the solution phase. This catalyst can use only O_2 as an oxidant to generate C_2 oxygenated compounds. By contrast, CO is usually required as a co-reactant (reductant) on other noble metal (Rh and Ir) modified zeolite catalysts. This study provides a strategy for the "holy grail" reaction in the field of methane selective conversion.

15.5 Summary and Outlook

In conclusion, revolutionary progresses have been developed by zeolite-based catalytic materials in the utilization of light alkanes, especially in ethane/PDH and methane selective oxidation. Although significant advances have been achieved by using zeolite-based catalysts, there are still several key issues that need to be addressed for their further commercial application.

A variety of metal@zeolite catalysts can catalyze PDH with conversions at the thermodynamic limit with enough propylene space–time yield. A challenge for these catalysts in replacing the current PDH catalysts would be their cycling stability under multiple redox conditions. The current Oleflex technology requires oxychlorination treatment to re-disperse the sintered Pt nanoparticles in each cycle for catalyst regeneration (coke removal in the air at high temperatures). The metal@zeolite structure would efficiently hinder the metal migration and ripening under redox conditions, which might improve the current processes by avoiding oxychlorination.

For EDH, the zeolite-based catalysts have efficiently catalyzed the reaction even with very rapid gas feeding rates. However, the one-pass conversion of ethane in non-ODH is still limited by thermodynamic equilibrium. The low ethane conversions in catalytic EDH ($<30\%$) are far from the commercial interest because of the high cost of separation and reflux. Improving the ethane conversion under mild reaction temperatures by selectively combusting the hydrogen to shift the equilibrium would be an important developing trend in this field.

The zeolite-based catalysts have catalyzed the selective oxidation of methane into various oxygen products, but the yield is still very low. Further research should be focused on further improvement of the catalysts and reaction system to enhance the methane conversion with sufficient product selectivity. In addition, most of the reaction systems still rely on precious metal catalysts, and efficient catalysts with earth-abundant metals are still highly required for methane oxidation reactions.

Due to the structural complexity of metal species in zeolites, the structure–performance relationship for most catalysts was still not fully understood. Some metal species in metallosilicates are not typically tetrahedrally coordinated with M-(O—Si)$_4$ structure in the zeolite framework, resulting in a variety of metallic sites in zeolites with diversified micro-structures and catalytic properties. For example, zeolites containing boron sites with predominantly B—O—Si coordinate structure were proved to be catalytically inactive in the ODH of propane [51]. In striking contrast, certain isolated boron sites which can be hydrolyzed forming B—OH…O(H)—Si groups are verified to be highly efficient active sites for this reaction [53, 87]. Some metallic active sites in zeolite – for example, the Zn sites for PDH [44] and the Ti sites for propylene epoxidation [28] – have been even demonstrated to be in form of dinuclear structures. These cases call for further in situ investigation of the fine structure of metal sites in zeolite under reaction conditions, which are potentially important for guiding the design of the next generation of alkane conversion catalysts in the future.

Acknowledgments

The present work was supported by the National Natural Science Foundation of China (U21B20101, 21932006, and 22202177), the China National Postdoctoral Program for Innovative Talents (BX2021255), and the China Postdoctoral Science Foundation (2021 M702807).

References

1 He, M., Sun, Y., and Han, B. (2013). Green carbon science: scientific basis for integrating carbon resource processing, utilization, and recycling. *Angew. Chem. Int. Ed.* 52: 9620–9633. https://doi.org/10.1002/anie.201209384.

2 Grant, J.T., Venegas, J.M., McDermott, W.P., and Hermans, I. (2018). Aerobic oxidations of light alkanes over solid metal oxide catalysts. *Chem. Rev.* 118: 2769–2815. https://doi.org/10.1021/acs.chemrev.7b00236.

3 Sun, M., Zhang, J., Putaj, P. et al. (2018). Catalytic oxidation of light alkanes (C1–C4) by heteropoly compounds. *Chem. Rev.* 118: 2769–2815. https://doi.org/10.1021/cr300302b.

4 Guo, Z., Liu, B., Zhang, Q. et al. (2014). Recent advances in heterogeneous selective oxidation catalysis for sustainable chemistry. *Chem. Soc. Rev.* 43: 3480–3524. https://doi.org/10.1039/C3CS60282F.

5 Morejudo, S.H., Zanón, R., Escolástico, S. et al. (2016). Direct conversion of methane to aromatics in a catalytic co-ionic membrane reactor. *Science* 353: 563–566. https://doi.org/10.1126/science.aag0274.

6 Schüth, F. (2019). Making more from methane. *Science* 363: 1282–1283. https://doi.org/10.1126/science.aaw7738.

7 Coors, W.G., Martínez, A., Norby, T. et al. (2016). Direct conversion of methane to aromatics in a catalytic co-ionic membrane reactor. *Science* 353: 563–566. https://doi.org/10.1126/science.aag0274.

8 Sabbe, M.K., Van Geem, K.M., Reyniers, M.-F., and Marin, G.B. (2011). First principle-based simulation of ethane steam cracking. *AIChE J.* 57: 482–496. https://doi.org/10.1002/aic.12269.

9 Chan, K.Y.G., Inal, F., and Senkan, S. (1998). Suppression of coke formation in the steam cracking of alkanes: ethane and propane. *Ind. Eng. Chem. Res.* 37: 901–907.

10 Guo, X., Fang, G., Li, G. et al. (2014). Direct, nonoxidative conversion of methane to ethylene, aromatics, and hydrogen. *Science* 344: 616–619. https://doi.org/10.1126/science.1253150.

11 Wang, P., Zhao, G., Wang, Y., and Lu, Y. (2017). $MnTiO_3$-driven low-temperature oxidative coupling of methane over TiO_2-doped Mn_2O_3-Na_2WO_4/SiO_2 catalyst. *Sci. Adv.* 3: e1603180.

12 Meng, X., Cui, X., Rajan, N.P. et al. (2019). Direct methane conversion under mild condition by thermo-, electro-, or photocatalysis. *Chem* 5: 1–30. https://doi.org/10.1016/j.chempr.2019.05.008.

13 Chen, S., Chang, X., Sun, G. et al. (2021). Propane dehydrogenation: catalyst development, new chemistry, and emerging technologies. *Chem. Soc. Rev.* 50: 3315–3354. https://doi.org/10.1039/D0CS00814A.

14 Sattler, J.J.H.B., Ruiz-Martinez, J., Santillan-Jimenez, E., and Weckhuysen, B.M. (2014). Catalytic dehydrogenation of light alkanes on metals and metal oxides. *Chem. Rev.* 114: 10613–10653. https://doi.org/10.1021/cr5002436.

15 Otroshchenko, T., Jiang, G., Kondratenko, V.A. et al. (2021). Current status and perspectives in oxidative, non-oxidative and CO_2-mediated dehydrogenation of propane and isobutane over metal oxide catalysts. *Chem. Soc. Rev.* 50: 473–527. https://doi.org/10.1039/D0CS01140A.

16 Schlögl, R. (2015). Heterogeneous catalysis. *Angew. Chem. Int. Ed.* 54: 3465–3520. https://doi.org/10.1002/anie.201410738.

17 Del Campo, P., Martínez, C., and Corma, A. (2021). Activation and conversion of alkanes in the confined space of zeolite-type materials. *Chem. Soc. Rev.* 50: 8511–8595. https://doi.org/10.1039/D0CS01459A.

18 Zhang, Q., Yu, J., and Corma, A. (2020). Applications of zeolites to C1 chemistry: recent advances, challenges, and opportunities. *Adv. Mater.* 32: 2002927. https://doi.org/10.1002/adma.202002927.

19 Wang, Y., Tao, Z., Wu, B. et al. (2015). Effect of metal precursors on the performance of Pt/ZSM-22 catalysts for n-hexadecane hydroisomerization. *J. Catal.* 322: 1–13. https://doi.org/10.1016/j.jcat.2014.11.004.

20 Chen, N., Wang, N., Ren, Y. et al. (2017). Effect of surface modification with silica on the structure and activity of Pt/ZSM-22@SiO_2 catalysts in hydrodeoxygenation of methyl palmitate. *J. Catal.* 345: 124–134. https://doi.org/10.1016/j.jcat.2016.09.005.

21 Pawelec, B., Mariscal, R., Navarro, R.M. et al. (2004). Simultaneous 1-pentene hydroisomerisation and thiophene hydrodesulphurisation over sulphided Ni/FAU and Ni/ZSM-5 catalysts. *Appl. Catal. A Gen.* 262: 155–166. https://doi.org/10.1016/j.apcata.2003.11.037.

22 Zhu, Q., Zhou, H., Wang, L. et al. (2022). Enhanced CO_2 utilization in dry reforming of methane achieved through nickel-mediated hydrogen spillover in zeolite crystals. *Nat. Catal.* 5: 1030–1037. https://doi.org/10.1038/s41929-022-00870-8.

23 Campelo, J.M., Lafont, F., and Marinas, J.M. (1995). Hydroisomerization and hydrocracking of n-heptane on Pt/SAPO-5 and Pt/SAPO-11 catalysts. *J. Catal.* 156: 11–18. https://doi.org/10.1006/jcat.1995.1226.

24 Geng, C.H., Zhang, F., Gao, Z.-X. et al. (2004). Hydroisomerization of n-tetradecane over Pt/SAPO-11 catalyst. *Catal. Today* 93–95: 485–491. https://doi.org/10.1016/j.cattod.2004.06.104.

25 Wang, H., Wang, L., and Xiao, F.-S. (2020). Metal@zeolite hybrid materials for catalysis. *ACS Cent. Sci.* 6: 1685–1697. https://doi.org/10.1021/acscentsci.0c01130.

26 Wang, L., Wang, L., Meng, X., and Xiao, F.-S. (2019). New strategies for the preparation of sinter-resistant metal-nanoparticle-based catalysts. *Adv. Mater.* 31: 1901905. https://doi.org/10.1002/adma.201901905.

27 Wu, Q., Xu, C., Zhu, L. et al. (2023). Recent strategies for synthesis of metallosilicate zeolites. *Catal. Today* 410: 2–12. https://doi.org/10.1016/j.cattod.2022.11.022.

28 Gordon, C.P., Engler, H., Tragl, A.S. et al. (2020). Efficient epoxidation over dinuclear sites in titanium silicalite-1. *Nature* 586: 708–713. https://doi.org/10.1038/s41586-020-2826-3.

29 Corma, A., Nemeth, L.T., Renz, M., and Valencia, S. (2001). Sn-zeolite beta as a heterogeneous chemoselective catalyst for Baeyer-Villiger oxidations. *Nature* 412: 423–425. https://doi.org/10.1038/35086546.

30 Rightor, E.G. and Tway, C.L. (2015). Global energy & emissions reduction potential of chemical process improvements. *Catal. Today* 258: 226–229. https://doi.org/10.1016/j.cattod.2015.02.023.

31 Corma, A., Melo, F.V., Sauvanaud, L., and Ortega, F. (2005). Light cracked naphtha processing: controlling chemistry for maximum propylene production. *Catal. Today* 107–108: 699–706. https://doi.org/10.1016/j.cattod.2005.07.109.

32 Fang, W., Wang, C., Liu, Z. et al. (2022). Physical mixing of a catalyst and a hydrophobic polymer promotes CO hydrogenation through dehydration. *Science* 377: 406–410. https://doi.org/10.1126/science.abo0356.

33 Wang, C., Yang, L., Gao, M. et al. (2022). Directional construction of active naphthalenic species within SAPO-34 crystals toward more efficient methanol-to-olefin conversion. *J. Am. Chem. Soc.* 144: 21408–21416. https://doi.org/10.1021/jacs.2c10495.

34 Zacharopoulou, V. and Lemonidou, A.A. (2018). Olefins from biomass intermediates: a review. *Catalysts* 8: 2. https://doi.org/10.3390/catal8010002.

35 Monai, M., Gambino, M., Wannakao, S., and Weckhuysen, B.M. (2021). Propane to olefins tandem catalysis: a selective route towards light olefins production. *Chem. Soc. Rev.* 50: 11503–11529. https://doi.org/10.1039/D1CS00357G.

36 Kytökivi, A., Jacobs, J.-P., Hakuli, A. et al. (1996). Surface characteristics and activity of chromia/alumina catalysts prepared by atomic layer epitaxy. *J. Catal.* 162: 190–197. https://doi.org/10.1006/jcat.1996.0276.

37 Kumar, M.S., Hammer, N., Rønning, M. et al. (2009). The nature of active chromium species in Cr-catalysts for dehydrogenation of propane: new insights by a comprehensive spectroscopic study. *J. Catal.* 261: 116–128. https://doi.org/10.1016/j.jcat.2008.11.014.

38 Motagamwala, A.H., Almallahi, R., Wortman, J. et al. (2021). Stable and selective catalysts for propane dehydrogenation operating at thermodynamic limit. *Science* 373: 217–222. https://doi.org/10.1126/science.abg7894.

39 Zhu, X., Wang, T., Xu, Z. et al. (2022). Pt-Sn clusters anchored at Al^{3+} penta sites as a sinter-resistant and regenerable catalyst for propane dehydrogenation. *J. Energy Chem.* 65: 293–301. https://doi.org/10.1016/j.jechem.2021.06.002.

40 Liu, L., Lopez-Haro, M., Lopes, C.W. et al. (2019). Regioselective generation and reactivity control of subnanometric platinum clusters in zeolites for high-temperature catalysis. *Nat. Mater.* 18: 866–873. https://doi.org/10.1038/s41563-019-0412-6.

41 Liu, L., Lopez-Haro, M., Lopes, C.W. et al. (2020). Structural modulation and direct measurement of subnanometric bimetallic PtSn clusters confined in zeolites. *Nat. Catal.* 3: 628–638. https://doi.org/10.1038/s41929-020-0472-7.

42 Sun, Q., Wang, N., Fan, Q. et al. (2020). Subnanometer bimetallic platinum–zinc clusters in zeolites for propane dehydrogenation. *Angew. Chem. Int. Ed.* 59: 19450–19459. https://doi.org/10.1002/anie.202003349.

43 Ryoo, R., Kim, J., Jo, C. et al. (2020). Rare-earth–platinum alloy nanoparticles in mesoporous zeolite for catalysis. *Nature* 585: 221–224. https://doi.org/10.1038/s41586-020-2671-4.

44 Zhao, D., Tian, X., Doronkin, D.E. et al. (2021). In situ formation of ZnO_x species for efficient propane dehydrogenation. *Nature* 599: 234–238. https://doi.org/10.1038/s41586-021-03923-3.

45 Cavani, F., Ballarini, N., and Cericola, A. (2007). Oxidative dehydrogenation of ethane and propane: how far from commercial implementation? *Catal. Today* 127: 113–131. https://doi.org/10.1016/j.cattod.2007.05.009.

46 Grant, J.T., Carrero, A., Goeltl, F. et al. (2016). Selective oxidative dehydrogenation of propane to propene using boron nitride catalysts. *Science* 354: 1570–1573. https://doi.org/10.1126/science.aaf7885.

47 Grant, J.T., McDermott, W.P., Venegas, J.M. et al. (2017). Boron and boron-containing catalysts for the oxidative dehydrogenation of propane. *ChemCatChem* 9: 3623–3626. https://doi.org/10.1002/cctc.201701140.

48 Yan, B., Li, W.-C., and Lu, A.-H. (2019). Metal-free silicon boride catalyst for oxidative dehydrogenation of light alkanes to olefins with high selectivity and stability. *J. Catal.* 369: 296–301. https://doi.org/10.1016/j.jcat.2018.11.014.

49 Sheng, J., Yan, B., Lu, W.-D. et al. (2021). Oxidative dehydrogenation of light alkanes to olefins on metal-free catalysts. *Chem. Soc. Rev.* 50: 1438–1468. https://doi.org/10.1039/D0CS01459A.

50 Love, A.M., Thomas, B., Specht, S.E. et al. (2019). Probing the transformation of boron nitride catalysts under oxidative dehydrogenation conditions. *J. Am. Chem. Soc.* 141: 182–190. https://doi.org/10.1021/jacs.8b08165.

51 Altvater, A.R., Dorn, R.W., Cendejas, M.C. et al. (2020). B-MWW zeolite: the case against single-site catalysis. *Angew. Chem. Int. Ed.* 59: 6546–6550. https://doi.org/10.1002/anie.201914696.

52 Venegas, J.M., McDermott, W.P., and Hermans, I. (2018). Serendipity in catalysis research: boron-based materials for alkane oxidative dehydrogenation. *Acc. Chem. Res.* 51: 2556–2564. https://doi.org/10.1021/acs.accounts.8b00330.

53 Zhou, H., Yi, X., Hui, Y. et al. (2021). Isolated boron in zeolite for oxidative dehydrogenation of propane. *Science* 372: 76–80. https://doi.org/10.1126/science.abe7935.

54 Dai, Y., Gao, X., Wang, Q. et al. (2021). Recent progress in heterogeneous metal and metal oxide catalysts for direct dehydrogenation of ethane and propane. *Chem. Soc. Rev.* 50: 5590–5630. https://doi.org/10.1039/D0CS01260B.

55 Yang, Y., Li, H., Zhou, H. et al. (2020). Coking-resistant iron catalyst in ethane dehydrogenation achieved through siliceous zeolite modulation. *J. Am. Chem. Soc.* 142: 16429–16436. https://doi.org/10.1021/jacs.0c07792.

56 Liu, L., Li, H., Zhou, H. et al. (2022). Rivet of cobalt in siliceous zeolite for catalytic ethane dehydrogenation. *Chem* https://doi.org/10.1016/j.chempr.2022.10.026.

57 Maeno, Z., Yasumura, S., Wu, X. et al. (2020). Isolated indium hydrides in CHA zeolites: speciation and catalysis for nonoxidative dehydrogenation of ethane. *J. Am. Chem. Soc.* 142: 4820–4832. https://doi.org/10.1021/jacs.9b13865.

58 Wang, C., Han, Y., Tian, M. et al. (2022). Main-group catalysts with atomically dispersed in sites for highly efficient oxidative dehydrogenation. *J. Am. Chem. Soc.* 144: 16855–16865. https://doi.org/10.1021/jacs.2c04926.

59 Gesser, H.D., Hunter, N.R., and Prakash, C.B. (1985). The direct conversion of methane to methanol by controlled oxidation. *Chem. Rev.* 85: 235–244. https://doi.org/10.1021/cr00068a001.

60 Hickman, D.A. and Schmidt, L.D. (1993). Production of syngas by direct catalytic oxidation of methane. *Science* 259: 343–346. https://doi.org/10.1126/science.259.5093.3.

61 Maier, L., Schädel, B., Delgado, K.H. et al. (2011). Steam reforming of methane over nickel: development of a multi-step surface reaction mechanism. *Top. Catal.* 54: 845. https://doi.org/10.1007/s11244-011-9702-1.

62 Stiles, A.B., Chen, F., Harrison, J.B. et al. (1991). Catalytic conversion of synthesis gas to methanol and other oxygenated products. *Ind. Eng. Chem. Res.* 30: 811–821. https://doi.org/10.1021/ie00053a002.

63 Gotti, A. and Prins, R. (1998). Basic metal oxides as cocatalysts for Cu/SiO_2 catalysts in the conversion of synthesis gas to methanol. *J. Catal.* 178: 511–519. https://doi.org/10.1006/jcat.1998.2167.

64 Liu, H.F., Liu, R.S., Liew, K.Y. et al. (1984). Partial oxidation of methane by nitrous oxide over molybdenum on silica. *J. Am. Chem. Soc.* 106: 4117–4121. https://doi.org/10.1021/ja00327a009.

65 Herman, R.G., Sun, Q., Shi, C. et al. (1997). Development of active oxide catalysts for the direct oxidation of methane to formaldehyde. *Catal. Today* 37: 1–14. https://doi.org/10.1016/S0920-5861(96)00256-8.

66 Parmaliana, A., Arena, F., Frusteri, F., and Mezzapica, A. (1998). High yields in the catalytic partial oxidation of natural gas to formaldehyde: catalyst development and reactor configuration. *Stud. Surf. Sci. Catal.* 119: 551–556. https://doi.org/10.1016/S0167-2991(98)80489-8.

67 Lou, Y., Wang, H., Zhang, Q., and Wang, Y. (2007). SBA-15-supported molybdenum oxides as efficient catalysts for selective oxidation of ethane to formaldehyde and acetaldehyde by oxygen. *J. Catal.* 247: 245–255. https://doi.org/10.1016/j.jcat.2007.02.011.

68 Wang, Y., Yang, W., Yang, L. et al. (2006). Iron-containing heterogeneous catalysts for partial oxidation of methane and epoxidation of propylene. *Catal. Today* 117: 156–162. https://doi.org/10.1016/j.cattod.2006.05.018.

69 Zhang, Q., Li, Y., An, D., and Wang, Y. (2009). Catalytic behavior and kinetic features of FeO_x/SBA-15 catalyst for selective oxidation of methane by oxygen. *Appl. Catal. A Gen.* 356: 103–111. https://doi.org/10.1016/j.apcata.2008.12.031.

70 Periana, R.A., Taube, D.J., Evitt, E.R. et al. (1993). A mercury-catalyzed, high-yield system for the oxidation of methane to methanol. *Science* 259: 340–343. https://doi.org/10.1126/science.259.5093.340.

71 Periana, R.A., Taube, D.J., Gamble, S. et al. (1998). Platinum catalysts for the high-yield oxidation of methane to a methanol derivative. *Science* 280: 560–564. https://doi.org/10.1126/science.280.5363.560.

72 Groothaert, M.H., Smeets, P.J., Sels, B.F. et al. (2005). Selective oxidation of methane by the bis (μ-oxo) dicopper core stabilized on ZSM-5 and mordenite zeolites. *J. Am. Chem. Soc.* 127: 1394–1395. https://doi.org/10.1021/ja047158u.

73 Banerjee, R., Proshlyakov, Y., Lipscomb, J.D., and Proshlyako, D.A. (2015). Structure of the key species in the enzymatic oxidation of methane to methanol. *Nature* 518: 431–434. https://doi.org/10.1038/nature14160.

74 Alayon, E.M., Nachtegaal, M., Ranocchiari, M., and Van Bokhoven, J.A. (2012). Catalytic conversion of methane to methanol over Cu–mordenite. *Chem. Commun.* 48: 404–406. https://doi.org/10.1039/C1CC15840F.

75 Sushkevich, V.L., Palagin, D., Ranocchiari, M., and van Bokhoven, J.A. (2017). Selective anaerobic oxidation of methane enables direct synthesis of methanol. *Science* 356: 523–527. https://doi.org/10.1126/science.aam9035.

76 Imbao, J., van Bokhoven, J.A., and Nachtegaal, M. (2021). On the promotional and inhibitory effects of water on Wacker-type ethylene oxidation over Pd–Cu/Zeolite Y. *ACS Catal.* 11: 8684–8691. https://doi.org/10.1021/acscatal.1c00507.

77 Imbao, J., van Bokhoven, J.A., Clark, A., and Nachtegaal, M. (2020). Elucidating the mechanism of heterogeneous Wacker oxidation over Pd-Cu/zeolite Y by transient XAS. *Nat. Commun.* 11: 1118. https://doi.org/10.1038/s41467-020-14982-x.

78 Huang, W., Zhang, S., Tang, Y. et al. (2016). Low-temperature transformation of methane to methanol on Pd1O4 single sites anchored on the internal surface of microporous silicate. *Angew. Chem. Int. Ed.* 55: 13441–13445. https://doi.org/10.1002/ange.201604708.

79 Min, J.S., Ishige, H., Misono, M., and Mizuno, N. (2001). Low-temperature selective oxidation of methane into formic acid with H_2–O_2 gas mixture catalyzed by bifunctional catalyst of palladium–heteropoly compound. *J. Catal.* 198: 116–121. https://doi.org/10.1006/jcat.2000.3117.

80 Freakley, S.J., He, Q., Harrhy, J.H. et al. (2016). Palladium-tin catalysts for the direct synthesis of H_2O_2 with high selectivity. *Science* 351: 965–968. https://doi.org/10.1126/science.aad5705.

81 Rahim, M.H.A., Forde, M.M., Jenkins, R.L. et al. (2013). Oxidation of methane to methanol with hydrogen peroxide using supported gold–palladium alloy nanoparticles. *Angew. Chem. Int. Ed.* 52: 1280–1284. https://doi.org/10.1002/anie.201207717.

82 Jin, Z., Wang, L., Zuidema, E. et al. (2020). Hydrophobic zeolite modification for in situ peroxide formation in methane oxidation to methanol. *Science* 367: 193–197. https://doi.org/10.1126/science.aaw1108.

83 Shan, J., Li, M., Allard, L.F. et al. (2017). Mild oxidation of methane to methanol or acetic acid on supported isolated rhodium catalysts. *Nature* 551: 605–608. https://doi.org/10.1038/nature24640.

84 Tang, Y., Li, Y., and Tao, F. (2022). Activation and catalytic transformation of methane under mild conditions. *Chem. Soc. Rev.* 51: 376–423. https://doi.org/10.1039/D1CS00783A.

85 Boetius, A., Ravenschlag, K., Schubert, C.J. et al. (2000). A marine microbial consortium apparently mediating anaerobic oxidation of methane. *Nature* 407: 623–626. https://doi.org/10.1038/35036572.

86 Qi, G., Davies, T.E., Nasrallah, A. et al. (2022). Au-ZSM-5 catalyses the selective oxidation of CH_4 to CH_3OH and CH_3COOH using O_2. *Nat. Catal.* 5: 45–54. https://doi.org/10.17035/d.2021.0142278187.

87 Yan, H., Alayoglu, S., Wu, W. et al. (2021). Identifying boron active sites for the oxidative dehydrogenation of propane. *ACS Catal.* 11: 9370–9376. https://doi.org/10.1021/acscatal.1c02168.

16

Design and Applications of Single-Site Photocatalysis Using Metallosilicates

Priyanka Verma[1,2] *and Hiromi Yamashita*[1]

[1] *Osaka University, Graduate School of Engineering, Division of Materials and Manufacturing Science, 2-1 Yamadaoka, Suita, Osaka 565-0871, Japan*
[2] *Department of Chemistry, Indian Institute of Technology Delhi, Hauz Khas, New Delhi 110016, India*

16.1 Introduction

The discovery of photocatalytic properties of TiO_2 in 1972 by Fujishima and Honda accelerated the research in various applications, such as water splitting [1], solar cells [2, 3], and environmental remediations [4]. The TiO_2-based multifunctional photocatalysts can display higher photocatalytic activities due to the formation of charge carriers (electron and holes) under UV-light excitation [5]. In recent years, the design and construction of photocatalysts have also advanced starting from bulk to molecular-based systems. The downsizing of semiconductor photocatalysts to nanometer size range can lead to the creation of unique properties by quantum size effect [6]. These smaller-sized nanoparticles (NPs) are capable of displaying unique photocatalytic activity, in some cases much higher than the conventional semiconductor species [7–13]. However, decreasing the size of NPs to achieve isolated single-atom species is still a difficult task because of the increase in the surface energy of the particles. Therefore, the use of suitable support material is essential in order to provide stability, preserve the original structure, and be able to strongly interact with the NPs and single-atomic species [14, 15]. For this purpose, the use of zeolites [10] and mesoporous silica (MS) materials [16] with the presence of micropores and mesopores, respectively, can be used as host materials to design multifunctional photocatalysts because of the following advantages: (i) structural tunability to control the pore size, (ii) easy accessibility of substrate molecules due to pore connectivity facilitating the diffusion, (iii) UV–vis light transparency providing an inactive surface for reaction, and finally, (iv) the chemical and thermal stability along with a large number of surface silanol groups providing an opportunity for further functionalization [17].

The highly dispersed tetrahedral coordinated transition metal ions (Ti, Cr, Mo, V, and W) within the silica-based porous solids (zeolites or MS) are known as *single-site photocatalysts* [18]. These are formed by introducing the metal precursor solution during the synthesis of zeolite or MS in which the "Si" atoms from the

Micro-Mesoporous Metallosilicates: Synthesis, Characterization, and Catalytic Applications, First Edition. Edited by Peng Wu and Hao Xu.

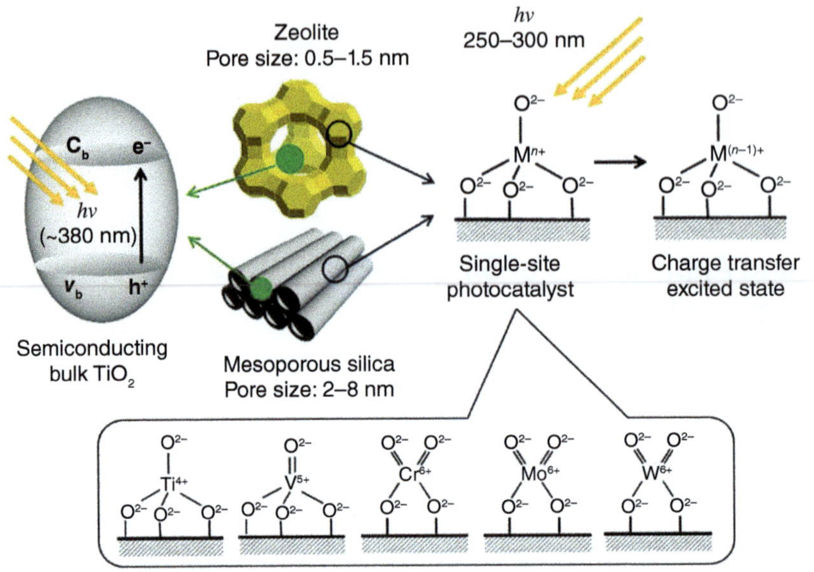

Figure 16.1 Titanium oxide photocatalysts from semiconducting bulk TiO_2 to isolated Ti-oxide molecular species implanted within silica matrices and the formation of a charge-transfer excited state with tetrahedrally coordinated metal oxide moieties under UV-light irradiation. Source: Reproduced with permission from Ref. [6]. Copyright 2018, The Royal Society of Chemistry.

framework are replaced with the heteroatoms [19, 20]. These single-site heterogeneous catalysts are capable of displaying a unique photocatalytic response due to the ligand to metal charge transfer (LMCT) process of the transition metal ion present in tetrahedral coordination (Figure 16.1) [14, 21]. Under light irradiation conditions, the formation of electron and hole pairs takes place due to the electron transfer from oxygen to metal ions. The positioning of these electron–hole pairs is much closer to each other when compared with the electron–hole pairs in semiconductor materials. These localized charge carriers have been reported to display enhanced photocatalytic activities in reactions such as NO reduction, CO_2 reduction, ethylene polymerization, and CO oxidation [8, 9, 22, 23]. The synthesis of single-site photocatalysts and their characterization was confirmed by several operando techniques, for example, in situ photoluminescence, X-ray absorption fine structure (XAFS), Fourier transform-infrared (FT-IR), UV–vis, and electron paramagnetic resonance (EPR) spectroscopy. The single-site photocatalysts are not only responsible for unique photocatalytic reactions but also can assist the synthesis of new functional materials [20, 24–27].

This chapter will focus on the introduction of metallosilicate photocatalysts with transition metal (Ti, Cr, V, Mo, W) incorporated porous materials exhibiting UV and visible-light responsiveness, the preparation of single-site thin films, and the synthesis of metal NPs deposition (monometallic, bimetallic, and alloys) on single-site photocatalysts by photo-assisted deposition (PAD) method. The photocatalytic systems of Ti, Cr, and V-oxide moieties incorporated in silica and zeolites are also discussed for their superior catalytic performance in NO decomposition, CO_2 reduction, partial oxidation of hydrocarbons, and alkene epoxidation under UV and visible-light irradiation. The structural characterization by operando techniques to understand the coordination environment and enhanced catalytic performance has also been elucidated.

16.2 Ti-Based Single-Site Photocatalysis Within Zeolites and Mesoporous Silica

The single-site Ti-oxide moieties have been incorporated within MS and zeolite frameworks to be used as efficient photocatalysts. The existence of isolated tetrahedrally coordinated Ti-oxide species was evidenced by EPR, UV–vis, and XAFS analysis. It was observed that the existence of single-site Ti-oxide species within the framework of zeolites and MS was able to display unique catalytic activity in the photocatalytic reduction of CO_2 with H_2O, when compared with bulk TiO_2 powder [8, 18, 22]. The reduction of CO_2 with water led to the formation of CH_3OH and CH_4 as the main products upon UV-light irradiation using TS-1 zeolite and Ti-containing MS. The obtained results are shown in Figure 16.2 where Ti-MCM-48 was found to exhibit a much higher catalytic response than Ti-MCM-41, TS-1, and bulk TiO_2 powder. The larger pore size along with the 3-D channel of the framework and better dispersion of Ti-oxide species contribute to the higher product yield when MCM-48 was used as the support framework. The superior

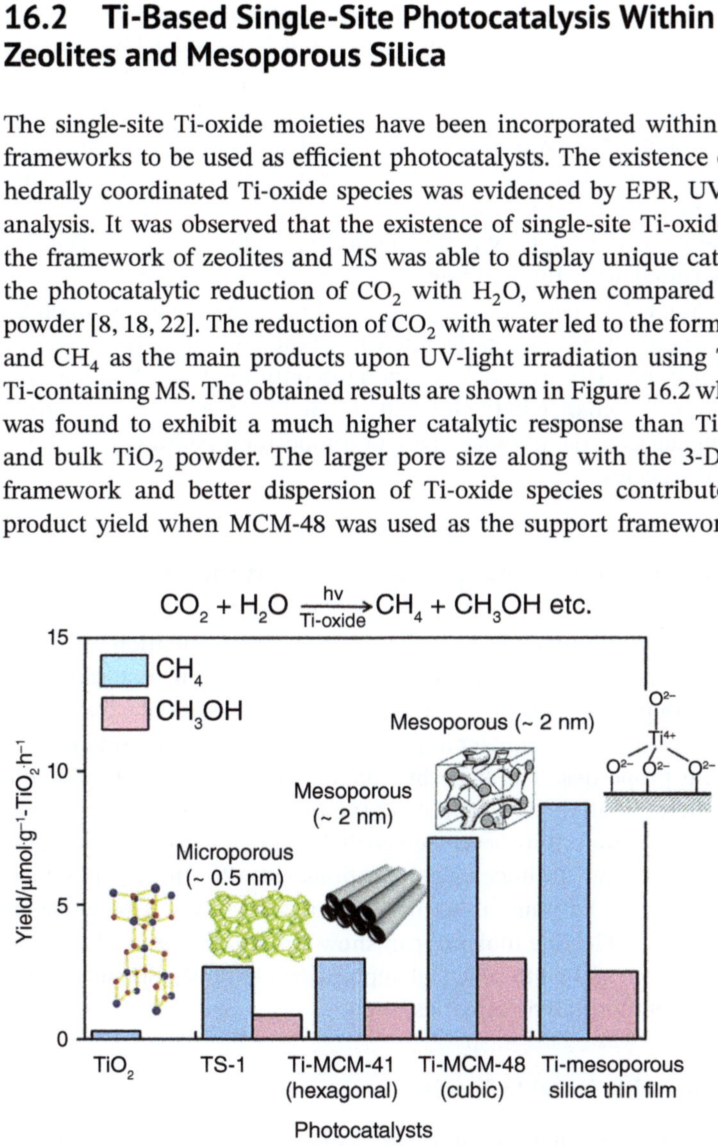

Figure 16.2 The photocatalytic reduction of CO_2 with H_2O on various Ti-oxide photocatalysts. Source: Reproduced with permission from Ref. [6]. Copyright 2018, The Royal Society of Chemistry.

catalytic response indicated the specific interaction between the charge-transfer excited state of single-site Ti-oxide and reactant molecules (CO_2 or H_2O). This was confirmed by the observation of photoluminescence spectra due to the radioactive decay of charge-transfer excited state of Ti-oxide $(Ti^{3+}\text{-}O^-)^*$ to their ground state upon light excitation at 250–280 nm. The interaction of Ti-oxide with the reactant molecules led to the quenching of the photoluminescence spectral peak, evidencing the interaction between the two components in a ground or excited state.

The single-site Ti-oxide species can also display unique photocatalytic activity in the NO decomposition reaction to form N_2, O_2, and N_2O under UV-light irradiation conditions. The single-site Ti-oxide moieties incorporated within zeolites (Ti-Y-zeolite) displayed a higher reactivity and selectivity for N_2 formation, whereas N_2O as the major component was observed in bulk TiO_2 powder and aggregated Ti-oxide catalyst with an octahedrally coordinated state [28]. The measurement of electron paramagnetic resonance (ESR) spectra was employed to evidence the role of Ti-oxide moieties in the photocatalytic NO decomposition reaction. The ESR spectra were measured with the addition of NO along with single-site Ti-oxide moieties at 77 K. A new ESR peak at a g value of 2.0015 was observed, and upon UV light irradiation, a gradual decrease in the intensity of the ESR signal was observed. The ESR signal returned to its original level upon discontinuing the illumination of UV-light irradiation. The change in the intensity of the ESR peak with and without light illumination suggests the photocatalytic decomposition pathway of NO over single-site Ti-oxide species.

16.3 Single-Site Photocatalysis Based Thin Films

The development of thin film-based photocatalysts has gained attention in antifogging, optical, electronic, and separation systems because of their easy separation when compared with powders [29–31]. The introduction of superhydrophilic character in thin films with a water contact angle of less than 5° is intriguing because of its self-cleaning properties. The superhydrophilic property of TiO_2 thin film was reported by Shirai et al. in 1974 [32]. After the first report on the amphiphilic properties of TiO_2, TiO_2 thin films have been coated on glass, mirror, and building walls for self-cleaning and antifogging applications. Ideally, the photoinduced surface properties should be maintained from 1 day to 1 week. The single-site photocatalysts involving Ti-oxide moieties can show similar properties, but unfortunately, it has not been applied in practical applications so far despite its efficient and highly selective photocatalytic activities.

16.3.1 Thin Films with Ti-Oxide Species

The addition of transition metal in the mesoporous silica thin films (MSTFs) can lead to the generation of unique active sites. The removal of the template creates a porous framework with more accessible active sites which can exhibit molecular shape selectivity and penetration effects. Figure 16.3 displays the schematic illustration of

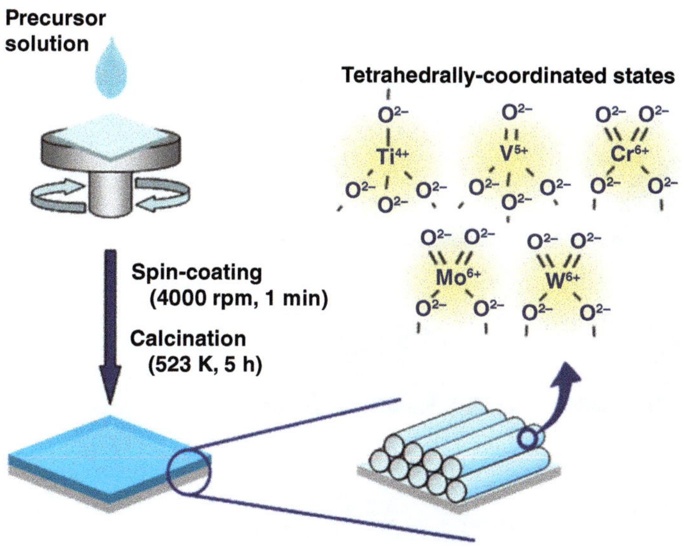

Figure 16.3 Schematic illustration of the synthesis procedure of single-site photocatalyst mesoporous silica thin film (SSP-MSTFs) containing transition metal ions. Source: Reproduced with permission from Ref. [19]. Copyright 2020, Wiley.

the synthesis procedure of MSTF-based single-site photocatalysts by a spin coating method using an appropriate precursor solution followed by calcination. For example, Ti-MSTF was fabricated on the quartz substrate (10 mm × 10 mm × 1 mm) by spin coating of the precursor solution, tetraethylorthosilicate for Si and tetraethylorthotitanate for Ti and BrijR30 as the template material, followed by calcination. The incorporation of various heteroatom oxides including V, Cr, Mo, and W within porous framework has been reported to investigate their photocatalytic applications as discussed in Section 16.3.2. The presence of porous framework was confirmed by XRD, TEM, and N_2 physisorption analysis. The hexagonal mesoporous structure was confirmed by the presence of a sharp peak at 3° in the XRD pattern due to the (100) plane of the mesoporous structure. Furthermore, a type IV isotherm in the N_2 physisorption analysis with an average pore diameter of 2.7 nm evidenced the formation of mesoporous structure. The local environment around the Ti-oxide moieties can be confirmed by UV–vis and XAS analyses. An absorption peak at around 215 nm was attributed to the ligand-to-metal charge-transfer (LMCT) transition from O^{2-} to Ti^{4+} in tetrahedrally coordinated single site Ti-oxide species. The XANES spectra of Ti-MSTF showed a sharp intense peak at 4967 eV, similar to the reference precursor tetrapropyl orthotitanate (TPOT), confirming the tetrahedral coordination of Ti-oxide species. The FT-EXAFS spectra also displayed a single peak at 1.4 Å corresponding to the presence of Ti—O bond. Hence, the above-mentioned characterizations can be used as a strategy for characterizing the successful incorporation of single-site moieties within the framework.

The wettability of Ti-MSTF was evaluated by measuring the water contact angle before and after UV light irradiation as shown in Figure 16.4. The high hydrophilic

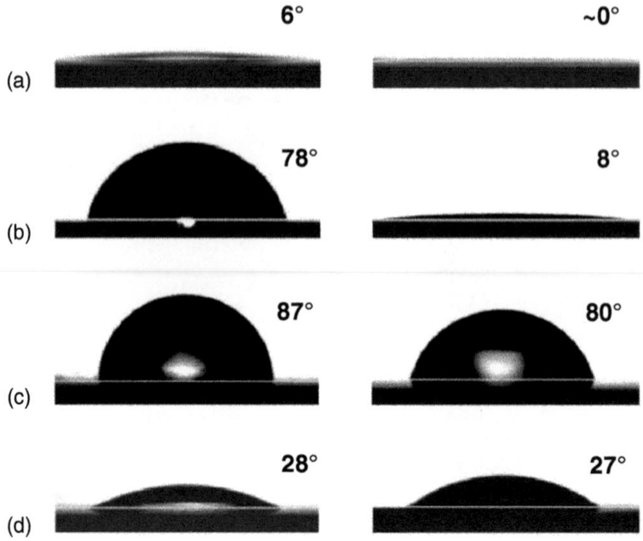

Figure 16.4 The images of water droplets and their corresponding water contact angles on (a) Ti-MSTFs, (b) TiO_2 thin film, (c) Ti-STFs, and (d) pure silica MSTFs before (left) and after (right) UV-light irradiation. Source: Reproduced with permission from Ref. [33]. Copyright 2011 Elsevier.

character of Ti-MSTF film was evidenced by the water contact angle of 6° when compared to 78° for the conventional TiO_2 film. The bare MSTF film also displayed a hydrophilic character with a contact angle of 28°. The introduction of Ti-oxide species led to the improved hydrophilic character due to the formation of a newly introduced adsorption site for water molecules. However, the Ti-STF thin film without mesoporous component displayed higher contact angle of 87° than Ti-MSTF. The lower hydrophilic character can be attributed to the lesser number of silanol groups present on the surface of the silica. Furthermore, the light illumination on Ti-MSTF led to the achievement of superhydrophilic character with a water contact angle of ~0°.

16.3.2 Thin Films with Various Metal Oxide Species (Mo, V, Cr, and W)

The design of single-site MSTFs doped with several transition metal ions (V, Cr, Mo, W) has been reported to design novel thin film-based photocatalysts [33]. For V-MSTF, Cr-MSTF, and Mo-MSTF, the NH_4VO_3, $Cr(NO_3)_6 \cdot 9H_2O$, and $(NH_4)_6Mo_7O_{24} \cdot 4H_2O$ solutions were used as the precursor solutions, respectively. The metal oxide doped thin film materials, denoted as M-MSTFs, were characterized to gain an understanding of the structural details. For example, the presence of a sharp diffraction peak in the low angle region (2.5–4°) confirmed the existence of a 2-D mesoporous structure in all M-MSTFs. The formation of single-site oxide moieties was also confirmed due to the absence of peaks corresponding to bulk metal oxides. The high dispersion of single-site metal oxides can also be confirmed

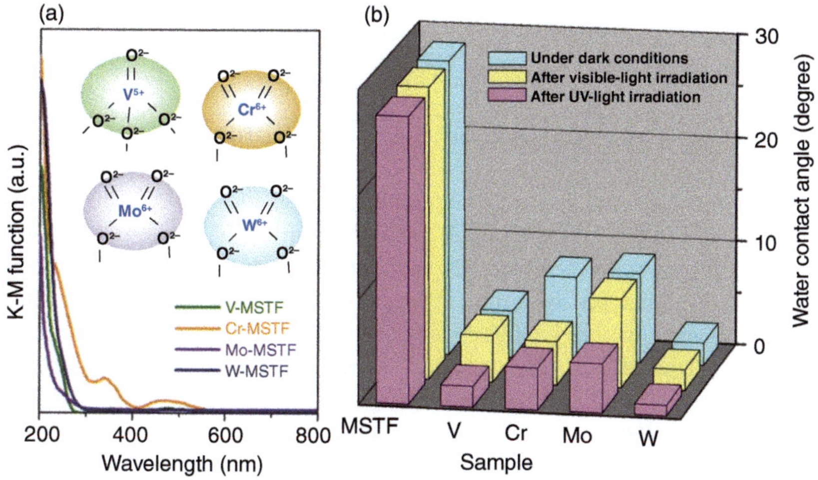

Figure 16.5 (a) UV–vis spectra of M-MSTFs (M = V, Cr, Mo, and W) and (b) water contact angles of pure silica MSTF and M-MSTFs (M = V, Cr, Mo, and W) before and after UV-light irradiation. Source: Reproduced from Ref. [33]/with permission of Elsevier.

by the UV–vis absorption spectra (Figure 16.5a) in which Cr-MSTF displayed three typical absorption peaks corresponding to LMCT transition from O^{2-} to Cr^{6+} species. The d–d transition absorption band at 550 nm corresponding to Cr_2O_3 was not observed. Furthermore, the metal oxides M-MSTF absorption bands also confirmed the LMCT transition from O^{2-} to M^{n+} and their tetrahedral coordination. Therefore, XRD and UV–vis spectra can confirm the formation of highly dispersed single-site metal oxide species for all metals (M = V, Cr, Mo, and W). The water contact angle measurements also revealed an enhancement in the hydrophilic properties in M-MSTFs when compared with bare MSTFs in dark, after visible-light irradiation and after UV light irradiation. This observation suggests that the incorporation of single-site metal oxide can lead to the formation of new adsorption sites due to electron delocalization. In particular, W-MSTF displayed the lowest water contact angle of 3° in dark conditions with the highest hydrophilicity (Figure 16.5b). Under UV irradiation, the water contact angle was further reduced, suggesting the creation of photo-induced superhydrophilicity in prepared samples. The Cr-MSTF can display a decrease in the water contact angle even under visible-light irradiation due to its light absorption property in the visible regime, whereas no such change was observed in other M-MSTFs composed of V, Mo, and W. The durability of hydrophilic character was also measured for the W-MSTF sample, and the obtained results revealed that it is capable of maintaining the superhydrophilic property for a period of consecutive 10 days.

The Cr-MSTF has also been reported to successfully synthesize polyethylene in the presence of ethylene gas under UV-light irradiation [34]. The formation of polyethylene was confirmed by the presence of two strong bands in the FT-IR spectra at 2584 and 2916 cm^{-1}, assigned to the symmetric and asymmetric stretching modes of CH_2 groups. Moreover, the formation of polyethylene on Cr-MSTF drastically changed the surface property from hydrophilic to hydrophobic character, as confirmed by the increased value of water contact angle after the reaction.

16.4 Visible-Light Sensitive Single-Site Photocatalysis

As described in the previous section, the Ti-based single-site photocatalysts can prove to display excellent photocatalytic activities under UV-light irradiation conditions. The doping with heteroatoms such as V, Cr, and Mo can also form photocatalysts with higher photocatalytic efficiency [7, 10, 35–39]. The single-site Cr-oxide moieties can display catalytic activities under visible-light irradiation due to the charge transfer process on Cr=O bond in tetrahedrally-coordinated Cr-oxide species [10]. The photoluminescence emission peak of Cr-oxide species appears at around 550–750 nm when excited at 450 nm. The insertion of Cr-oxide in MS (Cr-HMS) led to the formation of a fine structure in comparison to the case when zeolite frameworks were used as the support material. This observation is attributed to the clear energy separation between the Cr=O vibronic transition bands in the case of a mesoporous structure due to their more open structure. The formation of such unique Cr-oxide single-site species in mesoporous framework displayed enhanced photocatalytic reaction yields in the NO decomposition under visible-light irradiation ($\lambda > 450$ nm) with 97% selectivity in the N_2 formation. The zeolite framework immobilized Cr-oxide species did not display any significant catalytic response in UV or visible-light irradiation.

Interestingly, the creation of binary metal oxide combinations was possible with Ti, Cr-oxide and V, Ti-oxide species. These bimetal oxide species can be prepared by the PAD method and wet-ion implantation method. Yamashita et al. found that the incorporation of V-oxide with Ti-oxide can shift the optical absorption toward the visible region of the spectrum and is capable of displaying visible-light enhanced epoxidation of propene in the presence of molecular oxygen [24]. However, the bare Ti-oxide moiety did not display any photocatalytic activity. Similarly, the preparation of Cr, Ti-oxide binary oxide was done by the PAD method in which Ti^{4+} species were irradiated with UV-light ($\lambda > 200$ nm) to deposit Cr species as shown in Figure 16.6a. The absorption peaks arising from Ti^{4+} (220–260 nm) and Cr^{6+} were observed due to the LMCT absorption of the oxide molecules. The Ti–Cr bimetal oxide was found to be more active and selective in the propene oxidation to propene oxide (PO) when compared with monometallic oxides. A 2.5% PO yield was obtained with Cr–Ti/HMS in comparison to 0.5% yield with Cr-HMS photocatalyst. Moreover, the catalytic activity varied based on the preparation method of the photocatalyst. For example, the catalyst prepared by PAD method displayed much higher photocatalytic activities than the catalyst prepared by the wet-impregnation method (Figure 16.6b). This was attributed to the formation of an oxo-bridge between Cr–O–Ti within the MS framework. In the exploration of different preparation methods of single-site bimetal oxides, Kamegawa et al. employed chemical vapor deposition using MCM-41 silica support material [40]. The absorption spectra of Cr–Ti/MCM-41 displayed three distinct peaks but slightly

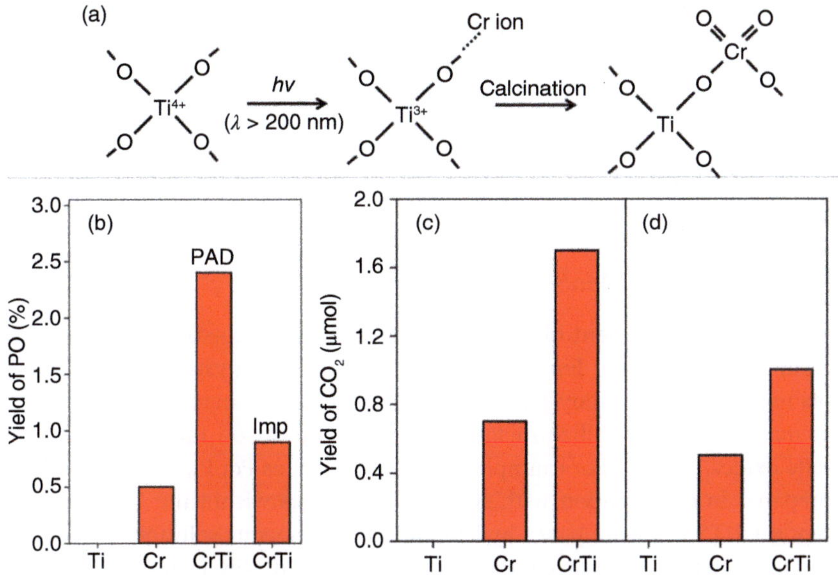

Figure 16.6 (a) Synthesis of Cr, Ti-binary oxide by the PAD method. (b) Photocatalytic oxidation of propene with O_2 to PO on Cr–Ti/HMS catalyst under visible-light irradiation ($\lambda > 450$ nm). Yields of CO_2 in the photocatalytic oxidation of CO with O_2 on Cr–Ti/MCM-41 catalyst (c) under UV-light irradiation and (d) under visible-light irradiation ($\lambda > 420$ nm). Source: (b) Reproduced from Ref. [24]/with permission of Springer Nature. (c, d) Reproduced from Ref. [40]/with permission of Springer Nature.

different from individual Cr/MCM-41 and Ti/MCM-41, implying the possible interactions between the two species. The photocatalytic activity in ethylene polymerization reaction was found to be superior (80% conversion) due to the possible interaction between Cr^{6+} and Ti^{4+} species. No activity was obtained in dark conditions. It could be possible that the interaction between Cr^{6+} and Ti^{4+} ions could change the environment around Cr^{6+} ions, making them easy to be reduced and hence leading to superior catalytic performance than individual Cr/MCM-41 and Ti/MCM-41. The same binary oxide Cr–Ti/MCM-41 was found to be active in the photocatalytic oxidation of CO to CO_2 under UV and visible-light irradiation conditions. A higher product yield of 1.7 µmol was obtained under UV light irradiation when compared to 1.0 µmol under visible-light irradiation (Figure 16.6c,d). A lower catalytic response was shown by Cr/MCM-41, and however, Ti/MCM-41 did not display any catalytic activity. The photoredox mechanism of Cr^{6+} to Cr^{4+} species allows the conversion of CO to CO_2 in the presence of O_2. The interaction between Cr^{6+} and Ti^{4+} could also have an influence in enhancing the catalytic activity of binary oxide. Therefore, the formation of binary oxides by PAD method played a crucial role in improving the catalytic response of single-site photocatalysts.

Other than Cr–Ti-based photocatalysts, V-based binary oxides have also been reported for their unique catalytic performances. For example, V–Ti/MCM-41 catalyst was found to be active in the propane oxidation to acetone with 72% product selectivity, much higher than V/MCM-41 (56% selectivity) [36]. The propylene epoxidation to propylene oxide (PO) was also synergistically catalyzed by binary V–Ti/MCM-41 catalyst with a reaction rate of 643 μmol g^{-1} h^{-1} [41].

16.5 Nano-Sized Metal Preparation Using Single-Site Photocatalyst

16.5.1 Preparation of Monometallic NPs

The design and development of methods to synthesize nano-sized metals within porous supports are crucial for preparing highly active catalysts [42–45]. In such attempts to prepare new catalysts, single-site photocatalysts have been used as the support material to immobilize metal precursor solutions on the excited state of single-site oxides [18, 26]. For example, metal NPs such as Pd, Pt, Ag, and Au have been prepared in combination with single-site catalysts within the MS or zeolite framework. Pd/TS-1 was prepared by irradiating the aqueous solution of TS-1 and $PdCl_2$ solution with UV-light irradiation to prepare Pd NPs (Pd/Ti = 0.6). The local environment and oxidation state of metal NPs were confirmed by FT-EXAFS analysis in which the presence of a single peak at 2.2 Å confirmed the existence of Pd in a metallic state. The comparison of the intensity of FT-EXAFS spectra revealed the efficiency of PAD method in the formation of much smaller-sized NPs, when comparing it with a wet-impregnation method. This method also proved its efficiency in the formation of much smaller NPs in the range of 4–5 nm for Pt/Ti-HMS prepared by PAD under UV-light irradiation. The sample prepared by the impregnation method formed larger-sized NPs in the range of 2–20 nm. The formation of highly dispersed metal NPs can be accounted by the use of PAD method and single-site photocatalysts as the support material.

Recently, the use of single-site photocatalysts has also been employed in the synthesis of uniformly dispersed plasmonic Ag metal NPs on Ti-SBA-15 support material [46]. Plasmonic catalysts exploit the oscillation of free electrons to establish a resonance condition by impinging light irradiation of suitable wavelength on the metal NPs [47–50]. Utilizing this concept, the preparation of plasmonic Ag NPs was achieved on a Ti-SBA-15 catalyst by microwave-assisted reduction method. The small-sized Ag NPs were uniformly deposited on the Ti-SBA-15 support with an average size of 4–5 nm. The catalyst Ag/Ti-SBA-15 displayed superior catalytic performance in the hydrogen generation from ammonia borane when compared with Ag/SBA-15 and Ag/TiO_2/SBA-15 catalysts in dark and under UV–vis light irradiation conditions. A reaction rate of 3.3 μmol min^{-1} was obtained over Ag/Ti-SBA-15 under UV–vis light irradiation when compared to 2.4 μmol min^{-1} in dark reaction conditions. The inferior reaction rate for Ag/TiO_2/SBA-15 catalyst could be due to the aggregated Ti-oxide moieties in octahedral coordination geometry. The enhanced catalytic performance was attributed to the improved electron transfer and the creation of heterojunction between the charge transfer excited state of Ti-oxide moieties and plasmonic Ag NPs (Figure 16.7).

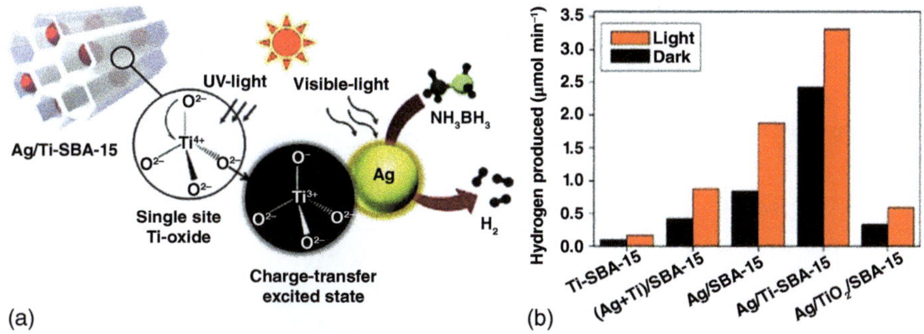

Figure 16.7 (a) Schematic illustration of a light-induced charge transfer pathway on the surface of the Ag/Ti-SBA-15 catalyst under UV–vis light irradiation. (b) Reaction rates for all catalysts in the ammonia borane dehydrogenation at room temperature in the dark and under light irradiation conditions. Source: Reproduced with permission from Ref. [46]. Copyright 2017, Wiley.

16.5.2 Preparation of Bimetallic Metal NPs

The deposition of bimetallic NPs in the form of alloy or bimetal core–shell NPs with single-site photocatalysts has also attracted much attention in order to improve the catalytic and photocatalytic activities [27]. The PAD of Pd and Au NPs was carried out by exciting the Ti-oxide species in TS-1 (Ti/Si = 0.017) under UV-light irradiation and utilizing the aqueous solution of metal precursors ($PdCl_2$ and $HAuCl_4$) [25, 51]. The TEM micrographs revealed the bimodal size distribution of metal NPs in which the smaller NPs (5–6 nm) were mainly composed of Pd metal and larger NPs (12–15 nm) had Au and Pd in 80 and 20 mol%, respectively. The FT-EXAFS spectra of Pd K-edge evidenced the formation of metallic Pd NPs, and Au L_{III}-edge FT-EXAFS spectra confirmed the formation of electron-rich Au NPs. The bimetal NPs formed in the Au core and Pd shell type morphology due to the more positive reduction potential of Au than Pd. The same catalyst, PdAu/TS-1, prepared by the PAD method was tested in the H_2O_2 synthesis from H_2 and O_2 and found to display the most active catalytic performance, when compared with the PdAu/TS-1 catalyst prepared by impregnation method (imp-PdAu/TS-1) and Pd-based monometallic catalyst [25]. The obtained results suggested the significance of PAD method and use of bimetallic catalytic system. A one-pot oxidation system of phenol was also developed in which the oxidant, H_2O_2, was generated in situ over PdAu/TS-1 catalyst [25]. The PAD-Pd/TS-1 displayed higher catalytic activity than imp-Pd/TS-1 due to the smaller size and higher dispersion of metal NPs. The PAD-PdAu/TS-1 catalyst displayed the highest catalytic activity when compared to imp-PdAu/TS-1, PAD-Pd/TS-1, and PAD-Au/TS-1 catalyst. The synergistic interaction of two metals along with the effectiveness of the PAD method led to the superior catalytic response of the prepared catalyst.

The deposition of bimetal Ni and Pd NPs can also be carried out on Ti-HMS silica by photoexcitation of tetrahedral coordinated Ti-oxide under UV-light

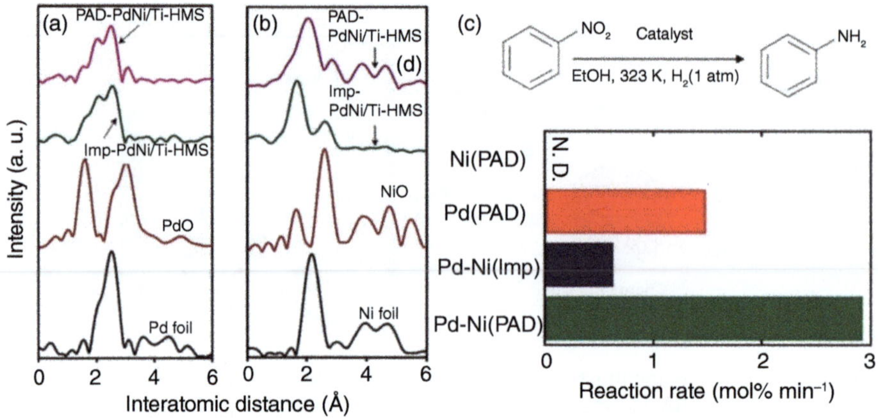

Figure 16.8 (a) Pd K-edge FT-EXAFS spectra, (b) Ni K-edge FT-EXAFS spectra, and (c) the hydrogenation activity of nitrobenzene to aminobenzene on different catalysts. Source: Adapted from Ref. [52].

irradiation in the presence of metal precursors ($NiSO_4$ and $PdCl_2$) and ammonia solution (pH = 10), followed by hydrogen reduction [52]. The synthesis of NPs by impregnation method was also done to compare the size and catalytic performance of bimetallic NPs prepared by PAD method. The smaller-sized metal NPs with a diameter ranging from 2 to 8 nm were obtained with PAD in comparison to a much larger particle size of about 14.6 nm prepared by the impregnation method. The prepared catalyst, PdNi/Ti-HMS catalyst, was characterized by FT-EXAFS analysis for Pd K-edge and Ni K-edge in order to confirm the oxidation state and coordination environment of metal NPs (Figure 16.8a,b). The formation of small-sized uniformly dispersed metal NPs was confirmed for PAD sample, whereas a core–shell type morphology was observed for the sample prepared by the impregnation method. The catalytic activity was tested in the hydrogenation of nitrobenzene to aminobenzene, and PAD PdNi/Ti-HMS displayed superior catalytic performance than pure Pd and Ni-base catalysts as shown in Figure 16.8c. The superior catalytic performance was attributed to the size and structural aspects of the Pd–Ni metal NPs. Especially, the core–shell morphology of Pd(core)–Ni(shell) with large-sized NPs obtained by impregnation method displayed inferior catalytic activity due to the shielding effect of Pd NPs via Ni shell. The higher catalytic performance of PAD PdNi/Ti-HMS was due to the synergistic presence of two metals with smaller size in a highly dispersed manner with a Ni/Pd molar ratio of 1.5.

16.6 Conclusions

In this chapter, we introduced single-site photocatalysts with heteroatom oxides incorporated within the frameworks of MS and zeolites, exhibiting UV and visible-light sensitivity to achieve superior catalytic performances. Table 16.1 has

Table 16.1 Single-site photocatalysts and their catalytic applications.

Single-site photocatalyst	Photocatalytic application reaction	Product yield	References
Ti-based catalyst			
Ti-MCM-48	Photocatalytic reduction of CO_2 to form CH_3OH and CH_4	Yield of CH_4 ~7 μmol g^{-1} h^{-1}	[6, 22]
Ti-MSTF	Superhydrophilicity measurements	Water contact angle under UV light <1°	[33]
Ti-Y-zeolite	NO decomposition to N_2, O_2, and N_2O under UV-light	91% N_2 selectivity	[28]
Cr-based catalyst			
Cr-HMS	NO decomposition to N_2, O_2, and N_2O under visible-light irradiation	97% N_2 selectivity under $\lambda > 450$ nm	[10]
	Epoxidation of propene with O_2 to propylene oxide; $\lambda > 450$ nm	~0.5% yield of PO	[24]
Cr–Ti/HMS	Epoxidation of propene with O_2 to propene oxide; $\lambda > 450$ nm	~2.5% yield of PO	[12, 24]
Cr–Ti/MCM-41	Photocatalytic oxidation of CO with O_2 to CO_2	Under UV light (1.7 μmol) and visible light (1.0 μmol) of CO_2	[40]
	Ethylene polymerization under UV light	80% conversion of C_2H_4	[40]
Cr-MS thin film	Ethylene polymerization under UV and visible light	Formation of polyethylene on thin film confirmed by FT-IR bands	[34, 40]
	Evaluation of hydrophilicity	Decrease in water contact angle under UV and visible light	[34]
V-based catalyst			
VS-1	NO decomposition to N_2 in the presence of propane under UV-light irradiation	40 μmol g-cat^{-1} yield of N_2 after 7 h	[53]
V/MCM-41	Photocatalytic oxidation of methane to methanol in the presence of NO and UV-light irradiation	Methanol selectivity 80%	[37]
	Oxidation of propane to acetone under UV-light irradiation	56% selectivity	[36]

(Continued)

Table 16.1 (Continued)

Single-site photocatalyst	Photocatalytic application reaction	Product yield	References
V-Ti/MCM-41	Oxidation of propane to acetone under UV-light irradiation	72% selectivity	[36]
	Epoxidation of propylene to propylene oxide	Reaction rate of 634 $\mu mol\ g^{-1}\ h^{-1}$ under UV-light irradiation	[41]
Metal NPs based			
PdAu/TS-1	H_2O_2 synthesis from H_2 and O_2	~80 mM H_2O_2	[25]
	One pot oxidation of phenol	13.3% product yield	[25]
PdNi/Ti-HMS	Nitrobenzene reduction to aminobenzene	Reaction rate ~3.0 mol% min^{-1}	[52]
Ag/Ti-SBA-15	Hydrogen generation from ammonia borane	H_2 production rate of 3.3 $\mu mol\ min^{-1}$	[46]

been created at the end to summarize all single-site-based photocatalysts introduced in this chapter. The use of single-site photocatalysts incorporated within mesoporous frameworks can offer various advantages when compared to bulk semiconductors, for example, (i) enhanced diffusion of substrate molecules due to the porous framework, (ii) tunability of the pore size and surface functionalization of the support framework with different functional groups, and (iii) enhanced photocatalytic performance under UV and UV–vis light irradiation due to the formation of charge-transfer excited state with charge carriers at a much closer distance. Within porous frameworks, the design of single-site photocatalysts can be carefully tuned with respect to particle size and dispersibility.

The future directions and applications of single-site porous photocatalysts include the ability for selective functionalization of specific functional groups on the surface of the framework to carry out the challenging reaction with better efficiency such as artificial photosynthesis. The efforts should be directed in order to increase the dopant concentration amount within the porous framework to carry out practical application reactions. Till now, the amount of heteroatom is limited, and therefore, sometimes a larger quantity of catalyst is required to achieve reasonable catalytic performances. For example, TS-1 zeolite with an approximate ratio of Si/Ti = 40 : 1 can be modified to increase the active site (Ti) concentration and enhance the catalytic performances. Another aspect of controlling the coordination structure and reaction environment around single-site moieties should be evaluated. The utilization of single-site photocatalysts is a promising approach, and it can be expected to be applied in several interdisciplinary fields to design unique photo-functional materials for diverse applications.

Acknowledgments

The present work was supported by Grants-in-Aid for Scientific Research from the Japan Society for the Promotion of Science (JSPS) [grant nos 19K15311, 19H00838, and 22K21326]. A part of this work was supported by the Cooperative Research Program of the "Network Joint Research Center for Materials and Devices" [grant nos 20201097 and 20211069]. H. Y. thanks the Elements Strategy Initiative of MEXT [grant no. JPMEXTP0112101003]. P. V. gratefully acknowledges the Japan Society for the Promotion of Science for International Postdoctoral Fellowship (P21042) at Osaka University, Japan.

References

1 Miyoshi, A., Nishioka, S., and Maeda, K. (2018). Water splitting on rutile TiO_2-based photocatalysts. *Chem. Eur. J.* 24: 18204–18219.

2 Zhao, D. and Yang, C.F. (2016). Recent advances in the TiO_2/CdS nanocomposite used for photocatalytic hydrogen production and quantum-dot-sensitized solar cells. *Renew. Sust. Energy Rev.* 54: 1048–1059.

3 Choi, W. (2006). Pure and modified TiO_2 photocatalysts and their environmental applications. *Catal. Surv. Asia* 10: 16–28.

4 Fujishima, A. and Honda, K. (1972). Electrochemical photolysis of water at a semiconductor electrode. *Nature* 238: 37–38.

5 Hashimoto, K., Irie, H., and Fujishima, A. (2006). TiO_2 photocatalysis: a historical overview and future prospects. *Jpn. J. Appl. Phys.* 44: 8269–8285.

6 Yamashita, H., Mori, K., Kuwahara, Y. et al. (2018). Single-site and nano-confined photocatalysts designed in porous materials for environmental uses and solar fuels. *Chem. Soc. Rev.* 47: 8072–8096.

7 Higashimoto, S., Matsuoka, M., Yamashita, H. et al. (2000). Effect of the Si/Al ratio on the local structure of V oxide/ZSM-5 catalysts prepared by solid-state reaction and their photocatalytic reactivity for the decomposition of NO in the absence and presence of propane. *J. Phys. Chem. B* 104: 10288–10292.

8 Ikeue, K., Yamashita, H., Anpo, M., and Takewaki, T. (2001). Photocatalytic reduction of CO_2 with H_2O on Ti−β zeolite photocatalysts: effect of the hydrophobic and hydrophilic properties. *J. Phys. Chem. B* 105: 8350–8355.

9 Yamashita, H., Ikeue, K., Takewaki, T., and Anpo, M. (2002). In situ XAFS studies on the effects of the hydrophobic-hydrophilic properties of Ti-Beta zeolites in the photocatalytic reduction of CO_2 with H_2O. *Top. Catal.* 18: 95–100.

10 Yamashita, H. and Anpo, M. (2003). Local structures and photocatalytic reactivities of the titanium oxide and chromium oxide species incorporated within micro- and mesoporous zeolite materials: XAFS and photoluminescence studies. *Curr. Opin. Solid State Mater. Sci.* 7: 471–481.

11 Yamashita, H., Kida, K., Ikeue, K. et al. (2003). Photocatalytic epoxidation of propene with molecular oxygen under visible light irradiation on V ion-implanted Ti-HMS and Cr-HMS mesoporous molecular sieves. *Stud. Surf. Sci. Catal.* 146: 597–600.

12 Yamashita, H., Chiyoda, O., Masui, Y. et al. (2005). Design of visible light sensitive (Cr,Ti)-containing mesoporous silica photocatalyst using a photo-assisted deposition (PAD) method. *Stud. Surf. Sci. Catal.* 158: 43–50.

13 Amano, F., Ito, T., Takenaka, S., and Tanaka, T. (2005). Selective photocatalytic oxidation of light alkanes over alkali-ion-modified V_2O_5/SiO_2: kinetic study and reaction mechanism. *J. Phys. Chem. B* 109: 10973–10977.

14 Qian, X., Fuku, K., Kuwahara, Y. et al. (2014). Design and functionalization of photocatalytic systems within mesoporous silica. *ChemSusChem* 7: 1528–1536.

15 Che, M., Mori, K., and Yamashita, H. (2012). Elaboration, characterization and properties of silica-based single-site heterogeneous photocatalysts. *Proc. R. Soc. A Math. Phys. Eng. Sci.* 468: 2113–2128.

16 Verma, P., Kuwahara, Y., Mori, K. et al. (2020). Functionalized mesoporous SBA-15 silica: recent trends and catalytic applications. *Nanoscale* 12: 11333–11363.

17 Verma, P., Kondo, Y., Kuwahara, Y. et al. (2021). Design and application of photocatalysts using porous materials. *Catal. Rev. Sci. Eng.* 63: 165–233.

18 Yamashita, H. and Mori, K. (2007). Applications of single-site photocatalysts implanted within the silica matrixes of zeolite and mesoporous silica. *Chem. Lett.* 36: 348–353.

19 Mori, K., Qian, X., Kuwahara, Y. et al. (2020). Design of advanced functional materials using nanoporous single-site photocatalysts. *Chem. Rec.* 20: 660–671.

20 Suib, S.L., Přech, J., Čejka, J. et al. (2020). Some novel porous materials for selective catalytic oxidations. *Mater. Today* 32: 244–259.

21 Navlani-García, M., Verma, P., Salinas-torres, D. et al. (2020). Single-site heterogeneous catalysts and photocatalysts for emerging applications. In: *Advanced Heterogeneous Catalysis Volume 2: Application to the Single-Atom Scale*, Chapter 7, vol. 2, 151–188.

22 Mori, K., Yamashita, H., and Anpo, M. (2012). Photocatalytic reduction of CO_2 with H_2O on various titanium oxide photocatalysts. *RSC Adv.* 2: 3165–3172.

23 Kamegawa, T., Morishima, J., Matsuoka, M. et al. (2007). Eliminating traces of carbon monoxide photocatalytically from hydrogen with a single-site, non-noble metal catalyst. *J. Phys. Chem. C* 111: 1076–1078.

24 Yamashita, H., Mori, K., Shironita, S., and Horiuchi, Y. (2008). Applications of single-site photocatalysts to the design of unique surface functional materials. *Catal. Surv. Asia* 12: 88–100.

25 Mori, K., Miura, Y., Shironita, S., and Yamashita, H. (2009). New route for the preparation of Pd and PdAu nanoparticles using photoexcited Ti-containing zeolite as an efficient support material and investigation of their catalytic properties. *Langmuir* 25: 11180–11187.

26 Shironita, S., Takasaki, T., Kamegawa, T. et al. (2009). Application of microwave-assisted deposition for the synthesis of noble metal particles on Ti-containing mesoporous silica. *Catal. Lett.* 129: 404–407.

27 Verma, P., Kuwahara, Y., Mori, K., and Yamashita, H. (2019). Design of silver-based controlled nanostructures for plasmonic catalysis under visible light irradiation. *Bull. Chem. Soc. Jpn.* 92: 19–29.

28 Yamashita, H., Ichihashi, Y., Anpo, M. et al. (1996). Photocatalytic decomposition of NO at 275 K on titanium oxides included within Y-zeolite cavities: the structure and role of the active sites. *J. Phys. Chem.* 100: 16041–16044.

29 Takeuchi, M., Yamashita, H., Matsuoka, M. et al. (2000). Photocatalytic decomposition of NO under visible light irradiation on the Cr-ion-implanted TiO_2 thin film photocatalyst. *Catal. Lett.* 67: 135–137.

30 Negishi, N., Takeuchi, K., and Ibusuki, T. (1998). Surface structure of the TiO_2 thin film photocatalyst. *J. Mater. Sci.* 3: 5789–5794.

31 Sunada, K. (1998). Bactericidal and detoxification effects of TiO_2 thin film photocatalysts. *Environ. Sci. Technol.* 32: 726–728.

32 Shirai, K. (1974). Studies on light-sensitive material of organic titanium compounds. *J. Soc. Photogr. Sci. Technol. Jpn.* 37: 192–198.

33 Horiuchi, Y. and Yamashita, H. (2011). Design of mesoporous silica thin films containing single-site photocatalysts and their applications to superhydrophilic materials. *Appl. Catal. A Gen.* 400: 1–8.

34 Yamashita, H., Nishio, S., Imaoka, S. et al. (2008). Photo-induced surface property on transparent mesoporous silica thin films containing single-site photocatalyst. *Top. Catal.* 47: 116–121.

35 Wen, M., Mori, K., Kuwahara, Y. et al. (2018). Design of single-site photocatalysts by using metal-organic frameworks as a matrix. *Chem. Asian J.* 13: 1767–1779.

36 Hu, Y., Wada, N., Tsujimaru, K., and Anpo, M. (2007). Photo-assisted synthesis of V and Ti-containing MCM-41 under UV light irradiation and their reactivity for the photooxidation of propane. *Catal. Today* 120: 139–144.

37 Hu, Y., Higashimoto, S., Takahashi, S. et al. (2005). Selective photooxidation of methane into methanol by nitric oxide over V-MCM-41 mesoporous molecular sieves. *Catal. Lett.* 100: 35–37.

38 Higashimoto, S., Matsuoka, M., Yamashita, H., and Anpo, M. (1999). Investigation of the local structure of vanadium silicalite catalyst (VS-1) using XAFS, FT-IR and photoluminescence spectroscopic methods. *Jpn. J. Appl. Phys.* 38: 47–50.

39 Yamashita, H., Yoshizawa, K., Ariyuki, M. et al. (2001). Photocatalytic reactions on chromium containing mesoporous silica molecular sieves (Cr-HMS) under visible light irradiation: decomposition of NO and partial oxidation of propane. *Chem. Commun.* 435–436.

40 Kamegawa, T., Shudo, T., and Yamashita, H. (2010). Preparation of Cr-Ti binary oxide anchored mesoporous silica by CVD method and their photocatalytic activities. *Top. Catal.* 53: 555–559.

41 Nguyen, V.H., Lin, S.D., and Wu, J.C.S. (2015). Synergetic photo-epoxidation of propylene over V-Ti/MCM-41 mesoporous photocatalysts. *J. Catal.* 331: 217–227.

42 Fukuoka, A. and Dhepe, P.L. (2009). Sustainable green catalysis by supported metal nanoparticles. *Chem. Rec.* 9: 224–235.

43 Li, X., Wang, X., and Antonietti, M. (2012). Mesoporous g-C_3N_4 nanorods as multifunctional supports of ultrafine metal nanoparticles: hydrogen generation from water and reduction of nitrophenol with tandem catalysis in one step. *Chem. Sci.* 3: 2170–2174.

44 Lewera, A., Timperman, L., Roguska, A., and Alonso-vante, N. (2011). Metal support interactions between nanosized Pt and metal oxides (WO_3 and TiO_2) studied using X-ray photoelectron spectroscopy. *J. Phys. Chem. C* 115: 20153–20159.

45 Li, W., Kovarik, L., Mei, D. et al. (2014). A general mechanism for stabilizing the small sizes of precious metal nanoparticles on oxide supports. *Chem. Mater.* 26: 5475–5481.

46 Verma, P., Kuwahara, Y., Mori, K., and Yamashita, H. (2017). Enhancement of Ag-based plasmonic photocatalysis in hydrogen production from ammonia borane by the assistance of single-site Ti-oxide moieties within a silica framework. *Chem. Eur. J.* 23: 3616–3622.

47 Verma, P., Kuwahara, Y., Mori, K. et al. (2022). New insights in establishing the structure-property relations of novel plasmonic nanostructures for clean energy applications. *EnergyChem* 4: 100070.

48 Verma, P., Mori, K., Kuwahara, Y. et al. (2021). Plasmonic nanocatalysts for visible-NIR light induced hydrogen generation from storage materials. *Mater. Adv.* 2: 880–906.

49 Verma, P., Mori, K., Kuwahara, Y. et al. (2020). Synthesis of plasmonic gold nanoparticles supported on morphology-controlled TiO_2 for aerobic alcohol oxidation. *Catal. Today* 352: 255–261.

50 Verma, P., Kuwahara, Y., Mori, K., and Yamashita, H. (2020). Visible-light-driven reduction of nitrostyrene utilizing plasmonic silver nanoparticle catalysts immobilized on oxide supports. *Catal. Today* 355: 620–626.

51 Yamashita, H., Miura, Y., Mori, K. et al. (2007). Synthesize of nano-sized Pd metal catalyst on Ti-containing zeolite using a photo-assisted deposition (PAD) method. *Catal. Lett.* 114: 75–78.

52 Fuku, K., Sakano, T., Kamegawa, T. et al. (2012). Enhanced hydrogenation activity of nano-sized Pd-Ni bimetal particles on Ti-containing mesoporous silica prepared by a photo-assisted deposition method. *J. Mater. Chem.* 22: 16243–16247.

53 Higashimoto, S., Matsuoka, M., Zhang, S.G. et al. (2001). Characterization of the VS-1 catalyst using various spectroscopic techniques and its unique photocatalytic reactivity for the decomposition of NO in the absence and presence of C_3H_8. *Microporous Mesoporous Mater.* 48: 329–335.

Index

Micro-Mesoporous Metallosilicates: Synthesis, Characterization, and Catalytic Applications, First Edition. Edited by Peng Wu and Hao Xu.

u

v